Principles of Electronics

Principles of Electronics

Colin D. Simpson
George Brown College

Prentice Hall
Englewood Cliffs, New Jersey   Columbus, Ohio

Library of Congress Cataloging-in-Publication Data
Simpson, Colin D. (Colin David)
    Principles of electronics / Colin D. Simpson.
       p.   cm.
    Includes index.
    ISBN 0-13-034406-0
    1. Electronics.   I. Title.
TK7816.S544   1996
621.3--dc20   95-18400
                                CIP

Cover photo: Allen Cheuvront Studios
Editor: David Garza
Production Editor: Mary Irvin
Design Coordinator: Julia Z. Van Hook
Cover Designer: Brian Deep
Production Manager: Pamela A. Bennett
Marketing Manager: Debbie Yarnell
Illustrations: Rolin Graphics
Production Supervision: Betty O'Bryant

This book was set in Times Roman by Clarinda Company and was printed and bound by R. R. Donnelley & Sons Company. The cover was printed by Phoenix Color Corp.

©1996 by Prentice-Hall, Inc.
A Simon & Schuster Company
Englewood Cliffs, New Jersey 07632

Photo Credits: *Chapter opening photos:* Chapters 1, 7, and 13—Courtesy of Tektronix, Inc.; Chapter 2—Scott Cunningham/Prentice Hall; Chapters 3 and 9—Courtesy of Toyota Motors Manufacturing; Chapters 4, 6, 11, and 14—Courtesy of International Business Machines Corporation; Chapters 5 and 10—Anne Vega; Chapters 8 and 18—Courtesy of Children's Hospital, Columbus, Ohio; Chapter 12—Courtesy of R. R. Donnelley & Sons Company; Chapter 15—Courtesy of Honeywell; Chapter 16–Courtesy of Honda of America Manufacturing, Inc.; Chapter 17—Courtesy of NASA; Chapter 19—Courtesy of Sony Electronics, Inc.; Chapter 20—Courtesy of the U.S. Navy, Dava Still Media Depository; Chapter 21—Courtesy of The Glidden Company, Member of ICI Paints; Chapters 22 and 23—Courtesy of Miller Brewing Company. (Credits continue on p. v.)

Printed in the United States of America
10 9 8 7 6 5 4 3 2 1

ISBN: 0-13-034406-0

Prentice-Hall International (UK) Limited, *London*
Prentice-Hall of Australia Pty. Limited, *Sydney*
Prentice-Hall of Canada, Inc., *Toronto*
Prentice-Hall Hispanoamericana, S. A., *Mexico*
Prentice-Hall of India Private Limited, *New Delhi*
Prentice-Hall of Japan, Inc., *Tokyo*
Simon & Schuster Asia Pte. Ltd., *Singapore*
Editora Prentice-Hall do Brasil, Ltda., *Rio de Janeiro*

Credits for interior figures: Figure 1–1—Scott Cunningham/Prentice Hall; Figures 1–2 and 1–4—Courtesy of International Business Machines Corporation; Figures 1–3, 2–8(b), and 2–22—Courtesy of Motorola Semiconductor Products Sector; Figure 1–5—Courtesy of Sony Electronics, Inc.; Figure 1–6—Courtesy of Children's Hospital, Columbus, Ohio; Figure 1–7—Courtesy of Motorola, Inc., used by permission; Figures 1–8, 7–22, and 12–14—Reproduced from Cook, Nigel (1993), *Introductory DC/AC Electronics,* 2nd ed. (Englewood, Cliffs, NJ: Prentice Hall), pp. 779, 705, and 739, respectively; Figure 1–9(a)—Courtesy of Allen-Bradley Co.; Figures 1–9(b), 1–11(a) and (b), 2–20, 2–21, and 14–8(c)—Courtesy of Dale Electronics, Inc.; Figures 1–9(c) and (d) and 2–27—Courtesy of Bourns, Inc.; Figure 1–10(a) and (d)—Courtesy of Mepco/Centralab; Figure 1–10(b)—Courtesy of Murata Erie North America; Figure 1–10(c) and (g)—Courtesy of Vishay Sprague; Figure 1–11(c)—Courtesy of Delevan; Figure 2–7—Courtesy of Eveready Battery Company, Inc.; Figure 2–8(a)—Courtesy of EG&G VACTEC, Inc.; Figure 2–9—Courtesy of Omega Engineering—An Omega Technologies Company; Figure 2–10—Prentice-Hall, Inc., Reliance Electric Company; Figures 2–11, 7–16(a) and (b), 11–1(b), and 20–18—Courtesy of B & K Precision; Figure 2–17—Courtesy of Stackpole Carbon Co.; Figures 2–19, 13–8, and 13–9—Reproduced from Floyd, T. (1995), *Electronic Fundamentals: Circuits, Devices, and Applications,* 3rd ed. (Englewood Cliffs, NJ: Prentice Hall), pp. 31, 351 and 353, respectively; Figure 2–28—Reproduced from Bell, D. (1995), *Electric Circuits: Principles, Applications, and Computer Analysis* (Englewood Cliffs, NJ: Prentice Hall), p. 74; Figures 4–30 and 4–33—Courtesy of Bussmann, Cooper Industries; Figure 4–34(a), (b), and (d)—Courtesy of Eaton Corp.; Figure 4–34(c) and (e)—Courtesy of Grayhill, Inc.; Figures 7–2, 7–6, and 7–12—Furnished by the Simpson Electric Company, Elgin, Ill.; Figures 7–18 and 12–11—Courtesy of Fluke Corporation; Figure 11–1(a)—Courtesy of General Electric; Figures 12–3 and 23–69—Courtesy of Tektronix, Inc.; Figure 13–6—Courtesy of Custom Electronics, Inc.; Figure 13–7—Carbide Electronics Corporation; Figure 14–8(a) and (b)—Courtesy of Delevan/American Precision; Figure 15–6—Courtesy of Superior Electric Co.; Figures 19–10 and 20–36—Reproduced from Webb, J., and Greshock, K. (1993), *Industrial Control Electronics,* 2nd ed. (Englewood Cliffs, NJ: Prentice Hall), pp. 89 and 221, respectively; Figure 19–24—Reproduced from Simpson, Colin (1993), *Introduction to Electronic Circuits and Machines* (Scarborough, Ontario: Prentice-Hall Canada, Inc.), p. 365; Figure 20–1(a) and (b)—Copyright of Motorola, Inc., used by permission; Figure 23–67—Reproduced from Floyd, T. (1994), *Digital Fundamentals*, 5th ed. (Englewood Cliffs, NJ: Prentice Hall), pp. 32–33.

Data Sheets: Courtesy of Motorola, Inc., Fairchild, and National Semiconductor, Inc.

In memory of Paul Simpson. My brother. My friend.

PREFACE

Principles of Electronics has been written primarily for use in electronic engineering technician and technology programs. The content is suitable for community colleges and technical institutes. The topics follow a typical sequence for two- and four-year electronics courses. For each subject, the basic principles are explained first, followed by applications and troubleshooting techniques. Students using this book should have an understanding of algebra up to quadratic equations and of trigonometry up to the simple properties of triangles. The availability of a scientific calculator is assumed. The calculator must handle trigonometric and exponential functions as a minimum. Basic principles, theorems, circuit behavior, and problem-solving procedures are presented so that the average student can get a clear understanding of essential concepts. The large number of example problems and exercises should make this book useful for self-study and review, and should serve as preparation for more advanced courses.

The introductory chapters cover essential concepts of electricity, including atomic structure, electric charges, Ohm's law, Kirchhoff's laws, energy, and power. Early in the book introductory circuit analysis techniques are emphasized to assist the student in developing an approach to solving problems related to electronic circuits. Thévenin's theorem, the superposition theorem, and the maximum power transfer theorem are covered in early chapters. In addition to electric circuit analysis, the book also covers topics such as magnetism, resonance, filters, semiconductor diodes, transistors, and electronic circuits. Electron current flow is used throughout the text, and, in accordance with the Institute of Electrical and Electronic Engineers (IEEE), the SI system of measurement is used wherever practical.

My primary objective in writing this book was to take what many perceive to be a difficult subject and make it easier to understand. In my experience, I have found that many students have difficulty grasping the basic principles of electronic circuits and that the most effective method of teaching these principles is to present them in a relatively nonmathematical format. Over the years, I have developed a method of teaching electronics that is straightforward and uses a minimum of mathematical equations. This is not to say that the depth or complexity of the subject matter has been sacrificed. It simply means that the book is more readable. Concepts are reinforced with practical applications as well as mathematical solutions. If the student can relate the circuit or device to a real-world application, the learning curve improves dramatically.

Every effort has been made to achieve accuracy and clarity throughout the text. Having written three books on the subject of electricity and electronics, I was able to apply many of the lessons learned from previous publications. I was also able to utilize a database of several hundred professors who have taught electronics to obtain their opinions on course content, depth of material, chapter organization, and accuracy. As always, I welcome all praise and criticism from students, professors, and people in industry. As an author, feedback from users of my books is tremendously helpful. Please feel free to write to me c/o George Brown College, P.O. Box 1015 Station B, Toronto, Ontario, Canada, M5T 2T9. My Internet address is csimpson@gbrownc.on.ca.

Each chapter contains numerous features that are designed to aid in the learning process and to reinforce key concepts and applications for the chapter topics.

- Learning objectives at the beginning of each chapter specify the chapter's goals.
- A practical application task is also introduced at the beginning of each chapter, and the final section provides a solution.
- Hundreds of worked examples throughout the text illustrate fundamental concepts.
- Review questions at the end of each section provide an immediate self-check of key concepts and principles, with answers given at the end of the chapter.
- A detailed summary at the end of each chapter emphasizes the important concepts covered in the chapter.
- The end-of-chapter problems test learning of chapter concepts and reinforce problem-solving skills.
- A comprehensive glossary at the end of the book serves as easy-to-use reference source for students.

The two-page chapter opener lists the chapter's learning objectives on the left page.

The right page of the chapter opener introduces a practical application task that is solved in the final section of the chapter.

5

Parallel Circuits

Learning Objectives

Upon completion of this chapter you will be able to

- Define a parallel circuit.
- Calculate resistance in parallel.
- Describe the flow of current in a parallel circuit.
- Express Kirchhoff's current law.
- Use the current divider rule.
- Apply Ohm's law for parallel circuit calculations.
- Calculate power in a parallel circuit.
- Describe the effect of connecting voltage sources in parallel.
- List some typical applications for parallel circuits.
- Troubleshoot parallel circuits.

Practical Application

As a technician for an audio-engineering company, you are required to design an audio system for a nightclub that requires eight speakers to be driven from a 1000 W amplifier (500 W per side). The specifications for the installation stipulate that four speakers are to be connected in parallel to each side of the amplifier. The minimum output resistance per side of the amplifier is rated at 4 Ω.

Your task is to sketch an installation diagram showing the speaker connections, determine the ohmic value of each speaker, and calculate the minimum power-handling capability of each speaker.

118

119

Hundreds of worked examples appear throughout the text.

Section review questions provide an immediate self-check at the end of each section.

The final section in each chapter provides a detailed solution for the practical application task introduced in the chapter opening.

A chapter summary that highlights key points and the answers to section reviews appear at the end of every chapter.

The back-of-book glossary provides easy access to key terms and definitions.

EXAMPLE 6–19

Solution The total resistance of the circuit with R_L opened consists of R_1 and R_2.

$$R_T = R_1 + R_2 = 18.2\ \Omega + 200\ \Omega$$

The voltage drops can now be found using the voltage divider rule.

$$V_{R1} = \left(\frac{R_1}{R_T}\right)E_T = \left(\frac{18.2\ \Omega}{218.2\ \Omega}\right)20\ V = 1.67\ V$$

$$V_{R2} = \left(\frac{R_2}{R_T}\right)E_T = \left(\frac{200\ \Omega}{218.2\ \Omega}\right)20\ V = 18.33\ V$$

Calculate the voltages across R_1 and R_2 for the circuit of Figure 6–37 (b).

Section Review

1. What happens to a series-parallel circuit when a short occurs in one of the components?
2. Why is it more difficult to provide overcurrent protection for a series-parallel circuit than a parallel circuit?
3. When an open-circuit condition occurs in a series-parallel circuit, the voltage drops in the circuit remain constant (True/False).

6–10
PRACTICAL APPLICATION SOLUTION

The design requirements for a dual-polarity power supply were outlined in the chapter opener. Your task was to design a power supply capable of supplying 200 mA for a +10 V and −10 V load. The following steps illustrate the method of solution for this practical application of series-parallel circuits.

STEP 1   Sketch a wiring diagram showing the reference point for the dual polarity supply (Figure 6–38). Three resistors are required for the circuit. Because the source is 24 V and the total voltage requirements are 20 V, a series-dropping resistor (R_1) must drop the difference (4 V). The other two resistors (R_2 and R_3) form a divider to provide the +10 V and −10 V values. The two load resistors are connected with R_2 and R_3 at the reference, or ground, point for the circuit.

FIGURE 6–38   Circuit for practical application.

FIGURE 9–10   Diagram of stop/start circuit with control relay.

was to energize and de-energize the DC motor. The following steps demonstrate the method of solution for this practical application of magnetism.

STEP 1   Draw a diagram of how the circuit is to be connected. Figure 9–10 shows a diagram of the low-voltage control circuit and high-voltage power circuit. The dashed lines between the windings of the electromagnet and contacts indicate that the electromagnet controls these contacts. When the start button is pressed, the contact in parallel with the start button closes and the contact in series with the DC motor and 100 V supply also closes. The relay remains energized after the start button is released because a path for the current is provided through the parallel contact. Because the stop button is in series with the start button and holding contact, when the stop button is pressed, current flow is interrupted to the electromagnetic coil and the contacts return to their normally open state.

STEP 2   Test the circuit. Thoroughly check the wiring for the circuit and verify that the installation is complete before starting the motor.

Summary

1. A moving electron carries a negative electrical charge. The spinning effect of the electron creates a magnetic field.
2. The total number of lines of force in a given region is called *magnetic flux*.
3. The attraction or repulsion between magnets varies directly with the product of their strengths and inversely with the square of the distance between them.
4. When a material is easy to magnetize, it is said to have a *high permeability*.
5. When an electric current is passed through a long straight conductor, a magnetic field is established in and around the conductor.
6. The magnetic field around a current-carrying conductor can be intensified by forming the conductor into a coil, or solenoid.
7. Solenoids are electromagnetic devices with a moveable iron core.
8. Relays are electromagnetic devices that use one or more sets of contacts to make or break control circuits.
9. Permanent magnets are often used with electromagnets in audio and video devices for the recording and playback of signals.
10. The Hall effect is based on the Hall voltage, which is a voltage developed across a magnetic field.

Answers to Section Reviews

Section 9–1
1. False
2. That the earth is a huge magnet with its magnetic north pole near the geographical north pole and the magnetic south pole located near the geographical south pole
3. The presence of iron

GLOSSARY

Absolute permittivity (ε) The flux produced with a vacuum as dielectric; also known as absolute capacitivity.

AC resistance The effective resistance of a conductor; includes factors such as skin effect and radiation loss.

Active device A device, such as a transistor, capable of controlling voltage or current.

Active filter A filter network that uses an active device to obtain the desired filtering effect.

Additive polarity A winding connection that causes the direction of counter emf in both windings to be the same.

Admittance (Y) The ease at which an AC current flows in a circuit; the reciprocal of impedance, measured in siemens (S).

Admittance triangle A right-angle triangle that relats conductance, susceptance, and admittance.

Air-core inductor An inductor that contains no magnetic iron and generally is wound on a tubular insulating material.

Air gap A part of the magnetic circuit of electromagnets, and used to increase reluctance.

Alternating current (AC) A current in which the magnitude and direction varies with time.

Alternating emf A voltage in which the magnitude and direction varies with time.

American Wire Gauge (AWG) A standard for manufacturing and numbering wires.

Ammeter A measuring instrument used to indicate electrical current in amperes.

Ampere (A) The SI base unit of electric current; the rate of electric charge flow when 1 C of charge passes a given point in 1 s.

Ampere's circuit law A law that states that the algebraic sum of the rises and drops of the mmf around a closed loop of a magnetic circuit is equal to zero.

Amplitude The maximum positive or negative value of an alternating current, voltage, or power; also known as peak value.

Analog meter A moving-coil measuring instrument.

Angular velocity The rate of change of a quantity, such as voltage, in an ac circuit.

Apparent power The product of the total rms voltage and current in a circuit, expressed in volt-amperes (VA).

Arcing A phenomenon caused by interrupting current, such as opening a switch, that produces a very high induced voltage because of the rapidly collapsing field.

The supporting ancillary package for *Principles of Electronics* includes the following:

- Test item file
- Computerized test bank
- Transparency package with transparency masters
- Student study guide
- Instructor's manual
- Laboratory manual
- Video library
- Electronics workbench software
- Bulletin board in America Online

Organization of the Book

The book is divided into 23 chapters. The chapter opener provides an insight into a practical application of electronic technology. A particular task is outlined at the beginning of the chapter and the solution is presented in the final numbered section. The hundreds of worked examples and problems in this book have been selected and written to illustrate fundamental concepts essential to the troubleshooting and design of electronic circuits. To promote immediate reinforcement of problem-solving skills, section reviews are provided at the end of each major topic. The answers for the section reviews are included at the end of the chapter.

Chapter 1 presents an introduction to electronics, including applications, components, and the principles of electric charge. Chapters 2 and 3 provide information on the fundamentals of current, voltage and resistance, as well as Ohm's law, power, and energy. Chapters 4 through 6 describe series, parallel, and series-parallel circuits with an emphasis on troubleshooting and practical application. Troubleshooting is reinforced in Chapter 7 with a presentation of DC measuring instruments and practices.

A thorough discussion of DC network theorems is provided in Chapter 8, including Thévenin's theorem, Norton's theorem, and the maximum power transfer theorem. Chapters 9 and 10 describe the principles of magnetism and the magnetic circuit. Chapter 11 begins by introducing the phenomenon known as alternating current and ends with a discussion of nonsinusoidal waveforms and harmonic frequencies. Troubleshooting with AC meters is emphasized in Chapter 12 using equipment such as oscilloscopes, frequency counters, and signal generators.

The principles of capacitors and inductors are presented in Chapters 13 and 14. Chapter 15 describes the fundamentals of transformers including isolation transformers, multiple-winding transformers, transformer polarity, pulse transformers, and troubleshooting. Chapter 16 provides a comprehensive treatment of alternating current circuits and describes the effects of AC on resistors, inductors, and capacitors. Resonance in series and parallel circuits is discussed in Chapter 17, including selectivity, bandwidth, and resonant frequency. Chapter 18 describes the basics of coupling and filtering circuits with an emphasis on practical application.

Chapter 19 begins the final stage in the organization of the book. This chapter on semiconductor fundamentals provides the foundation for the subsequent chapters on transistors, thyristors, amplifiers, op-amps, and digital circuits. Chapter 20 presents a discussion on the basic operating principles of BJTs, FETs, SCRs, triacs, and UJTs. Amplifier circuits are described in Chapter 21 with an emphasis on practical applications. Chapter 22 introduces the op-amp, and Chapter 23 presents an overview of digital electronics.

Acknowledgments: This book could not have emerged in its present form without the helpful suggestions made by many people during its preparation. I would like to thank my colleagues at George Brown College, particularly Jim Drennan, Jennifer Bolt, Ming Quon, and Hilde Zimmer. I would also like to thank the editorial staff at Prentice Hall, especially Dave Garza and Sylvia Huning. A special thanks goes to Betty O'Bryant, whose copyediting skills contributed greatly to the accuracy of this text, and to Monica Ohlinger for her photo research and editing. I would also like to express my sincere gratitude to Bud Skinner of Applied Physics Specialties for his insightful suggestions and immeasurable contributions to this book. Thanks are also extended to all my students for their eager desire to learn and for their feedback during the classroom testing of the manuscript. I also wish to thank the following reviewers who offered many helpful suggestions in the early stages of the manuscript's development: Toby Boydall, Conestoga College; Joseph Thomas, University of Wisconsin; Leonard Sokoloff, DeVry Institute of Technology; Mike Goulding, Fanshawe College; Stanley W. Lawrence, Utah Technical College; Richard D. Morris, Portland Community College; Chuck Hansman, Timroc Industries; Lee Rosenthal, Fairleigh Dickinson University; J.N. Tompkin, British Columbia Institute of Technology; H. Hayre, University of Houston; Brian Delaney, Laidlaw Industries; Richard Parker, Seneca College; Robert D. Thompson, ITT Technical Institute; Doug Fuller, Humber College; David Ingram, DeVry Institute of Technology; and Robert I. Eversoll, Western Kentucky University.

CONTENTS

Principles of Electronics

1

Introduction

Learning Objectives

Upon completion of this chapter you will be able to

- Understand the historical perspective of electricity and electronics.
- Describe some of the important areas where electronics technology is applied.
- List examples of common electronic components.
- Define the basic units of measurement.
- Describe the SI system of measurement.
- Be able to express numbers in scientific notation.
- Convert from one power of 10 to another.
- Define engineering notation.
- Describe basic atomic structure.
- Explain the principle of electric charge.
- Express Coulomb's law.

Practical Application

Each chapter in this book begins with a practical application of electronics. The purpose of the practical application is to illustrate how the material covered in a particular chapter relates to situations you may encounter as an electronics technician. Because the field of electronics is so diverse, every effort has been made to present a wide cross section of the employment possibilities available and a brief overview of the tasks expected of you as a technician.

The practical applications presented at the beginning of each chapter outline a task you are expected to perform as an electronic technician for companies involved in the manufacture, design, installation, and maintenance of electronic equipment. When you have completed a chapter, your knowledge of the material covered in the chapter should be sufficient to solve the task described in the practical application. The last section in each chapter describes the method of solution for a given task.

HISTORY OF ELECTRICITY AND ELECTRONICS

In 600 B.C. the Greeks discovered that certain substances, when rubbed with fur, caused other substances to be attracted to them. Thales of Miletus (640 B.C.–546 B.C.) is credited with having been among the first to observe the attraction of amber for small fibrous materials and bits of straw. Amber, a solidified tree sap, was used by these people for ornamental purposes. The Greek word for amber was *elektron,* which is the root word for *electricity.*

Although the electrification of amber by friction was known for many centuries, it was not until the beginning of the seventeenth century that Sir William Gilbert (1544–1603) announced the discovery that many substances could be electrified by friction. In Gilbert's book *De Magnete,* he also described how amber differs from magnetic lodestones in its attraction of certain materials. Gilbert demonstrated that lodestone always attracts iron or other magnetic bodies, but an electric object exerts its attraction only when it has been recently rubbed.

It was Gilbert's discoveries that led scientists to the realization that all fundamental properties of electricity and magnetism can be traced to the state or motion of something called *electric charge.* In 1733, a Frenchman named Charles F. DuFay (1698–1739) made an important discovery while experimenting with the conduction of electricity. DuFay found that a glass rod became electrified when rubbed with silk and that a rod made of wax would become electrified when rubbed with fur. What made DuFay's discovery so significant was that he determined that the rods possessed two different kinds of electrical "fluid."

In the eighteenth century, Benjamin Franklin (1706–1790) introduced the terms *positive* (+) and *negative* (−) to describe the two types of electricity. Franklin defined a glass rod that had been rubbed with silk as having a positive charge and a wax or rubber rod rubbed with fur as having a negative charge.

In 1785, physicist Charles Augustin de Coulomb (1736–1806) proved the laws of attraction and repulsion that exist between positive and negative electric charges. **Coulomb's law** states that the force acting between two charges is directly proportional to the product of the two charges and inversely proportional to the square of the distances between the charges.

A year later, in 1786, the Italian scientist Luigi Galvani made a startling discovery during his experiments with electricity. Galvani found that a frog's leg could be made to twitch if copper and iron were brought into contact with a nerve and a muscle. Galvani named his discovery *animal electricity.*

In 1796, Alessandro Volta furthered Galvani's research by proving that electricity could be produced if unlike metals separated by moistened paper were brought into contact. In his original design Volta stacked pairs of unlike metals on top of each other in order to increase the intensity of the electric charge. This arrangement became known as the *voltaic pile,* which was the first battery.

The first significant connection between magnetism and electricity was uncovered by Hans Christian Oersted (1777–1851) in 1820. Oersted accidently discovered that a current-carrying wire influenced the orientation of a nearby compass needle. Oersted's discovery tied the origin of magnetic fields to the motion of electric charges and became known as *electromagnetism.*

In 1826 German physicist Georg Simon Ohm (1787–1854) observed that the electrical resistance of metallic conductors remains constant over wide ranges of potential difference. This observation became known as **Ohm's law.**

Two scientists are given credit for the discovery of **electromagnetic induction:** Michael Faraday (1791–1867) and Joseph Henry (1797–1878). Since Faraday was the first to publish his findings, the laws of electromagnetic induction are called **Faraday's laws.**

The dawn of electronics began in 1868 when Heinrich Geissler (1814–1879) developed "Geissler tubes," in which electrical discharges in rarefied gases produced different colors. These tubes, made in diverse sizes, shapes, and colors, attracted the attention of physicists in the leading scientific institutions and universities of the world.

One of the scientists intrigued by Geissler tubes was Sir William Crookes (1832–1919), an English physicist and chemist. Crookes is widely credited with being the inventor of the first cathode ray tube.

In 1883 Thomas Edison (1847–1931) made an interesting discovery during his work on the incandescent light bulb. When Edison inserted an electrode in a glass bulb with a filament, he found that a current would flow if a positive potential was applied to the electrode and the filament was hot. Edison also noted that no current would flow if the filament was cold. This device was the first thermionic diode and eventually led to the invention of the electronic vacuum tube.

Sir John Fleming recognized the importance of Edison's discovery, and in 1904 he developed the Fleming valve, which is now called a vacuum-tube rectifier, or diode. The Fleming valve is still used in some circuits, such as radio and television transmitters and receivers, to change alternating current into direct current.

The first vacuum tube capable of boosting, or amplifying, small electrical signals was patented in 1907 by Lee de Forest (1873–1961). The vacuum tube used by de Forest was a triode tube.

The first tetrode tube was invented by the German scientist Walter Schottky in 1916. Schottky is credited with a great many inventions during his career, among them the first semiconductor diode in 1938.

In 1947 three American scientists, John Bardeen, Walter Brattain, and William Shockly, invented the transistor. It was this discovery that ushered in the era of solid-state electronics.

In 1958 the integrated circuit (IC) was developed by Jean Hoerni, Jack Kilby, Kurt Lehovec, and Robert Noyce. The integrated circuit can now contain thousands of transistors on a semiconductor chip the size of a thumbnail.

The development of the IC led to the invention of the microprocessor by Ted Hoff in 1971. Hoff's discovery revolutionized the computer industry, greatly reducing the size and cost of computing machines.

In 1977 three companies, Apple, Radio Shack, and Commodore, introduced personal computers for home and office use. The development of personal computers was furthered by the Motorola Corporation in 1979 when it began marketing a powerful 16-bit microprocessor.

In 1980, Microsoft introduced the MS-DOS disk-operating system for personal computers, which evolved into an industry standard for computers using INTEL microprocessors. Further developments in microprocessor technology led to the development of the 64-bit, 100 MHz, Pentium microprocessor by INTEL, which was made commercially available in 1993.

Section Review

1. What laws did Charles Augustin de Coulomb prove in 1785?
2. Who was it that discovered "animal electricity"?
3. What device was invented in the 1970s that led to the development of the personal computer?

APPLICATIONS OF ELECTRONICS

Before the invention of the integrated circuit, electronic equipment such as televisions were extremely expensive and required considerable maintenance. The first generation of televisions used vacuum tubes that were handmade and had relatively short lifespans. Telephone systems used thousands of electromechanical relays for interconnecting calls. Robots, cellular phones, and portable computers were seen only in science fiction movies. Thirty years ago, the idea of walking up to a bank machine to pay bills and withdraw money would have seemed absurd.

Electronics has had a tremendous impact on almost every aspect of life in the twentieth century. Without electronics there would be no radio, television, motion pictures, fluorescent lighting, computers, or long-distance telephone calls. Electronics has played a major role in the improvement of the quality of life through major advancements in the fields of science and medicine. In this section, we shall examine some of the general applications of electronics to provide some insight into this diverse and exciting field.

Communications

The purpose of any communication system is to convey a message from one point to another. Modern communication systems use electronic circuits and devices for the collection, processing, and storage of information before being transmitted. After the message has been transmitted, electronic circuits play a major role in the decoding, storage, and interpretation of this data. Thanks to satellites and digital electronics, radio, television, and telephone transmissions are fast, efficient, and relatively inexpensive for today's consumer. Recent developments in fiber optic communications now make it possible to transmit data thousands of miles at the speed of light. By 1990, over 10 million kilometers of fiber optic cable had been installed worldwide, allowing the transmission of telephone calls, computer data, and cable television signals.

As the field of electronics continues to grow, old technologies eventually become obsolete. Recent advancements in the development of integrated circuits such as microprocessors have created new markets for products such as fax machines and cellular telephones. The fax machine has replaced the telex machine and telegraph in the same way the push-button telephone replaced the rotary-dial. Fax machines allow the transmission of printed documents using conventional telephone lines. The cellular phone succeeded in rendering the mobile phone obsolete in much the same way that the fax machine eliminated the telex. Cellular phones use satellite systems to transmit phone calls with astounding speed and clarity. Figure 1–1 shows a satellite communications system.

Computers

The modern electronic computer started with the ENIAC (Electronic Numerical Integrator and Calculator) completed in 1946 by J. Pesper Eckert and John Mauchly at the University of Pennsylvania. This discovery created, for all practical purposes, a second industrial revolution. The personal computer and software industry is a multibillion dollar industry employing hundreds of thousands of people worldwide. Computers are used to calculate gas and electric bills, to determine the price of goods in supermarkets, and to control the financial and management activities of businesses; in fact, any decision pertaining to the sorting and calculating of data on a large scale will usually involve a computer.

FIGURE 1-1 *Satellite system.*

Scientists, engineers, and technicians use high-speed computers for calculations or data analysis that would otherwise be extremely tedious or even impossible because of the complexity or length of the calculations. Some of these computers are capable of performing one hundred million calculations per second. Personal computers have created another multibillion dollar market for small, inexpensive computers that can be used in the home or office. These machines use programs to perform word processing, business spreadsheets, video games, tutorial lessons, and a host of other consumer requirements.

The computer industry has spawned a massive software industry that employs thousands of computer programmers and software designers. These people are involved in the design of computer programs for almost every aspect of society, such as computer programs that land airplanes, teach children to read, run entire businesses, build automobiles, and predict earthquakes. The applications for computer programs are unlimited. Figure 1-2 shows a typical computer.

Microprocessors

In addition to being the single most important device in a computer system, microprocessors are found in a wide array of commercial and industrial applications. They are used in calculators, video games, and home appliances such as the microwave oven and VCR (video cassette recorder). Today, most homes contain one or more microprocessor-controlled appliances. Microprocessors are common in automobiles to control fuel injection, temperature, and cruising speed. Homes and offices use microprocessor-based systems to control temperature, humidity, lighting, and security. Almost all fire and burglar alarm systems count on a microprocessor to oversee all activities in the system. VCRs and video cameras also use microprocessor technology to record and reproduce video images. Figure 1-3 shows a typical microprocessor.

FIGURE 1–2   *Typical computer system.*

FIGURE 1–3   *Microprocessor.*

Automation

Electronic circuits and devices have radically altered the control of manufacturing processes. By using electronics it is now possible to automatically control everything from the manufacture of soup, soap, and gasoline, to the manufacture of engine blocks, sheet steel, and paper. Automation is also used in the control of modern weapons such as nuclear submarines, supersonic airplanes, missiles, and rockets. One reason automation has become extremely popular is convenience. For example, automatic speed control in an automobile is primarily for the convenience of the driver. Vending machines, home heating and air-conditioning systems, and many remotely controlled servomechanisms are also primarily for convenience.

Approximately two-thirds of the wealth created in the United States comes from manufacturing. In order to remain competitive in a world market, U.S. manufacturers have become more and more dependent on automation to increase production, reduce manufacturing costs, and improve quality. Electronic devices have played a major role in the automation of factories and have been partly responsible for the loss of certain jobs in the manufacturing sector. Fortunately, automation has also created a demand for the design, construction, and maintenance of electronic components used in automated manufacturing. An automated manufacturing facility is shown in Figure 1–4.

Consumer Products

The entire home entertainment industry is a direct result of electronics. Since the early days of radio and television, a wide range of consumer products has been developed, such as CD (compact disk) players, VCRs, and color televisions. Audio equipment for home use has changed radically in the last half of the twentieth century. Long-playing records (LPs) have been replaced by CDs, providing superior sound quality and durability. The VCR has significantly affected the way we watch television, and the video camera has made it possible to record events and play them back using a television set.

FIGURE 1–4 *Automated manufacturing.*

FIGURE 1–5 *Electronic consumer products.*

The calculator and home computer have also brought tremendous change to the consumer's life. Now, almost anyone can afford computing machines that decades ago would have cost hundreds of thousands of dollars. Video games and educational programs have provided both entertainment and a learning environment that was previously unheard of in a typical household. Digital clock radios, answering machines, electronic typewriters, microwave ovens, and remote-controlled television sets have all provided convenience and comfort to today's consumer. Some typical electronic consumer products are pictured in Figure 1–5.

Medicine and Science

Medical electronics is a combination of electronics and biology. In the field of medical electronics, various types of electronic equipment, such as shown in Figure 1–6, are used to monitor and diagnose sickness, disease, and injury. Electrocardiograph (ECG) machines are used to diagnose heart ailments; electroencephalograph (EEG) machines record brain

FIGURE 1–6 *Medical electronic equipment.*

activity. X-ray machines and surgical lasers owe their existence to electronics. Virtually every hospital in the world contains electronic equipment of some sort. Diagnostic imaging uses a computer to create cross-sectional planes of the human body.

In the science laboratory, electronic measuring equipment is invaluable. These instruments can measure particles invisible to the human eye and can access and compare data at incredible speeds. In scientific evaluation, computers and medical software have played a major role in uncovering some of the mysteries of sickness and disease. Machines performing billions of computations per second have greatly reduced the time and cost involved in scientific research and development.

Section Review

1. Electronics has played a major role in the improvement of quality of life through advancements in the fields of science and medicine (True/False).
2. How much of the wealth created in the United States comes from manufacturing?
3. In what year was the ENIAC computer completed?

1-3
ELECTRONIC COMPONENTS

In this section we shall examine several types of electronic components that you will study in detail later in this text. There are surprisingly few types of electronic components, considering the wide variety of applications. One of the most popular types of electronic components in use is the **semiconductor.** These devices include everything from diodes and transistors to integrated circuits (ICs). Figure 1–7 shows some typical semiconductor devices.

Another common electronic component is the visual display device, of which there are several types, including the **cathode-ray tube (CRT)** for television sets, **liquid-crys-**

(a)

(b)

FIGURE 1–7   *Semiconductor devices. (a) Diodes and transistors.*
(b) Integrated circuits.

FIGURE 1–8 *Light-emitting diode. (a) Schematic symbol.*
(b) Packaging. (c) Seven-segment display. (d) Display styles.

tal display (LCD) for calculators and wristwatches, and the **light-emitting diode (LED).** A single LED may be used to indicate an ON-OFF status and an array of LEDs is capable of producing all alphanumeric characters as shown in Figure 1–8. LEDs are very popular in consumer products such as clock radios because of their high visibility in the dark.

Resistors are electronic components that are found in almost every electronic circuit. These devices are made either of carbon-composite materials or wound with special resistance wire and wrapped around a ceramic-core form. Resistors come in a wide variety of shapes and sizes and are rated by their ability to resist the flow of current and to dissipate heat. Some typical resistors are shown in Figure 1–9.

Another common electronic component is the **capacitor.** The capacitor is capable of storing an electrical charge and is popular in electronic circuits where filtering is required. Often the filter requirements include blocking DC current and passing AC current, or smoothing out voltages produced by rectifiers and power supplies. Figure 1–10 shows some typical capacitors.

Inductors, another popular type of electronic component, are used in a wide variety of applications. Also known as coils, or chokes, inductors are used to store energy in an electromagnetic field. Chokes are common in audio-electronic circuits because of their excellent filtering capabilities. Some typical inductors are shown in Figure 1–11.

FIGURE 1–9   *Typical resistors. (a) Carbon-composition resistors. (b) Wire-wound resistors. (c) Variable resistors or potentiometers. (d) Resistor networks.*

FIGURE 1–10   *Typical capacitors.*

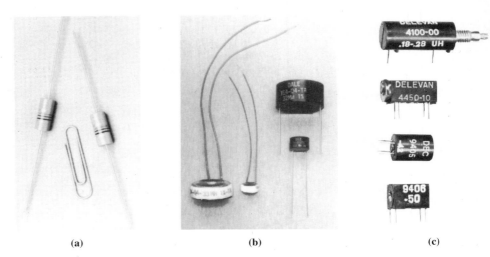

(a)   (b)   (c)

FIGURE 1-11 *Typical inductors.*

Section Review

1. Which electronic device is made from carbon-composite materials or wound with special resistance wire?
2. Capacitors are used to store energy in an electromagnetic field (True/False).
3. List three types of visual display devices.

1-4
UNITS

There are at least three quantities that can be proven to exist without the use of any measuring instruments. These three quantities are **length, mass,** and **time.** These are known as **fundamental units.** Other quantities that are obtained from the fundamental units are referred to as **derived units.** Over the years measurement of these three fundamental units has led to at least three systems of measurement: the MKS system, the CGS system, and the English system of units. The International System of Units, or **SI,** represents a rationalized selection of the MKS system and is commonly referred to as the *metric system.*

In recent history, the systems of units most commonly used were the English and metric. The English system of units developed in a rather natural way based on physical objects used to convey the dimensional information. English units of length consist of the foot, rod, and chain; English units of mass consist of the stone and slug. The metric, or SI, system was developed scientifically using precise measuring instruments. In the SI system the unit of length is the **meter, m,** the unit of mass is the **kilogram, kg,** and the unit of time is the **second, s.**

In October 1965, **IEEE** (the Institute of Electrical and Electronics Engineers) adopted the SI system as a standard for all engineering and scientific literature. The SI system uses unit prefixes to form multiples and submultiples of a unit. The United States officially recognized the SI system in 1975 by passing the Metric Conversion Act. The International System of Units has seven base units: length, mass, time, electric current, temperature, luminous intensity, and molecular substance.

There are many situations in which base or supplementary units of measure are not suitable and derived units are required. The SI derived units are formed from the previously defined SI base units. For example, the base unit of time has the following derived

TABLE 1-1 *Partial listing of SI derived units.*

Quantity	Quantity Symbol	Unit	Unit Symbol
Capacitance	C	farad	F
Conductance	G	siemens	S
Electric charge	Q	coulomb	C
Electromotive force	E	volt	V
Energy, work	W	joule	J
Force	F	newton	N
Frequency	f	hertz	Hz
Inductance	L	henry	H
Magnetic flux	Φ	weber	Wb
Magnetic flux density	B	tesla	T
Power	P	watt	W
Resistance	R	ohm	Ω
Reactance	X	ohm	Ω
Impedance	Z	ohm	Ω

units: frequency, velocity, and acceleration. Units derived from the base unit of mass deal with work, force, power, pressure, and other physical science measurements. There are 16 derived units, of which 14 are considered essential for the study of electronics. Table 1–1 lists these 14 derived units with special names.

Section Review

1. Length, mass, and time are known as fundamental units (True/False).
2. Without referring to Table 1–1, list as many derived units as possible.
3. Which system of units is also referred to as the metric system?

1-5
SCIENTIFIC NOTATION

Numbers that are very large or very small can be more conveniently written as a number multiplied by 10 and raised to a power. This method of writing numbers is known as **scientific notation.** A number written in scientific notation is expressed as the product of a number greater than or equal to 1 and less than 10, and is a power of 10. To express a number in scientific notation, the decimal point is moved until there is one significant digit to the left of the decimal point. The result is then multiplied by the appropriate power of 10 to return the quantity to its original value.

--- EXAMPLE 1-1 ---

Solution

(a)

(b)

$$72,300 = 7.23 \times 10^4$$
$$0.0057 = 5.7 \times 10^{-3}$$

Express the following numbers in scientific notation.

(a) 72,300
(b) 0.0057

TABLE 1-2 *Prefixes for use with SI units.*

Prefix	Symbol	Scientific Notation	Value
tera	T	10^{12}	1,000,000,000,000
giga	G	10^{9}	1,000,000,000
mega	M	10^{6}	1,000,000
kilo	k	10^{3}	1,000
milli	m	10^{-3}	0.001
micro	μ	10^{-6}	0.000001
nano	n	10^{-9}	0.000000001
pico	p	10^{-12}	0.000000000001
femto	f	10^{-15}	0.000000000000001

Since some of these powers of 10 appear frequently, various multiples and submultiples are assigned prefixes and symbols. Those that are most frequently used in electronic calculations are listed in Table 1-2.

_____ **EXAMPLE 1-2** _____

Express 5600 grams in a more appropriate form using SI unit prefixes.

Solution   Moving the decimal point three places to the left results in a figure of 5.6, which now must be multiplied by 1000, or 10^{3}, to return the quantity to its original value.

$$5600 \text{ g} = 5.6 \times 10^{3} \text{ g}$$

From Table 1-2, the value of 10^{3} can be replaced by using the prefix kilo on the root unit.

$$5600 \text{ g} = 5.6 \times 10^{3} \text{ g} = 5.6 \text{ kg}$$

To convert a fraction to scientific notation, it is necessary to first divide the fraction out into decimal form. The decimal point is now moved to the right until there is one significant digit to the left of the decimal.

_____ **EXAMPLE 1-3** _____

Express 1/500 meters in scientific notation.

Solution   Divide the fraction out into decimal form.

$$1/500 \text{ m} = 0.002 \text{ m}$$

The decimal point is now moved three places to the right, and the result is multiplied by 10^{-3}.

$$0.002 \text{ m} = 2.0 \times 10^{-3} \text{ m}$$

or, using SI unit prefixes,

$$2.0 \times 10^{-3} \text{ m} = 2.0 \text{ mm}$$

To change a number from scientific notation to ordinary notation, the procedure is reversed. The following examples illustrate this method.

EXAMPLE 1–4

Solution   The decimal point must be moved five places to the right. Therefore, additional zeroes must be included for the proper location of the decimal point.

Convert 7.84×10^5 to ordinary notation.

$$7.84 \times 10^5 = 784{,}000$$

To add or subtract numbers expressed in scientific notation, it is necessary to convert numbers to a common power of 10.

EXAMPLE 1–5

Solution

Calculate the sum of (3.4×10^5) + (5.9×10^6).

$$(0.34 \times 10^6) + (5.9 \times 10^6) = 6.24 \times 10^6$$

EXAMPLE 1–6

Solution

Find the sum of (8.4×10^3) − (4.7×10^2).

$$(8.4 \times 10^3) - (0.47 \times 10^3) = 7.93 \times 10^3$$

To multiply numbers expressed in scientific notations, the exponents are added; to divide, the exponents are subtracted.

EXAMPLE 1–7

Solution

Calculate the product of (4×10^5) × (2.2×10^2).

$$(4 \times 10^5) \times (2.2 \times 10^2) = 4 \times 2.2 \times 10^{(5+2)}$$
$$= 8.8 \times 10^7$$

EXAMPLE 1–8

Solution

Find the result of (6×10^6)/(3×10^2).

$$\frac{6 \times 10^6}{3 \times 10^2} = \frac{6}{3} \times 10^{(6-2)}$$
$$= 2 \times 10^4$$

Numbers that are written in powers of 10 notation having exponents that are multiples of 3 are written in **engineering notation.** This technique calls for the use of numbers between 1.0 and 999 times the appropriate third power of 10. The values shown in Table 1–2 would be considered to be in engineering notation because these prefixes all represent powers of 10 notation that have exponents with multiples of 3. The following rules are used to express a decimal number in engineering notation.

1. Express the number in scientific notation.
2. If required, make the exponent a multiple of 3 by subtracting 1 or 2 from the exponent.
3. The decimal point is moved one or two places to the right to correspond to the change in exponent.

EXAMPLE 1–9

Write 645,000 in engineering notation.

Solution   First, express the number in scientific notation.

$$645,000 = 6.45 \times 10^5$$

Subtract 2 from the exponent.

$$6.45 \times 10^{5-2}$$

Move the decimal point two places to the right.

$$645 \times 10^3$$

The number 645,000 could also be expressed in engineering notation by converting the number to scientific notation and moving the decimal point one place to the left.

$$645,000 = 6.45 \times 10^5$$

Add 1 to the exponent.

$$= 6.45 \times 10^{5+1}$$

Move decimal point one place to the left.

$$= 0.645 \times 10^6$$

With some calculators, it is possible to fix a readout to either SCI (scientific) or ENG (engineering) notation, as well as selecting the number of decimal places required. The EXP, or EE, key is used on calculators to enter a number expressed in scientific notation. The following example illustrates how to enter numbers expressed in scientific notation on a calculator.

EXAMPLE 1–10

Enter 220 kΩ on a calculator with an EXP key.

Solution

Step 1. Press the keys 2 2 0.
Step 2. Press the EXP key.
Step 3. Enter the number 3.

The display should now read

$$220 \ 03$$

Section Review

1. Convert 5.35×10^2 to ordinary notation.
2. What notation uses exponents that are multiples of 3?
3. What is the EXP or EE key used for on a calculator?

1-6

MOLECULES AND ATOMIC STRUCTURE

In the seventeenth century Galileo (1564–1642) discovered that when no force is exerted on a body, it stays at rest or it moves with constant velocity. This became known as Galileo's law of inertia. **Mass** is a measure of inertia. Mass is defined as the quantity of inertia possessed by a substance that occupies space. The mass of an object is an indication of the quantity of matter that it possesses. In other words, the more massive an object, the greater its inertia. Mass is often mistaken for weight, although there is a distinct difference between the two. The weight of a body is the force with which the body is pulled vertically downward by gravity. Mass is a universal constant equal to the ratio of a body's weight to the gravitational acceleration due to its weight.

Anything that occupies space and has mass is called **matter.** All matter is composed of small particles called **atoms.** There are presently over 100 known types of atoms, referred to as **elements.** An element is a substance that cannot be chemically decomposed and contains atoms of one kind only. Atoms can combine to form **molecules.** For example, one molecule of water contains two atoms of hydrogen and one atom of oxygen. Historically, molecules were regarded as being formed by the association of individual atoms. When atoms become attached to each other, a chemical *bond* has occurred.

The nature and structure of atoms was a principal subject of study by physicists and chemists during the first half of the twentieth century. Before the 1900s, there was considerable speculation concerning the nature of the atom. In 1808 John Dalton (1766–1844) presented his version of atomic theory in a treatise entitled *A New System of Chemical Philosophy.* Among other things, Dalton noted that matter consists of individual atoms, atoms are unchangeable, and each element consists of a characteristic kind of identical atoms. While chemists were applying Dalton's atomic theory, a discovery by Humphry Davy and Michael Faraday proved that electricity and matter are closely related. Their research in electrochemistry, now called *electrolysis,* helped to establish the fact that electricity is atomic in character and that the "atoms" of electricity are part of the atoms of matter.

Early attempts to explain the structure of an atom were based on a model developed by J.J. Thomson, the discoverer of the electron. A discovery in 1911 by an English scientist named Ernest Rutherford established an atomic structure. His model was based on the principle of the atom being primarily an open space with all the mass concentrated in a central core, called the **nucleus.**

Rutherford's model of the atom was further extended by Niels Bohr. In 1913 Bohr proposed the theory that the simplest atom, hydrogen, consisted of a nucleus with a positively charged particle, or **proton,** and a planetary negative **electron** revolving in a circular orbit. The charge of a proton is equal in magnitude to that of an electron. Since the number of protons in the nucleus is equal to the electrons orbiting the nucleus, the atom is considered to be electrically neutral in charge.

Bohr's model of a hydrogen atom with one revolving electron is shown in Figure 1–12. This planetary atom theory may seem rather naive compared with more recent quan-

FIGURE 1–12 *Bohr model of a hydrogen atom.*

tum-mechanical theories, but in fact, considering what it was designed to do, Bohr's theory is an excellent example of a successful physical model.

As each electron is constantly in motion about the nucleus, it has energy of motion, or **kinetic energy.** Electrons travel at very high velocities and have charges that are equal, but opposite, to the charge of a proton. It is this attraction between the proton and electron that causes an equilibrium with the centrifugal force acting upon it.

Neutrons are particles of mass that do not have an electrical charge. The name is derived from the word *neutral,* meaning in this case neither positive nor negative. Neutrons are about the same size as protons and have approximately the same mass.

Most of the mass of an atom is located in the nucleus because each proton has 1837 times as much mass as an electron. Because neutrons are neither positively or negatively charged, the net charge of the nucleus is positive, due to the presence of protons. The neutron does not contribute to the flow of electricity, but it does contribute significantly to the mass of an atom. Table 1–3 shows the masses and charges of the proton, neutron, and electron.

In addition to having kinetic energy, the electron also has **potential energy.** This energy is the result of an attraction between the nucleus and the electron, which tends to "draw" the electron towards the nucleus. The potential energy increases as the distance from the nucleus increases. The closer an electron is to the nucleus, the greater its kinetic energy. However, the farther an electron is from the nucleus, the greater its potential energy. The potential energy of an electron is opposed to its kinetic energy. The net energy of an electron is the difference in magnitudes between the two energies.

The electrons very close to the nucleus are tightly held in their orbit and are called **bound electrons.** The outermost orbit, or **electron shell,** is known as the **valence orbit.** Electrons in the valence orbit are referred to as *valence electrons.* Because these electrons are farther from the nucleus, the attracting forces are not as great as in the inner orbit. A shell that contains its full complement of electrons is said to be *filled.* A given shell will

TABLE 1–3 *Some properties of three particles.*

Particle	Symbol	Charge	Mass
Proton	p	+e	$1.6725485 \times 10^{-27}$ kg
Neutron	n	0	$1.6749543 \times 10^{-27}$ kg
Electron	e^-	−e	9.109534×10^{-31} kg

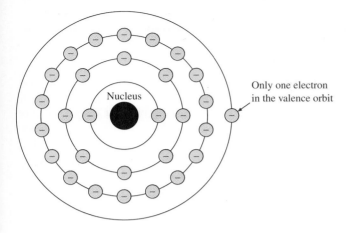

Only one electron
in the valence orbit

Nucleus

contain electrons only if shells closer to the nucleus are filled. Therefore, only outer shells can be partially filled, or incomplete. Figure 1–13 is a diagram of a copper atom with a single valence electron in the outer shell.

By adding energy, such as heat or light, it is possible for a valence orbit electron to absorb this energy and move into the valence orbit of a neighboring atom. An electron freed from its valence state is called a **free electron.** Because a single cubic inch of copper contains approximately 1.4×10^{24} free electrons, a veritable cloud of free electrons moves about in a piece of copper. An atom that has lost one or more valence electrons has a net positive charge and is called a **positive ion.** This concept of free electrons is extremely important because free electrons make an electric current possible.

As an electron moves from the valence orbit of one atom into the valence orbit of another atom, it leaves behind a space that it used to occupy. This space is often referred to as a hole. By definition, a **hole** is the absence of a negative electron and cannot be electrically neutral. Consequently, holes are considered to be positively charged and equal in magnitude to the negative charge of an electron.

Orbits, or shells, are designated by either numbers or letters. These shells are called *permissible energy levels.* Every electron that orbits the nucleus of an atom must exist at a permissible energy level. When atoms are in close proximity, as in a solid, the energy levels that existed for single isolated atoms are split to form *bands* of energy levels. In the study of electronics, the total possible number of energy levels in the outer subshells of a solid are divided into two classes: **valence band** and **conduction band.** The valence band of electron energies contains all the energy levels available to the valence electrons in the material. The electrons that are in the valence band do not move readily from atom to atom. The conduction band contains electrons with an energy level so high that the electrons are not attached or bound to any atom; instead, they are mobile and capable of being influenced by an external force.

Section Review

1. What is matter composed of?
2. According to Bohr's theory, atoms consist of a nucleus with a negatively charged particle called an electron and a planetary negative electron revolving in a circular orbit (True/False).
3. Which particles have mass but do not have an electrical charge?

__1-7__

ELECTRIC CHARGES

If a neutral atom were to acquire an extra electron, it would become negatively charged. Conversely, if it were to lose an electron, it would become positively charged. An electric charge must have either negative or positive polarity, labeled $-Q$ or $+Q$. **Charge** is one of the fundamental properties of matter. Whenever one atom exchanges charge with another, the charge gained by one is exactly equal to the charge lost by the other. Because of this, the total net charge of any isolated system never changes. This fact is expressed in terms of the **law of conservation of electric charge,** which is stated as follows:

> The algebraic sum of all electric charges in any isolated system is a constant.

It is possible to transfer an electric charge by causing friction between such materials as glass and silk. In these types of materials, charges do not move as freely as they would in metal. For this reason, they are referred to as **static charges,** or **static electricity.** A piece of glass rubbed with silk obtains a positive charge that does not easily escape, because the glass is a good insulator. If a copper rod is also rubbed with silk, it will become charged, but the charge will rapidly be conducted away, due to the poor insulation qualities of copper.

Whenever a body becomes charged, it is said to have energy. The energy is not in the charge itself but in an **electrostatic field** surrounding the body. The magnitude of the field is a function of the strength of the electric charge on the body. The electrostatic field is also referred to as a **dielectric field.** Since the ability of an electric charge to attract or repel another charge is a physical force, the forces between charged bodies are called **electrostatic forces.** These fields of force spread out in the space surrounding their point of origin and, generally, diminish in proportion to the square of the distance from their source.

Electric fields are often shown as lines of force, as in Figure 1–14. These lines are imaginary and are used merely to represent the direction and strength of the field. The arrowheads show the direction in which the field acts. The closeness or density of the lines indicates the strength of the field.

If an electron were placed within the electrostatic field of a positively charged body, the electron would be drawn toward the positive charge. If an electron were placed within a negative electrostatic field, the lines of force surrounding the charged body would repel the electron. In summary,

> Like charges repel each other, and unlike charges attract each other.

Figure 1–15 (a) represents the repulsion of two positively charged bodies and their associated fields. The force of attraction between two opposite-charged bodies is shown in Figure 1–15 (b). When two materials are charged with opposite charges and placed near

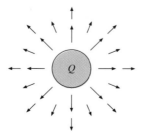

FIGURE 1–14 *Electric field around a stationary charge Q.*

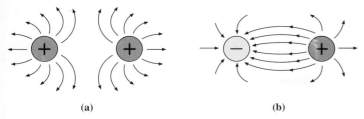

(a)   (b)

FIGURE 1–15 *Electrostatic lines of force. (a) Like charges repel. (b) Unlike charges attract.*

one another, the excess electrons on the negatively charged material will be pulled toward the positively charged material.

Whenever two charged bodies are joined by any conducting medium, they always try to neutralize each other. If the two materials were connected by a conductor, a path would be provided for the electrons of the negative charge to cross over to the positive charge, and the charges would neutralize each other. If the two materials contained large amounts of electric charge and were brought close together, the electrons may jump from the negative charge to the positive charge before the two materials actually touch each other. In this case, you would actually see the discharge in the form of an arc. With very strong charges, static electricity can discharge across large gaps, causing massive arcs. Lightning is an example of the discharge of static electricity resulting from the accumulation of a static charge in a cloud as it moves through the air.

The basic unit for measuring the quantity of electric charge is the **coulomb, C,** named after Charles Augustine de Coulomb. One coulomb represents the collective charge found on 6.24×10^{18} electrons. The number of electrons in one coulomb of charge is important because this number is the basis for the unit of electric current. The reciprocal of 6.24×10^{18} gives the number of coulombs in an electron or proton. Therefore, the equation for defining a charge in coulombs is

$$Q = \frac{n}{6.24 \times 10^{18}} \qquad\qquad (1.1)$$

where Q = charge, in coulombs
n = number of electrons

_____ **EXAMPLE 1–11** _____

Solution

$$Q = \frac{42.71 \times 10^{12} \text{ electrons}}{6.24 \times 10^{18} \text{ C}}$$

$$= 6.845 \ \mu C$$

What charge in microcoulombs does a body possess if it loses 42.71×10^{12} electrons?

_____ **EXAMPLE 1–12** _____

Solution

$$n = 3 \text{ C} \times 6.24 \times 10^{18} \text{ electrons}$$

$$= 18.72 \times 10^{18} \text{ electrons}$$

Three coulombs of negative charge were transferred from body A to body B. How many electrons changed bodies?

Section Review

1. What is static electricity?
2. State the law of conservation of electric charge.
3. Like charges repel each other, and unlike charges attract each other (True/False).
4. What is the basic unit for measuring the quantity of electric charge?

___ Summary ___

1. The SI system uses unit prefixes to form multiples and submultiples of a unit.
2. Numbers written in scientific notation with exponents that are multiples of 3 are also written in engineering notation.
3. Anything that occupies space and has mass is called *matter*.
4. All matter is composed of small particles called *atoms*.
5. Electrons have charges that are equal, but opposite, to the charge of a proton.
6. Holes are considered to be positively charged and equal in magnitude to the negative charge of an electron.
7. The algebraic sum of all electric charges in any isolated system is a constant.
8. Like charges repel each other, and unlike charges attract each other.
9. The basic unit for measuring the quantity of electric charge is the coulomb, C.

___ Answers to Section Reviews ___

Section 1–1

1. The laws of attraction and repulsion that exist between positive and negative electric charge
2. Luigi Galvani
3. The microprocessor

Section 1–2

1. True
2. Two-thirds
3. 1946

Section 1–3

1. Resistors
2. False
3. Liquid-crystal displays (LCDs), light-emitting diodes (LEDs), cathode ray tubes (CRTs)

Section 1–4

1. True
2. Compare your list with Table 1–1
3. The SI system

Section 1–5

1. 535
2. Engineering notation
3. To enter a number expressed in scientific notation

Section 1–6

1. Atoms
2. False
3. Neutrons

Section 1–7

1. The electric charge often transferred by friction between materials such as glass and silk
2. The algebraic sum of all electric charges in any isolated system is a constant
3. True
4. The coulomb, C

Review Questions

Multiple Choice Questions

1–1 The electrification of amber by friction was first noted in
(a) 400 B.C.   **(b)** 600 B.C.   **(c)** 6000 B.C.   **(d)** 800 B.C.

1–2 Who first described how amber differs from magnetic lodestone in its attraction of certain materials?
(a) Thales of Miletus   **(b)** Charles Augustin de Coulomb
(c) William Gilbert   **(d)** Hans Christian Oersted

1–3 All fundamental properties of electricity and magnetism can be traced to a state or motion called
(a) Force   **(b)** Gravity   **(c)** Charge   **(d)** Electromagnetism

1–4 The first battery was invented by
(a) Thomas Edison   **(b)** Luigi Galvani
(c) Benjamin Franklin   **(d)** Alessandro Volta

1–5 Hans Christian Oersted discovered
(a) Positive and negative fluids   **(b)** Animal electricity
(c) Electromagnetism   **(d)** Electromagnetic induction

1–6 Who noted that the electrical resistance of metallic conductors remains constant over wide ranges of potential difference?
(a) Georg Simon Ohm   **(b)** Benjamin Franklin
(c) Michael Faraday   **(d)** Joseph Henry

1–7 The laws of electromagnetic induction are named after
(a) William Crookes   **(b)** Michael Faraday
(c) Thomas Edison   **(d)** Hans Christian Oersted

1–8 The cathode ray tube was invented by
(a) William Crookes   **(b)** Thomas Edison
(c) Sir John Fleming   **(d)** Lee de Forest

1–9 Who invented the semiconductor diode?
(a) Walter Schottky   **(b)** William Schockly
(c) Sir John Fleming   **(d)** Lee de Forest

1–10 In what year was the first ENIAC computer put to use?
(a) 1926   **(b)** 1938   **(c)** 1946   **(d)** 1948

1–11 Which of the following is not an electronic component?
(a) Resistor   (b) Capacitor
(c) Computer   (d) Semiconductor device

1–12 Three fundamental units are
(a) Second, kilogram, and meter   (b) Length, mass, and time
(c) Frequency, velocity, and acceleration   (d) Energy, mass, and gravity

1–13 The number 52,400 expressed in engineering notation would be
(a) 5.24×10^4 (b) 0.524×10^5 (c) 524×10^3 (d) 52.4×10^3

1–14 Express 1/400 seconds in engineering notation
(a) 2.5×10^{-3} ms   (b) 0.25×10^{-3} ms   (c) 2.5 ms   (d) 25.0 μs

1–15 Mass is a measure of
(a) Velocity   (b) Inertia   (c) Gravity   (d) Weight

1–16 The law of inertia is named after
(a) Newton   (b) Coulomb   (c) Galileo   (d) Dalton

1–17 Anything that occupies space and has mass is called
(a) Element   (b) Molecule   (c) Atom   (d) Matter

1–18 When atoms become attached to each other, what has occurred?
(a) Bonding   (b) Electric charge   (c) Velocity   (d) Magnetism

1–19 The electron was discovered by
(a) Ernest Rutherford   (b) Newton   (c) J. J. Thomson   (d) Galileo

1–20 Electrons have
(a) Potential energy only   (b) Kinetic energy only
(c) No energy   (d) Both kinetic and potential energy

1–21 Electrons close to the nucleus are called
(a) Bound electrons   (b) Free electrons   (c) Valence electrons   (d) Shells

1–22 An atom that has lost a valence electron is called a
(a) Negative ion   (b) Positive ion   (c) Hydrogen atom   (d) Negative hole

1–23 Shells are also known as
(a) Bound electrons   (b) Valence electrons
(c) Ions   (d) Permissible energy levels

1–24 If an electron were placed within the electrostatic field of a positively charged body, the electron would be
(a) Repelled by the positive charge   (b) Not affected by the positive charge
(c) Alternately attracted and repelled   (d) Drawn toward the positive charge

1–25 An arc is caused by
(a) Two positively charged bodies brought close together
(b) A positive charge and a negative charge brought close together
(c) Negatively charged bodies brought close together
(d) Two positively charged bodies touching each other

1–26 The basic unit for measuring the quantity of electric charge is the
(a) Electron   (b) Volt
(c) Proton   (d) Coulomb

Practice Problems

1–27 Express the following numbers as powers of 10.
(a) 0.0001   (b) 100,000   (c) 0.00000001   (d) 100,000,000

1–28 Put the following numbers into scientific notation form.
(a) 220,000   **(b)** 0.00318   **(c)** 4,570,000   **(d)** 0.0002755

1–29 Express 1/2750 meters in scientific notation.

1–30 Express the following voltages in engineering notation.
(a) 5000 V   **(b)** 60,000 V   **(c)** 0.0000025 V   **(d)** 0.072 V

1–31 Convert the following numbers to ordinary notation.
(a) 6.77×10^6 **(b)** 3.25×10^{-3} **(c)** 8.46×10^8 **(d)** 1.73×10^{-4}

1–32 Find the sum of the following numbers.
(a) $(8.25 \times 10^3) + (3.57 \times 10^4)$ **(b)** $(9.75 \times 10^6) + (1.33 \times 10^5)$
(c) $(6.25 \times 10^3) - (0.38 \times 10^3)$ **(d)** $(4.38 \times 10^6) - (2.65 \times 10^5)$

1–33 Determine the products as powers of 10.
(a) $0.001 \times 1,000,000$ **(b)** 1000×100 **(c)** $0.01 \times 0.0001 \times 10^{-3}$
(d) $100 \times 10,000 \times 10^6$ **(e)** $10^6 \times 10^5 \times 10^{-8}$

1–34 Find the quotients as powers of 10.
(a) $1000/0.01$ **(b)** $(0.001)^2/10^2$
(c) $(1000)^2/0.01$ **(d)** $(0.00001)^3/(0.0001)^2$

1–35 Calculate the product of the following numbers.
(a) $(4.38 \times 10^6) \times (7.75 \times 10^3)$ **(b)** $(1.77 \times 10^5) \times (6.11 \times 10^{-8})$
(c) $(9.51 \times 10^{-3}) \times (2.28 \times 10^{-6})$ **(d)** $(8.75 \times 10^9) \times (3.15 \times 10^6)$

1–36 Find the quotient of the following numbers.
(a) $(3.25 \times 10^6)/(1.61 \times 10^4)$ **(b)** $(8.35 \times 10^2)/(2.44 \times 10^5)$
(c) $(3.71 \times 10^{-5})/(7.52 \times 10^3)$ **(d)** $(6.25 \times 10^{-3})/(9.75 \times 10^{-6})$

1–37 How many coulombs are represented by each of the following quantities of electrons?
(a) 1.248×10^{20} **(b)** 1.56×10^{18} **(c)** 1.872×10^{19} **(d)** 3.12×10^{19}

Essay Questions

1–38 What significance did amber have in the discovery of electricity?

1–39 Define electric charge.

1–40 How was the vacuum tube invented?

1–41 Describe four applications of electronics.

1–42 Why was the microprocessor such an important invention?

1–43 List three common semiconductor devices.

1–44 What is the difference between a base unit and a derived unit?

1–45 Explain the advantage of using scientific notation in electronic calculations.

1–46 What is the difference between engineering notation and scientific notation?

1–47 Describe Bohr's model of the hydrogen atom.

1–48 How does an electron move from the valence orbit of one atom to the valence orbit of another atom?

2

Current, Voltage, and Resistance

Learning Objectives

Upon completion of this chapter you will be able to

- Define electric current.
- Describe electron flow and conventional flow.
- Discuss electric potential and voltage.
- List the five main types of voltage sources.
- Differentiate between a voltage source and a current source.
- Explain the difference between a dependent source and independent source.
- Define resistance.
- Describe the relationship between temperature and resistance.
- List various types of resistors.
- Utilize the resistor color code.

Practical Application

As an engineering technician for a television station, you are required to select and install a new cable for a satellite transmitter located 800 ft from the station. The installation requirements are as follows:

1. The cable must be made of copper.
2. The total resistance of the cable must be less than 2 Ω.
3. The diameter of the conductor must be less than 2 mm.

In Section 2–19 you will be able to complete this task.

2-1
THE MOTION OF ELECTRIC CHARGE

In Chapter 1, it was stated that the outer-shell electron in a conductor such as copper is loosely held to its nucleus. This enables the outer-shell electrons to migrate through the material. When no external electric field is present, the outer-shell electrons move, or drift, in a *random motion* as shown in Figure 2–1. The random motion of the free electrons from atom to atom is normally equal in all directions so that electrons are not lost or gained by any particular part of the material. What the outer-shell electrons need, if they are to move in a specific direction, is the influence of an electric field. When most of the electron movement takes place in the same direction, so that one part of the material loses electrons while another part gains electrons, the net electron movement or flow is called **current.**

Figure 2–2 shows a typical battery with an excess of electrons at its negative terminal and a shortage of electrons at its positive terminal. When a copper conductor is connected across the battery, the excess electrons are drawn to the positive terminal. Because electrons repel each other and are attracted by positive charges, they always tend to move, or drift, from a point having an excess of electrons toward a point having a lack of electrons. If the excess and shortage of electrons at the two ends of the conductor were fixed at a definite quantity, in a very short time all the excess electrons would have traveled through the conductor toward the positive end. However, the battery continues to furnish excess electrons at one terminal and continuously removes electrons from the other terminal, so that the two terminals remain negative and positive for the life of the battery.

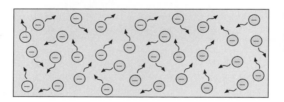

FIGURE 2–1 *Random movement of free electrons in a material.*

FIGURE 2–2 *Flow of electrons through a conductor.*

Section Review

1. When no external electric field is present, electrons will move in a fixed direction (True/False).
2. If one part of a material loses electrons while another part gains electrons, what is the net electron movement called?
3. Batteries continually furnish excess electrons at one terminal and remove electrons from the other terminal (True/False).

2-2
ELECTRIC CURRENT

Solid materials are composed of large numbers of atoms. In certain materials, called **conductors,** it is easy to remove some of the electrons from individual atoms. A conductor is a material that has the ability to transfer charge from one object to another. Materials that are poor conductors have electrons that are tightly bound to individual atoms. These materials are called **insulators.** An insulator is a material that resists the flow of charge. A class of materials called **semiconductors** is intermediate between conductors and insulators in its ability to transfer charge. A detailed discussion of semiconductors is presented in Chapter 19.

Electric current is defined as the rate of flow of charged particles through a conductor in a specified direction. The charged particle may be an electron, a positive ion, or a negative ion. In a solid, such as copper wire, the charged particle is the electron. In a copper wire, the ions are rigidly held in place by the atomic structure of the material. Therefore, ions cannot be current carriers in solid material. In a semiconductor, the current is a movement of electrons in one direction and a movement of positively charged holes in the opposite direction.

Electric current, or the rate of flow of charge, is measured in the units of coulombs per second. The SI unit of electric current is the **ampere, A.** It can then be stated that one ampere is equal to one coulomb of electric charge passing a certain point in an electric circuit in one second.

Amperes and coulombs per second express the same unit. The symbol for electric current is **I.** In equation form, current can be represented by the statement

$$I = \frac{Q}{t}$$ (2.1)

where I = current, in amperes
Q = charge transferred, in coulombs
t = time, in seconds

EXAMPLE 2-1

If 1.80 coulombs of charge pass a certain point in a conductor every minute, what is the electric current?

Solution

$$I = \frac{Q}{t}$$

$$= \frac{1.80 \text{ C}}{60.0 \text{ s}}$$

$$= 30 \text{ mA}$$

_____ EXAMPLE 2–2 _____

If a movement of 8 C creates a current of 10 A, in what time does this movement of charge take place?

Solution Since $I = Q/t$,

$$t = \frac{Q}{I}$$

$$= \frac{8\text{ C}}{10\text{ A}}$$

$$= 0.8\text{ s}$$

Section Review

1. What is the definition of electric current?
2. Insulators have the ability to transfer charge easily from one object to another (True/False).
3. What is the symbol for electric current?
4. One ampere is equal to one coulomb of electric charge passing a certain point in an electric circuit in one millisecond (True/False).

_____ 2–3 _____

ELECTRON VELOCITY

Although all free electrons in a conductor begin moving almost instantaneously when an external electric field is applied, the actual velocity, or drift, is extremely slow. One electron will move only a fraction of an inch in a second. However, the effective velocity is nearly 186,000 miles per second, the speed of light. This instantaneous velocity occurs because electrons moving at one end of a conductor will cause the electrons at the other end to move at the same instant. This chain reaction is similar to the effect of the cars of a long train starting and stopping together. If one car of a train moves, it causes all the cars of the train to move by the same amount. Free electrons behave the same way in a conductor. As each electron moves slightly, it exerts a force on the next electron, causing it to move slightly and, in turn, to exert a force on the next electron.

Section Review

1. What is the effective velocity of electrons in a conductor?
2. The drift velocity of electrons in a conductor is extremely slow (True/False).
3. Free electrons in a conductor behave in the same manner as the cars of a long train starting and stopping together (True/False).

_____ 2–4 _____

THE DIRECTION OF ELECTRIC CURRENT

The direction that current flows in a conductor is purely arbitrary as long as it is used consistently. Current can be thought of as a flow of negative charges in a conductor or a flow of positive charges. In a metallic conductor, the flow of current consists entirely of electrons, or a flow of negative charges. The flow of charge caused by an electric field in a gas or liquid consists of a flow of positive **ions** in the direction of the field or a flow of electrons opposite to the field direction. In semiconductors, there is another type of current carrier, called a **hole.** These holes flow in a direction opposite to that of electrons.

Historically, scientists were under the assumption that a positive charge represented a larger amount of electricity than a negative charge. As a result, it was taken for granted that current flowed from positive to negative. Since that time, it was determined that current flow in a typical conductor, such as copper, is actually the flow of electrons from negative to positive.

Electron Flow

The flow of electrons through a circuit is from the negative terminal of the power supply to the positive terminal. When the flow of current is in this direction, it is called **electron flow.** Since the electron is the lightest of the charged particles we have discussed, it would stand to reason that this particle could be most easily forced into directed motion as an electric current. In this text, we shall use electron flow to represent the flow of current through a device or circuit.

Conventional Current Flow

Conventional current is defined as the direction in which positive charge carriers flow through a circuit. If electrons flow one way in a material, the conventional electric current is in the opposite direction. This conventional flow is taken to imply that the electric current flows from the positive terminal of the power supply; it then travels through the external circuit and returns to the negative terminal of the power supply.

In the study of electronics, it is often necessary to be able to think in terms of both conventional current direction and electron flow. For example, all schematic symbols for electronic devices will use arrowheads that indicate conventional current flow.

Hole Flow

The current consisting of electrons jumping from one position to another is called **hole current** or **hole flow.** Whenever an atom has lost an electron, a nearby bound electron may receive enough energy from some external source to jump into the position left by the lost electron. It is easier for an electron to move into such a vacant position than it is to break completely clear of the atom. Consequently, another nearby electron may move into the vacancy created by the previous move. As each electron fills a hole, it must also leave a hole to be filled by another electron. The direction of motion of the hole is opposite to the direction of movement of the electrons.

Section Review

1. The direction in which current flows in a conductor is purely arbitrary as long as it is used consistently (True/False).
2. In what direction is the flow of electrons in a conductor?
3. What is conventional current flow?
4. What type of flow is caused by electrons jumping from one position to another?

DIRECT CURRENT AND ALTERNATING CURRENT

A current is considered to be unidirectional if it always maintains the same direction of flow; it is bidirectional if it changes direction. When a unidirectional current is unchanging or changes negligibly, it is referred to as **direct current (DC).** This type of current is a constant value that does not change with time nor cross the zero axis. Figure 2–3 shows a direct current signal with no variation over time.

If the magnitude of the unidirectional current varies with time, it is called a **pulsating current.** Figure 2–4 shows an example of a pulsating direct current. This type of waveform is similar to the output waveshape for a rectifier. A **rectifier** is a device that converts alternating current to direct current.

If the magnitude and direction of the current varies with time, it is referred to as **alternating current (AC).** When a conductor carries an alternating current, all of the moving charges continually change their directions of motion, and they vibrate about their average positions in the conductor. Figure 2–5 shows an example of an alternating current waveform.

Section Review

1. What is direct current?
2. If the magnitude of a unidirectional current varies with time, what is it called?
3. A rectifier converts direct current to alternating current (True/False).
4. What is the name for a current whose magnitude and direction vary with time?

FIGURE 2–3   *Direct current, DC.*

FIGURE 2–4   *Pulsating DC.*

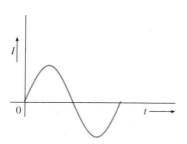

FIGURE 2–5   *Alternating current, AC.*

ELECTRIC POTENTIAL AND VOLTAGE

Potential energy is defined as energy possessed by a system by virtue of position. For example, a lifted object or a compressed spring both have the potential for doing work because of position or condition. It is energy that can be stored for long periods of time in its present form. Potential energy is capable of doing work when it is converted from its stored form into another form, such as kinetic energy. The potential energy of a charge at a given point is directly proportional to the charge itself.

Work is the amount of energy converted from one form to another, as a result of motion or conversion of energy from potential to kinetic. In order for work to be done, three things must occur.

1. A force must be applied.
2. The force must act through a certain distance.
3. The force must have a component along the displacement.

In the SI system, the unit of work is the **joule, J.** One joule is equal to the work done by a force of one newton acting through a distance of one meter. The **electric potential,** or potential difference, between two points in an electric circuit is the amount of work required to move a unit charge from one point to another. In other words, electric potential is a measure of work per unit charge. When a difference in potential exists between two charged bodies connected by a conductor, electrons will flow along the conductor until the two charges are equal and the potential difference no longer exists.

The units of electric potential are expressed in joules per coulomb. Because this quantity is so frequently employed, its units have been given the name **volts, V.** One volt is the potential difference between two points in an electric circuit, when one joule of energy is required to move one coulomb of electric charge from one point to the other. The relationship between voltage, energy, and charge can be expressed as

$$E = \frac{W}{Q} \qquad (2.2)$$

where E = the potential difference, in volts
      W = energy, in joules
      Q = quantity of electric charge, in coulombs

_____ EXAMPLE 2-3 _____

Solution

$$E = \frac{W}{Q}$$

$$Q = \frac{W}{E}$$

$$= \frac{0.44\ \text{J}}{2.6\ \text{V}}$$

$$= 0.169\ \text{C}$$

What value of charge is moved between two points if 0.44 J of energy is used and a potential of 2.6 V is developed?

_____ EXAMPLE 2–4 _____

Determine the energy expended moving a charge of 80 μC through a potential difference of 9 V.

Solution

$$W = QE$$
$$= (80 \times 10^{-6} \text{ C}) (9 \text{ V})$$
$$= 720 \text{ μJ}$$

Voltage, or a difference in potential, exists between any two charges that are not exactly equal to each other. Even an uncharged body has a potential difference with respect to a charged body; it is positive with respect to a negative charge and negative with respect to a positive charge. Voltage also exists between two unequal positive charges or between two unequal negative charges. Therefore, voltage is purely relative and is not used to express the actual amount of charge, but rather to compare one charge to another and indicate the electromotive force between the two charges being compared.

To maintain a constant current in a conductor, it is necessary to maintain a steady potential difference across the conductor. This potential difference can be supplied only if some device converts another form of energy into electric energy. Such a device is called a source of **electromotive force, emf.** As charges pass through a source of electrical energy, work is done on them. The emf of the source is the work done per coulomb on the charges. A typical car battery has an emf of 12 volts, which means that 12 joules of work are done on each coulomb of charge that passes through the battery.

Section Review

1. Work is the amount of energy converted from one form to another (True/False).
2. What is the SI unit of work?
3. What is the equation for calculating voltage if the amount of energy and quantity of electric charge are known?
4. Voltage exists between any two charges that are equal to each other (True/False).

___ 2–7 ___

VOLTAGE DROP

In the previous section, a source of emf was defined as a device that produces voltage. In other words, emf represents a **voltage rise,** or the energy imparted to the free electrons for the development and maintenance of electric current. Any device that contributes energy to an electrical or electronic circuit is called an **active device.** A device that contributes no energy or removes energy is called a **passive device.**

A **voltage drop** represents the energy used by the free electrons while engaged in current flow. Passive devices, such as resistors, create voltage drops in a circuit because they absorb energy and dissipate it in the form of heat.

In order to distinguish between a source of voltage in a circuit and a voltage loss across a dissipative component, the following notation will be used in this text:

E = voltage source (volts)

V = voltage drop (volts)

Section Review

1. A source of emf is a device that produces voltage (True/False).
2. What is a passive device?
3. The energy used by the free electrons while engaged in current flow is called a voltage rise (True/False).

2–8

VOLTAGE AND CURRENT SOURCES

A voltage source is a device capable of converting one form of energy into electrical potential energy. A voltage **cell** is considered to be a voltage source because it is a source of stored chemical energy that may be converted to electrical energy as required. A **battery** consists of one or more cells connected in series. A **primary cell** provides electrical energy to the limit of its chemical energy. Primary cells are nonrenewable voltage sources. A **secondary cell** is capable of being recharged. The electrical energy returned to the secondary cell during its charging process restores the chemicals in the battery to their original state.

The schematic symbols used for voltage sources are shown in Figure 2–6. Figure 2–6(a) represents a fixed DC voltage source, and implies that the voltage remains constant and does not vary with time. Figure 2–6(b) shows a variable DC voltage source, which allows the output voltage of the source to be manually adjusted. Figure 2–6(c) is the symbol used for an AC voltage source.

There are five main types of voltage sources.

1. Chemical sources. These sources convert chemical energy into electrical energy. Examples of such sources are primary cells, secondary cells, and fuel cells. Fuels cells operate on the principle that when hydrogen and oxygen are combined, water is produced and electrical energy is released. Hydrogen is available from hydrocarbon sources such as ammonia, petroleum, propane, and natural gas. Figure 2–7 shows some common chemical voltage sources.

2. Solar and photovoltaic cells. The photovoltaic effect, which is the basis of all solar cells, was discovered in 1893 by Edmond Becquerel. Using a pair of electrodes immersed in an electrolyte, Becquerel observed a flow of current when the cell was illuminated with sunlight. A solar cell is a semiconductor device consisting of a thin layer of heavily doped P-type silicon on a heavily doped N-type silicon wafer. Incident light passes through the thin layer of P-type silicon to the junction region where it creates a large number of free electron-hole pairs. A strong electric field causes the free electrons to flow to the negative terminal, while free holes move to the positive output terminal. The net result is a potential difference created in the device when it is irradiated with light. Some examples of solar and photovoltaic cells are shown in Figure 2–8.

(a)   **(b)**   **(c)**

FIGURE 2–6 *Schematic symbols for voltage sources. (a) Fixed DC supply. (b) Variable DC supply. (c) AC supply.*

FIGURE 2–7 *Typical chemical voltage sources. (a) Cutaway of alkaline cell. (b) Primary cells.*

3. Thermoelectric generation. Thermoelectric generation is based on the principle that if a metal rod is heated at one end, negatively charged electrons flow from the hot end to the cooler end to reduce their energy. Examples of such sources are thermocouples and semiconductor thermoelectric engines. Figure 2–9 shows a photograph of a thermocouple.

4. Electromagnetic generation. This type of device is capable of converting mechanical energy to electrical energy. A generator, or dynamo, is a device in which mechanical energy is used to rotate conductors in a magnetic field to produce an emf. Examples of such sources are DC generators, as shown in Figure 2–10.

5. Electrical conversion. A **power supply** is a device that converts one type of electric potential or current to another. A DC power supply converts an alternating signal into a signal of a fixed magnitude. In addition to DC power supplies, other examples of electrical conversion sources are rotary converters and motor-generator sets. A typical power supply is shown in Figure 2–11.

An ideal voltage source provides a constant voltage across its terminals, regardless of the size of load connected to it. This type of source is also referred to as an *ideal independent voltage source,* or *constant voltage source.*

A **current source** is similar to a voltage source. An ideal, or constant current source, provides a specified value of current through its terminals regardless of the voltage across the terminals. This type of source is also referred to as an ideal independent current source. Ideal current sources are generally represented as a circle with an arrow symbol, as shown in Figure 2–12.

Section Review

1. What is a voltage source?
2. Primary cells are rechargeable (True/False).
3. Name five types of voltage sources.

40-W, high density solar module
100-mm x 100-mm (4" x 4") square cells are
used to provide maximum power in a minimum of
space. The 33 series cell module provides a strong
12-V battery charging current for a wide range of
temperatures (−40°C to 60°C)

(a)   **(b)**

FIGURE 2–8 *Typical solar and photovoltaic cells. (a) Photo conductive
cells. (b) Solar module.*

FIGURE 2–9 *Typical thermocouple.*

FIGURE 2–10 *DC generator.*

FIGURE 2–11 *Typical electronic power supplies.*

FIGURE 2–12 *Schematic symbol for an ideal current source.*

2–9

RESISTANCE

The ease with which electric current flows through a material depends on whether there are relatively large numbers of free electrons. A material with few electrons per unit volume would have a substantial opposition to current flow. If an electric potential is applied to a conductor, the electrons are given increased kinetic energy and collide more frequently with atoms. This friction caused by atoms colliding increases the temperature of the conductor. Therefore, when electric current flows in a conductor, some of the electrical potential energy is converted to heat energy. Because of this heat dissipation, **resistance** is not only in opposition to current flow, but is also a producer of heat energy in a conductor.

The unit of measurement of resistance is the **ohm, Ω,** the capital Greek letter omega. The standard international ohm is defined as the resistance at zero degrees Celsius (0°C) of a column of mercury of uniform cross section having a mass of 14.4521 grams and a length of 106.3 centimeters. The resistance of any material with a uniform cross-sectional area is determined by the following four factors:

1. The kind of material
2. The length
3. The cross-sectional area
4. The temperature

As the cross-sectional area of the conductor increases, its resistance decreases. As the length of the current path through the conductor increases, its resistance increases. This statement can be expressed in the form of the following basic rule.

> The resistance of a conductor is directly proportional to its length, and inversely proportional to its cross-sectional area.

By using a constant of proportionality, ρ (the Greek lowercase letter rho), this rule can be expressed by the following mathematical equation:

$$R = \rho \frac{l}{A} \qquad (2.3)$$

where R = resistance of material, in ohms
l = length of the current path through the conductor, in meters
A = cross-sectional area of the conductor, in meters squared
ρ = constant of proportionality between R and l/A

The resistance of a conductor having unit length and unit cross-sectional area is defined as the specific resistance, or **resistivity.** The constant of proportionality in the above equation would represent the resistivity of the conductor. Resistivity is used to compare the inherent resistance characteristics of different materials. A material with the lowest resistivity will be the best conductor and the poorest insulator. Conductors are defined as materials having resistivities from $10^{-6}\,\Omega \cdot m$ to $10^{-8}\,\Omega \cdot m$. Because the SI base unit of length and cross section is the meter, the resistivity of a material is equal to the resistance of a piece of the material one meter in length and one square meter in cross-sectional area. A comparison of resistivities for materials used as conductors, insulators, and semiconductors is given in Table 2–1.

TABLE 2–1 *Resistivities of common materials.*

Material	Resistivity, Ω · m at 20°C	Description
Aluminum	2.83×10^{-8}	Conductor
Copper	1.72×10^{-8}	Conductor
Germanium	47×10^{-2}	Semiconductor
Gold	2.45×10^{-8}	Conductor
Mica	2.02×10^{10}	Insulator
Silicon	6.4×10^{2}	Semiconductor
Silver	1.64×10^{-8}	Conductor
Teflon	3×10^{15}	Insulator

EXAMPLE 2–5

For the uniform copper conductor shown in Figure 2–13, determine the resistance of the conductor if it is 100 m in length and 0.0005 m² in cross-sectional area.

Copper conductor

FIGURE 2–13 *Uniform copper conductor.*

Solution

$$R = \rho \frac{l}{A}$$

$$= (1.72 \times 10^{-8} \, \Omega \cdot m) \, \frac{100 \, m}{0.0005 \, m^2}$$

$$= 3.44 \, m\Omega$$

EXAMPLE 2–6

What is the resistance of a piece of silicon 0.4 cm long with a cross-sectional area of 1 cm²?

Solution

$$R = \rho \frac{l}{A}$$

$$= (6.4 \times 10^2 \, \Omega \cdot m) \, \frac{0.004 \, m}{0.0001 \, m^2}$$

$$= 25.6 \, k\Omega$$

EXAMPLE 2–7

The printed circuit board shown in Figure 2–14 uses strips of copper foil as conductors. One of these conductors is 0.18 m long, 0.75 mm wide, and 0.05 mm thick. Determine the resistance of the copper foil conductor at ambient temperature (20° C)

FIGURE 2–14 *Copper conductors on printed circuit board.*

Solution

$$\text{Area} = \text{width} \times \text{thickness}$$

$$= 0.75 \, mm \times 0.05 \, mm$$

$$= 37.5 \times 10^{-9} \, m^2$$

$$R = \rho \frac{l}{A}$$

$$= (1.72 \times 10^{-8} \, \Omega \cdot m) \, \frac{0.18 \, m}{37.5 \times 10^{-9} \, m^2}$$

$$= 82.56 \, m\Omega$$

For a circular wire conductor, it is convenient to use circular units to denote the area of a material. The unit chosen to represent the relatively small diameter of a wire is the mil (0.001 in.). The **circular mil, CM,** is the area of a circular cross section having a diameter of 1 mil. The standard of reference for resistivity is, therefore, defined as the resistance of one circular mil-foot of conductor. The equivalent area of one circular mil in square inches is

$$1 \text{ CM} = 7.854 \times 10^{-7} \text{ in.}^2$$

To express the cross-sectional area of a circular conductor in circular mils, multiply the inch value of the diameter by 1000 and square the result. In equation form, the area represented by a circular mil is given as

$$\text{area in CM} = d^2 \qquad \qquad (2.4)$$

EXAMPLE 2–8

Solution

How many circular mils in the circular cross section of a material 2 in. in diameter?

$$2 \text{ in.} = 2 \text{ in.} \times 1000 \text{ mils/in.}$$
$$= 2 \times 1000 \text{ mils}$$
$$= 2 \times 10^3 \text{ mils}$$
$$A = d^2 = (2 \times 10^3)^2$$
$$= 4 \times 10^6 \text{ CM}$$

Section Review

1. The resistance of a conductor is inversely proportional to its length and directly proportional to its cross-sectional area (True/False).
2. What is the unit of measurement and symbol for resistance?
3. What is the main purpose of determining the resistivity of a conductor?
4. One circular mil is equal to what value in square inches?

2–10

RELATIONSHIP BETWEEN TEMPERATURE AND RESISTANCE

By increasing the temperature of a conductor, an increase in atomic motion occurs in the conductor. As free electrons flow through a metallic conductor, their motion is impeded when they interact with the atomic ion cores. When a conductor is heated, its atoms increase their energy and vibrate with larger amplitudes. The higher the temperature, the faster the electrons move about between atoms and the more the atoms vibrate. Consequently, the higher the temperature of a conducting material, the more likely it is that collisions will occur between the electrons drifting along the conductor and the atoms of the material. If the number of collisions increase, the flow of current will decrease.

Generally, the resistance of a conductor will increase with temperature. When pure metals are used within normal temperature ranges, the resistance increases in a fairly linear progression. For copper, the resistance at $100°C$ is nearly 43% greater than at $0°C$, and the resistance increases proportionately between these temperatures.

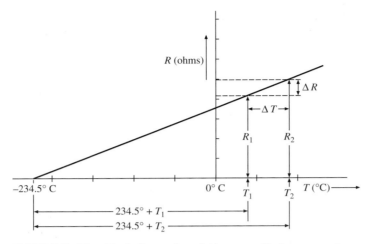

FIGURE 2–15 *Variation of resistance with temperature for copper.*

Because of the linear relationship between the resistance and temperature of a conductor, the slope $\Delta R/\Delta T$ is constant, as shown in Figure 2–15. The increase in resistance for most metals is approximately linear when compared with temperature changes. The proportional change in resistance to the change in temperature can be expressed in terms of the original resistance R as follows:

$$R = \frac{\Delta R}{\alpha \Delta T}$$

where ΔR represents the change in resistance and ΔT the change in temperature. The proportionality constant α is called the **temperature coefficient of resistance** and is defined as the change in resistance per unit resistance per degree change in temperature. Since the resistance of all pure metals increases as the temperature increases, metals are said to have a *positive temperature coefficient.*

The temperature intercepts and coefficients for common conducting materials are shown in Table 2–2. The temperature of $-234.5°C$ is called the **inferred zero-resistance temperature** of copper. Nichrome wire has a very small temperature coefficient, which means that its straight line graph of resistance versus temperature has a very small slope.

TABLE 2–2 *Temperature coefficients of resistance and conductor materials.*

Material	Temperature Coefficient, α	Inferred Zero-Resistance Temperature (°C)
Aluminum	0.0039	−236
Copper	0.00393	−234.5
Nichrome	0.0004	−2480
Nickel	0.006	−147
Platinum	0.003	−310
Silver (99.98% pure)	0.0038	−243
Steel, soft	0.0042	−218

Within normal ranges of temperature, the resistance (R_2) of a conductor at any temperature (t), is given in terms of the resistance (R_1) at a given temperature. This can be expressed by the following equation:

$$R_2 = R_1(1 + \alpha \Delta T) \qquad (2.5)$$

where

$$\Delta T = t_2 - t_1$$

EXAMPLE 2–9

Solution

What is the resistance of a copper wire at 30° C if the resistance at 20° C is 4.31 Ω and if the proportionality constant α = 20° C = 0.00393?

$$R_2 = R_1(1 + \alpha \Delta T)$$
$$= 4.31\ \Omega[1 + 0.00393(30°\ C - 20°\ C)]$$
$$= 4.31\ \Omega(1 + 0.0393)$$
$$= 4.48\ \Omega$$

Section Review

1. By increasing the temperature of a conductor, an increase in atomic motion results in the conductor (True/False).
2. Does the resistance of a conductor increase or decrease with an increase in temperature?
3. What is the definition of the temperature coefficient of resistance?
4. The temperature of −234.5° C is called inferred zero-resistance temperature (True/False).

2–11
CONDUCTANCE

Conductance can be defined as the reciprocal of resistance—the ease with which current flows through a circuit. The SI unit of measurement is **siemens, S,** and is represented by the symbol G.

$$G = \frac{1}{R} \qquad (2.6)$$

EXAMPLE 2–10

Solution

A given conductor has a resistance of 200 Ω. What is its conductance?

$$G = \frac{1}{200}$$
$$= 0.005\ S \quad \text{or} \quad 5\ mS$$

The reciprocal of resistivity is **conductivity,** or specific conductance, which is assigned the symbol σ. Conductivity is frequently used instead of resistivity when discussing the properties of semiconductor materials. The relationship between resistivity and conductivity is expressed by the following equation:

$$\sigma = \frac{1}{\rho} \qquad (2.7)$$

Section Review

1. What is the reciprocal of resistance?
2. Define conductance.
3. What is the SI unit of measurement for conductance?

___ 2-12 ___
WIRE SIZES

Wires are manufactured in sizes numbered according to a standard called the **American Wire Gauge (AWG).** By analyzing Table 2–3, the following statements can be made.

A lower AWG number indicates a greater cross-sectional area in circular mils. A decrease of one gauge number represents an increase in the cross-sectional area by approximately 25%. An increase in three gauge numbers multiplies the resistance by a factor of 2 and decreases the area by a factor of 2. A decrease in three gauge numbers doubles the area and halves the resistance. A change of ten wire gauge numbers represents a ten-to-one change in the cross-sectional area.

For such considerations as cost, weight, and bulk, it is desirable to use wire of the smallest diameter consistent with the minimum resistance that can be tolerated. Table 2–3 lists the information for copper wire only. Wire is also available in many other metals, such as aluminum, silver, platinum, and iron.

As Table 2–3 illustrates, the wire diameters become smaller as the gauge numbers become larger. Larger and smaller sizes of conductors than those shown in the table are

TABLE 2–3 *American Wire Gauge (AWG) wire sizes of commercial solid copper conductors at 20° C.*

AWG	Diameter		Area		Resistance	
	mil	mm	CM	mm^2	Ω/kft	Ω/km
0	324.9	8.252	105,530	53.49	0.0983	0.316
1	289.3	7.348	83,694	42.41	0.1240	0.406
2	257.6	6.543	66,373	33.62	0.1563	0.511
3	229.4	5.827	52,634	26.67	0.1970	0.645
4	204.3	5.189	41,742	21.15	0.2485	0.813
5	181.9	4.620	33,102	16.77	0.313	1.028
6	162.0	4.115	26,250	13.30	0.395	1.29
7	144.3	3.665	20,816	10.55	0.498	1.63
8	128.5	3.264	16,509	8.367	0.628	2.06
9	114.4	2.9065	13,094	6.631	0.792	2.59
10	101.9	2.588	10,381	5.261	0.998	3.27
11	90.74	2.305	8,234	4.170	1.260	4.10
12	80.81	2.0525	6,529.9	3.310	1.588	5.20
13	71.96	1.828	5,178.4	2.630	2.003	6.55
14	64.08	1.628	4,106.8	2.08	2.525	8.26
15	57.07	1.450	3,256.7	1.650	3.184	10.4
16	50.82	1.291	2,582.9	1.310	4.016	13.1
17	45.26	1.150	2,048.2	1.040	5.064	16.6
18	40.30	1.024	1,624.3	0.823	6.385	21.0
19	35.89	0.912	1,288.1	0.653	8.051	26.3
20	31.96	0.812	1,021.5	0.519	10.15	33.2

available. Very large conductors are measured in thousand circular mils, or kCM. The Roman symbol M is also used to express 1000 CM. For example, 250,000 circular mils could be expressed as either 250 MCM, or 250 kCM. Table 2–3 can be used when making calculations of a conductor's resistance.

_____ **EXAMPLE 2–11** _____

Solution

Calculate the resistance of 600 ft of No. 18 AWG wire at 20° C.

$$R = \text{length} \times \frac{\Omega}{1000 \text{ ft}}$$

$$= 600 \text{ ft} \times \frac{6.385 \ \Omega}{1000 \text{ ft}}$$

$$= 3.83 \ \Omega$$

Section Review

1. A lower AWG number indicates a lower cross-sectional area in circular mils (True/False).
2. An increase in three gauge numbers has what effect on wire resistance?
3. Table 2–3 lists the information for copper and aluminum wire (True/False).

_____ **2–13** ___

LOAD RESISTANCE

A **load** is the name given to any device connected across an energy source. Common loads are light bulbs, loudspeakers, motors, electrical appliances, and so forth. **Load resistance** represents the amount of opposition to current flow offered by the load. Often this value of load resistance is quite small. For example, a typical loudspeaker has a load resistance of 8 Ω. The term *load* actually refers to the amount of current drawn by a device. Therefore, a circuit with a small resistance would have a large load because of the greater value of current. The expression small load or large load is also relative to the type of circuit where the term is applied. For example, one ampere might be a very large load in an electronic circuit where it is common to have microampere values, but it would be a very small load in an industrial electronic circuit with hundreds of amperes flowing.

Loads always represent voltage drops in a circuit because they absorb electrical energy and dissipate it as heat, sound, light, or mechanical energy. The polarity of a voltage drop across a load is indicated by + and − symbols. Figure 2–16 shows an electric circuit with a source (E) connected to a load (R). The flow of current through the load resistance is indicated, as is the polarity of the voltage drop across the load. The current in Figure 2–16 travels a path from the source to the load through conductors with very small

FIGURE 2–16 *Load resistance connected to source.*

or negligible resistance. Therefore, the source voltage and the load voltage should be approximately equal in this circuit, because the load is the only device offering opposition to current flow.

Section Review

1. A load is the name given to any device connected across an energy source (True/False).
2. What is load resistance?
3. Loads can represent voltage drops or voltage rises in a circuit (True/False).

— 2–14 _____

RESISTORS

Resistors are devices that conduct electricity but also dissipate electric energy as heat. By adding resistance, the supply voltage may be reduced or current limited. Resistors are made from low-conductance materials such as carbon and special metal alloys. The power or wattage rating of the resistor determines its size and shape. As the power rating increases, the size of the resistor also increases. Generally, resistors with high resistance ratings will have low wattage ratings since the current passing through the resistor will be small.

There are two basic types of resistors: *fixed* and *variable*. Fixed resistors have permanent ohmic values that cannot be changed. Variable resistors have resistances that can be changed manually or by applying energy such as heat or light. Fixed value resistors come in standard ohmic values called the *preferred resistance values*. These values for resistors were standardized by the electronics industry and are available in 1/8-, 1/4-, 1/2-, 1-, 2-, and 5-watt sizes. Appendix A provides a complete table of the preferred resistance values for resistors. Fixed-type resistors are manufactured to be accurate within a certain degree, or **tolerance.** The tolerance value for standard resistors can be between 1% and 20%, depending on the manufacturer's specification. For example, a 100 Ω resistor with a tolerance of 5% would have an actual value somewhere between 95 Ω and 105 Ω. Resistors with tolerances of 1% and under are called **precision resistors.**

In electronics circuits, resistors serve very specific purposes. When they are used to limit the flow of current to a safe value, they are called **current limiting resistors.** When a resistor divides the voltage into different values from a single source, it is called a **bleeder resistor.** If it performs the function of applying a load to a circuit, it is called a **load resistor.** In transistor circuits, the term load resistor is often applied to the resistor that acts as both a current limiting resistor and a bleeder resistor. Another type of resistor that is popular in electronic applications is the **fusible resistor.** The fusible resistor is designed to burn open when the power rating of the resistor is exceeded. Fusible resistors provide both resistance and overcurrent protection to electronic circuits.

One of the more unusual resistors found in electronic circuits is the **zero-ohm resistor.** A resistor with an ohmic value of 0 Ω might seem like a useless component, but it actually serves a very important function in printed circuit board applications. Most printed circuit boards have components such as resistors inserted automatically by machine. Whenever a piece of jumper wire is to be installed between two points on the circuit board, the machine will automatically insert a zero-ohm resistor. This saves a tremendous amount of time and cost in the manufacture of printed circuit boards. If a zero-ohm resistor were not inserted, jumper wires would have to be installed manually, causing large delays in the manufacturing process.

Section Review

1. Resistors are made from high-conductance materials such as carbon and special metal alloys (True/False).
2. What is the term for a resistor that divides the voltage into different values from a single source?
3. What is a precision resistor?
4. Fixed-type resistors typically have tolerances from 1% to 50% (True/False).

2-15

FIXED RESISTORS

Fixed resistors are available in a variety of sizes and shapes, depending on the wattage rating and type of resistor. The power rating of a resistor is based on the assumption that the resistor is mounted in free air at room temperature. The dissipated heat then has no obstruction to prevent its free radiation, convection, and conduction from the body of the component. If the temperature exceeds a certain specified value, the resistor may be damaged. To prevent a resistor from overheating, the device must be physically large enough to have the necessary minimum external surface to properly dissipate the heat. At rated power, the temperature of a resistor is in the range of 300° C to 400° C. To avoid such high temperatures, the power loss in a resistor should be limited to approximately one-half its rated power. Figure 2–17 shows several of standard resistors of various power ratings.

The schematic symbol for a fixed-type resistor is shown in Figure 2–18. This symbol is also used on some drawings to indicate a load resistance. The resistor symbol is almost always shown with an ohmic value and, in some cases, will have a power rating shown. The tolerance of each resistor in a circuit is not usually indicated beside the resistor itself

(a)

(b)

FIGURE 2–17 *Comparison of typical fixed resistors.*

FIGURE 2–18 *Symbol for fixed resistor.*

but, in most cases, is mentioned in the specifications as a general statement such as "all resistors 5% tolerance."

There are four general categories for fixed-type resistors: carbon-composition, film, wire-wound, and chip resistors.

Carbon-composition resistors are the most common resistors used in electronic circuits. They are relatively inexpensive and are available in power ratings of 1/10 watts to 5 watts. Figure 2–19 shows the basic construction of a fixed molded-carbon composition resistor. The resistive element contains a mixture of graphite powder and silica as well as a binding compound, which is molded under a combination of heat and pressure. Since graphite is a conductor and silica is an insulator, higher resistance values are obtained by increasing the amount of silica in proportion to the graphite. The carbon-composition resistor is used to a great extent in low-power applications (under 1 watt) and has typical tolerances of 5% or 10%.

Film resistors are available in three basic types: carbon film, metal film, and metal-oxide film. Examples of film-type resistors are shown in Figure 2–20. *Carbon-film resistors* have a thin coating of resistive material on a ceramic insulator. The main advantage of using carbon-film resistors compared to carbon-composition resistors is that carbon-film resistors generate less noise because of random electron motion. This is particularly noticeable in audio and communication electronic circuits. These types of resistors are also manufactured to more exacting standards, allowing for tolerances of between 2% and 5%.

Metal-film resistors are often used where a high degree of ohmic accuracy is required. These resistors are capable of providing very precise ohmic values and are often used as precision resistors with tolerances of less than 1%. Metal-film resistors are manufactured by spraying a relatively thin layer of metal on a ceramic cylinder.

Metal-oxide film resistors are made by oxidizing tin chloride on a heated-glass substrate. The ratio of the oxide insulator to the tin conductor will determine the resistance of the device. These resistors have low noise and excellent temperature characteristics. Metal-oxide film resistors are popular where precision resistors of high ohmic value are required.

Wire-wound resistors are considerably different from composition resistors. They are made from alloys of relatively high resistivity and drawn into wire with precisely controlled characteristics. This wire is then wrapped around a ceramic-core form. Different wire alloys are used to provide various resistor ranges. Wire-wound resistors are characterized by their high stability and power ratings up to 250 watts. Unlike carbon-composition resistors, wire-wound resistors do not use a color code to identify the resistance rating.

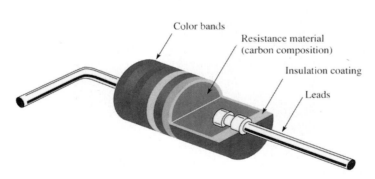

FIGURE 2–19 *Construction of carbon-composition resistor.*

FIGURE 2–20 *Typical thick-film resistors.*

FIGURE 2–21 *Typical wire-wound resistors.*

The ohmic value of wire-wound resistors is usually stamped on the resistor case. Some examples of wire-wound resistors are shown in Figure 2–21. Applications of wire-wound resistors include voltage dropping, bleeder resistors, bias resistors for transistor circuits, and power-supply filter devices.

Low-power resistors are also available in packages that resemble integrated circuit chips and, as a result, are called *chip resistors.* These resistors are small, high-quality resistors manufactured on small, rectangular ceramic chips. Chip packages are fabricated from thin-film and thick-film metals to exacting tolerances. Figure 2–22 shows two standard chip packages. Chip resistors are designed for printed circuit board installations and are very popular in *surface mount technology* (SMT). Chip resistors used in SMT applications have no leads and are mounted directly to the printed circuit board's surface.

Cermet-film resistors are manufactured by covering a ceramic substrate with a coating of metal alloy and placing it into a ceramic metal casing, or *cermet.* Cermet-film resistors are available in either a chip package or in a traditional tubular design. Figure 2–23 shows a cermet in a dual-inline package (DIP) for printed circuit board applications.

Section Review

1. List the four general categories for fixed-type resistors.
2. What is the most common type of resistor used in electronic circuits?
3. What type of resistor is designed for surface-mount, printed circuit board applications?

FIGURE 2–22 *Standard chip packages.*

(a)   **(b)**

FIGURE 2–23   *Cermet resistor. (a) IC package with cutaway showing resistive element. (b) Resistive network connection diagram showing seven separate resistors in the package.*

___ 2–16 _____

VARIABLE RESISTORS

There are three basic types of variable resistors: manual, heat, and optical. Manual variable resistors are extremely popular in electronic circuits for a wide variety of applications. The two types of manual variable resistors are the **rheostat** and the **potentiometer.** The rheostat is a two-terminal variable resistor that consists of a resistance element wound around a circular insulated form. Inside this form is a contact that is rotated to vary the resistance.

Rheostats are used to control the circuit current by varying the amount of resistance in the resistance element. This is accomplished by "tapping" the resistance element. Usually the resistance element is on a circular track, and a contact, attached to a dial, rolls or sweeps from one end of the resistance to the other. The most familiar example is the volume control on a TV set. Figure 2–24 shows the two schematic symbols used to represent a rheostat. The amount of resistance in the circuit is the section between the fixed end and the movable contact.

The relationship between the angle of rotation and resistance, called the **taper,** is frequently linear, although logarithmic tapers are also quite popular. If a very fine division of the total resistance is required, the resistance element will be deposited as a helix and the tracking contact must both rotate and translate. This type of variable resistor may require ten or more rotations of the dial before the entire resistance has been traversed.

The main difference between a potentiometer and a rheostat is the way that they are used in a circuit. Any potentiometer can be used as a rheostat by eliminating one terminal. The potentiometer, as shown in Figure 2–25(a), is a variable resistor having three electrical

FIGURE 2–24   *Alternative symbols for variable resistor or rheostat.*

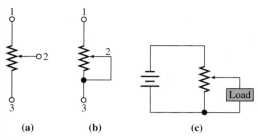

(a)   (b)   (c)

FIGURE 2–25 *Symbols for (a) potenti-ometer, (b) potentiometer connected as a rheostat, and (c) potentiometer connected to vary circuit resistance and load voltage.*

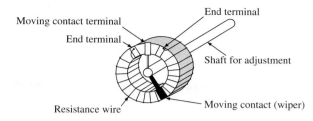

FIGURE 2–26 *Typical wire-wound potentiometer.*

connections, the center one being the movable contact, or **wiper.** Figure 2–25(b) shows a potentiometer connected as a rheostat. Both ends of the resistance element of the poten-tiometer are connected in the circuit. In Figure 2–25(c), the wiper, or movable contact, is used to control the voltage applied across the circuit load.

Potentiometers may have carbon, film, or wire as the resistive element. Figure 2–26 shows a typical wire-wound potentiometer with a moving contact that slides to vary the resistance of the wire.

Trimmer resistors are variable resistors that are used where small and infrequent adjustments of a resistance are necessary to maximize circuit performance. The adjustment on a trimmer resistor is usually made with a screwdriver. These miniature potentiometers, also known as *trim pots,* are normally found inside a piece of equipment and not intended for frequent adjustment. Figure 2–27 shows an example of a trimmer resistor.

FIGURE 2–27 *Typical trimmer resistors.*

FIGURE 2–28 *Decade resistance box.*

Another type of adjustable resistor is the **decade box.** This device has a set of dials that can be rotated to provide a desired value of resistance. The decade box is popular in low-power laboratory measurements such as bridge circuits, where high accuracy (0.1%) is required. Decade boxes are constructed of precision wire-wound resistors that provide a resistance range from 1 kΩ to 999 GΩ, with an accuracy of 9 significant figures. Figure 2–28 shows an example of a decade resistance box.

Resistance substitution boxes should not be confused with decade boxes, although they do serve the same basic function; both insert resistance in a circuit. The main difference is resistance substitution boxes can handle larger amounts of power, but are less accurate. A typical resistance substitution box has a range of between 15 Ω and 10 MΩ, with power handling up to 1 W, and a tolerance of ±10%. An example of a resistance substitution box is shown in Figure 2–29.

A **thermistor** is a resistor whose resistance varies with temperature. Consequently, thermistors are classified as nonlinear resistors. They are usually made of semiconductor material and are extremely sensitive to changes in temperature. Thermistors can have either a positive temperature coefficient (PTC) or a negative temperature coefficient

FIGURE 2–29 *Resistance substitution box.*

FIGURE 2–30
*Symbol for a
thermistor.*

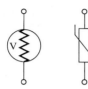

FIGURE 2–31
*Alternate
symbols for a
varistor.*

FIGURE 2–32
*Alternate
symbols for
photoresistors.*

(NTC), depending on the application. The temperature at which the resistance rises sharply is called the *switching point*. Figure 2–30 shows the schematic symbol for a thermistor.

A **varistor** is a voltage-dependent, metal-oxide material that has the property of rapidly decreasing resistance with increasing voltage. The term varistor is derived from the words *variable* and *resistor*. Varistors and thermistors are both nonlinear semiconductor resistors. Varistors are often referred to as *metal-oxide varistors* (MOVs) and are quite popular in protection circuits where voltage surges occur. The electronic symbols for a varistor are shown in Figure 2–31.

A *ballast resistor* is used in electronic circuits to maintain a constant value of current through a wide range of temperatures. The characteristics of the ballast resistor are such that its resistance increases as the current through the device increases. Therefore, if a ballast resistor is connected in series with a load, and the load current increases, the resistance of the ballast resistor will also increase, dropping the load back to normal. Ballast resistors are popular in circuits where loads have a low resistance when cold compared to their normal operating temperature.

Photoresistors are made of semiconductor materials that change resistivity as the level of light around the semiconductor changes. As the light level increases, the resistance of the material decreases. The type of semiconductor material used determines the sensitivity of the device to light. Typical dark-to-light resistance ratios are from 100 : 1 to 10,000 : 1. The semiconductor cadmium sulfide (CdS) exhibits a spectral response that closely matches that of the human eye. The CdS can switch from a high value of resistance to a low resistance in less than 100 ms. Highly sensitive, it is popular in consumer products such as camera light meters. Figure 2–32 shows two types of schematic symbols used to represent photoresistors.

Figure 2–33 shows a cutaway view of a photoresistor. When light shines through the device window onto the photoresistive material, electrons are released from their valence

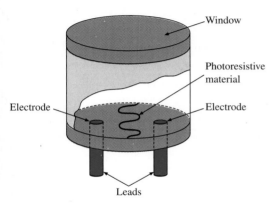

FIGURE 2–33 *Photoresistor construction.*

orbit. These electrons become available as current flow, and the resistance between the two electrodes decreases.

Section Review

1. List the three basic types of variable resistors.
2. The rheostat is a two-terminal variable resistor that consists of a resistance element wound around a circular insulated form (True/False).
3. What is the center terminal of a variable resistor called?
4. What is a trimmer resistor?
5. Varistors and thermistors are both linear semiconductor resistors (True/False).

___ 2–17 ___

RESISTOR COLOR CODE

The resistance value of a general purpose carbon-composition resistor is specified by a set of four **color code bands** on the resistor housing. General purpose resistors are classified as having tolerances between 5% and 20%. To determine the ratings of the resistor, it is held so that the bands are on the left, and the reliability designation is on the right. The first and second bands produce a two-digit number, the third band indicates the multiplier, and the fourth band designates the percent tolerance, or accuracy, of the resistor's rating. The four resistor color bands are shown in Figure 2–34.

Table 2–4 contains the color code for general purpose carbon-composition resistors. Three color bands near one end of the resistor indicate the resistance value, and an added gold or silver band means the accuracy is ±5% or ±10% of the amount designated by the three bands. When the resistor contains only three bands, the tolerance is taken as ±20%.

— First significant digit
— Second significant digit
— Multiplier
— Tolerance

FIGURE 2–34 *Resistor color bands.*

___ EXAMPLE 2–12 ___

Determine the rating of a resistor having the following four color bands:

Band 1—Red
Band 2—Violet
Band 3—Orange
Band 4—Gold

Solution

$$Red = 2 \quad \text{(first significant digit)}$$
$$Violet = 7 \quad \text{(second significant digit)}$$
$$Orange = 10^3 \quad \text{(multiplier in third band)}$$
$$Gold = \pm 5\% \quad \text{(tolerance)}$$
$$Resistor = 27 \times 10^3 \ \Omega \pm 5\%$$
$$= 27{,}000 \ \Omega \pm 5\%$$

TABLE 2–4 *Color code for general purpose carbon-composition resistors.*

Color	First Band First Significant Digit	Second Band Second Significant Digit	Third Band Multiplier	Fourth Band Tolerance
Black	—	0	10^0	—
Brown	1	1	10^1	—
Red	2	2	10^2	—
Orange	3	3	10^3	—
Yellow	4	4	10^4	—
Green	5	5	10^5	—
Blue	6	6	10^6	—
Violet	7	7	10^7	—
Gray	8	8	10^8	—
White	9	9	10^9	—
Gold	—	—	10^{-1}	±5%
Silver	—	—	10^{-2}	±10%
None	—	—	—	±20%

_____ EXAMPLE 2–13 _____

Solution

Orange = 3	(first significant digit)
White = 9	(second significant digit)
Gold = 0.1	(multiplier in third band)
None = ±20%	(tolerance)

Resistor = $39 \times 0.1 \ \Omega \ \pm 20\%$

= 3.9 Ω ±20%

Find the rating of a resistor with the following color bands:

Band 1—Orange

Band 2—White

Band 3—Gold

Band 4—No color

Some general purpose resistors also have a fifth color band, which indicates the **reliability factor.** The fifth band gives the percentage of failure per 1000 hours of use. For example, a 1% failure would indicate that 1 out of every 100 resistors will not lie within the tolerance range after 1000 hours of operation at its rated power. Tolerances for fifth-band general purpose resistors are as follows:

Brown = 1%   Orange = 0.01%

Red = 0.1%   Yellow = 0.001%

Precision resistors with tolerances of less than 2% use five color bands, as shown in Table 2–5. The first three bands of a precision resistor indicate the three significant digits. The fourth band is the multiplier and the fifth band indicates tolerance.

TABLE 2–5 *Color code for five-band precision resistors.*

Color	First Band First Significant Digit	Second Band Second Significant Digit	Third Band Third Significant Digit	Fourth Band Multiplier	Fifth Band Tolerance
Black	—	0	0	10^0	—
Brown	1	1	1	10^1	±1%
Red	2	2	2	10^2	±2%
Orange	3	3	3	10^3	—
Yellow	4	4	4	10^4	—
Green	5	5	5	10^5	±0.5%
Blue	6	6	6	10^6	±0.25%
Violet	7	7	7	10^7	±0.1%
Gray	8	8	8	10^8	—
White	9	9	9	10^9	—
Gold	—	—	—	10^{-1}	—
Silver	—	—	—	10^{-2}	—

EXAMPLE 2–14

What is the resistance and tolerance of the precision resistor shown in Figure 2–35?

Brown
Green
Red
Gold
Blue

FIGURE 2–35 *Resistor for Example 2–14.*

Solution

Band 1 = Brown = (first significant digit) = 1

Band 2 = Green = (second significant digit) = 5

Band 3 = Red = (third significant digit) = 2

Band 4 = Gold = (multiplier) = 10^{-1}

Band 5 = Blue = (tolerance) = ± 0.25%

$152 \times 0.1 = 15.2\ \Omega \pm 0.25\%$

Section Review

1. What is the tolerance range of general purpose resistors?
2. Resistor color bands indicate ohmic value, tolerance, and power ratings (True/False).
3. What is the reliability factor?

A common problem in electronic circuits is fixed-resistor failure. Resistor failure is often due to excess heat caused by too large a value of current flowing through the resistor. When the maximum power rating of a resistor is exceeded, the interior of the device will either burn out or melt. When the interior is burned out, it produces an **open circuit** so that no current can flow. Figure 2–36 shows an example of an open-circuit condition. In this circuit, the flow of current has been interrupted because of a break in the normal current path.

Open circuits can also be caused by cold solder joints. A *cold solder joint* occurs when the resistor is not properly soldered to a printed circuit board. To prevent cold solder joints, always observe proper soldering techniques when inserting any electronic device on a circuit board.

Another mode of failure occurs when the resistor becomes short circuited. A **short circuit** means that the resistor value has effectively fallen to zero ohms. Figure 2–37 shows a short-circuit condition. This circuit is very dangerous because it implies that there is no resistance between the two terminals of the power supply.

Resistors can also become short circuited when the device is being installed in a printed circuit board. If the resistor leads are poorly soldered in place, excess solder may produce a *bridge* around the component. Solder bridges are particularly dangerous because of their ability to allow large values of current flow. Figure 2–38 shows a short-circuit condition where the resistor is now bypassed by a bridge. Since current always takes the path of least resistance, the flow of current is "shunted" around the resistor and through the short-circuit bridge.

Mechanically variable resistors are also capable of failure, although problems with these types of resistors are usually limited to the rotating arm or to the element of the resistor. A condition referred to as a *dirty pot* is very common with potentiometers that have

FIGURE 2–36 *Open circuit.*

FIGURE 2–37 *Short circuit.*

Solder bridge

FIGURE 2–38 *Short circuit produced by a solder bridge.*

FIGURE 2–39 *Resistance measurement using an ohmmeter.*

reached a certain age or are used in dusty conditions. The contact assembly oxidizes and can cause a noticeable noise in audio circuits such as volume controls or filters such as equalizers. Although oxidizing often can be eliminated by spraying the potentiometer with contact cleaner, it is usually necessary to replace the potentiometer once oxidization has occurred.

The most popular instrument used in troubleshooting resistors is the **ohmmeter.** An ohmmeter is a resistance-measuring instrument that has a battery as an internal power source. Ohmmeters are available as either single-function meters or as part of a multimeter that is capable of also measuring current and voltage. Figure 2–39 shows an example of a digital multimeter (DMM) connected to measure the ohmic value of a resistor.

The ohmmeter should *never* be used on a circuit that is energized. The power always must be turned off when attempting to measure resistance with an ohmmeter. To properly read the resistance of a resistor in a circuit, at least one lead of the resistor should be disconnected from the circuit. Otherwise, any other devices that are connected to the resistor may contribute to the resistance being read by the ohmmeter. A detailed discussion of ohmmeters is presented in Chapter 7.

Section Review

1. What is the main cause of resistor failure?
2. Open circuits can be caused by cold solder joints (True/False).
3. What instrument is used to test resistance?

___ 2–19 ___

PRACTICAL APPLICATION SOLUTION

In the chapter opener, your task was to select and install an 800-ft cable. The specifications for the installation were that the cable must be made of copper, with a resistance of less than 2 Ω and a diameter of 2 mm. The following steps illustrate the method of solution for this practical application.

STEP 1   Determine the wire size. Referring to Table 2–3, the largest copper conductor less than 2 mm in diameter is a #13 AWG. The next step is to determine if this size conductor has a suitable resistance for the length given.

STEP 2 Determine the resistance of the conductor.

$$R = \text{length} \times \frac{\Omega}{1000 \text{ ft}}$$

$$= 800 \text{ ft} \times \frac{2.003}{1000 \text{ ft}}$$

$$= 1.6 \ \Omega$$

STEP 3 Install the conductor. From the data obtained in Steps 2 and 3, it is apparent that a #13 AWG conductor meets the installation specifications for the practical application outlined.

Summary

1. The net electron movement in one direction is called *current flow.*
2. A conductor is a material that has the ability to transfer charge from one object to another.
3. Electric current is the rate of flow of charged particles through a conductor in a specified direction.
4. The direction in which current flows in a conductor is arbitrary as long as it is used consistently.
5. A source of electromotive force (emf) converts another form of energy into electrical energy.
6. A voltage drop represents the energy used by the free electrons while engaged in current flow.
7. A power supply is a device that converts one type of electrical energy to another type of electrical energy.
8. The resistance of a conductor is directly proportional to its length and inversely proportional to its cross-sectional area.
9. Conductance is the reciprocal of resistance.
10. Resistors conduct electricity but also dissipate electric energy as heat.
11. The resistance of a typical resistor is specified by a set of four color code bands on the resistor housing.
12. The most popular instrument used in troubleshooting resistors is the ohmmeter.

Answers to Section Reviews

Section 2-1

1. False
2. Current flow
3. True

Section 2-2

1. The rate of flow of charged particles through a conductor in a specified direction
2. False
3. *I*
4. False

Section 2–3

1. 186,000 miles per second
2. True
3. True

Section 2–4

1. True
2. From negative to positive
3. The direction in which positive charge carriers flow through a circuit
4. Hole flow

Section 2–5

1. A unidirectional current that is unchanging or changes negligibly
2. Pulsating current
3. False
4. Alternating current (AC)

Section 2–6

1. True
2. The joule
3. $E = W/Q$
4. False

Section 2–7

1. True
2. A device that contributes no energy or removes energy
3. False

Section 2–8

1. A device capable of converting one form of energy into electrical potential energy
2. False
3. Chemical sources, solar cells, thermoelectric generation, electromagnetic generation, electrical conversion

Section 2–9

1. False
2. Ohm, Ω
3. To compare the inherent resistance characteristics of different materials
4. 7.854×10^{-7} in.^{2}

Section 2–10

1. True
2. Increase
3. The change in resistance per unit resistance per degree change in temperature
4. True

Section 2–11

1. Conductance
2. The ease with which current flows through a circuit
3. Siemens, S

Section 2–12

1. False
2. The resistance doubles
3. False

Section 2–13

1. True
2. The amount of opposition to current flow offered by the load
3. False

Section 2–14

1. False
2. Bleeder resistor
3. A resistor with a tolerance of 1% or under
4. False

Section 2–15

1. Carbon-composition, film, wire-wound, and chip resistors
2. Carbon-composition resistors
3. Chip resistors

Section 2–16

1. Manual, heat, and optical
2. True
3. The wiper
4. A variable resistor that is used where small and infrequent adjustments of a resistance are necessary
5. False

Section 2–17

1. 5% to 20%
2. False
3. An indication of the percentage of failure per 1000 hours of use

Section 2–18

1. Excess heat
2. True
3. Ohmmeter

Review Questions

Multiple Choice Questions

2–1 When no external field is present,
(a) Electrons move in a fixed pattern
(b) Electrons move in a random pattern
(c) Electrons flow from the positive to the negative terminal
(d) There is no electron movement

2–2 Current flow is defined as
(a) The random motion of electric charge
(b) The motion of charged particles through a conductor in a random direction

(c) The flow of charged particles through a conductor in a fixed direction

(d) None of the above

2–3 If 2.46 C of charge pass a certain point in a conductor every minute, what is the electric current?

(a) 0.041 mA   (b) 0.41 mA   (c) 41 mA   (d) 4.1 mA

2–4 Hole flow is

(a) Opposite to electron flow   (b) Opposite to conventional flow

(c) The same as electron flow   (d) None of the above

2–5 Electron flow is

(a) From the positive terminal of a power supply to the negative

(b) Arbitrary and cannot be determined

(c) The direction positive charge carriers flow in a circuit

(d) From the negative terminal of a power supply to the positive

2–6 When a unidirectional current is unchanging, it is called

(a) Pulsating current   (b) Alternating current

(c) Electron flow   (d) Direct current

2–7 In the SI system, the unit of work is

(a) Watt   (b) Volt   (c) Joule   (d) Coulomb

2–8 The units of electric potential are expressed in

(a) Joules per second   (b) Coulombs per second

(c) Joules per coulomb   (d) Volts per second

2–9 A source of emf is called a

(a) Voltage drop   (b) Passive device

(c) Current source   (d) Voltage rise

2–10 The unit of measurement of resistance is the

(a) Ohm   (b) Volt   (c) Ampere   (d) Watt

2–11 A material with low resistivity will be a(n)

(a) Poor conductor   (b) Semiconductor

(c) Insulator   (d) Good conductor

2–12 When the temperature of a conductor increases,

(a) Its resistance decreases   (b) Its resistance remains constant

(c) Its resistance increases   (d) None of the above

2–13 Conductance is

(a) Measured in ohms   (b) The reciprocal of voltage

(c) The reciprocal of resistance   (d) The opposition to current flow

2–14 Resistors are devices that

(a) Conduct electricity and dissipate energy as heat

(b) Completely block the flow of current

(c) Are not designed to dissipate heat

(d) Act as short circuits

2–15 Zero-ohm resistors are

(a) Not practical   (b) Poor conductors

(c) Used as jumpers on printed circuit boards   (d) Made by accident

2–16 To prevent resistors from being damaged, the power loss should be

(a) 1/4 of its rated value   (b) Equal to its rated value

(c) Double its rated value   (d) 1/2 its rated value

2–17 The relationship between angle and resistance of a variable resistor is called
(a) Taper   (b) Wiper   (c) Tolerance   (d) Trimmer

2–18 A thermistor is a resistor whose resistance varies
(a) With light intensity   (b) Mechanically
(c) With temperature   (d) With voltage

2–19 General purpose resistors have tolerances
(a) Between 5% and 10%   (b) Below 5%
(c) Between 2% and 20%   (d) Between 5% and 25%

2–20 If a resistor is open circuited, its internal resistance would be
(a) 0 Ω   (b) Infinite
(c) Equal to its rated value   (d) Low

Practice Problems

2–21 If a movement of 1.5 C creates a current of 4 A, in what time does this movement of charge take place?

2–22 Determine the resistance of an annealed-copper bus bar that is 0.75 m long and 1.5 cm × 3.5 cm wide.

2–23 Find the resistance of a round annealed-copper conductor 0.22 cm in diameter and 750 m long.

2–24 Calculate the resistance of 250 m of aluminum conductor that has a cross-sectional area of 3.5 mm^2.

2–25 A 1 kΩ carbon resistor has a diameter of 1.75 mm and a length of 1.55 cm. Determine the resistivity of its material.

2–26 Calculate the area in circular mils for the following diameters.
(a) 0.0375 in.   (b) 0.062 in.   (c) 0.166 in.   (d) 0.003 in.

2–27 Find the diameter in inches for the following areas.
(a) 270 CM   (b) 5500 CM   (c) 1200 CM   (d) 820 CM

2–28 A cable contains four conductors 0.025 in. in diameter, six conductors 0.0035 in. in diameter, and five conductors 0.055 in. in diameter. What is the amount of circular mils of conductor in this cable?

2–29 A conductor made of annealed copper has a resistance of 4.45 Ω at 20° C. What will the resistance of the conductor be at 60° C?

2–30 A length of pure silver wire has a resistance of 2.4 Ω at 20° C. Calculate its resistance at 40° C.

2–31 Determine the resistance of an aluminum conductor at 50° C if the resistance is 1.86 Ω at 20° C.

2–32 A coil of copper wire has a resistance of 120 Ω. Determine its conductance.

2–33 Calculate the resistance of 400 ft of No. 16 AWG wire at 20° C.

2–34 What is the resistance of 750 ft of No. 8 wire at 20° C?

2–35 Determine the range of resistance for a carbon-composition resistor having color bands red, black, orange, and gold?

2–36 What is the resistance rating of a carbon-composition resistor having color bands yellow, violet, orange, and silver.

2–37 A resistor has color bands in the order of blue, yellow, and gold. What is the ohmic value and tolerance of this resistor?

2–38   Determine the color bands for a 1.5 Ω, ±10% resistor.

2–39   Find the color bands for a 6.8 MΩ, ±5% resistor.

Essay Questions

2–40   Define current flow.

2–41   List the three types of electrical current flow and briefly describe each type.

2–42   Explain the difference between direct current and alternating current.

2–43   Describe the three things that must occur in order for work to be done.

2–44   What is the difference between a voltage drop and a voltage rise?

2–45   How does a primary cell differ from a secondary cell?

2–46   Define resistance.

2–47   What four factors affect the resistance of any material with a uniform cross-sectional area?

2–48   What is resistivity?

2–49   Explain what is meant by the terms *load* and *load resistor.*

2–50   Describe the two basic types of resistors.

2–51   List three functions of resistors in electronic circuits.

2–52   What is a zero-ohm resistor used for?

2–53   Name the four general categories for fixed-type resistors.

2–54   What is a chip resistor?

2–55   Describe the two types of mechanically variable resistors.

2–56   Explain the difference between a thermistor and a varistor.

2–57   Why do some resistors have five color bands?

2–58   What causes a resistor to fail?

2–59   Define the term *open circuit.*

2–60   What is the difference between an open circuit and a short circuit?

3

Ohm's Law, Power, and Energy

Learning Objectives

Upon completion of this chapter you will be able to

- Define Ohm's law.
- Utilize Ohm's law to determine current, voltage, or resistance.
- Describe the linear relationship between current and voltage.
- Differentiate between work and energy.
- Define power.
- Determine the efficiency of an electrical device.
- Calculate power consumption in terms of kilowatt-hours.

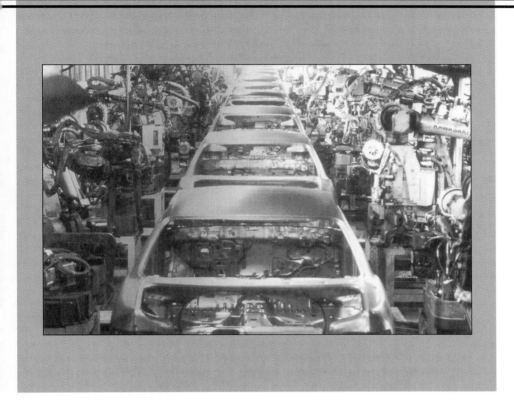

Practical Application

As a technician for a company specializing in robotic equipment, you are asked to assist in the design of a robotic arm with a 48 V, 1/2 hp, DC servo-motor that is to be controlled by a microprocessor. Your task is to determine the efficiency and current-handling capability of the electric motor if the motor produces 7/16 hp when full-load current flows. You are also required to calculate the power consumed by the motor over a 30-day period at a cost of 8 cents ($0.08) per kW · h. To complete this practical application, you will

1. Determine the power used by the motor.
2. Calculate the current drawn by the motor based on the rated power and voltage.
3. Determine the efficiency of the motor.
4. Calculate the power used by the robotic arm over a 30-day period.

3-1

OHM'S LAW

In 1827 a German scientist named Georg Simon Ohm discovered a relationship between the voltage that existed across a simple electric circuit and the current through that circuit. He determined that the magnitude of a current is, in general, proportional to the magnitude of the emf that produces it. When he doubled the potential difference, he found that the current was doubled; when he tripled the potential difference, the current was tripled; and so on. Ohm discovered that as the voltage increased, the current increased in direct proportion to the applied voltage, maintaining a constant ratio of voltage to current. **Ohm's law** is stated as follows:

> The current produced in a given conductor is directly proportional to the difference of potential between its end points.

The linear relationship between voltage and current is illustrated in Figure 3–1. The straight line that results means that the resistance of the conductor is the same regardless of the magnitude of applied voltage. Assuming the temperature of a conductor is constant, the slope of the curve, (k), is the value of the voltage divided by the current at any given instant. This relationship can be expressed mathematically as

$$K = \frac{E}{I} \tag{3.1}$$

The slope (k) is resistance; therefore, Ohm's law is expressed in equation form as

$$R = \frac{E}{I} \tag{3.2}$$

Recall from Chapter 2 that the symbol for a voltage source is considered to be E and the symbol for a voltage drop is represented by V. The curve shown in Figure 3–1 would be applicable to either a voltage source or voltage drop. Consequently, E and V are interchangeable in Equations 3.1 and 3.2.

By manipulating Ohm's law, any one of the three variables can be determined, if two are known. Figure 3–2 shows what is commonly referred to as the *Ohm's law triangle*. In Figure 3–2(a), the unknown current is solved for by dividing the voltage by the resistance. In Figure 3–2(b), the unknown voltage is solved for by multiplying the current by the resistance; and in Figure 3–2(c), the unknown resistance is determined by dividing the voltage by the current. The Ohm's law triangle illustrates that the current in a circuit is directly proportional to the applied voltage and inversely proportional to the circuit resistance.

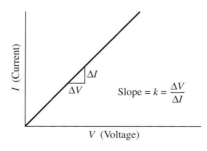

FIGURE 3–1 *Linear current versus voltage curve.*

FIGURE 3-2 *Ohm's law triangle. (a) Solving for current. (b) Solving for voltage. (c) Solving for resistance.*

(a) (b) (c)

EXAMPLE 3-1

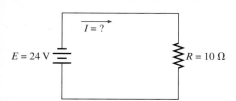

FIGURE 3-3 *Circuit for Example 3-1.*

Referring to Figure 3–3, if the resistance of the circuit is 10 Ω and the applied voltage is 24 V, determine the value of current flowing in the circuit.

Solution According to Ohm's law, the current flowing in the circuit must be equal to the voltage divided by the resistance.

$$I = \frac{E}{R}$$
$$= \frac{24V}{10\ \Omega}$$
$$= 2.4\ A$$

EXAMPLE 3-2

Solution

$$R = \frac{E}{I}$$
$$= \frac{120\ V}{0.833\ A}$$
$$= 144.06\ \Omega$$

An incandescent lamp is connected to a 120 V supply; current flow is 0.833 A. What is the resistance of the lamp?

EXAMPLE 3-3

Solution

$$V = I \times R$$
$$= 160\ mA \times 0.25\ \Omega$$
$$= 40\ mV$$

A current of 160 mA is measured through a conductor whose resistance is 0.25 Ω. Determine the voltage drop across the conductor.

_____ EXAMPLE 3–4 _____

A 48 V source is connected to a circuit with a resistance of 15 Ω. Determine the current drawn from the source.

Solution

$$I = \frac{E}{R}$$
$$= \frac{48 \text{ V}}{15 \text{ }\Omega}$$
$$= 3.2 \text{ A}$$

Section Review

1. The current produced in a given conductor is inversely proportional to the difference of potential between its end points (True/False).
2. What is the Ohm's law equation for finding voltage?
3. According to Ohm's law, what is the relationship between current, voltage, and resistance?
4. If $E = 5$ V and $I = 25$ mA, what is R equal to?

_____ 3–2 _____
WORK AND ENERGY

Whenever a force acts so as to produce motion in a body, work is done. Unless a force acts through a distance, no work is done, regardless of how great the force. If the force and the body's direction are perpendicular to each other, no work is done, even though the body has moved. An example of this is motion parallel to the earth's surface, where the force of gravity, which pulls downward, is perpendicular to all horizontal displacements. Motion must take place in a mechanical consideration of work, such as a rotating electric motor. The SI unit for work is the **joule, J.** One joule is the work accomplished when an object is moved a distance of one meter against an opposing force of one newton.

When energy is converted from mechanical energy to electrical energy, work is said to be done. This implies that **energy** can be defined as the *ability to do work.* There are two kinds of energy: kinetic and potential. **Kinetic energy** is possessed by a body by virtue of its motion. **Potential energy** is possessed by a system by virtue of its position, or condition. Examples of kinetic energy are a moving car or a bullet. Examples of potential energy are a compressed spring or a lifted object.

Figure 3–4 shows an example of a hydroelectric system that uses both potential energy and kinetic energy. In this system, a lake holds a large volume of water, maintained high above sea level by a dam. The water would be considered to have potential energy, or the energy of position. The water is moved from the dam to the generating station through pipes known as penstocks. The force of gravity forces the water through the penstocks. Consequently, the falling water is said to have kinetic energy, or the energy of motion. The kinetic energy of the water causes the shafts of the generators to rotate and produce electrical energy. This electrical energy is then distributed to homes, offices, factories, and so forth. In a consumer's home, a heater would convert the electrical energy to heat energy, or a light bulb would convert the electrical energy to light energy.

Because work is performed when electrical energy is converted to mechanical energy and when mechanical energy is converted to its electrical counterpart, the joule is also used as a unit measurement of electric energy. Consequently, the joule is used as the

FIGURE 3-4 *Hydroelectric system.*

SI unit for both electric energy and work. The definition of the joule can be stated in electrical terms as

> One joule of electric energy is required to raise one coulomb of electric charge through a potential difference of one volt.

$$1 \text{ joule of energy} = 1 \text{ coulomb} \times 1 \text{ volt}$$

$$W = Q \times E \qquad \qquad (3.3)$$

where W = energy, in joules
Q = charge, in coulombs
E = potential difference, in volts

EXAMPLE 3-5

Solution

$$W = QE$$
$$= (2.55 \times 10^5 \text{ C})\ 12 \text{ V}$$
$$= 30.6 \times 10^5 \text{ J}$$

A standard 12 V car battery can store 2.55×10^5 coulombs of electrons. How much energy does this battery possess?

EXAMPLE 3-6

Solution

$$Q = \frac{W}{E}$$
$$= \frac{30.6 \times 10^5 \text{ J}}{6 \text{ V}}$$
$$= 5.1 \times 10^5 \text{ C}$$

How much charge would a 6 V battery have to store in order to possess the same electrical energy as the 12 V battery in Example 3-5?

Therefore, the 6 V battery would have to store twice as much charge as the 12 V battery in order to produce the same amount of energy.

Section Review

1. When energy is converted from mechanical energy to electrical energy, work is said to be done (True/False).
2. What are the two types of energy?
3. What is the SI unit of measurement for mechanical energy and electrical energy?

__3-3__

POWER

Power is an indication of how much work can be accomplished in a specified amount of time. From this statement, **power** can be defined as the *rate at which work is done* or *energy expended.* This definition is true for both mechanical power and electrical power. Because work is measured in joules, J, and time in seconds, *t,* power is, therefore, measured in *joules per second.* The electrical unit of measurement for power is the **watt, W.** One watt is the rate of doing work when one joule of work is done in one second. Expressing this in equation form,

$$P = \frac{W}{t} \qquad (3.4)$$

where P = power, in watts
W = work, in joules
t = time, in seconds

As stated, the unit of measurement of power is the watt, which is derived from the surname of James Watt, the developer of the steam engine. This unit was originated in 1782 by Watt as a method of specifying how much power his new steam engines would develop in terms of the horses they were intended to replace. In the British Engineering System, power may be expressed in foot-pounds per second, but more often it is given in **horsepower, hp.** The horsepower, watt, and foot-pound are related in the following manner:

$$1 \text{ horsepower} = 550 \text{ foot-pounds/second} = 746 \text{ watts} \qquad (3.5)$$

Electric power may also be defined as

> One watt is the power expended when one ampere of direct current flows through a resistance of one ohm.

One watt is also the power dissipated when there is one ampere as the result of the application of one volt.

In the previous chapter, the relationship between current and voltage was discussed. Voltage was defined as being a ratio between energy and charge.

$$\text{electric potential } (E) = \frac{\text{energy (joules)}}{\text{charge (coulombs)}}$$

Electric current is given as a ratio between charge and time.

$$\text{electric current } (I) = \frac{\text{charge (coulombs)}}{\text{time (seconds)}}$$

If the current and voltage were multiplied together, the product of $E \times I$ would result in a ratio of energy/time, or joules per second, which, according to Equation 3.4, is the formula for power. Therefore, in a direct current circuit the amount of power dissipated can be determined by the following equation:

$$P = EI \qquad (3.6)$$

where P = power dissipated, in watts
E = applied voltage, in volts
I = current flow, in amperes

By direct substitution of Ohm's law, the equation for power can be obtained in two other forms:

$$\begin{aligned} P &= EI \\ &= E\left(\frac{E}{R}\right) \\ &= \frac{E^2}{R} \end{aligned} \qquad (3.7)$$

and

$$\begin{aligned} P &= EI \\ &= (IR)I \\ &= I^2 R \end{aligned} \qquad (3.8)$$

The interrelationship between Equations 3.7 and 3.8 is shown in Figure 3–5. The circle is divided into four sections: current, resistance, voltage, and power. As long as two quantities are known, the third quantity can be solved for by applying the appropriate formula.

EXAMPLE 3–7

Solution

$$\begin{aligned} P &= EI \\ &= (24\ \text{V})\,(8\ \text{mA}) \\ &= 192\ \text{mW} \end{aligned}$$

What is the power delivered to a DC circuit when the input current is 8 mA and the supply voltage is 24 V?

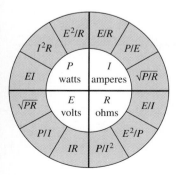

FIGURE 3–5   *Wheel showing relationships between voltage, current, resistance, and power.*

_____ **EXAMPLE 3–8** _____

What is the power dissipated by an incandescent light bulb with a rated voltage of 120 V and a resistance of 240 Ω?

Solution

$$P = \frac{E^2}{R}$$

$$= \frac{120^2 \text{ V}}{240 \text{ Ω}}$$

$$= 60 \text{ W}$$

_____ **EXAMPLE 3–9** _____

An electronic device operates at 12 V and is supplied with 200 mW of power. Determine the current drawn by the device.

Solution

$$P = EI$$

$$I = \frac{P}{E}$$

$$= \frac{200 \text{ mW}}{12 \text{ V}}$$

$$= 16.67 \text{ mA}$$

_____ **EXAMPLE 3–10** _____

Determine the voltage across a 2 kΩ resistor receiving a power of 50 mW.

Solution

$$V = \sqrt{PR}$$
$$= \sqrt{(50 \times 10^{-3} \text{ W})(2 \times 10^3 \text{ Ω})}$$
$$= 10 \text{ V}$$

_____ **EXAMPLE 3–11** _____

What is the resistance of a device that dissipates 5 W of power when 30 V is applied?

Solution

$$R = \frac{E^2}{P}$$

$$= \frac{30^2}{5}$$

$$= 180 \text{ Ω}$$

_____ **EXAMPLE 3–12** _____

A 100 W light bulb has a measured resistance of 144 Ω. Determine the amount of current drawn by the bulb.

Solution

$$I = \sqrt{\frac{P}{R}}$$

$$= \sqrt{\frac{100 \text{ W}}{144 \text{ Ω}}}$$

$$= 0.833 \text{ A}$$

Section Review

1. What is the electrical unit of measure for power?
2. One watt is the rate of doing work when one joule of work is done in one second (True/False).
3. One horsepower is equal to how many watts?
4. What is the equation for calculating power if current and resistance are known?

POWER DISSIPATION AND RATING OF CIRCUIT COMPONENTS

Whenever current flows through a component, heat is produced. The power dissipated by a resistor is actually a conversion from electric energy to heat energy. As heat is developed in a resistor, it will be distributed in the surrounding atmosphere. If a resistor gives off as much heat as is developed within it by the current, its temperature will remain fairly constant. But if the resistor develops heat more rapidly than it is capable of giving off, its temperature will rise. If the temperature rises too high, the material of the resistance may change its composition—expand, contract, or burn due to the heat. For this reason, all types of circuit components are rated for a maximum wattage. This rating may be in terms of watts or in terms of maximum voltage and current, which effectively give the rating in watts.

Resistors are rated in watts in addition to ohms of resistance. This wattage rating indicates the amount of power a resistor can handle safely. For example, carbon resistors are commonly manufactured in wattage ratings of 1/4, 1/2, 1, and 2 watts. The larger the size of carbon resistor, the higher the wattage rating, because a larger amount of material will absorb and give off heat more easily. Generally, a *safety factor* of at least 100% is used when determining power requirements for a circuit component. In other words, if it is known that a resistor in a given circuit will dissipate 1/2 watt, it is recommended that a 1-watt resistor be used. This will ensure that in the event of a momentary surge in current through the device, its rating will not be exceeded. When resistors of wattage ratings greater than 2 watts are required, wire-wound resistors are often used. These resistors are made in ranges between 5 watts and 200 watts.

EXAMPLE 3–13

The maximum current expected to flow through a 1 kΩ resistor is 20 mA. Determine the wattage rating of the resistor using a 100% safety factor.

Solution   The wattage of the resistor is

$$P = I^2 R$$
$$= (0.02 \text{ A})^2 (1000 \text{ }\Omega)$$
$$= 400 \text{ mW}$$

To have a 100% safety factor, this resistor should have a power-dissipation rating of 400 mW × 2 = 800 mW. Therefore, a 1 W resistor would be required.

Section Review

1. Whenever current flows through a component, heat is produced (True/False).
2. List four common wattage ratings for carbon resistors.
3. What value of safety factor is generally used when determining power requirements for a circuit component?

3–5
EFFICIENCY

Efficiency is a measure of how completely the power put into a circuit is used as output. Some of the energy put into a circuit is lost either in overcoming friction or in some other way. When any system or device converts one form of energy to another, some energy is lost or wasted, usually in the form of heat. The resistance of electrical conductors causes the production of heat when current flows. This energy is lost when it occurs in a power supply, such as a generator or battery. Electronic circuits are particularly susceptible to heat losses when semiconductor devices are used to control large amounts of power. In these circuits, a substantial amount of the input power is lost as waste heat. To overcome these losses of power, the input power must be greater than the output power, or

$$\text{power input} = \text{power output} + \text{power losses}$$

Efficiency is defined as the ratio of useful output energy to total input energy. In electronic circuits, such as amplifiers, the efficiency represents a ratio between the AC output power and the DC input power. The letter symbol for efficiency is the Greek letter η (eta). Efficiency is expressed by the equation

$$\eta = \frac{P_{out}}{P_{in}} \times 100\% \qquad (3.9)$$

where η = efficiency
        P_{out} = useful output power, in watts
        P_{in} = useful input power, in watts

EXAMPLE 3–14

Determine the efficiency of a circuit that has a useful output power of 15 W for each 20 W of input power.

Solution

$$\eta = \frac{P_{out}}{P_{in}} \times 100\%$$
$$= \frac{15\text{ W}}{20\text{ W}} \times 100\%$$
$$= 75\%$$

EXAMPLE 3–15

A motor has an efficiency of 85%. What current does it draw from a 120 V source when its output is 3/4 hp?

Solution

$$P_{out} = \frac{3}{4} \times 746 \quad \text{(W/hp)}$$
$$= 559.5\text{ W}$$

$$\eta = \frac{P_{out}}{P_{in}}$$

$$P_{in} = \frac{P_{out}}{\eta}$$
$$= \frac{559.5\text{ W}}{0.85} = 658.24\text{ W} \quad \text{(input power)}$$

$$P = EI$$

$$I = \frac{P}{E}$$

$$= \frac{658.24 \text{ W}}{120 \text{ V}}$$

$$= 5.49 \text{ A}$$

Section Review

1. Efficiency is a measure of how completely the power put into a circuit is used as output (True/False).
2. What is the letter symbol for efficiency?
3. If the input power is 100 W and the output power is 90 W, what is the efficiency?

3–6

THE KILOWATT-HOUR

The basic unit of electric energy is the joule. Because this unit is too small for typical problems, a more practical unit of electrical energy is required. Since

$$P = \frac{W}{t} \quad \text{then} \quad W = Pt$$

If P is power in watts and t is time in seconds, then W must be work in watt-seconds or joules. Therefore, if P is power in kilowatts, and t is time in hours, then W must be work in **kilowatt-hours.**

The watt-hour meter is an instrument for measuring the energy supplied to the residential or commercial user of electricity.

EXAMPLE 3–16

A generator delivers 4 kW to a customer for 60 h. How much energy does the customer receive in this time?

Solution

$$W = Pt$$
$$= (4 \text{ kW}) (60 \text{ h})$$
$$= 240 \text{ kW} \cdot \text{h}$$

EXAMPLE 3–17

At 8 cents ($0.08) per kW · h, how much will it cost to use a 60 W lamp for 20 days?

Solution

$$\text{cost} = \frac{8¢}{\text{kW} \cdot \text{h}} \times 60 \text{ W} \times 20 \text{ days} \times \frac{24 \text{ h}}{\text{day}}$$

$$= \frac{8 \times 60 \times 20 \times 24}{1000} ¢$$

$$= 230 ¢$$

$$= \$2.30$$

_____ EXAMPLE 3–18 _____

A light-emitting diode (LED) is used as an indicating light in a fire alarm panel. The LED is continuously on, drawing 30 mA from a 6 V supply. Determine the yearly cost to maintain its operation at $0.12 per kW · h.

Solution

$$P = EI = (6)(0.03) = 0.18 \text{ W}$$

$$\text{cost} = \frac{12¢}{\text{kW} \cdot \text{h}} \times 0.18 \text{ W} \times 365 \text{ days} \times \frac{24 \text{ h}}{\text{day}}$$

$$= \frac{12 \times 0.18 \times 365 \times 24}{1000} ¢$$

$$= 18.92 ¢ = \$0.19$$

Section Review

1. Describe the purpose of a watt-hour meter.
2. How much power is delivered by a generator supplying 1000 kW · h of electricity for 24 h.

_____ 3-7 _____

PRACTICAL APPLICATION SOLUTION

In the chapter opener, you were asked to assist in the design of a robotic arm with a 48 V, 1/2 hp motor controlled by a microprocessor. The following steps outline the solution for the given assignment.

STEP 1   Determine the power used by the motor.

$$1 \text{ hp} = 746 \text{ W}$$

Therefore,

$$1/2 \text{ hp} = 373 \text{ W}$$

STEP 2   Calculate the current drawn by the motor.

$$I = \frac{P}{E} = \frac{373 \text{ W}}{48 \text{ V}} = 7.77 \text{ A}$$

STEP 3   Determine the efficiency of the motor. The input power was determined in Step 1 as 373 W. In the chapter opener, the output power was indicated as 7/16 hp.

$$1 \text{ hp} = 746 \text{ W}$$

Therefore,

$$7/16 \text{ hp} = 326.38 \text{ W}$$

$$\eta = \frac{P_{\text{out}}}{P_{\text{in}}} \times 100\% = \frac{326.38}{373} \times 100\% = 87.5\%$$

STEP 4   Calculate the power used over a 30-day period. The cost of power was outlined in the chapter opener as 8 cents per kW · h.

$$\text{cost} = \frac{\$0.08}{\text{kW} \cdot \text{h}} \times 373 \text{ W} \times 30 \text{ days} \times \frac{24 \text{ h}}{\text{day}}$$

$$= \frac{8 \times 373 \times 30 \times 24}{1000} ¢$$

$$= 2148.5 ¢ = \$21.49$$

1. Georg Simon Ohm discovered that the current produced in a given conductor is directly proportional to the difference of potential between its end points.
2. The Ohm's law triangle demonstrates that the current in a circuit is directly proportional to the applied voltage and inversely proportional to the circuit resistance.
3. One joule of electric energy is required to raise one coulomb of electric charge through a potential difference of one volt.
4. Power is an indication of how much work can be accomplished in a specified amount of time.
5. One watt is the rate of doing work when one joule of work is done in one second.
6. Whenever current flows through a component, heat is produced.
7. Resistors are rated in watts, in addition to ohms of resistance.
8. Efficiency is a measure of how completely the power put into a circuit is used as output.
9. To overcome losses such as heat, the input power must be greater than the output power.

Answers to Section Reviews

Section 3–1

1. False
2. $E = I \times R$
3. That current in a circuit is directly proportional to the applied voltage and inversely proportional to the circuit resistance.
4. $R = E/I = 5$ V/25 mA $= 200\ \Omega$

Section 3–2

1. True
2. Kinetic energy and potential energy
3. The joule, J

Section 3–3

1. The watt, W
2. True
3. 1 hp $= 746$ W
4. $P = I^2 R$

Section 3–4

1. True
2. 1/4 W, 1/2 W, 1 W, and 2 W
3. Generally 100%

Section 3–5

1. True
2. The Greek letter η (eta)
3. $\eta = (P_{out}/P_{in}) \times 100\% = (100$ W/90 W$) \times 100\% = 90\%$

Section 3–6

1. An instrument for measuring the energy supplied to the residential or commercial user of electricity
2. $P = W/t = 1000$ kW $\cdot$ h/24 h $= 41.67$ kW

Review Questions

Multiple Choice Questions

3–1 Georg Simon Ohm discovered a relationship between
(a) Voltage and power
(b) Resistance and power
(c) Voltage and current
(d) Power and current

3–2 Which is the correct equation for Ohm's law?
(a) $E = I/R$
(b) $I = E \times R$
(c) $R = E \times I$
(d) $E = I \times R$

3–3 The Ohm's law triangle is used to solve for
(a) Voltage, current, and resistance
(b) Voltage, current, and power
(c) Voltage, resistance, and power
(d) Current, resistance, and power

3–4 Work is measured in
(a) Newtons
(b) Meters
(c) Joules
(d) Amperes

3–5 Energy is defined as the
(a) Amount of current produced in a given conductor
(b) Ability to oppose current
(c) Ability to do work
(d) Rate at which work is done

3–6 Kinetic energy is
(a) Possessed by a body by virtue of its position
(b) Dependent on gravity
(c) Possessed by a body by virtue of its motion
(d) The rate at which work is done

3–7 When rating circuit components, what safety factor is used?
(a) 50%
(b) 100%
(c) 150%
(d) 200%

3–8 Efficiency is a measure of
(a) The rate at which work is done
(b) The current produced in a given conductor
(c) How completely the power put into a circuit is used as output
(d) The power expended when one ampere of direct current flows through a resistance of one ohm

Practice Problems

3–9 What is the resistance of a lamp that draws 2.15 A from a 120 V supply?

3–10 A resistor in a television receiver must have a 2.5 V drop across it when its current flow is 4 mA. What value of resistance is required?

3–11 If the current in a circuit is 0.5 A and the resistance is 12 Ω, what is the voltage?

3–12 How much voltage must be applied across a 30 Ω load to cause a current of 4A to flow?

3–13 An electric heater draws 12 A and its terminal voltage is 120 V. If the total resistance of the supply conductors is 0.75 Ω, what is the voltage drop in the supply conductors?

3–14 A radio receiver has an "internal" resistance of 225 Ω. How much current will the receiver draw when 120 V is applied?

3–15 If the voltage across a resistance doubles, what is the effect on the current flowing through the resistance?

3–16 A 1 MΩ resistor in a television receiver has a voltage drop of 5 V. When the resistor heats up, its resistance changes to 1.12 MΩ, but the voltage is held at 5 V by a voltage regulator. What value of current flows when the resistor is heated?

3–17 What is the power supplied to a DC motor when the input current is 6 A and the supply voltage is 240 V?

3–18 A soldering iron requires 0.33 A at 120 V. How much power is dissipated?

3–19 An electric range will draw 80 A at 220 V when on the maximum heat setting. What is the power requirement of the range?

3–20 Determine the maximum voltage and current ratings for a 2 kΩ, 5 W resistor.

3–21 Find the current and resistance of a 120 V, 100 W lamp.

3–22 At what rate must a 60 Ω soldering iron dissipate heat if the current through it is 0.64 A?

3–23 A 60 W light bulb has a measured resistance of 240 Ω. Determine the amount of current drawn by the bulb.

3–24 The maximum current expected to flow through a 600 Ω resistor is 12 mA. Determine the wattage rating of the resistor using a 100% safety factor.

3–25 A motor requires 6.4 A when operating from a 120 V source. If the efficiency of the motor is 81%, what is the horsepower output?

3–26 What is the input power in watts to a 2 hp motor operating at 96% efficiency?

3–27 One kilowatt-hour is equivalent to how many joules?

3–28 A microwave oven, rated at 700 W, operates for 20 minutes. How much electric energy is used
(a) In kW · h?   **(b)** In joules?

3–29 A generator delivers 100 kW to a building for 30 days. How many kW · h were supplied to the building?

3–30 If the cost per kW · h is 6 cents, how much would it cost to supply 80 kW over a period of 15 days?

3–31 Assuming a rate of $0.08 per kW · h, determine the cost of operating the following electrical equipment for 30 days:
(a) 600 W iron for 2 h/day
(b) 260 W TV receiver for 5 h/day
(c) Four 60 W, two 100 W, and three 40 W lamps for 6 h/day
(d) 750 W microwave oven for 30 min/day

3–32 A 30 hp pump motor is used for an average of 8 h/day. If the motor has an efficiency of 88%, what would be the cost of operating the motor over a 20-day period if the industrial rate is 6 cents/kW · h?

Essay Questions

3–33 Define Ohm's law.

3–34 What two factors is work dependent on?

3–35 Explain why the joule can be used as a unit of measure for both electric energy and work.

3–36 Why is current directly proportional to voltage?

3–37 What is the Ohm's law triangle and what three quantities can be solved for using this triangle?

3–38 Define the two types of energy and list three examples of each type.

3–39 What is the definition of power?

3–40 Explain the relationship between the horsepower, watt, and foot-pound.

3–41 List three formulas that can be used to determine power in a circuit.

3–42 Why is it necessary to use a safety factor when rating circuit components?

3–43 Explain why efficiency is important in the study of electronic circuits.

3–44 What is a kilowatt-hour, and why is it used?

4

Series Circuits

Learning Objectives

Upon completion of this chapter you will be able to

- Describe how voltages are distributed around a series circuit.
- Explain the purpose of double-subscript notation.
- Define Kirchhoff's voltage law.
- Express the voltage divider rule and understand where it can be applied.
- Determine the polarity of emfs and voltage drops.
- Explain the meaning of positive ground and negative ground.
- Calculate power in a series circuit.
- Define internal resistance.
- Explain the purpose of fuses and switches.
- Troubleshoot open-circuit and short-circuit conditions in a series circuit.

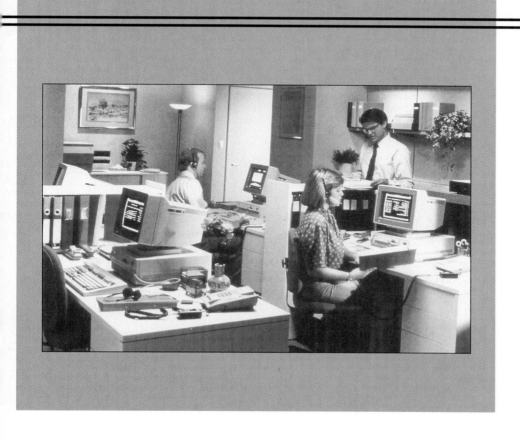

Practical Application

As a technician for a telecommunications company, you are required to troubleshoot a local area network (LAN) consisting of six computers connected in a configuration known as a ring network. In this type of LAN, each computer is connected to its nearest neighbor until all computers are connected in a closed path or ring. Data is transmitted in only one direction and is read by each computer until it reaches its destination. Each computer uses a device called a *repeater* to relay information from one computer to another

Your task is to determine why the network is not functioning properly. Your supervisor has informed you that he/she believes that the problem is caused by one of the repeaters in the network because a similar problem occurred previously in this system. To accomplish this practical application, you will utilize troubleshooting techniques based on the following data:

1. All computers in the network are not functioning and are presently incapable of transmitting or receiving data.
2. When functioning properly, each repeater in the LAN has a rated resistance of 65 Ω. The measured resistance of each repeater is 65 Ω.
3. The resistance of each conductor interconnecting the six computers is 2 Ω. The measured resistance of all conductors but one is 2 Ω. One of the conductors between the third and fourth computer appears to have an infinite resistance.

__ 4-1

INTRODUCTION

The coverage of this chapter is confined to direct-current circuits. That is, we will consider those circuits in which there is only direct current without variation and in which there is only the effects of resistance. Later, in the study of alternating-current circuits (Chapter 11), many of the principles contained in this chapter will be found to be applicable to AC circuits.

A **series circuit** can be defined as a circuit in which there is only one current path and all components are connected end to end along this path. If any component in the chain becomes open, it interrupts the flow of current through all the elements in the circuit. Consequently, all devices connected in series depend on each other for the continuous flow of current through the circuit. A common example of this dependence is a series string of Christmas-tree lights in which all of the lamps go out when one lamp stops working. When the filament in one lamp is burned out, it creates an open-circuit condition, preventing electrons from flowing through the rest of the circuit.

Section Review

1. A series circuit has only one current path and all components are connected end to end along this path (True/False).
2. If six lamps are connected in series and one lamp goes out, what happens to the other lamps?

__ 4-2

RESISTANCE IN SERIES CIRCUITS

In a series circuit, the total resistance is the sum of the individual resistances. The more resistance in the circuit, the more opposition to current flow. Figure 4–1(a) shows an example of three series-connected resistors. The only path for the current to flow in the circuit is through all three resistors. If one of the resistors became open circuited, as shown in Figure 4–1(b), the resistance of the circuit would rise to infinity. If one of the resistors became short circuited, Figure 4–1(c), the circuit resistance would decrease to a value equal to the sum of the remaining resistors.

The total resistance of a series circuit is determined as follows:

$$R_T = R_1 + R_2 + R_3 + \cdots + R_n \qquad (4.1)$$

where R_T = total ohmic value
R_n = highest numbered resistor in the circuit

FIGURE 4–1 *Total resistance in a series circuit. (a) Three series-connected resistors. (b) Open circuit. (c) Resistor R_2 short circuited.*

EXAMPLE 4–1

Solution

Determine the total resistance of the circuit shown in Figure 4–2.

$$R_T = R_1 + R_2 + R_3 + R_4$$
$$= 100 \text{ k}\Omega + 3300 \text{ k}\Omega + 470 \text{ k}\Omega + 97.6 \text{ k}\Omega$$
$$= 3967.6 \text{ k}\Omega \quad \text{or} \quad 3.97 \text{ M}\Omega$$

FIGURE 4–2 *Circuit for Example 4–1.*

EXAMPLE 4–2

Figure 4–3 shows an assembly diagram with six resistors. Connect the resistors together to form a series circuit.

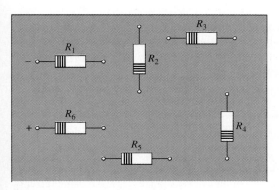

FIGURE 4–3 *Assembly drawing.*

Solution   The resistors should be connected together, end to end, as shown in Figure 4–4.

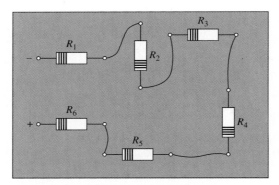

FIGURE 4–4   Solution to Example 4–2.

By using two or more resistors connected in series, it is possible to design a specific resistance that is close to the value required for a circuit. For example, suppose you need a 56 kΩ resistor but only have a 33 kΩ and a 22 kΩ. By combining the two resistors, you would have a 55 kΩ that would be close to the desired value. Because all resistors have a tolerance rating, a slight variation from ideal will always be present.

EXAMPLE 4–3

A 68 kΩ resistor is required for a circuit. The resistors available are as follows: 10 kΩ, 12 kΩ, 18 kΩ, 22 kΩ, 27 kΩ, 39 kΩ, and 47 kΩ. Which pair of resistors would make the closest substitute?

Solution

$$22 \text{ k}\Omega + 47 \text{ k}\Omega = 69 \text{ k}\Omega$$

The 22 kΩ and the 47 kΩ resistors connected in series would produce the closest approximation for the available resistors.

Section Review

1. How is the total resistance found in a series circuit?
2. What is the total resistance of three 100 Ω resistors connected in series?
3. If a 22 kΩ resistor is required for a circuit, what resistance would have to be combined in series with a 10 kΩ resistor?

— 4–3

CURRENT IN SERIES CIRCUITS

As the amount of resistance in a series circuit increases, the opposition to the flow of current must also increase. Conversely, when the amount of resistance decreases, the current flow will increase. Consider the circuit of Figure 4–5(a). In this circuit a 20 V supply is

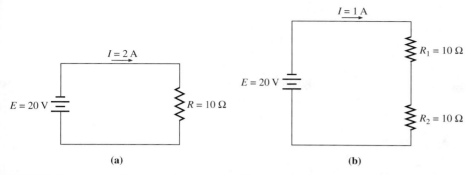

$I = 1$ A

$R_1 = 10\ \Omega$

$E = 20$ V

$R_2 = 10\ \Omega$

$I = 2$ A

$E = 20$ V

$R = 10\ \Omega$

(a)

(b)

FIGURE 4–5 *(a) Circuit with 2 A of current. (b) Doubling the resistance decreases current by 50%.*

connected in series with a 10 Ω resistor. According to Ohm's law, the current flowing through the circuit must equal the applied voltage divided by the resistance,

$$\frac{20\ \text{V}}{10\ \Omega} = 2\ \text{A}$$

In Figure 4–5(a), 2 A of current are flowing due to a circuit resistance of 10 Ω. In Figure 4–5(b) another 10 Ω resistor is connected in series with the original 10 Ω resistor. This produces a total resistance of

$$10\ \Omega + 10\ \Omega = 20\ \Omega$$

In Figure 4–5(b) the flow of current has now been reduced to

$$\frac{20\ \text{V}}{20\ \Omega} = 1\ \text{A}$$

Therefore, if the resistance of a series circuit doubles, its current is decreased by one-half.

The same amount of current must flow in every part of a series circuit. This is because the same number of electrons must pass through the source per second as pass through any other part of the series circuit in the same time interval. Mathematically, this is expressed as

$$I_T = I_1 = I_2 = I_3 = \cdots = I_n \qquad \textbf{(4.2)}$$

Ammeters are instruments used to measure current in electrical and electronic circuits. Since the ammeter's purpose is to show the magnitude of the common current, it must be connected in *series* with the circuit. This will allow the current being measured to flow through the device. The internal resistance of a typical ammeter is low. Ideally, the resistance of the ammeter should be zero ohms in order to have no effect on the circuit. If the internal resistance is high, it can have an effect on the current level it is measuring. Figure 4–6 shows a digital multimeter set to function as an ammeter. The wiring connection diagram illustrated in Figure 4–6(a) shows the ammeter connected in series between the resistor and the DC power supply.

In a circuit with a fixed voltage source, the current will be constant as long as the resistance does not change. To calculate the current in a series circuit, we first determine the total resistance. Once we determine the total resistance of a series circuit, we then can solve for the common current by using Ohm's law.

(a)

(b)

FIGURE 4–6 *(a) Wiring diagram for current measurement. (b) Schematic diagram.*

___ **EXAMPLE 4–4** ___

For the circuit shown in Figure 4–7, determine the amount of current that will flow.

$R_1 = 220\ \Omega$

$E_T = 24$ V

$R_2 = 150\ \Omega$

$R_3 = 330\ \Omega$

FIGURE 4–7 *Circuit for Example 4–4.*

Solution

$$R_T = R_1 + R_2 + R_3$$
$$= 220\ \Omega + 150\ \Omega + 330\ \Omega = 700\ \Omega$$

$$I_T = I_1 = I_2 = I_3 = \frac{E_T}{R_T}$$

$$= \frac{24\ \text{V}}{700\ \Omega}$$

$$= 34.29\ \text{mA}$$

Section Review

1. As the amount of resistance in a series circuit decreases, the amount of current flow in the circuit also decreases (True/False).
2. If two resistors are connected in series and 5 A flow through one resistor, how much current flows through the other resistor if it has twice as much resistance?
3. If the resistance of a series circuit doubles, its current is increased by 1/2 (True/False).
4. How are ammeters connected to read current?

4-4

VOLTAGE IN SERIES CIRCUITS

Whenever current flows in a circuit, a voltage drop will occur across a resistive device. Figure 4–8(a) shows an example of a circuit with a mechanical switch and two voltmeters. **Voltmeters** are voltage measuring instruments with very high internal resistances. The voltmeter is *always* connected across the load to be measured, never in series. It must also be connected with the negative terminal of the meter connected on the negative side of the power supply.

(a)

(b)

(c)

FIGURE 4–8 *Voltage drops around a series circuit. (a) Wiring connections. (b) Closed circuit. (c) Open circuit.*

In Figure 4–8(b), when the switch is closed, current flows and a voltage drop is developed across the resistor. The voltage drop across the switch drops to 0 V and the voltage drop across the resistor rises to the applied voltage. The voltage drop across the load resistor is solved by Ohm's law: $V = IR$. Because there is only one resistor in Figure 4–8(b), the entire circuit voltage is dropped across the device, and $E_T = V_R$.

Voltage drops are often called *IR drops* because of the relationship between current and resistance in the Ohm's law equation. In Figure 4–8(c), an open-circuit condition is present and the flow of current is interrupted. If there is no current, there is no voltage drop across the resistor. The voltage is now being dropped across the open circuit.

The following examples illustrate how the voltage drops in a series circuit are calculated using Ohm's law.

EXAMPLE 4–5

The circuit of Figure 4–9 shows two series-connected resistors. Each of these resistors is rated at 10 kΩ, and the supply voltage is 20 V. Determine the voltage drops across the individual resistors by using Ohm's law.

FIGURE 4–9   *Circuit for Example 4–5.*

Solution   First, the total resistance of the circuit is calculated.

$$R_T = R_1 + R_2$$
$$= 10 \text{ k}\Omega + 10 \text{ k}\Omega$$
$$= 20 \text{ k}\Omega$$

The total current can now be determined.

$$I = \frac{E_T}{R_T}$$
$$= \frac{20 \text{ V}}{20 \text{ k}\Omega}$$
$$= 1 \text{ mA}$$

The individual voltage drops are now solved for using Ohm's law. Since $I_T = I_1 = I_2 = I_3$, then

$$V_{R1} = IR_1$$
$$= (1 \text{ mA}) (10 \text{ k}\Omega)$$
$$= 10 \text{ V}$$

$$V_{R2} = IR_2$$
$$= (1 \text{ mA}) (10 \text{ k}\Omega)$$
$$= 10 \text{ V}$$

Because the resistors of the circuit shown in Figure 4–9 are of equal ohmic value, the voltage drops are the same across both resistors.

EXAMPLE 4–6

FIGURE 4–10   *Circuit for Example 4–6.*

Figure 4–10 has three series-connected resistors. The values are $R_1 = 330\ \Omega$, $R_2 = 560\ \Omega$, and $R_3 = 470\ \Omega$, and the power supply $(E_T) = 100$ V. What is the voltage drop across resistor R_2?

Solution

$$R_T = R_1 + R_2 + R_3$$
$$= 330\ \Omega + 560\ \Omega + 470\ \Omega$$
$$= 1360\ \Omega$$

$$I_T = \frac{E_T}{R_T} = \frac{100\text{ V}}{1360\ \Omega} = 73.53\text{ mA}$$

$$V_{R2} = IR_2 = (73.53\text{ mA})(560\ \Omega) = 41.18\text{ V}$$

EXAMPLE 4–7

Solution   The current flowing through R_2 is equal to the total current flowing in the circuit. Therefore,

$$I_T = \frac{V_{R2}}{R_2}$$
$$= \frac{8.5\text{ V}}{3.3\text{ k}\Omega}$$
$$= 2.58\text{ mA}$$

A series circuit consisting of four resistors has the following values: $R_1 = 2.2$ kΩ, $R_2 = 3.3$ kΩ, $R_3 = 4.7$ kΩ, and $R_4 = 5.6$ kΩ. If the voltage drop across R_2 is 8.5 V, what is the total voltage of the circuit?

The total resistance of the circuit is equal to the sum of the individual resistances.

$$R_T = R_1 + R_2 + R_3 + R_4$$
$$= 2.2\text{ k}\Omega + 3.3\text{ k}\Omega + 4.7\text{ k}\Omega + 5.6\text{ k}\Omega$$
$$= 15.8\text{ k}\Omega$$

The total voltage is now found using Ohm's law.

$$E_T = I_T R_T$$
$$= (2.58\text{ mA})(15.8\text{ k}\Omega)$$
$$= 40.76\text{ V}$$

Section Review

1. How is a voltmeter connected to read a load voltage?
2. Why are voltage drops known as *IR* drops?
3. In a series circuit, if there is no current, there is no voltage drop across a resistor (True/False).

4-5

POLARITY OF VOLTAGES

To calculate currents and voltages correctly, it is necessary to determine the polarity of the emfs and voltage drops in a circuit. As mentioned earlier, a source of emf is called an active circuit element, because it generates electric energy. A rise of potential occurs in an active element, such as a voltage source, when electrons flow from the positive (+) to the negative (−) terminals. In passive circuit elements (which consume electric energy) electrons flow from the − to the + terminals. The polarity of a voltage drop in a resistance is always such that, if it were an emf, it would oppose the current producing it.

The polarity of a voltage source or drop will often be taken with respect to **ground.** An electrical ground utilizes the earth or an equipment chassis as a reservoir of charge. Originally, the term *ground* meant a point in a circuit directly connected to the earth, either for safety in power systems or for efficient communication in radio transmission and reception. Although this meaning is still true, ground is now frequently applied to a circuit point connected to a relatively large metal object, such as the chassis, frame, or cabinet supporting the equipment. The terms *ground, common,* and *reference* all apply to chassis-ground systems and are represented schematically by the symbols shown in Figure 4–11.

The symbol shown in Figure 4–11(a) represents an earth ground connection. Figure 4–11(b) shows the symbol for a chassis ground. Another symbol used to indicate a common connection is shown in Figure 4–11(c). Although this symbol is occasionally used to represent a ground connection, it is frequently used to show a common connection in a complicated circuit, that is, a circuit with a large number of connections to a particular device or conductor.

The ground provides a source for electrons and can also act as a conductor for the supply of charge to the system. In an automotive electrical system, the negative terminal of the battery is connected to the chassis of the automobile. In this type of system, the chassis acts as a conductor between the battery and all electrical and electronic devices in the system.

The polarity of voltage sources is considered in terms of which terminal is grounded. If the positive terminal is connected to ground, it is called **positive ground.** If the negative terminal is grounded, it is referred to as **negative ground.** Most automobiles and electronic systems use negative ground. Figure 4–12(a) shows a circuit with negative ground and a 12 V supply. The polarity across the devices is shown; because the negative terminal is grounded, all voltages are considered positive with respect to ground. Figure 4–12(b)

FIGURE 4–11
Ground symbols.
(a) Earth ground.
(b) Chassis
ground. (c)
Common-
connection
symbol.

(a)　　　　　　　(b)

FIGURE 4–12　*(a) Negative ground circuit.*
(b) Positive ground circuit.

FIGURE 4–13 *Voltage sources in series. (a) Series aiding. (b) Series opposing.*

shows an example of a positive ground circuit. In this network, all voltages are considered negative with respect to ground.

The polarity of voltage sources is extremely important in situations where more than one source is used in a circuit. Figure 4–13(a) shows an example of two batteries connected together in series. The positive terminal of E_1 is connected to the negative terminal of E_2. This connection is called **series aiding** because the flow of electrons inside the two sources flows from positive to negative to positive to negative and are additive. Essentially, we have created one large voltage source from two smaller voltage sources. The series-aiding connection of voltage sources is very popular in small consumer products using batteries. For example, a flashlight requiring 3 V may use two 1.5 V batteries, or a portable CD player may require four 1.5 V batteries to produce 6 V.

When two voltage sources are connected as shown in Figure 4–13(b), the flow of electrons from one source cannot pass through to the second source. This configuration is called **series opposing** because it actually opposes the flow of electrons through the two sources. The net result of a series-opposing connection is that the voltage sources "subtract" from each other. For example, if a 9 V battery and a 6 V battery were connected series opposing, the net voltage would be 3 V. If two series-opposing voltages are of equal value, they will cancel each other out and the net voltage will be zero. When the voltages are not equal, the polarity of the resultant voltage is determined by the larger of the two voltage sources.

EXAMPLE 4–8

Determine the terminal voltages for the three voltage source connections shown in Figure 4–14.

FIGURE 4–14 *Voltage sources in series for Example 4–8.*

Solution

(a) The two voltage sources are connected series aiding. Therefore, the voltages are additive.

$$E_T = E_1 + E_2 = 6\ \text{V} + 6\ \text{V} = 12\ \text{V}$$

(b) The two voltage sources are connected series opposing, and the two voltages are subtractive.

$$E_T = E_1 - E_2 = 6\text{ V} - 3\text{ V} = 3\text{ V}$$

(c) In Figure 4–14(c) there is a combination of series-aiding and series-opposing voltages. The terminal voltage is a combination of the two. Voltage sources E_1 and E_2 are connected series opposing, while E_2 and E_3 are connected series aiding.

$$E_T = E_1 - E_2 + E_3 = 6\text{ V} - 9\text{ V} + 6\text{ V} = 3\text{ V}$$

In a DC circuit, the direction of current indicates the polarity of a device and whether the device is absorbing power or delivering it. According to the theory of electron flow, the direction of current in Figure 4–15 through resistor R_1 is shown as top to bottom. This means that the top side of R_1 is more negative than the bottom side, or point A is more negative than point B.

When current flows through a passive element, such as resistor R_1 in Figure 4–15, a drop of potential occurs in the direction of current. Since the current is flowing from point A to point B, the voltage drop of R_1 is referred to as V_{AB}. The notation V_{AB} has two subscripts and is called **double-subscript notation.** Using this type of notation, the second subscript indicates the reference point. Since the voltage drop V_{AB} represents the flow of current in one direction, voltage drop V_{BA} must represent current flow in the opposite direction. Therefore, for the voltage drop shown in Figure 4–15, $-V_{AB} = V_{BA}$.

The use of double-subscripts identifies not only which points are being measured in a circuit but also the relative polarity of the voltage between these two points. The relative polarity is always taken in terms of a reference point and is defined as the voltage at one point in a circuit compared to the voltage at a reference point. The reference can be any point in a circuit. The reference point is often chosen as ground, although it can also be at a potential above or below actual ground (0 V).

If a single subscript is indicated, the voltage at that point is considered to be referenced to ground. Figure 4–16 shows a positive ground circuit with both single- and double-subscripts used. The voltage drop V_{AB} represents the amount of voltage dropped between points A and B, or across resistor R_1. The voltage drop V_A represents the voltage from point A to ground, and V_B represents the voltage from point B to ground (across R_2).

Section Review

1. What is an electrical ground?
2. Why is the negative terminal of a battery connected to the chassis of an automobile?
3. One large voltage source can be made by connecting two smaller voltage sources in a series-opposing configuration (True/False).
4. In a DC circuit, what does the direction of current indicate?

FIGURE 4–15 *Double-subscript to indicate voltage drop.*

FIGURE 4-16 *Voltage drops using subscripts.*

_____ **4-6** __

KIRCHHOFF'S VOLTAGE LAW

In 1845 Gustav Robert Kirchhoff developed a law dealing with the distribution of voltages around a circuit. **Kirchhoff's voltage law (KVL)** states

> In any closed loop, the algebraic sum of the voltage drops and rises equals zero.

Any closed path is a **loop** and may be defined as a continuous connection of branches where current flow leaving a point in one direction travels through a loop in one direction and returns to its point of origin. A *loop equation* is used to specify the voltages present in a closed loop. Voltage sources producing the assumed current direction around the loop are positive algebraic terms. Voltage drops across resistors are negative algebraic terms in the equation.

Figure 4–17 shows a series circuit with two voltage drops, V_{R1} and V_{R2}. The KVL equation for this circuit is as follows:

$$E - V_{R1} - V_{R2} = 0$$

Kirchhoff's voltage law also may be stated as

> The algebraic sum of all the voltage drops in a circuit must equal the applied voltage.

Mathematically, Kirchhoff's voltage law is expressed as

$$E_{\text{supply}} = V_{R1} + V_{R2} + V_{R3} + \cdots + V_{Rn} \qquad (4.3)$$

According to Equation 4.3, the KVL equation for the circuit shown in Figure 4–17 could also be written as

$$E = V_{R1} + V_{R2}$$

FIGURE 4-17 *Series circuit with two voltage drops.*

EXAMPLE 4-9

Calculate the applied voltage, E_T, for the circuit shown in Figure 4–18.

FIGURE 4–18 *Circuit for Example 4–9.*

Solution   The KVL equation for Figure 4–18 is

$$E_T = V_{R1} + V_{R2} + V_{R3}$$
$$= 8\ V + 6\ V + 10\ V$$
$$= 24\ V$$

EXAMPLE 4-10

Solve for the unknown voltage drop shown in Figure 4–19.

FIGURE 4–19 *Circuit for Example 4–10.*

Solution   According to Kirchhoff's voltage law, the algebraic sum of the voltages around the loop must equal 0. Therefore,

$$E_1 - E_2 - V_{R1} - V_{R2} = 0$$

Since all values except V_{R2} are known, the KVL equation appears as follows:

$$18\ V - 12\ V - 3.5\ V - V_{R2} = 0$$

which reduces to

$$2.5\ V - V_{R2} = 0$$

Rearranging terms results in

$$V_{R2} = 2.5\ V$$

EXAMPLE 4–11

FIGURE 4–20   Circuit for Example 4–11.

Write the KVL equation for the circuit shown in Figure 4-20. Use the Ohm's law to prove the validity of the equation.

Solution   The sum of the voltage rises must equal the sum of the voltage drops in any closed loop. Therefore,

$$E = IR_1 + IR_2 + IR_3$$

Substituting values, the KVL equation yields

$$24\ V = I(1\ k\Omega) + I(3.3\ k\Omega) + I(4.7\ k\Omega)$$

The validity of the above KVL equation can be proven by solving for I and inserting the value in the equation.

$$R_T = R_1 + R_2 + R_3 = 1\ k\Omega + 3.3\ k\Omega + 4.7\ k\Omega = 9\ k\Omega$$

$$I = \frac{24\ V}{9\ k\Omega} = 2.67\ mA$$

The value of I (2.67 mA) is now inserted in the KVL equation and the sum of the voltage drops is proven to equal the applied voltage.

$$24\ V = (2.67\ mA)\ (1\ k\Omega) + (2.67\ mA)\ (3.3\ k\Omega) + (2.67\ mA)\ (4.7\ k\Omega)$$
$$= 2.67\ V + 8.80\ V + 12.53\ V$$
$$= 24\ V$$

Section Review

1. The algebraic sum of all voltage drops in a circuit must equal the applied voltage (True/False).
2. What is the purpose of a loop equation?
3. State Kirchhoff's voltage law as a mathematical equation for three series-connected resistors.

4–7
VOLTAGE DIVIDERS

In a series circuit, the voltage across a given resistance has the same relation to the total voltage as the resistance has to the total resistance. This proportional method of determining voltage drop is known as the **voltage divider rule.** The voltage divider rule also may be stated as

> The ratio between any two voltage drops in a series circuit is the same as the ratio of the two resistances across which these voltage drops occur.

FIGURE 4–21 *Potentiometer as a voltage divider.*

A useful application of the voltage divider rule is for *voltage divider circuits.* A voltage divider circuit is used to allow a load to operate at a voltage that is different from the supply voltage. **Voltage dividers** are popular in electronic circuits to bias transistors, in integrated circuits for proper operation, and in power supply circuits to establish preset voltage levels.

Mathematically, the voltage divider rule is stated in equation form as follows:

$$V_{Rx} = E\left(\frac{R_x}{R_T}\right) \tag{4.4}$$

where V_{Rx} = voltage drop across a given resistance
E = applied voltage
R_x = given resistance
R_T = total resistance

The simplest type of voltage divider is the potentiometer. If a voltage source is placed across the potentiometer, an output voltage is obtained at the wiper. This output voltage represents a portion of the input voltage. Figure 4–21 shows a potentiometer connected across a voltage source. The output voltage is taken either from point A to C, or from point C to B. According to Kirchhoff's voltage law, the sum of these two drops must equal the applied voltage.

___ **EXAMPLE 4–12** ___

The wiper arm of a 10 kΩ potentiometer is set as shown in Figure 4–22. Determine the voltage measured from point A to C.

FIGURE 4–22 *Circuit for Example 4–12.*

Solution The resistance between point A and C is 6 kΩ. The voltage drop across this resistance is found by using the voltage divider rule.

$$V_{Rx} = E\left(\frac{R_x}{R_T}\right)$$

$$= 12 \text{ V}\left(\frac{6 \text{ k}\Omega}{10 \text{ k}\Omega}\right)$$

$$= 7.2 \text{ V}$$

EXAMPLE 4–13

Use the voltage divider rule to find the voltage drops across the resistors in the circuit shown in Figure 4–23.

FIGURE 4–23 *Circuit for Example 4–13.*

Solution

$$R_T = R_1 + R_2 + R_3$$
$$= 75\ \Omega + 33\ \Omega + 47\ \Omega$$
$$= 155\ \Omega$$

$$V_{R1} = E\left(\frac{R_1}{R_T}\right)$$
$$= 24\ V\left(\frac{75\ \Omega}{155\ \Omega}\right)$$
$$= 11.61\ V$$

$$V_{R2} = E\left(\frac{R_2}{R_T}\right)$$
$$= 24\ V\left(\frac{33\ \Omega}{155\ \Omega}\right)$$
$$= 5.11\ V$$

$$V_{R3} = E\left(\frac{R_3}{R_T}\right)$$
$$= 24\ V\left(\frac{47\ \Omega}{155\ \Omega}\right)$$
$$= 7.28\ V$$

To check:

$$E_T = V_{R1} + V_{R2} + V_{R3}$$
$$= 11.61\ V + 5.11\ V + 7.28\ V$$
$$= 24\ V$$

When a series resistor is used to reduce the supply voltage to a value suitable for a load, a voltage divider circuit is designed using a **series-dropping resistor.** The voltage across the series-dropping resistor accounts for the unwanted voltage in the circuit. The following example illustrates the application of a series-dropping resistor.

EXAMPLE 4–14

A 12 V, 2 A radio receiver is to be installed in a circuit with a 18 V supply (Figure 4–24). Calculate the ohmic value of a series-dropping resistor that will limit the current to the rated value of 2 A.

FIGURE 4–24 *Circuit for Example 4–14.*

Solution   The total resistance of the circuit is found by Ohm's law.

$$R_T = \frac{E_T}{I} = \frac{18 \text{ V}}{2 \text{ A}} = 9 \ \Omega$$

The internal resistance of the radio receiver is also determined by Ohm's law.

$$R_{\text{int}} = \frac{V_T}{I} = \frac{12 \text{ V}}{2 \text{ A}} = 6 \ \Omega$$

The value of R is now found.

$$R_T = R + R_{\text{int}}$$

$$9 \ \Omega = R + 6 \ \Omega$$

$$R = 3 \ \Omega$$

Section Review

1. What rule states the ratio between any two voltage drops in a series circuit is the same as the ratio of the two resistances across which these voltage drops occur?
2. A voltage divider circuit allows a load to operate at a voltage that is different from the supply voltage (True/False).
3. Which resistor in a voltage divider circuit accounts for the unwanted voltage in the circuit?

4–8

POWER IN SERIES CIRCUITS

The total power dissipated in a circuit is the sum of the individual amounts of power expended by the individual resistances. The total power of the circuit shown in Figure 4–25 is the sum of the power at each point in the network. In equation form, this is expressed as

$$P_T = P_1 + P_2 + P_3 + \cdots + P_n \tag{4.5}$$

FIGURE 4–25   *Power in a series circuit.*

EXAMPLE 4–15

FIGURE 4–26   *Circuit for Example 4–15.*

A 9 V battery is to be used to supply power to a portable CD player that operates on 6 V. Determine the power dissipated by the series-dropping resistor and the power used by the CD player. Assume a load current of 250 mA, as shown in Figure 4–26.

Solution   According to Kirchhoff's voltage law, the voltage across the series-dropping resistor must be

$$9\text{ V} - 6\text{ V} = 3\text{ V}$$

Since the current is known, the power dissipated by the series-dropping resistor is

$$P = IE = (250\text{ mA})\,(3\text{ V}) = 750\text{ mW}$$

The power used by the CD player is found using the same equation.

$$P = IE = (250\text{ mA})\,(6\text{ V}) = 1.5\text{ W}$$

EXAMPLE 4–16

Calculate the total power, as well as the power used in each resistor, shown in Figure 4–27.

FIGURE 4–27   *Circuit for Example 4–16.*

Solution

$$I_T = I_1 = I_2 = I_3 = 4\text{ A}$$

$$R_2 = \frac{V_{R2}}{I_2} = \frac{11\text{ V}}{4\text{ A}} = 2.75\ \Omega$$

$$V_{R1} = I_1\,R_1 = (4\text{ A})\,(5.6\ \Omega) = 22.4\text{ V}$$

$$V_{R3} = E_T - (V_{R1} + V_{R2}) = 60\text{ V} - (22.4\text{ V} + 11\text{ V}) = 26.6\text{ V}$$

$$R_3 = \frac{V_{R3}}{I_3} = \frac{26.6 \text{ V}}{4 \text{ A}} = 6.65 \text{ }\Omega$$

$$R_T = R_1 + R_2 + R_3 = 5.6 \text{ }\Omega + 2.75 \text{ }\Omega + 6.65 \text{ }\Omega = 15 \text{ }\Omega$$

$$P_1 = I_T^2 R_1 = (4^2 \text{ A}) (5.6 \text{ }\Omega) = 89.6 \text{ W}$$

$$P_2 = \frac{V_{R2}^2}{R_2} = \frac{11^2 \text{ V}}{2.75 \text{ }\Omega} = 44 \text{ W}$$

$$P_3 = I_T^2 R_3 = (4^2 \text{ A}) (6.65 \text{ }\Omega) = 106.4 \text{ W}$$

$$P_T = P_1 + P_2 + P_3 = 89.6 \text{ W} + 44 \text{ W} + 106.4 \text{ W} = 240 \text{ W}$$

___ **EXAMPLE 4–17** ___

Two lamps with ratings of 100 W at 80 V and 60 W at 100 V are connected in series across a 120 V supply. Calculate the voltage drop of each lamp.

Solution

$$P = \frac{E^2}{R}$$

$$R = \frac{E^2}{P}$$

$$R_1 = \frac{80^2 \text{ V}}{100 \text{ W}} = 64 \text{ }\Omega$$

$$R_2 = \frac{100^2 \text{ V}}{60 \text{ W}} = 166.7 \text{ }\Omega$$

$$R_T = R_1 + R_2 = 64 \text{ }\Omega + 166.7 \text{ }\Omega = 230.6 \text{ }\Omega$$

$$I = \frac{E_T}{R_T} = \frac{120 \text{ V}}{230.6 \text{ }\Omega} = 0.52 \text{ A}$$

$$V_{R1} = IR_1 = (0.52 \text{ A}) (64 \text{ }\Omega) = 33.3 \text{ V}$$

$$V_{R2} = IR_2 = (0.52 \text{ A}) (166.7 \text{ }\Omega) = 86.7 \text{ V}$$

Section Review

1. The total power dissipated in a circuit is the product of the individual amounts of power expended by the individual resistances (True/False).
2. Three resistors in series dissipate the following powers: 1 W, 2 W, and 5 W. What is the total power?
3. Two 10 kΩ resistors are connected in series with a total current flow of 50 mA. What is the total power dissipated?

___ 4–9 ___

INTERNAL RESISTANCE

A series-connected resistance that produces a voltage drop inside a source of emf is called **internal resistance.** The internal voltage drop that occurs as a result of this resistance represents a loss of voltage between the input and output terminals. Consequently, an internal resistance in a voltage source tends to reduce the voltage and current available at the load.

FIGURE 4–28 *Practical voltage source with internal resistance.*

Figure 4–28 shows a circuit illustrating the effect of internal resistance, R_i, on the terminal voltage of a battery. Because R_i is in series with the voltage source, E, the voltage drop IR_i is subtracted from E and the terminal voltage E_T is obtained.

$$E_T = E - IR_i \qquad (4.5)$$

As the internal voltage drop increases, the terminal voltage decreases. This is known as a **loading effect.** In Figure 4–28 the internal voltage drop can be present only when a load, such as load resistor R_L, is connected at the terminals of the supply. When a load is connected, the terminal voltage will decrease in proportion to the value of current flowing in the circuit. Ideal voltage sources have no internal resistance and the no-load voltage and full-load voltage are equal. However, this is seldom the case in practical applications. All voltage sources have internal losses, although many are small enough to be considered negligible in the circuit. For example, the internal resistance of a fully charged lead-acid battery cell is 0.01 Ω, so the internal voltage drop is extremely small.

Voltage sources such as batteries have internal resistances that vary with charge. When a battery is fully charged, it has a very low value of internal resistance. As the charge decreases, the internal resistance increases causing a lower value of terminal voltage. According to Ohm's law, if the load voltage decreases and the resistance remains the same, the current flowing through the load must also decrease. The value of internal resistance of a power supply is readily obtained by combining Kirchhoff's voltage law and Ohm's law:

$$R_i = \frac{E_{nl} - E_{fl}}{I}$$

where R_i = internal resistance of source
 E_{nl} = no-load terminal voltage
 E_{fl} = terminal voltage with load connected
 I = load current

_____ **EXAMPLE 4–18** _____

Solution

$$R_i = \frac{E_{nl} - E_{fl}}{I}$$

$$= \frac{25.5\ \text{V} - 24\ \text{V}}{2.5\ \text{A}}$$

$$= 0.6\ \Omega$$

Determine the internal resistance of a power supply shown in Figure 4–28 if the no-load voltage of 25.5 V is reduced to 24 V when the full-load current of 2.5 A flows.

The power losses in a voltage source are attributed to the power dissipated by the internal resistance of the source. The overall efficiency of a source, such as a power supply, is determined by the internal losses, output power, and input power. As mentioned in Chapter 3, these losses are in the form of heat.

EXAMPLE 4–19

Using the values given in Example 4–16, calculate the efficiency of the power supply.

Solution

$$P_{out} = E_{fl} I_{fl}$$
$$= (24 \text{ V}) (2.5 \text{ A})$$
$$= 60 \text{ W}$$

$$P_{in} = P_{out} + P_{losses}$$
$$= (24 \text{ V}) (2.5 \text{ A}) + (1.5 \text{ V}) (2.5 \text{ A})$$
$$= 63.75 \text{ W}$$

The efficiency is now found using Equation 3.9.

$$\eta = \left(\frac{P_{out}}{P_{in}} \right) 100\%$$
$$= \left(\frac{60 \text{ W}}{63.75 \text{ W}} \right) 100\%$$
$$= 94.1\%$$

Section Review

1. Internal resistance produces a voltage rise inside a source of emf (True/False).
2. As the internal resistance increases, what happens to the terminal voltage?
3. When a battery is fully charged, is the internal resistance very high or very low?
4. What three factors affect the overall efficiency of a source such as a power supply?

— 4–10
FUSES AND SWITCHES

A **fuse** is a series-connected protection device designed to create an open circuit in the event of excess current flow in the circuit. A fuse consists of a strip of metal that will melt at a relatively low temperature. As the current through the fuse rises, the heating effect causes the metal to melt, creating an open circuit and cutting off the flow of current in the circuit. The schematic symbol for a fuse is shown in Figure 4–29.

Figure 4–30 shows some examples of typical fuses. Generally, the fuse is made of zinc or of an alloy of tin and lead and has current ratings from 2 mA to several hundred

FIGURE 4–29
Symbol for fuse.

(a)

(b)

FIGURE 4–30 *Typical fuses. (a) Power fuses. (b) Fuses and fuse holders.*

FIGURE 4–31 *Fuse-
protecting load resistor.*

FIGURE 4–32
*Schematic
symbols for circuit
breakers.*

amperes. In addition to a current rating, fuses also have a voltage rating. This rating should be at least as high as the circuit voltage, because it represents the highest voltage at which a particular fuse can safely interrupt the current.

A standard fuse will usually melt when a 10% overcurrent occurs. This type of fuse is classified as a *fast-blow fuse,* since it will blow quickly on a relatively low value of excess current. Circuits that use electromagnetic principles, such as motors, have large current values when started and then fall to rated levels after a certain length of time. A *slow-blow fuse* is designed to allow an overcurrent condition for a number of seconds before finally melting. A typical slow-blow fuse will allow an overcurrent of 300% for over 3 seconds.

Figure 4–31 shows an example of a fuse in a circuit. In this circuit, the fuse is rated at 2 A. If the input current increases above that value, the inside of the fuse will melt and an open circuit is created. The resulting open circuit interrupts the flow of current and protects the load from current greater than 2 A.

Although fuses are extremely popular in electronic circuits, there are situations where the replacement of blown fuses is neither practical or economical. For this reason a circuit breaker is frequently used instead of a fuse. A **circuit breaker** is also an overcurrent protection device. The main difference between a circuit breaker and a fuse is a circuit breaker can be reset after the overcurrent problem has been corrected, but a fuse must be replaced after every failure. Two examples of schematic symbols that are used to represent circuit breakers are shown in Figure 4–32.

Figure 4–33 shows some examples of typical circuit breakers. Although there are several types of circuit breakers available, the most popular are the ones that operate on the thermal principle. In these devices, when an overload occurs, the excess current flowing through the bimetallic strip will heat the strip above normal temperature. As a result, the

FIGURE 4–33 *Typical circuit
breakers.*

FIGURE 4–34 *Standard switches. (a) Toggle switches. (b) Rocker switches. (c) Rocker switches in dual-in-line packages. (d) Push-button switches. (e) Rotary switches.*

end of the strip curves upwards releasing a latch and pulling the contact points apart. Once the current stops flowing, the strip will cool down and straighten up, reclosing the contact points inside the breaker. Circuit breakers are generally classified as slow-blow devices and are not suitable in circuits that are easily damaged by short-term overcurrent.

A **switch** is a device that mechanically interrupts the flow of current by opening a set of contacts. When the contacts are closed, the switch is on and current flows through the device. When the switch is off, the contacts are open and current flow is interrupted. Most switches in electronic circuits are simple two-position ON/OFF switches with one set of contacts. Some examples of common switches are shown in Figure 4–34.

Figure 4–35 shows the schematic symbol for a two-position ON/OFF switch. Figure 4–35(a) shows the switch in the ON position, conducting current. Figure 4–35(b) shows the switch in the OFF position, blocking current.

The term **pole** is used to describe the number of completely isolated circuits that are allowed to pass through the switch at a given time. A single pole switch only allows the current from one circuit to pass through, while a two-pole switch conducts the current from two different circuits.

The term **throw** refers to the total number of individual circuits that each pole is capable of controlling. In other words, it is quite common to have a single pole controlling more than one circuit. If there were two possible paths for the current to take, depending on the position of the switch, and the switch were fed from only one circuit, then the switch would be considered a single-pole, double-throw switch. Figure 4–36 shows an example of such a switch. When the switch is set at position A, resistor R_1 receives current. If the switch is moved to position B, resistor R_2 is now connected in the circuit, and R_1 is disconnected.

(a)   **(b)**

FIGURE 4–35 Two-position switch. (a) ON position. (b) OFF position.

FIGURE 4–36 Single-pole, double-throw switch.

(a)   **(b)**

FIGURE 4–37 Push-button. (a) Normally open. (b) Normally closed.

Figure 4–37 shows the schematic symbol for a push button. A **push button** is a momentary device that will either allow or interrupt current flow when its contacts are closed or open. In Figure 4–37(a) the push button's contacts are *normally open*. That is, at rest the device will not pass current. When the push button is pressed, current flows through the device. The push button shown in Figure 4–37(b) is called *normally closed* because its contacts are normally resting in the closed state and will not open until the button is pressed. This type of push button is very popular in circuits that require a "stop" function. For example, if a normally closed stop button were connected in series with an alarm bell, the alarm bell would stop ringing whenever the stop button was pressed. Normally closed switches are always connected in series to the load, since a parallel connection would cause a short circuit.

Section Review

1. A fuse is a series-connected protection device designed to create an open circuit in the event of excess current flow in a circuit (True/False).
2. What is the basic construction of a fuse?
3. What percentage of overcurrent typically causes a fast-blow fuse to melt?
4. What percentage of overcurrent is allowed by a slow-blow fuse for 3 seconds?
5. A switch is a device that mechanically interrupts the flow of current by closing a set of contacts (True/False).

4–11

TROUBLESHOOTING SERIES CIRCUITS

The ability to locate faulty components in electronic circuits is a skill that must be learned by all electronic technicians. This ability is gained by combining proper troubleshooting procedures with a knowledge of the theory of the circuits. Essentially, troubleshooting is a practical application of theory. It is a logical, systematic process of proceeding from "effect" to "cause." It is also a process of elimination. For the most part, troubleshooting procedures are based on narrowing the problem by identifying what parts of the circuit work and what parts do not work. To properly troubleshoot an electronic circuit, the technician must know "what should be there." Any service manuals and technical literature available for the circuit will help in assessing the problem and arriving at a solution.

Depending on the circumstances, electronic circuits can be either tested when energized (switched on) or de-energized (switched off). Great care should be taken when troubleshooting energized equipment. The hazards of electric shock are always present when working on "live" circuits. Circuits such as television sets can produce thousands of volts and maintain charge long after being unplugged. Ohmmeters should *never* be used to trou-

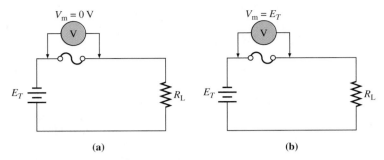

(a)   (b)

FIGURE 4–38 *Testing a fuse with a voltmeter. (a) Fuse is good. (b) Fuse is blown.*

bleshoot circuits that are energized. Because ohmmeters have their own internal voltage source, inaccurate readings as well as possible damage to the instrument may result.

The most common failure in a series circuit is an open-circuit condition. Open circuits can be caused by a variety of problems including burned out resistors, broken conductors, poor soldering, blown fuses, and faulty switches. Figure 4–38 shows how a fuse is tested on an energized circuit using a voltmeter. Since the internal resistance of a fuse is almost 0 Ω, the voltage drop, V_m, across a working fuse should be 0 V (Figure 4–38a). If the fuse is the cause of the open-circuit condition, the applied voltage of the circuit should equal the voltage drop across the fuse (Figure 4–38b).

Fuses can also be checked for failure with an ohmmeter. Figure 4–39 shows how a fuse can be checked for failure when it has been removed from the circuit. If the fuse is good, the ohmmeter should read very close to 0 Ω. If the fuse is blown, the ohmmeter should read infinity. When using a digital multimeter (DMM) the symbol for infinity is often displayed by flashing the number 1999 repeatedly.

Open-circuit conditions in series circuits can also be caused by defective switches. If the switch contacts become pitted, or oxidized, the switch could behave as an open circuit even though it is switched ON. A switch can be checked using the same methods employed

(a)   (b)

FIGURE 4–39 *Testing a fuse with an ohmmeter. (a) Fuse is good. (b) Fuse is blown.*

for testing a fuse. Using an ohmmeter, in the ON position the meter should read 0 Ω. In the OFF position, the meter should read infinity.

Switches can be tested using a voltmeter or ohmmeter in the same manner that fuses are tested. When using a voltmeter, if the voltage drop across the switch is approximately 0 V, the switch is in the ON position. If the voltage drop is equal to the applied voltage, the contacts are open and the switch is OFF.

Short circuits primarily occur in resistive components as a result of excess current and heat. The effect of a short circuit is exactly opposite to the effect of an open circuit. When a short-circuit condition occurs, the total circuit resistance decreases and current flow increases. Resistors and semiconductor devices are the most likely components to cause short-circuit conditions. Since fuses and switches have no resistance, these devices are not associated with short-circuit conditions.

The following steps represent a practical approach to troubleshooting electronic circuits.

STEP 1 *Look for symptoms of trouble.* Determine what is not working in the circuit, and understand what the circuit should be doing when functioning properly.

STEP 2 *Look for obvious problems.* Check fuses and circuit breakers to ensure that the circuit is receiving power. If troubleshooting a printed circuit board, look for signs of damage such as burned components from excess heat, damaged solder traces, loose connections, and so on.

STEP 3 *Localize the problem.* Isolate the trouble to as small an area as possible in the circuit. When the problem has been isolated, measure voltage and resistance and compare the results with "what should be there."

STEP 4 *Check each component systematically.* Devices such as semiconductors and resistors will have to be at least partially removed to obtain accurate readings with an ohmmeter on a de-energized circuit. A voltmeter can be used to test for shorts and opens while the circuit is energized and the components still in the circuit, although this is not feasible in some troubleshooting situations.

Section Review

1. What is the most common failure in a series circuit?
2. Name three causes of open circuits.

—————————————————————— 4–12 ——

PRACTICAL APPLICATION SOLUTION

The chapter opener stated that a LAN system connected in a ring network was malfunctioning. Your task was to determine what was wrong with the system based on the information provided. The following steps illustrate how a solution to this practical application is reached.

STEP 1 Sketch a diagram of the circuit (Figure 4–40). From the description of the system in the chapter opener, it is apparent that a ring network is basically a series circuit.

STEP 2 Further evaluation of the data provided in the chapter opener indicates that the measured resistance of each repeater corresponds to the rated resistance. This leads to the conclusion that the repeaters may in fact be functioning properly and that it is some other problem in the LAN.

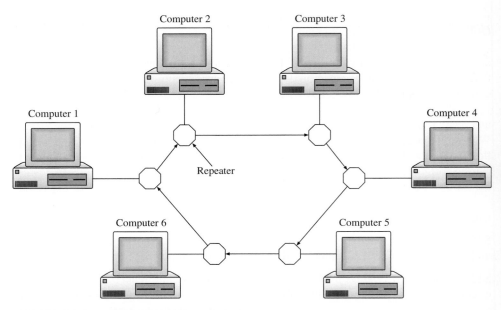

FIGURE 4–40 *Diagram of ring network.*

STEP 3   The fact that the resistance of one conductor is different from the others is cause for suspicion. Further inspection of the cable interconnecting computers 3 and 4 indicates a kink in the cable that has broken the conductor, producing an open-circuit condition. By replacing this cable, the problem has been corrected and the system is back in operation.

Summary

1. A series circuit is a circuit in which there is only one current path and all components are connected end to end along this path.
2. The same amount of current must flow in every part of a series circuit.
3. In a DC circuit, the direction of current indicates the polarity of a device and whether the device is absorbing power or delivering it.
4. In any closed loop, the algebraic sum of the voltage drops and rises equals zero.
5. The ratio between any two voltage drops in a series circuit is the same as the ratio of the two resistances across which these voltage drops occur.
6. The total power dissipated in a circuit is the sum of the individual amounts of power expended by the individual resistances.
7. A series-connected resistance that produces a voltage drop inside a source of emf is called *internal resistance.*
8. The power losses in a voltage source are attributed to the power dissipated by the internal resistance of the source.
9. A fuse is a series-connected protection device designed to create an open circuit in the event of excess current flow in the circuit.
10. The most common failure in a series circuit is an open-circuit condition.

Section 4–1
1. True
2. All the lamps will be off

Section 4–2
1. The total resistance is the sum of the individual resistances.
2. 300 Ω
3. 12 kΩ

Section 4–3
1. False
2. 5 A
3. False
4. Ammeters are always connected in series with the circuit.

Section 4–4
1. Voltmeters are always connected across the load to be measured, never in series.
2. Due to the relationship between current and resistance in the Ohm's law equation
3. True

Section 4–5
1. An electrical ground utilizes the earth or an equipment chassis as a reservoir of charge.
2. The chassis acts as a conductor between the battery and electrical devices in the system.
3. False
4. The polarity of a device and whether the device is absorbing power or delivering it

Section 4–6
1. True
2. To specify the voltages present in a closed loop
3. $E_T = V_{R1} + V_{R2} + V_{R3}$

Section 4–7
1. Voltage divider rule
2. True
3. The series-dropping resistor

Section 4–8
1. False
2. 8 W
3. $P_T = (50 \text{ mA})^2(10 \text{ k}\Omega) + (50 \text{ mA})^2(10 \text{ k}\Omega) = 50 \text{ W}$

Section 4–9
1. False
2. The terminal voltage decreases.
3. The internal resistance is very low.
4. Internal losses, output power, and input power

Section 4–10
1. True
2. A fuse consists of a strip of metal that will melt at a relatively low temperature.

3. 10% overcurrent
4. 300% overcurrent
5. False

Section 4-11

1. Open-circuit conditions are the most common.
2. Blown fuses, burnt out resistors, faulty switches

___ *Review Questions* ___

Multiple Choice Questions

4-1 A series circuit is a circuit with
(a) More than one current path   **(b)** No current path
(c) No voltage source   **(d)** Only one current path

4-2 In a series circuit, the total resistance is
(a) The product of the individual resistances
(b) The sum of the individual resistances
(c) The square root of the individual resistances
(d) Equal to the smallest resistance in the circuit

4-3 As the amount of resistance increases in a series circuit, the flow of current
(a) Increases   **(b)** Decreases   **(c)** Stays the same   **(d)** Cannot be determined

4-4 Voltage drops are also called
(a) *IE* drops   **(b)** *IV* drops   **(c)** *I/R* drops   **(d)** *IR* drops

4-5 Voltmeters are instruments
(a) With low internal resistance
(b) With high internal resistance
(c) Connected in series with the load to be measured
(d) Connected with the positive terminal on the negative side of the power supply

4-6 A voltage rise occurs
(a) In an active element when electrons flow from the + to − terminals
(b) In a passive element when electrons flow from the + to − terminals
(c) In an active element when electrons flow from the − to the + terminals
(d) In a passive element when electrons flow from the − to the + termminals

4-7 The term *ground* means
(a) A point in the circuit directly connected to earth   **(b)** A reservoir of charge
(c) A circuit point connected to a large metal chassis   **(d)** All of the above

4-8 If the positive terminal of a voltage source is connected to ground, it is called
(a) A short circuit   **(b)** Positive ground   **(c)** Negative ground   **(d)** Reverse polarity

4-9 When the positive terminal of a voltage source is connected in series with the negative terminal of another source, it is called
(a) Positive ground   **(b)** Negative ground   **(c)** Series opposing   **(d)** Series aiding

4-10 When using double-subscript notation, the second subscript implies
(a) The voltage is measured with respect to ground   **(b)** There is no potential difference
(c) The reference point is not at ground   **(d)** None of the above

4-11 A voltage divider circuit
(a) Allows a load to operate at a higher value of current than the source
(b) Allows a load to operate at a lower value of resistance than the source

(c) Allows the load to operate at a lower value of voltage than the source

(d) Is not practical in electronic circuit applications

4–12 The total power in a series circuit is

(a) A product of the power expended in individual resistors

(b) The square root of the power expended in individual resistors

(c) Equal to the power of any one resistor

(d) Equal to the sum of the power expended in individual resistors

4–13 A series-connected resistance that produces a voltage drop inside a source of emf is called

(a) Internal resistance   **(b)** Series-dropping resistance

(c) Load resistance   **(d)** Divider resistance

4–14 A fuse is a

(a) Series-connected resistive component designed to produce a short circuit

(b) Parallel-connected device designed to produce a short circuit

(c) Series-connected device designed to produce an open circuit

(d) Parallel-connected device designed to produce an open circuit.

4–15 When a short circuit occurs in a series circuit, the total circuit

(a) Voltage increases   **(b)** Current increases

(c) Resistance increases   **(d)** Current stays the same

Practice Problems

4–16 The following resistors are connected in series across a 24V supply: $R_1 = 30\ \Omega$, $R_2 = 100\ \Omega$, $R_3 = 70\ \Omega$. Determine the total current in this circuit.

4–17 A 50 Ω resistor is connected in series with another resistor whose value is not known. If the current is 0.75 A when 120V is applied, determine the unknown resistance.

4–18 A series circuit containing three resistors has a total current of 2 A when 120 V is applied. If $R_1 = 20\Omega$ and $R_2 = 15\ \Omega$, calculate the ohmic value of R_3.

4–19 What is the current through each resistor in a series circuit if the total resistance is 1.3 MΩ and the total voltage is 24 V.

4–20 The current in a 40 V circuit is 0.44 A. One of three series resistors has a value of 40 Ω, and another resistor has a value of 22 Ω. How much power is dissipated by the unknown resistor?

4–21 A series circuit containing four resistors has the following values: $R_1 = 2\ k\Omega$, $R_2 = 1\ k\Omega$, $R_3 = 5\ k\Omega$, $R_4 = 1.5\ k\Omega$. The voltage across the 2 kΩ resistor is 5.5 V. Determine the total applied voltage of the circuit.

4–22 Three resistors with values of 1 kΩ, 5 kΩ, and 10 kΩ are connected in series across a 120 V supply. What power is dissipated by each resistor, and what is the total power dissipated by the circuit?

4–23 Using the voltage divider rule, determine the voltage drop across the 51 Ω resistor for the circuit shown in Figure 4–41.

27 Ω

FIGURE 4–41

$E_T = 100\ V$

51 Ω

43 Ω

4–24 A 24 V, 1 A radio receiver is to be installed in a circuit with a 30 V supply. Calculate the ohmic value of a series-dropping resistor that will limit the current to the rated value of 1 A.

4–25 The following values have been measured in a series circuit containing three resistors: $R_1 = 11.5\ \Omega$, $I_{R2} = 3.2$ A, $V_{R3} = 18$ V. The applied voltage of the circuit is 80 V. Calculate the total power, as well as the power used in each resistor.

4–26 In a series circuit, R_1 is to have one-third the voltage drop of R_2. If R_2 has a resistance of 1000 Ω, what value is R_1?

4–27 Two lamps with ratings of 75 W at 100 V and 40 W at 50 V are connected in series across a 60 V supply. Calculate the voltage drop of each lamp.

4–28 A certain series electronic circuit requires a current-limiting resistor to lower the supply voltage to an acceptable level. The load voltage is 8 V and load current is 20 mA. The supply voltage is 18 V. Calculate the ohmic value and power rating of the resistor required for this circuit.

4–29 A bank of 10 series-connected light emitting diodes are to be supplied by a 24 V source. Each LED has a voltage drop of 1.6 V and an internal resistance of 600 Ω. Calculate the resistance and power rating of the necessary voltage-dropping resistor.

4–30 Two resistances in series have ohmic values that result in the power dissipated by R_1 being four times greater than the power dissipated by R_2. What is the voltage ratio of R_1 and R_2?

4–31 For a power supply with a no-load voltage of 14.5 V, a full-load voltage of 12 V, and a full-load current of 2 A, determine the internal resistance.

4–32 Using the values given in Problem 4–31, calculate the efficiency of the power supply.

Essay Questions

4–33 Define a series circuit.

4–34 Explain how resistance is calculated in a series circuit.

4–35 When measuring current in a series circuit, how is the ammeter connected?

4–36 Define the terms *voltage polarity* and *ground.*

4–37 What is the difference between a positive ground system and a negative ground?

4–38 Explain series-aiding and series-opposing connections of voltage sources.

4–39 What is double-subscript notation, and why is it used when measuring voltages?

4–40 Define Kirchhoff's voltage law.

4–41 Explain the principle of a voltage divider.

4–42 What is a series-dropping resistor?

4–43 Explain the relationship between the individual amounts of power expended and the total power in a series circuit.

4–44 What is meant by the term *internal resistance?*

4–45 What are the two most common types of faults in a series circuit?

4–46 Explain the four steps taken in troubleshooting a series circuit.

5

Parallel Circuits

Learning Objectives

Upon completion of this chapter you will be able to

- Define a parallel circuit.
- Calculate resistance in parallel.
- Describe the flow of current in a parallel circuit.
- Express Kirchhoff's current law.
- Use the current divider rule.
- Apply Ohm's law for parallel circuit calculations.
- Calculate power in a parallel circuit.
- Describe the effect of connecting voltage sources in parallel.
- List some typical applications for parallel circuits.
- Troubleshoot parallel circuits.

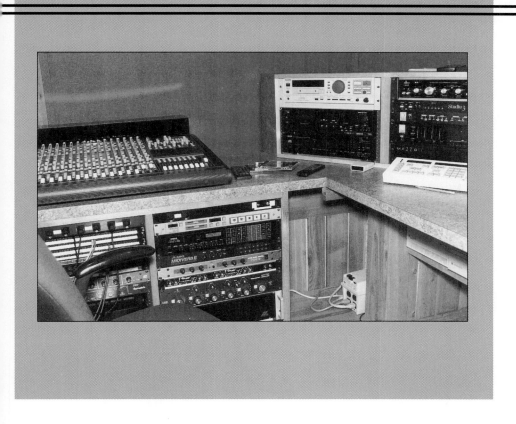

Practical Application

As a technician for an audio-engineering company, you are required to design an audio system for a nightclub that requires eight speakers to be driven from a 1000 W amplifier (500 W per side). The specifications for the installation stipulate that four speakers are to be connected in parallel to each side of the amplifier. The minimum output resistance per side of the amplifier is rated at 4 Ω.

Your task is to sketch an installation diagram showing the speaker connections, determine the ohmic value of each speaker, and calculate the minimum power-handling capability of each speaker.

INTRODUCTION

A **parallel circuit** can be defined as a circuit that provides more than one current path. When resistances, or other circuit components, are connected so that they have the same pair of terminal points, or **nodes,** the resistances are said to be in *parallel*. A node is any point in a circuit where two or more circuit paths intersect. Parallel circuits are extremely popular in electrical and electronic systems. For example, all lights, receptacles, and appliances in every home are connected in parallel with each other so that they can be operated independently. The headlights, tail lights, radio, and all other accessories in a car are also connected in parallel with each other. Whenever a device is required to be switched off without affecting other devices, it is wired in parallel. An example of a parallel circuit with four resistors interconnected on a breadboard is shown in Figure 5–1.

Section Review

1. A node is any point in a circuit where two or more circuit paths intersect (True/False).
2. What is a parallel circuit?
3. Parallel circuits are rarely used in electronic systems (True/False).

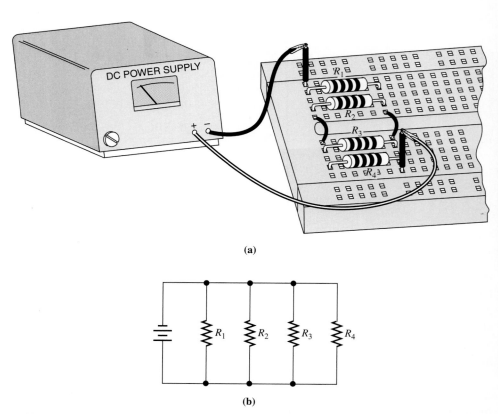

(a)

(b)

FIGURE 5–1 *Parallel circuit with four resistors and a power supply. (a) Breadboard connections. (b) Schematic diagram.*

FIGURE 5-2 *(a) Resistors in parallel. (b) Open-circuit voltage.*

5-2

VOLTAGE IN PARALLEL CIRCUITS

Because each of the resistors in Figure 5–1 are connected directly across the terminals of the power supply, the parallel circuit may be described as a circuit having a common voltage across its components. The voltage across any branch of a parallel combination is equal to the voltage across each of the other branches in parallel. This fact may be expressed mathematically in the following manner:

$$E = V_{R1} = V_{R2} = V_{R3} \qquad\qquad (5.1)$$

When a load is connected in parallel with a voltage source, the voltage measured across the load will not change if the load opens. Figure 5–2(a) shows two resistors connected in parallel. If R_2 in Figure 5–2(b) were open, V_{oc}, a voltmeter connected across R_2 would still measure the source voltage.

Section Review

1. A parallel circuit has a common voltage across its components (True/False).
2. What is the equation to express the relationship between a supply voltage, E, and the voltage across three parallel-connected resistors?
3. In a parallel circuit, the voltage measured across the load will not change if the load opens (True/False).

5-3

RESISTANCES IN PARALLEL

For resistors to be connected in parallel, they must be connected together so they have common nodes. On a circuit diagram this interconnection may be quite obvious. However, the actual physical wiring of the resistors may not be as readily apparent. The following example illustrates how resistors are wired together to form a parallel circuit.

_____ **EXAMPLE 5-1** _____

Draw the wiring connec-
tions for the assembly diagram shown in Figure 5–3.

FIGURE 5–3 *Assembly diagram for Example 5–1.*

Solution   One terminal of each resistor is wired so that it is common to the negative terminal of the source and one terminal of each resistor is wired to the positive
terminal of the source, as shown in Figure 5–4.

FIGURE 5–4 *Resistors wired in parallel.*

The wiring shown in Figure 5–4 constitutes a parallel circuit since the terminals of
any resistance can be traced back to the terminals of the source without having to go
through any other resistances. This corresponds to our definition of a parallel circuit that
each terminal of any given resistance connects directly to a corresponding terminal of
every other resistance.

Resistances in parallel are like water pipes in parallel. When two water pipes of the
same size are placed side by side, they will carry twice as much water as a single pipe. Two
resistors of equal value when connected in parallel will allow twice as much current to
flow. When resistances of unequal value are connected in parallel, the opposition to current
flow is not the same for each branch of the circuit. A large resistor is like a small-diameter
water pipe because it restricts the flow of current. Conversely, a small resistor has low
opposition to current flow, so it acts like a large-diameter water pipe.

As the number of resistors in parallel increases, the paths for the current to flow in
the circuit also increases, and the overall resistance must decrease. Therefore, the total, or

equivalent, resistance of resistors in parallel is always less than the resistance of any of the individual resistances in the circuit.

Parallel circuits are made up either of resistors of equal value or of resistors of unequal value. These two types of parallel circuits are most easily solved for by using the following two rules.

1. When only two resistors are connected in parallel, the resistance of the parallel combination of two resistances is equal to the product of the individual resistances divided by their sum. This rule is also known as the **product-over-sum rule.** In equation form,

$$R_T = \frac{R_1 \times R_2}{R_1 + R_2} \tag{5.2}$$

2. When two resistors of equal value are connected in parallel, twice as much current flows. To find the resistance of resistors of equal value connected in parallel, divide the value of one resistance by the number of resistances connected.

$$R_T = \frac{R_x}{N} \tag{5.3}$$

where R_T = equivalent resistance of parallel circuit
  R_x = value of one resistance
  N = number of resistors in circuit

_____ **EXAMPLE 5–2** _____

FIGURE 5–5   *Circuit for Example 5–2.*

Using the product-over-sum rule, calculate the total resistance of the circuit of Figure 5–5.

Solution

$$R_T = \frac{R_1 \times R_2}{R_1 + R_2}$$

$$= \frac{300\ \Omega \times 430\ \Omega}{300\ \Omega + 430\ \Omega}$$

$$= 176.71\ \Omega$$

_____ **EXAMPLE 5–3** _____

FIGURE 5–6   *Circuit for Example 5–3.*

Using Equation 5.3, determine the total resistance of the circuit shown in Figure 5–6.

Solution

$$R_T = \frac{R_x}{N}$$

$$= \frac{200\ \Omega}{3}$$

$$= 66.67\ \Omega$$

EXAMPLE 5–4

For the circuit shown in Figure 5–7, two resistors of equal ohmic value are connected in parallel. The total resistance of this parallel combination is 8 Ω. Determine the resistance of R_1 and R_2.

FIGURE 5–7 *Circuit for Example 5–4.*

Solution

$$R_T = \frac{R_x}{N}$$

$$R_x = R_T N$$

$$= (8\ \Omega)(2)$$

$$= 16\ \Omega$$

When determining the total resistance of a parallel circuit containing three or more resistors of unequal value, the inverse resistance, or conductance, is often used. The formula for conductance states that the reciprocal of the total resistance is equal to the sum of the reciprocal of each resistance and is written as

$$\frac{1}{R_T} = \frac{1}{R_1} + \frac{1}{R_2} + \frac{1}{R_3} + \cdots + \frac{1}{R_n} \tag{5.4}$$

As mentioned in Chapter 2, conductance is a term used to describe a circuit or component's ability to pass current. The SI unit of conductance is siemens (S). The ease with which current flows through a circuit is said to be the circuit's conductance. As the number of resistors connected in parallel increases, the ease with which current flows also increases. Therefore, when resistances are connected in parallel, the total conductance increases, or

$$G_T = G_1 + G_2 + G_3 + \cdots + G_n \tag{5.5}$$

Since resistance is the exact opposite, or inverse, of conductance, Equations 5.4 and 5.5 are combined to form Equation 5.6.

$$G_T = \frac{1}{R_T} = \frac{1}{R_1} + \frac{1}{R_2} + \frac{1}{R_3} + \cdots + \frac{1}{R_n} \tag{5.6}$$

The conductance of a parallel circuit and the resistance of a series circuit are often called the **duals** of each other because it is possible to develop some of the relationships for the parallel circuit directly from those of a series circuit by simply interchanging R and G. There are many other duals that exist between parallel and series circuits. For example, in a series circuit the current is the same, and in a parallel circuit the voltage is the same.

EXAMPLE 5–5

Calculate the total resistance of the circuit shown in Figure 5–8 using Equation 5.4.

$R_1 = 39\,\Omega$ $R_2 = 47\,\Omega$ $R_3 = 56\,\Omega$

FIGURE 5–8   *Circuit for Example 5–5.*

Solution

$$\frac{1}{R_T} = \frac{1}{R_1} + \frac{1}{R_2} + \frac{1}{R_3}$$

$$= \frac{1}{39\,\Omega} + \frac{1}{47\,\Omega} + \frac{1}{56\,\Omega}$$

$$= 0.06447\,\text{S}$$

$$R_T = \frac{1}{1/R_T}$$

$$= \frac{1}{0.06447}$$

$$= 15.44\,\Omega$$

EXAMPLE 5–6

Calculate the total conductance of the circuit shown in Figure 5–9.

$R_1 = 330\,\Omega$ $R_2 = 270\,\Omega$ $R_3 = 560\,\Omega$

FIGURE 5–9   *Circuit for Example 5–6.*

Solution

$$G_T = \frac{1}{R_T} = \frac{1}{R_1} + \frac{1}{R_2} + \frac{1}{R_3}$$

$$= \frac{1}{330\,\Omega} + \frac{1}{270\,\Omega} + \frac{1}{560\,\Omega}$$

$$= 8.52\,\text{mS}$$

Parallel resistors are often shown in circuit equations as two parallel vertical lines ∥. When these lines are shown between two resistors, such as R_1 and R_2, the notation would be

$$R_1 \parallel R_2$$

Often when calculating equivalent resistances, it is convenient to use an *approximation* of the actual value. Because fixed resistors are only available in specific ohmic values, approximations are frequently necessary. For example, suppose a resistance calculation requires a 9.99 kΩ resistor. According to the Table of Standard Resistors (Appendix A), the nearest size resistor available is 10 kΩ. This resistor could be used even though it is not the exact value. As all resistors have tolerance ratings, there is always a slight amount of variation in the true resistance of the circuit. In most electronic circuit calculations, if two quantities are within 1%, they are considered approximately equal to each other. The symbol for *approximately equal to* is ≈.

One method of approximation used in circuits with two parallel resistances is the **ten-to-one rule.** When two resistors are connected in parallel, if one resistor is ten or more times greater than the other resistor, then the greater value resistor may be ignored. The following example illustrates the approximation rule.

_____ **EXAMPLE 5–7** _____

Calculate the total resistance of the circuit shown in Figure 5–10 using the ten-to-one rule, and compare this value to the actual circuit resistance.

FIGURE 5–10 *Circuit for Example 5–7.*

Solution Using the approximation rule, the equivalent resistance is equal to the lowest resistance in the circuit. Therefore,

$$R_T \approx 1 \text{ k}\Omega$$

The true resistance of the circuit is found by the product-over-sum rule.

$$R_T = \frac{R_1 R_2}{R_1 + R_2}$$
$$= \frac{(33 \text{ k}\Omega)\,(1 \text{ k}\Omega)}{33 \text{ k}\Omega + 1 \text{ k}\Omega}$$
$$= 970.59 \ \Omega$$

The difference between the true value and the approximate value is 29.41 Ω, which represents a percentage of error of less than 3%.

Section Review

1. Does the overall resistance increase or decrease as the number of resistors in parallel is increased?
2. What is the product-over-sum rule?
3. The conductance of a parallel circuit and the resistance of a series circuit are often called the duals of each other (True/False).

FIGURE 5-11   *Flow of current in parallel circuit.*

_____ **5-4** ____

KIRCHHOFF'S CURRENT LAW

During his experiments with voltage in 1845, Kirchhoff also discovered that the algebraic sum of all currents entering and leaving a node is zero. This became known as **Kirchhoff's current law (KCL).** Kirchhoff defined the currents entering a node as positive and the currents leaving the node as negative. KCL also can be stated as

> The total current flowing from the source is the sum of the individual branch currents.

Therefore, in equation form,

$$I_T = I_1 + I_2 + I_3 + \cdots + I_n \qquad (5.7)$$

Figure 5-11 shows an example of Kirchhoff's current law in a parallel circuit with two resistors. The total circuit current I_T is shown flowing into the node. At the node, the current splits into two paths I_1 and I_2. This is because, in a parallel circuit, the total circuit current divides, with part of the total current flowing through each possible path. Each resistance is rated to pass a certain maximum current, although the total circuit current can be greater than this individual rated value.

_____ **EXAMPLE 5-8** _____

Two resistances shown in Figure 5-12 are connected in parallel. The current through one resistance is 750 mA, and the current through the other resistance is 2.15 A. Determine the total current supplied to the circuit.

FIGURE 5-12   *Circuit for Example 5-8.*

Solution

$$I_T = I_1 + I_2$$
$$= 0.75 \text{ A} + 2.15 \text{ A}$$
$$= 2.90 \text{ A}$$

EXAMPLE 5–9

For the circuit of Figure 5–13, solve for the branch current I_2 and the total current I_T in the circuit.

FIGURE 5–13   *Circuit for Example 5–9.*

Solution   Since R_2 is double the resistance of R_1, current I_1 must be twice as high as current I_2. Therefore,

$$\frac{R_2}{R_1} = \frac{I_1}{I_2} = \frac{30\ \Omega}{15\ \Omega} = \frac{100\ \text{mA}}{I_2}$$

$$I_2 = \frac{I_1}{2} = \frac{100\ \text{mA}}{2} = 50\ \text{mA}$$

$$I_T = I_1 + I_2 = 100\ \text{mA} + 50\ \text{mA} = 150\ \text{mA}$$

Any equation based on Kirchhoff's current law is referred to as a *nodal equation.* When nodal equations are written, the flow of current in one direction is considered additive, and the flow of current in the opposite direction is considered subtractive. The following example illustrates using nodal equations to solve for unknown current.

EXAMPLE 5–10

Solve for the unknown current shown in Figure 5–14 using a nodal equation.

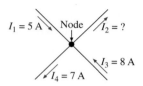

FIGURE 5–14   *Diagram for Example 5–10.*

Solution   If we assume that the current shown entering the node is positive and the current leaving the node is negative, the nodal equation is written as

$$I_1 - I_2 + I_3 - I_4 = 0$$

Rearranging terms and solving for I_2 results in

$$I_2 = I_1 + I_3 - I_4$$
$$= 5\ \text{A} + 8\ \text{A} - 7\ \text{A}$$
$$= 6\ \text{A}$$

Section Review

1. The total current flowing from the source is the product of the individual branch currents (True/False).
2. State the Kirchhoff equation for the current flowing in three parallel resistors.
3. How does the flow of current affect nodal equations?

5-5
CURRENT DIVIDERS

It is often necessary to find the individual branch currents in a circuit from the resistances and total current, but without knowing what the applied voltage of the circuit is. This type of problem can be solved by using the fact that currents divide inversely as the resistances of their paths. Because current *always* takes the path of least resistance, current is always greatest through the path of least opposition. A small value of resistance offers less opposition and, consequently, the smaller resistances in parallel circuits pass more current than the larger resistances.

The **current divider rule** is stated as follows:

> The amount of current in one of two parallel resistances is calculated by multiplying their total current by the other resistance and dividing by their sum.

The formulas for calculating currents through two branch circuits are

$$I_1 = \left(\frac{R_2}{R_1 + R_2} \right) I_T \qquad (5.8)$$

$$I_2 = \left(\frac{R_1}{R_1 + R_2} \right) I_T \qquad (5.9)$$

EXAMPLE 5-11

$I_T = 25$ mA

$R_1 = 6\ \Omega$ $R_2 = 9\ \Omega$

FIGURE 5-15 *Circuit for Example 5-11.*

Using the current divider rule, calculate the currents flowing through branches R_1 and R_2 of Figure 5-15.

Solution

$$I_1 = \left(\frac{R_2}{R_1 + R_2} \right) I_T$$

$$= \left(\frac{9\ \Omega}{6\ \Omega + 9\ \Omega} \right) 25 \text{ mA}$$

$$= 15 \text{ mA}$$

$$I_2 = \left(\frac{R_1}{R_1 + R_2} \right) I_T$$

$$= \left(\frac{6\ \Omega}{6\ \Omega + 9\ \Omega} \right) 25\ \text{mA}$$

$$= 10\ \text{mA}$$

To check:
$$I_T = I_1 + I_2$$

$$= 15\ \text{mA} + 10\ \text{mA}$$

$$= 25\ \text{mA}$$

It should be noted that Equations 5.8 and 5.9 are applicable only for *two* resistances. When a parallel circuit contains more than two resistances, the following equation may be used to find current in an individual branch:

$$I_x = \left(\frac{R_T}{R_x} \right) I_T \tag{5.10}$$

where I_x = current through an individual branch
         R_x = resistance of an individual branch

____ EXAMPLE 5–12 ____

For the circuit of Figure 5–16, determine the current through resistor R_3 using the current divider rule.

FIGURE 5–16   *Circuit for Example 5–12.*

Solution

$$\frac{1}{R_T} = \frac{1}{R_1} + \frac{1}{R_2} + \frac{1}{R_3}$$

$$= \frac{1}{510\ \Omega} + \frac{1}{270\ \Omega} + \frac{1}{430\ \Omega}$$

$$= 0.00799\ \text{S}$$

$$R_T = \frac{1}{1/R_T}$$

$$= \frac{1}{0.00799\text{S}}$$

$$= 125.16\ \Omega$$

$$I_3 = \left(\frac{R_T}{R_3} \right) I_T$$

$$= \left(\frac{125.16\ \Omega}{430\ \Omega} \right) 300\ \text{mA}$$

$$= 87.32\ \text{mA}$$

Section Review

1. The current divider rule is used to find the individual branch currents in a circuit without knowing the applied voltage (True/False).
2. Two resistors, 2.2 kΩ and 4.7 kΩ, are connected in parallel to a 100 mA supply. What is the current through the 2.2 kΩ resistor?
3. The current divider rule is used for two or more parallel resistors (True/False).

5–6
OHM'S LAW IN PARALLEL CIRCUITS

Unknown quantities of resistance, current, and voltage in parallel circuits also may be calculated using Ohm's law. As long as two quantities in a parallel circuit are known, the third can be determined. For example, the voltage of a parallel circuit can be determined if current and resistance are known for any branch. Since the voltage across each resistor is equal in a parallel circuit, the circuit voltage is found by multiplying together the known resistance and current values for any of the parallel paths.

Ohm's law for parallel circuits is stated as follows:

> Divide the common voltage across the parallel resistances by the total current of all the branches.

$$R_T = \frac{E}{I_T}$$

where I_T = the sum of all the branch currents
E = supply voltage
R_T = the equivalent resistance of all the parallel branches

EXAMPLE 5–13

FIGURE 5–17 *Circuit for Example 5–13.*

For the circuit shown in Figure 5–17, use Ohm's law to find the

(a) Total voltage.
(b) Current through R_2.
(c) Ohmic value for resistor R_3.

Solution

(a)
$$E_T = IR_1$$
$$= (225 \text{ mA}) (100 \text{ }\Omega)$$
$$= 22.5 \text{ V}$$

(b)
$$I_2 = \frac{E_T}{R_2}$$
$$= \frac{22.5 \text{ V}}{150 \text{ }\Omega}$$
$$= 150 \text{ mA}$$

(c)
$$R_3 = \frac{E_T}{I_3}$$
$$= \frac{22.5 \text{ V}}{250 \text{ mA}}$$
$$= 90 \text{ }\Omega$$

EXAMPLE 5–14

Calculate the value of resistor R_2 shown in Figure 5–18.

FIGURE 5–18 *Circuit for Example 5–14.*

Solution   Using Ohm's law, the total resistance of the circuit is found.

$$R_T = \frac{E_T}{I_T} = \frac{12 \text{ V}}{114.3 \text{ mA}} = 105 \text{ }\Omega$$

The value of resistor R_2 is now found using Equation 5.4

$$\frac{1}{R_T} = \frac{1}{R_1} + \frac{1}{R_2} + \frac{1}{R_3}$$

$$\frac{1}{R_2} = \frac{1}{R_T} - \frac{1}{R_1} - \frac{1}{R_3}$$

$$= \frac{1}{105 \text{ }\Omega} - \frac{1}{330 \text{ }\Omega} - \frac{1}{430 \text{ }\Omega}$$

$$= 0.00417 \text{ S}$$

$$R_2 = \frac{1}{1/R_2}$$

$$= \frac{1}{0.00417 \text{ S}}$$

$$= 240 \text{ }\Omega$$

EXAMPLE 5–15

Determine the amount of current flowing in branch R_3 of the circuit shown in Figure 5–19.

FIGURE 5–19 *Circuit for Example 5–15.*

Solution   Use Ohm's law to solve for the current through each branch.

$$I_1 = \frac{E}{R_1} = \frac{12\ V}{30\ \Omega} = 0.4\ A$$

$$I_2 = \frac{E}{R_2} = \frac{12\ V}{24\ \Omega} = 0.5\ A$$

The total current is now determined using the current divider rule.

$$I_T - (I_1 + I_2 + I_3) = 0$$
$$I_3 = I_T - (I_1 + I_2) = 1.6\ A - (0.4\ A + 0.5\ A) = 0.7\ A$$

Section Review

1. State Ohm's law for a parallel circuit.
2. Two 560 Ω resistors are connected in parallel to a 12 V source. What is the total circuit current?
3. Ohm's law cannot be applied to parallel circuits with more than three resistors (True/False).

5-7

VOLTAGE SOURCES IN PARALLEL

When voltage cells are connected in parallel to form a battery, they provide more than one current path through the battery with each cell furnishing only a part of the total battery current. This configuration is popular in electronic circuits that require large values of current. For example, Figure 5–20 shows two 12 V batteries with identical current ratings connected in parallel. Each battery is rated at 5 A and, according to Kirchhoff's current law, the total current supplied to the load, R_L, is 10 A. Voltage sources are connected in parallel to increase the current and power rating of a combination of batteries. Since the internal resistances of voltage cells in parallel will decrease, the overall efficiency of the total voltage source improves as voltage cells are added.

The only types of voltage sources that can be connected safely in parallel are sources with identical voltage ratings and internal resistances. Voltage sources of different voltage levels should never be connected in parallel with each other. When two unequal voltage sources are connected in parallel as in Figure 5–21, current flows rapidly from the negative terminal of the larger source to the negative terminal of the smaller source. The difference in potential between the two sources is −6 and −12 V, or 6 V. Essentially, the circuit of Fig-

FIGURE 5–20   Voltage sources in parallel.

FIGURE 5–21
Unequal voltage sources in parallel.

ure 5–21 represents a 6 V source connected across a short circuit, since the only opposition to current flow is the very small internal resistance of E_2. The current will flow away from the higher and toward the lower negative potential, as indicated by the arrow. This current direction between sources is because free electrons will be repelled with more force by the greater negative charge.

When voltage sources have identical voltage ratings and different internal resistances, the source having the highest internal resistance will supply the least amount of current. As mentioned earlier, when a battery ages, its internal resistance increases. Consequently, the flow of current supplied by the battery *decreases* as the internal resistance increases (due to aging). When a battery is said to be discharged, it implies that the current available from the battery is very close to zero. The principle of charging a battery is based on connecting a source, such as a battery charger or another battery, in parallel with a discharged, or dead, battery. Electrons are then drawn from the charged source to the discharged source, until the internal resistance falls to its rated value.

Section Review

1. Two 6 V batteries are connected in parallel. If each battery is rated at 300 mA, what is the total current supplied by the battery?
2. Why should batteries with different voltage ratings never be connected in parallel?
3. When voltage sources have different internal resistances and identical voltage ratings, the source having the lowest internal resistance will supply the least amount of current (True/False).

5–8

POWER IN PARALLEL CIRCUITS

As in any other circuit, power dissipation in a parallel circuit is the sum of the power dissipated in the individual resistances. The power expended by each component is calculated using Equations 3.6, 3.7, and 3.8:

$$P = EI = I^2R = \frac{E^2}{R}$$

In a parallel circuit, the smallest resistor passes the largest amount of current. Because the power of a resistor equals I^2R, the greatest amount of heat is dissipated by the smallest resistor. This characteristic is the opposite of that in a series circuit, although the total power supplied by the source is still equal to the sum of the power dissipated by all the components, or

$$P_T = P_1 + P_2 + P_3 + \cdots + P_n$$

EXAMPLE 5–16

For the circuit of Figure 5–22, determine the power dissipated by resistor R_1.

FIGURE 5–22 *Circuit for Example 5–16.*

Solution

$$P_T = E_T I_T$$
$$= (24 \text{ V}) (3 \text{ A})$$
$$= 72 \text{ W}$$

$$P_1 = P_T - P_2 - P_3$$
$$= 72 \text{ W} - 14 \text{ W} - 32 \text{ W}$$
$$= 26 \text{ W}$$

EXAMPLE 5–17

$E_T = 80$ V $R_1 = 39 \, \Omega$ $R_2 = 56 \, \Omega$

FIGURE 5–23 *Circuit for Example 5–17.*

Determine the power dissipated in each resistance and the total power expended in the circuit of Figure 5–23.

Solution

$$P_1 = \frac{E^2}{R_1} = \frac{80^2 \text{ V}}{39 \, \Omega} = 164.1 \text{ W}$$

$$P_2 = \frac{E^2}{R_2} = \frac{80^2 \text{ V}}{56 \, \Omega} = 114.29 \text{ W}$$

$$R_T = \frac{R_1 R_2}{R_1 + R_2} = \frac{(39 \, \Omega) (56 \, \Omega)}{39 \, \Omega + 56 \, \Omega} = 22.9895 \, \Omega$$

$$P_T = \frac{E^2}{R_T} = \frac{80^2 \text{ V}}{22.9895 \, \Omega} = 278.39 \text{ W}$$

To check: $P_T = P_1 + P_2 = 164.1 \text{ W} + 114.29 \text{ W} = 278.39 \text{ W}$

Section Review

1. Three 100 W lamps are connected in parallel. What is the total power dissipated by the lamps?
2. Power is calculated differently for parallel circuits compared to series circuits (True/False).
3. Does the smallest resistor in a parallel circuit dissipate the greatest amount of heat or the smallest?

5–9

APPLICATIONS OF PARALLEL CIRCUITS

Parallel circuits are used extensively in house wiring and consumer products. All electric outlets in a home are connected in parallel to a 120 V source supplied by the local electric utility company. Figure 5–24 shows an example of typical parallel circuits used in residential wiring. The electrical outlets allow consumer products to

FIGURE 5-24 *Parallel circuit for residential wiring.*

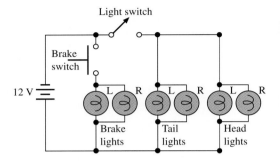

FIGURE 5-25 *Example of parallel circuits in automobiles.*

operate independently of each other. The three switches shown, Sw_1, Sw_2, and Sw_3, are used to mechanically disconnect the lamps from the applied voltage.

Parallel circuits are also popular in automobile electrical systems. All headlights, turn signals, and brake light systems are designed for parallel operation. For example, if one headlight burns out, the other headlight is still operational. Or if the brake pedal is pressed, only the brake lights come on. Figure 5–25 shows a simplified wiring diagram of an automobile lighting system.

Consumer products such as home audio systems use parallel circuits for connecting loudspeakers together. Figure 5–26 shows the wiring diagram for connecting four loudspeakers in parallel. The left channel has two 8 Ω speakers that, when connected in parallel, have a resistance of 4 Ω. The right channel also has two 8 Ω speakers connected in parallel and also has a total resistance of 4 Ω.

Section Review

1. Parallel circuits are used extensively in house wiring and consumer products (True/False).
2. What are three examples of parallel circuits in automobile electrical systems?

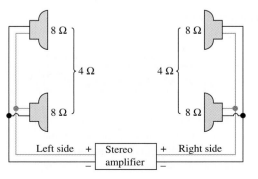

FIGURE 5-26 *Parallel connections for loudspeakers.*

TROUBLESHOOTING PARALLEL CIRCUITS

The same principles of troubleshooting that were applied to series circuits can also be applied to parallel circuits. The two main types of faults in parallel circuits are short circuits and open circuits. Unlike series circuits, an open-circuit condition does not necessarily interrupt power to an entire parallel network. Figure 5–27(a) shows an open circuit that interrupts current flow in the entire circuit. The circuit of Figure 5–27(b) shows an open circuit that interrupts current to resistor R_2 but still allows current through resistor R_1.

When an open circuit occurs in a branch of a parallel network, the resistance of the branch rises, and the total resistance of the circuit increases. An increase in the total resistance results in a decrease in the total current. Consider the circuit shown in Figure 5–28. In this circuit, resistor R_2 has become open circuited. It would be pointless to use a voltmeter to determine which resistor was destroyed because the voltmeter would provide the same reading across a good resistor as it would a bad resistor. To determine which resistor is faulty, the currents would have to be determined either by use of an ammeter or an ohmmeter and making calculations.

Figure 5–29 shows an example of using ammeters to detect an open circuit. In this circuit, ammeters A_1 and A_2 are reading current, while ammeter A_3 is reading 0 A. Since no current is flowing in this branch, it is apparent that resistor R_3 is open circuited.

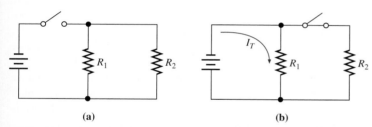

(a)   **(b)**

FIGURE 5–27 *Open in a parallel circuit. (a) Interrupting all current. (b) Interrupting part of the current.*

FIGURE 5–28 *Open-circuit voltage equals applied voltage.*

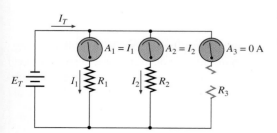

FIGURE 5–29 *Troubleshooting a parallel circuit using ammeters.*

FIGURE 5–30 *Short-circuit voltage equals 0 V.*

In a parallel circuit, a short in one branch will always result in no current flowing in any other branch. This is because a short has no resistance. Since current always takes the path of least resistance, all the circuit current will flow through the short and bypass all other circuit components. Shorts in parallel circuits are particularly dangerous because the voltage source is essentially shorted out, allowing massive amounts of current to flow through the loop.

Figure 5–30 shows an example of a short in a parallel circuit. In this circuit, resistor R_3 has become short circuited, so the voltage across R_3 is 0 V. Because the voltage is the same across all parallel components, this implies that the terminal voltage is also 0 V, and the current flow from the source is now limited only by the internal resistance of the source. Without proper overcurrent protection, such as a fuse, short circuits can cause extensive damage to voltage sources, conductors, and any other components connected in the path of the short-circuit current.

When troubleshooting parallel circuits, excess current flow will often cause noticeable heat damage, such as burned components and conductors. Blown fuses can also imply a short circuit of some type. Ohmmeters are excellent instruments for troubleshooting shorts in parallel circuits. Because a short produces 0 Ω of resistance, a short-circuit condition should be quite obvious when testing with an ohmmeter.

When a short exists in a parallel circuit, voltmeter and ammeter readings can become quite difficult to obtain on an energized circuit because of large values of short-circuit current. The longer a short-circuit current flows, the greater the potential for damage because of excess heat. When a circuit has proper overcurrent protection, a fuse or circuit breaker will open in the event of a short circuit. Therefore, it is often necessary to test for short-circuit conditions using an ohmmeter with the power disconnected to the circuit.

Section Review

1. The two main types of faults in parallel circuits are short circuits and excess current (True/False).
2. If two resistors are connected in parallel and one resistor becomes open circuited, what happens to the voltage supplied to the other resistor?
3. In a parallel circuit, a short in one branch will always result in no current flowing in any other branch (True/False).

__ 5–11

PRACTICAL APPLICATION SOLUTION

The system requirements stated in the chapter opener described a 1000 W amplifier driving four speakers per side. Your task was to sketch an installation diagram showing the speaker connections and to determine the ohmic value and power-handling capabilities of each speaker. The following steps outline the method of solution for this practical application of parallel circuits.

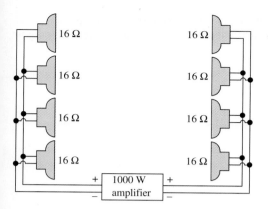

FIGURE 5–31 *Diagram of practical application.*

STEP 1   Figure 5–31 shows a sketch of the eight speakers. According to the information provided in the chapter opener, four speakers are to be connected in parallel to each side of the amplifier.

STEP 2   Because each side of the amplifier is rated at a minimum of 4 Ω, and four speakers are required to be in parallel, the ohmic value of each speaker is found using the following equation:

$$\frac{1}{R_T} = \frac{1}{R_1} + \frac{1}{R_2} + \frac{1}{R_3} + \frac{1}{R_4}$$

Therefore,
$$\frac{1}{4\ \Omega} = \frac{1}{16\ \Omega} + \frac{1}{16\ \Omega} + \frac{1}{16\ \Omega} + \frac{1}{16\ \Omega}$$

The ohmic value of each speaker must be 16 Ω.

STEP 3   Each side of the 1000 W amplifier is rated at 500 W. Because each side is to provide power to four speakers, the power-handling capabilities of each speaker is found as follows:

$$P_1 + P_2 + P_3 + P_4 = \frac{P_T}{4} = \frac{500\ \text{W}}{4} = 125\ \text{W}$$

Therefore,
$$P_1 = P_2 = P_3 = P_4 = \frac{P_T}{4} = \frac{500\ \text{W}}{4} = 125\ \text{W}$$

_____ **Summary** ___

1. A parallel circuit is defined as a circuit that provides more than one current path.
2. The voltage across any branch of a parallel circuit is equal to the voltage across each of the other branches in parallel.
3. The total resistance of parallel-connected resistors is always less than the resistance of any of the individual resistances in the circuit.
4. The algebraic sum of all currents entering and leaving a node is zero.
5. The amount of current in one of two parallel resistances is calculated by multiplying their total current by the other resistance and dividing by their sum.
6. As long as two quantities in a parallel circuit are known, the third can be determined.
7. Power dissipation in a parallel circuit is the sum of the power dissipated in the individual resistances.

8. Parallel circuits are used extensively in house wiring and consumer products.
9. In a parallel circuit, a short in one branch will always result in no current flowing in any other branch.

Answers to Section Reviews

Section 5-1
1. True
2. A circuit that provides more than one current path
3. False

Section 5-2
1. True
2. $E = V_{R1} = V_{R2} = V_{R3}$
3. True

Section 5-3
1. Decrease
2. A rule that states the resistance of the parallel combination of two resistances is equal to the product of the individual resistances divided by their sum
3. True

Section 5-4
1. False
2. $I_T = I_1 + I_2 + I_3$
3. The flow of current in one direction is considered additive while the flow in the other direction is subtractive.

Section 5-5
1. True
2. $I_1 = 4.7 \text{ k}\Omega/(2.2 \text{ k}\Omega + 4.7 \text{ k}\Omega) \times 100 \text{ mA} = 68.12 \text{ mA}$
3. True

Section 5-6
1. Divide the common voltage across the parallel resistances by the total current of all the branches.
2. $I_T = 12 \text{ V} (1/560 \text{ }\Omega + 1/560 \text{ }\Omega) = 42.86 \text{ mA}$
3. False

Section 5-7
1. 600 mA
2. Because current will flow rapidly from the negative terminal of the larger source to the negative terminal of the smaller one
3. False

Section 5-8
1. 300 W
2. False
3. The greatest amount of heat

Section 5–9

1. True
2. Headlights, brake lights, turn signals

Section 5–10

1. False
2. The voltage across the undamaged resistor would not be affected.
3. True

Review Questions

Multiple Choice Questions

5–1 A parallel circuit is defined as a circuit with
(a) Only one current path
(b) The same value of current in all parts of the circuit
(c) More than one current path
(d) Two or more devices connected in series

5–2 In a parallel circuit, if the load opens, the load voltage will
(a) Increase
(b) Decrease
(c) Be 0 V
(d) Not change

5–3 When two resistors of equal value are placed in parallel, the total current
(a) Triples
(b) Doubles
(c) Is reduced by 1/2
(d) Is reduced by 1/3

5–4 As the number of resistors in parallel increases, the total resistance
(a) Decreases
(b) Increases
(c) Stays the same
(d) Cannot be determined

5–5 The product-over-sum rule is used for
(a) Three resistors in parallel
(b) Two resistors in parallel
(c) Calculating current in a branch
(d) Calculating voltage in a branch

5–6 Conductance is
(a) The ease with which current flows
(b) The opposition to current flow
(c) Measured in ohms
(d) Measured in coulombs

5–7 Using the ten-to-one rule, one of two parallel resistors can be ignored if it
(a) Is 1/10 the value of the other resistor
(b) Has 10 times as much current as the other resistor
(c) Is 10 times greater than the other resistor
(d) Drops 10 times as much voltage as the other resistor

5–8 According to Kirchhoff's current law the
(a) Algebraic sum of the currents entering and leaving a node must equal the applied current
(b) Sum of the voltage drops equals the applied voltage
(c) Currents entering a node are negative and currents leaving a node are positive
(d) Algebraic sum of the currents entering and leaving a node is zero

5–9 Any equation based on Kirchhoff's current law is referred to as
(a) A voltage equation
(b) A loop equation
(c) A nodal equation
(d) None of the above

5–10 The current divider rule is used to find
(a) Total current
(b) Branch currents
(c) Total voltage
(d) Branch resistance

5–11 When voltage sources are connected in parallel the
(a) Total current increases
(b) Total voltage increases
(c) Total internal resistance increases
(d) Overall efficiency decreases

5–12 The only types of voltage sources that can safely be connected in parallel are sources with

(a) Different voltage ratings and identical internal resistances
(b) Identical voltage ratings and different internal resistances
(c) Identical current ratings
(d) Identical voltage ratings and internal resistances

5–13 In a parallel circuit, the smallest resistor dissipates

(a) The greatest amount of heat   (b) The smallest amount of heat
(c) All of the heat   (d) None of the above

5–14 When troubleshooting open circuits in a parallel network, which meter would be the least effective?

(a) Ohmmeter   (b) Ammeter
(c) Voltmeter   (d) All of the above

Practice Problems

5–15 What is the total resistance of the three parallel-connected resistors shown in Figure 5–32?

FIGURE 5–32

5–16 Find the equivalent resistance of a 20 Ω resistor in parallel with a 68 Ω resistor.

5–17 How much resistance must be connected in parallel with a 10 kΩ resistor to have a total resistance of 5 kΩ?

5–18 What value of resistance must be connected in parallel with a 470 Ω resistor to have an equivalent resistance of 220 Ω?

5–19 Determine the amount of current flowing in branch R_1 of the circuit shown in Figure 5–33.

FIGURE 5–33

5–20 Two parallel-connected resistors have values of $R_1 = 220$ Ω and $R_2 = 470$ Ω. The supply voltage is 12 V. Calculate the current flowing through each resistor as well as the total current supplied by the source.

5–21 Figure 5–34 shows two resistors connected in parallel. The resistors are supplied from a source rated at 20 A. Use the current divider rule to find the value of current in each branch.

FIGURE 5–34

5–22 Use the current divider rule to determine the branch currents in a circuit containing the following three parallel-connected resistors: $R_1 = 1$ kΩ, $R_2 = 2.2$ kΩ, $R_3 = 3$ kΩ. The supply current for the circuit is 40 mA.

5–23 Two circuit devices are connected in parallel. One component has a conductance of 480 μS and the other has a conductance of 800 μS. Find the total resistance of the circuit.

5–24 Three components with values of 1500 μS, 200 Ω, and 2500 μS are connected in parallel. If the total current is 81 mA, determine the voltage drop across the parallel components, as well as the current through each branch.

5–25 Use the product-over-sum rule to find the equivalent resistance of the parallel combination shown in Figure 5–35.

FIGURE 5–35

5–26 Four 10 kΩ resistors are connected in parallel. What is the total equivalent
(a) Resistance
(b) Conductance

5–27 Two resistors, $R_1 = 20$ Ω and $R_2 = 8$ Ω, are connected in parallel. If the circuit is supplied by a 24 V source, what is the amount of power dissipated by each resistance.

5–28 Three 60 W lamps (L) rated at 120 V are connected in parallel, as shown in Figure 5–36. What is the total resistance of these lamps?

FIGURE 5–36

5–29 A 12 Ω and 30 Ω resistor are connected in parallel. If the applied voltage is 12 V, determine the power dissipated by each resistor and the total power dissipated.

5–30 Three resistors, $R_1 = 1$ kΩ, $R_2 = 2$ kΩ, and $R_3 = 4$ kΩ, are connected in parallel to a 24 V supply. Find
(a) Total resistance   **(b)** Total current   **(c)** Current through each resistor
(d) Total power   **(e)** Power dissipated by each resistor

5–31 Four 120 V lamps (L) are connected in parallel to a 120 V source. The lamps are rated as follows: $L_1 = 25$ W, $L_2 = 60$ W, $L_3 = 40$ W, and $L_4 = 100$ W. Find the

(a) Amount of current flowing through each lamp

(b) Total resistance of the circuit

(c) Total power dissipated

5–32 Three resistors, $R_1 = 2.2$ kΩ, $R_2 = 4.7$ kΩ, $R_3 = 10$ kΩ, are connected in parallel. If each resistor is rated at 1/4W, find the maximum total current that would not cause any resistor to overheat.

5–33 Two resistors, $R_1 = 1$ kΩ and $R_2 = 2.2$ kΩ, are connected in parallel. The current in the first resistor is 5 mA. Find the

(a) Power dissipated by each resistor   (b) Total power drawn from the supply

Essay Questions

5–34 Define a parallel circuit.

5–35 Explain the relationship of voltages in a parallel circuit.

5–36 How is the total resistance found in a parallel circuit containing three equal resistors?

5–37 Define conductance.

5–38 When is it acceptable to use approximations?

5–39 What is the ten-to-one rule?

5–40 Explain Kirchhoff's current law.

5–41 What is a nodal equation?

5–42 Define the current divider rule.

5–43 Explain the effect of connecting voltage sources in parallel.

5–44 How is power determined in a parallel circuit?

5–45 List three applications of parallel circuits.

5–46 What are the two most common faults in parallel circuits?

6

Series-Parallel Circuits

Learning Objectives

Upon completion of this chapter you will be able to

- Define a series-parallel circuit.
- Determine the total resistance in a series-parallel circuit.
- Apply Kirchhoff's current and voltage laws to a series-parallel circuit.
- Calculate voltage drops and power.
- Recognize the various configurations of series-parallel networks.
- Explain the purpose of loaded voltage dividers.
- Describe the effects of open and short circuits on series-parallel resistor networks.
- Determine the total voltage of series-parallel voltage sources.

As a technician for a computer company, you are required to design a power supply to produce output voltages of +10 V and −10 V from a 24 V source. The power supply must be able to provide 200 mA at both the positive and negative 10 V terminals. A bleeder current of 10% of the total load is specified by your supervisor.

To complete this practical application you will

1. Sketch a wiring diagram, showing the reference, or ground, point for the power supply.
2. Determine the ohmic value of each resistor in the circuit.
3. Calculate the power consumption of each resistor required for the power supply.

6-1

INTRODUCTION

Circuits that contain both series and parallel combinations are called **series-parallel circuits.** A series-parallel circuit may be defined as a network in which some portions of the circuit have the characteristics of a simple series circuit and other portions have the characteristics of a simple parallel circuit.

In many cases, a complete series-parallel combination can be replaced by a single resistor that is electrically equivalent to the entire network. When calculating values of voltage, current, and resistance, it is often useful to simplify the circuit as these values are solved for. For example, three resistors in parallel can be redrawn as one resistor with an ohmic value equal to the equivalent value of the three resistors.

Series circuit laws and relations are applied to the components of a circuit that are connected in series, and parallel circuit laws and relations are applied to the components of a circuit that are connected in parallel. When solving problems relating to series-parallel circuits, it is advisable to begin on those components where two of the three Ohm's law factors are known. A series-parallel circuit can be analyzed using the following steps:

STEP 1   Reduce all parallel circuits to series equivalent resistances.

STEP 2   Combine all the branches containing more than one resistance in series into a single resistance.

STEP 3   Redraw the resulting **equivalent circuit.**

STEP 4   When the series-parallel circuit has been simplified, solve for total resistance, current, or voltage, as required.

STEP 5   To obtain a complete solution for a series-parallel circuit, use values obtained in the equivalent circuit and apply them to the original circuit. Calculate voltage, current, and power as required.

Section Review

1. Series-parallel circuits can often be replaced by a single resistor, which is electrically equivalent to an entire network (True/False).
2. What is the first step when solving problems relating to series-parallel circuits?

6-2

RESISTANCES IN SERIES-PARALLEL CIRCUITS

To determine the total resistance of a series-parallel circuit, it is necessary to determine which resistors form series circuits and which are parallel networks. Consider the circuit of Figure 6–1(a). In this circuit, resistor R_1 is connected in series with the parallel combination R_2 and R_3. Because R_2 and R_3 are connected in parallel, these two resistors can be reduced to a single equivalent value found by using the product-over-sum rule:

$$R_x = \frac{R_2 R_3}{R_2 + R_3}$$

(a)   (b)   (c)

FIGURE 6–1 *(a) Series-parallel circuit. (b) Simplified circuit. (c) Equivalent circuit.*

Figure 6–1(b) shows the equivalent circuit for Figure 6–1(a). The network now consists of two series-connected resistors, R_1 and R_x. The total resistance is equal to the sum of these two values, as shown in Figure 6–1(c).

$$R_T = R_1 + R_x$$

EXAMPLE 6–1

Determine the total resistance of the circuit shown in Figure 6–2.

FIGURE 6–2 *Circuit for Example 6–1.*

Solution   The two parallel resistors, R_2 and R_3, are solved by using the product-over-sum rule.

$$R_2 \parallel R_3 = \frac{R_2 R_3}{R_2 + R_3}$$

$$= \frac{(510\ \Omega)\,(330\ \Omega)}{510\ \Omega + 330\ \Omega}$$

$$= 200.36\ \Omega$$

The total resistance is now found by adding the equivalent resistance to resistor R_1.

$$R_T = R_2 \parallel R_3 + R_1$$

$$= 200.36\ \Omega + 470\ \Omega$$

$$= 670.36\ \Omega$$

The circuit shown in Figure 6–3(a) consists of two series networks connected in parallel. The circuit is reduced to a simple parallel circuit by adding resistors R_1 and R_2 together and adding R_3 and R_4 together, as shown in Figure 6–3(b). The

FIGURE 6–3 *(a) Two series networks connected in parallel. (b) Reduced simple parallel circuit. (c) Equivalent resistance.*

equivalent resistance, shown in Figure 6–3(c) is found by using the product-over-sum rule.

$$R_T = \frac{R_x R_y}{R_x + R_y}$$

EXAMPLE 6–2

Determine the total resistance of the circuit shown in Figure 6–4.

FIGURE 6–4 *Circuit for Example 6–2.*

Solution Resistors R_1 and R_2 are added together, as are resistors R_3 and R_4. The circuit is now reduced to a parallel network, as shown in Figure 6–5.

$$R_x = R_1 + R_2 = 12 \ \Omega + 22 \ \Omega = 34 \ \Omega$$

$$R_y = R_3 + R_4 = 15 \ \Omega + 27 \ \Omega = 42 \ \Omega$$

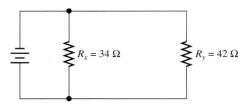

FIGURE 6–5 *Reduced parallel network.*

The total resistance is now found by using the product-over-sum rule.

$$R_T = \frac{R_x R_y}{R_x + R_y} = \frac{(34 \ \Omega)\,(42 \ \Omega)}{34 \ \Omega + 42 \ \Omega} = 18.79 \ \Omega$$

EXAMPLE 6-3

FIGURE 6-6   Circuit for Example 6-3.

Determine the total resistance of the circuit shown in Figure 6-6.

Solution   Resistors R_3 and R_4 are connected in parallel with resistor R_2. Therefore,

$$R_x = R_3 + R_4 = 47\ \Omega + 36\ \Omega = 83\ \Omega$$

The circuit is now redrawn as shown in Figure 6-7. The equivalent resistance of the parallel combination R_2 and R_x is found using the product-over-sum rule.

$$R_y = \frac{R_2 R_x}{R_2 + R_x} = \frac{(22\ \Omega)\ (83\ \Omega)}{22\ \Omega + 83\ \Omega} = 17.39\ \Omega$$

FIGURE 6-7   Redrawn circuit for Example 6-3.

FIGURE 6-8   Solution circuit for Example 6-3.

The total resistance is now found by adding the two series-connected resistors s together (Figure 6-8).

$$R_T = R_y + R_1 = 17.39\ \Omega + 33\ \Omega = 50.39\ \Omega$$

Section Review

1. To determine the total resistance of a series-parallel circuit, the voltage divider rule cannot be used (True/False).
2. What must be identified in order to calculate the total resistance of a series-parallel circuit?

6-3

CURRENT IN SERIES-PARALLEL CIRCUITS

The flow of current in a series-parallel circuit must always obey Kirchhoff's current law. Figure 6-9 illustrates the flow of current through a series-parallel network. The total current flows through resistor R_1 and divides at node A, with some of the current passing

FIGURE 6–9 *Current flow in a series-parallel circuit.*

Parallel Bank A

Parallel Bank B

FIGURE 6–10 *Series-parallel circuit.*

through R_2 and the remainder through R_3. The two branch currents regroup at node B, and the total current passes through R_4.

A group of parallel-connected resistors is often referred to as a *bank* of resistors. The circuit in Figure 6–10 shows two banks of resistors connected in series with each other. Bank A consists of resistors R_1, R_2, and R_3. Bank B consists of resistors R_4, R_5, and R_6. Bank A is in series with bank B, since the total current of bank A must go through bank B. Therefore, the total current, I_T, in each of the parallel banks is the same ($I_T = I_1 + I_2 + I_3$ and $I_T = I_4 + I_5 + I_6$).

_____ EXAMPLE 6–4 _____

Determine the values of current flowing through resistors R_3 and R_1 for the circuit shown in Figure 6–11.

FIGURE 6–11 *Circuit for Example 6–4.*

Solution According to Kirchhoff's current law, the sum of the currents entering a node must equal the sum of the currents leaving. Therefore, the current flowing through resistor R_5 must be the same as the current flowing through resistor R_1.

$$I_1 = I_5 = 11 \text{ A}$$

The total current flowing in resistors R_2, R_3, and R_4 must equal the current flowing into and out of the nodes of the resistor bank.

$$I_1 = I_5 = I_2 + I_3 + I_4$$

$$11\,A = 2\,A + I_3 + 5\,A$$

Rearranging terms,

$$I_3 = 11\,A - (2\,A + 5\,A)$$

$$= 4\,A$$

When solving for unknown currents in the branches of series-parallel circuits, the current divider rule is often useful. The following example illustrates solving for current using this rule.

EXAMPLE 6–5

FIGURE 6–12   Circuit for Example 6–5.

Determine the values of currents I_A and I_B for the circuit shown in Figure 6–12.

Solution

$$R_A = R_1 + R_2 = 30 + 56 = 86\ \Omega$$

$$R_B = R_3 + R_4 = 47 + 68 = 115\ \Omega$$

The circuit is now redrawn, as shown in Figure 6–13, and the currents solved for using the current divider rule.

FIGURE 6–13   Redrawn circuit for Example 6–5.

$$I_A = I_T \left(\frac{R_B}{R_A + R_B} \right) = 3\,A \left(\frac{115\ \Omega}{86\ \Omega + 115\ \Omega} \right) = 1.72\,A$$

$$I_B = I_T \left(\frac{R_A}{R_A + R_B} \right) = 3\,A \left(\frac{86\ \Omega}{86\ \Omega + 115\ \Omega} \right) = 1.28\,A$$

FIGURE 6–14 *Series-parallel circuit.*

Section Review

1. The flow of current in a series-parallel circuit must always obey Kirchhoff's current law (True/False).
2. What term is often applied to a group of parallel resistors in a series-parallel circuit?
3. When solving for unknown currents in the branches of series-parallel circuits, the current divider rule is often useful (True/False).

— 6–4

VOLTAGE IN SERIES-PARALLEL CIRCUITS

For any individual resistance in a branch, the current in the branch multiplied by the resistance equals the *IR* voltage drop across that particular resistance. The sum of the series *IR* drops in the branch equals the voltage drop across the entire branch. For the circuit of Figure 6–14,

$$E_T = V_{R1} + V_{R2} \quad \text{or} \quad E_T = V_{R1} + V_{R3}$$

According to Kirchhoff's voltage law, the sum of the voltage drops must always equal the applied voltage. Therefore, since resistors R_2 and R_3 are connected in parallel, the total voltage for the circuit of Figure 6–14 is equal to the voltage across this parallel combination plus the voltage drop across R_1. The voltage divider rule is once again useful for solving voltage drops when resistance is known. The following examples illustrate how Kirchhoff's voltage law and the current divider rule are used to determine voltage in a series-parallel circuit.

— EXAMPLE 6–6

Determine the voltage drop across resistor R_1 for the circuit of Figure 6–15.

FIGURE 6–15 *Circuit for Example 6–6.*

Solution The equivalent resistance of the parallel combination R_2 and R_3 must be determined first.

$$R_{AB} = \frac{R_2 R_3}{R_2 + R_3} = \frac{(18\ \Omega)\ (27\ \Omega)}{18\ \Omega + 27\ \Omega} = 10.8\ \Omega$$

FIGURE 6–16 *Equivalent circuit for Example 6–6.*

The circuit is now redrawn, as shown in Figure 6–16, as a simple series network with R_1 in series with R_{AB}. The voltage drop across R_1 is determined using the voltage divider rule.

$$V_{R1} = \left(\frac{R_1}{R_1 + R_{AB}} \right) E_T = \left(\frac{13 \ \Omega}{13 \ \Omega + 10.8 \ \Omega} \right) 24 \ \text{V} = 13.11 \ \text{V}$$

EXAMPLE 6–7

FIGURE 6–17 *Circuit for Example 6–7.*

What is the voltage drop across resistor R_5 in the circuit of Figure 6–17?

Solution R_A represents Bank A, and R_B represents Bank B.

$$R_A = \frac{R_1 R_2}{R_1 + R_2} = \frac{(10 \ \text{k}\Omega)\,(18 \ \text{k}\Omega)}{10 \ \text{k}\Omega + 18 \ \text{k}\Omega} = 6.43 \ \text{k}\Omega$$

$$\frac{1}{R_B} = \frac{1}{R_3} + \frac{1}{R_4} + \frac{1}{R_5} = \frac{1}{22 \ \text{k}\Omega} + \frac{1}{25 \ \text{k}\Omega} + \frac{1}{15 \ \text{k}\Omega}$$

$$R_B = 6.57 \ \text{k}\Omega$$

The circuit is now redrawn as shown in Figure 6–18.

FIGURE 6–18 *Redrawn circuit for Example 6–7.*

Using the voltage divider rule,

$$V_{RB} = E_T \left(\frac{R_B}{R_A + R_B} \right) = 30 \text{ V} \left(\frac{6.57 \text{ k}\Omega}{6.43 \text{ k}\Omega + 6.57 \text{ k}\Omega} \right) = 15.16 \text{ V}$$

The voltage at point A is 15.16 V

Section Review

1. For any individual resistance in a branch, the current in the branch times the resistance equals the *IR* voltage drop across that particular resistance (True/False).
2. The sum of the voltage drops in a series-parallel circuit must equal the applied voltage (True/False).

6-5

APPLYING OHM'S LAW IN SERIES-PARALLEL CIRCUITS

When determining unknown voltages, currents, and resistances in a series-parallel circuit, if two quantities are known, the other is easily solved for using Ohm's law. The following examples illustrate how Ohm's law is applied to series-parallel circuits.

EXAMPLE 6-8

For the circuit of Figure 6–19, find the

(a) Total current.
(b) Total voltage.

FIGURE 6–19   *Circuit for Example 6–8.*

Solution

(a) The sum of the currents flowing in R_2 and R_3 must equal the current flowing in R_1. According to KCL, the total current is

$$I_T = I_1 = I_2 + I_3$$

The currents flowing through R_2 and R_3 are found by Ohm's law.

$$I_2 = \frac{V_{R2}}{R_2} = \frac{8.5 \text{ V}}{150 \text{ }\Omega} = 56.67 \text{ mA}$$

$$I_3 = \frac{V_{R3}}{R_3} = \frac{8.5 \text{ V}}{100 \text{ }\Omega} = 85 \text{ mA}$$

The values of I_2 and I_3 are now inserted in the KCL equation, and the value of total current is solved for.

$$I_T = 56.67 \text{ mA} + 85 \text{ mA} = 141.67 \text{ mA}$$

(b) The total current flowing in the circuit is equal to the current flowing through R_1. The voltage drop V_{R1} is found by applying Ohm's law. Once V_{R1} is determined, the total voltage is found by applying KVL.

$$V_{R1} = I_1 R_1 = (141.67 \text{ mA})(120 \text{ }\Omega) = 17 \text{ V}$$

$$E_T = V_{R1} + V_{R2} = 17 \text{ V} + 8.5 \text{ V} = 25.5 \text{ V}$$

_____ EXAMPLE 6–9 _____

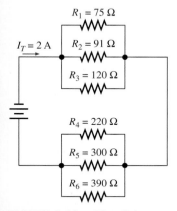

FIGURE 6–20   Circuit for Example 6–9.

For the circuit shown in Figure 6–20, resistors R_1, R_2, and R_3 form one group and resistors R_4, R_5, and R_6 form another group. Find the current through resistor R_2, and the voltage drop across each group.

Solution

$$\frac{1}{R_A} = \frac{1}{R_1} + \frac{1}{R_2} + \frac{1}{R_3}$$

$$= \frac{1}{75 \text{ }\Omega} + \frac{1}{91 \text{ }\Omega} + \frac{1}{120 \text{ }\Omega}$$

$$R_A = 30.62 \text{ }\Omega$$

$$\frac{1}{R_B} = \frac{1}{R_4} + \frac{1}{R_5} + \frac{1}{R_6}$$

$$= \frac{1}{220 \text{ }\Omega} + \frac{1}{300 \text{ }\Omega} + \frac{1}{390 \text{ }\Omega}$$

$$R_B = 95.76 \text{ }\Omega$$

The circuit is now redrawn as shown in Figure 6–21.

$I_T = 2 \text{ A}$ $R_A = 30.62 \text{ }\Omega$

$R_B = 95.76 \text{ }\Omega$

FIGURE 6–21   Redrawn circuit for Example 6–9.

$$E_{RA} = I_T R_A = (2 \text{ A}) (30.62 \ \Omega) = 61.24 \text{ V}$$

$$E_{RB} = I_T R_B = (2 \text{ A}) (95.76 \ \Omega) = 191.52 \text{ V}$$

$$I_2 = \frac{E_{RA}}{R_2} = \frac{61.24 \text{ V}}{91 \ \Omega} = 0.67 \text{ A}$$

EXAMPLE 6–10

Calculate the total resistance and total current for the circuit shown in Figure 6–22.

FIGURE 6–22   *Circuit for Example 6–10.*

Solution   The circuit can first be redrawn as the simplified diagram of Figure 6–23.

FIGURE 6–23   *Redrawn circuit for Example 6–10.*

There are now, basically, two groups of resistors in the circuit. The equivalent resistors can now be calculated:

$$R_A = \frac{(22 \text{ k}\Omega) (9 \text{ k}\Omega)}{22 \text{ k}\Omega + 9 \text{ k}\Omega} + 7 \text{ k}\Omega = 13.39 \text{ k}\Omega$$

$$R_B = \frac{(18 \text{ k}\Omega) (7 \text{ k}\Omega)}{18 \text{ k}\Omega + 7 \text{ k}\Omega} + 5 \text{ k}\Omega = 10.04 \text{ k}\Omega$$

The circuit is now simplified further, as shown in Figure 6–24.

FIGURE 6–24   *Circuit containing equivalent resistors R_A and R_B.*

$$R_T = \frac{R_A R_B}{R_A + R_B} = \frac{(13.39 \text{ k}\Omega)(10.04 \text{ k}\Omega)}{13.39 \text{ k}\Omega + 10.04 \text{ k}\Omega} = 5.74 \text{ k}\Omega$$

$$I_T = \frac{E_T}{R_T} = \frac{100 \text{ V}}{5.74 \text{ k}\Omega} = 17.42 \text{ mA}$$

Section Review

1. Two 10 kΩ resistors are connected in parallel. A 30 kΩ resistor is connected in series with the two parallel resistors to form a series-parallel circuit. If the applied voltage is 40 V, what is the voltage across the two 10 kΩ resistors?
2. What is the total current flowing in the circuit described in question 1 above?

6-6

POWER IN SERIES-PARALLEL CIRCUITS

Power in a series-parallel circuit is determined in the same manner as for simple series and simple parallel circuits. The total power dissipated in any circuit is always the sum of the quantities of power dissipated by the individual resistances. The power for each resistance is determined using the basic power equations.

$$P = I^2 R \qquad P = \frac{V^2}{R} \qquad P = IV$$

For the circuit shown in Figure 6–25, the power dissipated by each resistor is determined by the product of its current squared and the ohmic value of the resistance. The total power dissipation of the circuit is equal to the sum of the power dissipated by each resistor. That is,

$$P_T = P_1 + P_2 + P_3$$

FIGURE 6–25 *Power in a series-parallel circuit.*

EXAMPLE 6–11

Solution

(a) The value of current through R_3 is solved by KCL.

$$I_3 = I_1 - I_2$$
$$= 14 \text{ mA} - 4 \text{ mA}$$
$$= 10 \text{ mA}$$

For the circuit of Figure 6–25, assume $I_1 = 14$ mA and $I_2 = 4$ mA. Determine the

(a) Power dissipated by resistor R_3.
(b) Total power of the circuit.

The power dissipated by R_3 is now found.

$$P_3 = I_3^2 R_3$$
$$= (10 \text{ mA})^2 (12 \text{ k}\Omega)$$
$$= 1.2 \text{ W}$$

(b)
$$P_1 = I_1^2 R_1$$
$$= (14 \text{ mA})^2 (15 \text{ k}\Omega)$$
$$= 2.94 \text{ W}$$

$$P_2 = I_2^2 R_2$$
$$= (4 \text{ mA})^2 (180 \text{ k}\Omega)$$
$$= 2.88 \text{ W}$$

$$P_T = P_1 + P_2 + P_3$$
$$= 2.94 \text{ W} + 2.88 \text{ W} + 1.2 \text{ W}$$
$$= 7.02 \text{ W}$$

_____ **EXAMPLE 6–12** _____

Determine the power dissipated by resistor R_4 for the circuit shown in Figure 6–26.

FIGURE 6–26 *Circuit for Example 6–12.*

Solution   The voltage across R_4 can be found using the voltage divider rule. Because R_3 and R_4 are connected in parallel, the equivalent value of these two resistors, R_A, is calculated using the product-over-sum rule.

$$R_A = R_3 \| R_4 = \frac{R_3 R_4}{R_3 + R_4} = \frac{(200 \ \Omega)(300 \ \Omega)}{200 \ \Omega + 300 \ \Omega} = 120 \ \Omega$$

$$V_{R4} = \left(\frac{R_A}{R_A + R_5} \right) E_T = \left(\frac{120 \ \Omega}{120 \ \Omega + 100 \ \Omega} \right) 48 \text{ V} = 26.18 \text{ V}$$

The power dissipated by R_4 can now be determined.

$$P_4 = \frac{V_{R4}^2}{R_4} = \frac{(26.18 \text{ V})^2}{300 \ \Omega} = 2.28 \text{ W}$$

Section Review

1. The total power dissipated in a series-parallel circuit is the sum of the quantities of power dissipated by the individual resistances (True/False).
2. How is the power dissipated by each resistor in a series-parallel circuit determined?

LOADED VOLTAGE DIVIDERS

A specific application of the series-parallel circuit is the loaded voltage divider. As mentioned in Chapter 4, the main purpose of voltage divider circuits is to reduce the voltage fed to some components, or a section, below the available power supply value. When different voltage levels are required from a common power supply, the supply is designed for the highest voltage requirement, and the voltage divider is connected across the voltage supply. The number of taps or sections used depends on the number of loads supplied.

Figure 6–27 shows a typical loaded voltage divider circuit. Three loads are connected to a common source, with each load requiring different voltage and current for satisfactory operation. The current flowing through resistor R_3 is frequently called the *bleeder current,* which is the difference between the total current and the load current. Note that this current does not flow through any of the loads; it is merely "bled" from the power supply. The bleeder current, which is generally 10% to 25% of the total load, is considered as "loss" because it flows in the divider and not in a connected load. The amount of bleeder current that flows represents a design compromise; if the current is too low, small variations in load current will seriously affect the load voltage; if it is too high, large amounts of power will be dissipated by the divider, resulting in an inefficient system.

The methods used to find current, voltage, and resistance earlier in this chapter also can be applied to loaded voltage divider circuits. The following examples illustrate some basic problem solving using Ohm's law and Kirchhoff's laws.

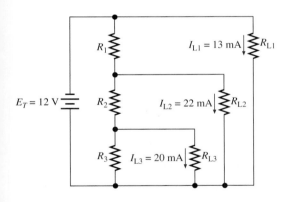

FIGURE 6–27 *Loaded voltage divider circuit.*

EXAMPLE 6–13

Solution To determine the values of R_1, R_2, and R_3, the value of current flowing through each resistor must be determined. The voltage drop across each resistor is found by Kirchhoff's voltage law.

$$V_{R1} = E_T - V_{RL2} = 12 \text{ V} - 5 \text{ V} = 7 \text{ V}$$

$$V_{R2} = V_{RL2} - V_{RL3} = 5 \text{ V} - 3 \text{ V} = 2 \text{ V}$$

$$V_{R3} = V_{RL3} = 3 \text{ V}$$

In Figure 6–27 three loads are connected to a common source. The three load voltages are $V_{RL3} = 3$ V, $V_{RL2} = 5$ V, and $V_{RL1} = 12$ V, and the three load currents are 13 mA, 22 mA, and 20 mA. Determine the values of resistors R_1, R_2, and R_3. Assume a bleeder current of 2 mA.

Since the bleeder current was given as 2 mA, the value of R_3 can be solved using Ohm's law.

$$R_3 = \frac{V_{R3}}{I_{R3}} = \frac{3\text{ V}}{2\text{ mA}} = 1500\ \Omega$$

According to Kirchhoff's current law, the current flowing through R_2 must be the sum of the current flowing in R_3 and R_{L3}.

$$I_{R2} = I_{R3} + I_{L3} = 2\text{ mA} + 20\text{ mA} = 22\text{ mA}$$

The value of R_2 is now found by Ohm's law.

$$R_2 = \frac{V_{R2}}{I_{R2}} = \frac{2\text{ V}}{22\text{ mA}} = 90.9\ \Omega$$

The current through R_1 is equal to $I_{R2} + I_{L2}$.

$$I_{R1} = I_{R2} + I_{L2} = 22\text{ mA} + 22\text{ mA} = 44\text{ mA}$$

$$R_1 = \frac{V_{R1}}{I_{R1}} = \frac{7\text{ V}}{44\text{ mA}} = 159.1\ \Omega$$

—— **EXAMPLE 6–14** ——

Determine the values for all resistances in the loaded voltage divider circuit of Figure 6–28.

FIGURE 6–28 *Circuit for Example 6–14.*

Solution   First, find the total current.

$$I_T = I_1 = I_2 + I_L = 8.5\text{ mA} + 85\text{ mA} = 93.5\text{ mA}$$

The values of R_1, R_2, and R_L are now solved by Ohm's law.

$$R_1 = \frac{V_{R1}}{I_1} = \frac{26\text{ V}}{93.5\text{ mA}} = 278.1\ \Omega$$

$$R_2 = \frac{V_{R2}}{I_2} = \frac{24\text{ V}}{8.5\text{ mA}} = 2.82\text{ k}\Omega$$

$$R_L = \frac{V_{RL}}{I_L} = \frac{24\text{ V}}{85\text{ mA}} = 282.4\ \Omega$$

In some situations, a loaded voltage divider is required that provides both positive and negative voltages with respect to ground by connecting the ground at some tap along the divider. Since the ground is the common reference point, all loads now connect back to ground instead of the end of the voltage divider. When both positive and negative load voltage are to be supplied from a voltage divider, it is essential that the principles of relative polarity be remembered.

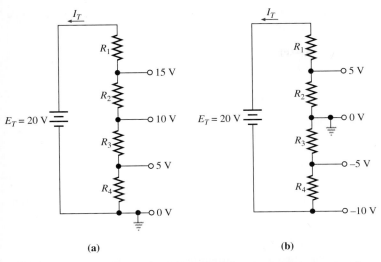

(a)   **(b)**

FIGURE 6-29 *(a) Voltage divider with negative ground.*
(b) Dual-polarity voltage divider.

Figure 6–29(a) shows an example of a single-polarity voltage divider with three taps and a negative ground as a reference. Consequently, all voltages are considered as positive with respect to this point. In Figure 6–29(b), the ground has now been moved and two polarities are available. The voltages of the divider are measured from this new reference. The voltage drops across each of the four resistors remains unchanged, and the method of determining ohmic values for R_1, R_2, R_3, and R_4 is also unchanged.

The position of the ground point in Figure 6–29(b) allows one value of positive voltage and two values of negative voltage to be available at the output. If the ground point was moved to between resistors R_1 and R_2, three negative voltages would be available. Since the ground is the common reference point, all loads connect back to ground instead of the end of the voltage divider. The voltages of the taps are obtained by subtracting the value of the potential at one end of the resistor from the potential at the other end:

$$15 \text{ V} - 10 \text{ V} = 5 \text{ V}$$

$$5 \text{ V} - 10 \text{ V} = -5 \text{ V}$$

$$0 \text{ V} - 10 \text{ V} = -10 \text{ V}$$

_____ **EXAMPLE 6–15** _____

FIGURE 6-30 *Circuit for Example 6-15.*

The voltage divider circuit of Figure 6–30 is designed to provide both a negative and positive voltage for the two loads. Load R_{L1} is to be supplied with –20 V at 50 mA and load R_{L2} is to supply +20 V at 100 mA. Determine the values of resistors R_1, R_2, and R_3. Assume a bleeder current of 10%.

Solution   In Figure 6–30, there is a total of 40 V between the −20 V and the +20 V outputs. Therefore, the voltage drop across resistor R_1 is 10 V.

$$V_{R1} = 50 \text{ V} - 40 \text{ V} = 10 \text{ V}$$

$$V_{R2} = V_{RL1} = 20 \text{ V}$$

$$V_{R3} = V_{RL2} = 20 \text{ V}$$

The current flowing through R_3 is 10% of I_{L2}, or 10 mA. The value of R_3 can be found by Ohm's law.

$$R_3 = \frac{V_{R3}}{I_3} = \frac{20 \text{ V}}{10 \text{ mA}} = 2 \text{ k}\Omega$$

The total current flowing in the circuit is the sum of bleeder current I_3 and the current through load R_{L2}.

$$I_T = I_3 + I_{L2} = 10 \text{ mA} + 100 \text{ mA} = 110 \text{ mA}$$

This value of current is the same as the current flowing through resistor R_1. Therefore, the resistance of R_1 can be determined by Ohm's law.

$$R_1 = \frac{V_{R1}}{I_T} = \frac{10 \text{ V}}{110 \text{ mA}} = 90.9 \text{ }\Omega$$

The current through R_2 is equal to the difference between I_T and I_{L1}.

$$I_2 = I_T - I_{L1} = 110 \text{ mA} - 50 \text{ mA} = 60 \text{ mA}$$

and

$$R_2 = \frac{V_{R2}}{I_2} = \frac{20 \text{ V}}{60 \text{ mA}} = 333.33 \text{ }\Omega$$

Up to this point, we have only examined voltage dividers with fixed loads. However, in many electronic circuits the load will actually vary. For example, the current drawn by a transistor varies with the change in the resistance of the device. Because the transistor is a semiconductor device, its internal resistance will vary with temperature. The introduction of a varying load complicates the calculations in a voltage divider circuit. Although, with an understanding of the changes brought on by the varying load condition, these problems can be reasoned out with a fair degree of accuracy.

Section Review

1. What happens to the output voltage of a loaded voltage divider if the voltage across the series-dropping resistor decreases?
2. What is the term applied to the lowest current in any of the components in a voltage divider?
3. What is a dual-polarity voltage divider?

— 6–8

WHEATSTONE BRIDGE

One of the most useful series-parallel configurations in electronics circuits is the *bridge*. The bridge circuit is used in a wide variety of applications, including rectifiers and mea-

FIGURE 6–31
Wheatstone bridge.

suring instruments. One such measuring instrument is the **Wheatstone bridge,** invented in 1850 by Charles Wheatstone. The Wheatstone bridge is an instrument that measures resistance with a much greater degree of accuracy than that obtainable from the typical ohmmeter. The accuracy of an ohmmeter is at best only 5% to 10%, but a Wheatstone bridge will measure resistance values from 0.01 Ω to 10 MΩ with an accuracy greater than 1%.

The circuit of the wheatstone bridge is shown in Figure 6–31. Three variable resistors (R_1, R_3, R_4) and the unknown resistor R_2 are connected in the form of a diamond. Connections are made from a battery to the two opposite corners of the diamond. A galvanometer is connected across the other two corners. A **galvanometer** is a meter that measures small currents in either direction and is zero at center scale.

The resistors are adjusted until the galvanometer reads zero. This operation is known as balancing the bridge. When the bridge is balanced, there is no current flow through the galvanometer. Therefore, the two points, A and B, must be at the same potential, and the currents $I_1 = I_3$ and $I_2 = I_4$.

Because points A and B in Figure 6–31 are at the same potential, the voltage drops across R_1 and R_2 must be equal, so

$$I_1 R_1 = I_2 R_2 \qquad (6.1)$$

also

$$I_3 R_3 = I_4 R_4 \qquad (6.2)$$

The equation relating the unknown resistor, R_2, to the known resistors can now be derived by applying the law of proportionality, which results in

$$R_2 = \frac{R_1 R_4}{R_3} \qquad (6.3)$$

_____ **EXAMPLE 6–16** _____

Solution

$$R_2 = \frac{R_1 R_4}{R_3}$$

$$= \frac{(56 \times 10^3)\,(100 \times 10^3)}{69.66 \times 10^3}$$

$$= 80.39 \text{ k}\Omega$$

In the wheatstone bridge in Figure 6–31, $R_1 =$ 56 kΩ, and $R_4 = 100$ kΩ. Determine the value of R_2 if R_3 must be adjusted to 69.66 kΩ to balance the bridge.

FIGURE 6–32 *Bridge circuit with photoresistor.*

In addition to measuring resistance, Wheatstone bridges are also used to measure quantities such as temperature and light. Figure 6–32 shows a Wheatstone bridge with a photoresistor. Resistor R_1 allows the bridge to be set for a certain operating temperature. When the bridge is balanced, the output of the bridge is 0 V. If the light intensity changes, the bridge is no longer balanced and a voltage is produced at the output of the bridge that is proportional to the change in light intensity.

Section Review

1. Bridge circuits are used in rectifiers and measuring instruments (True/False).
2. What instrument measures resistance with a high degree of accuracy?
3. What is a galvanometer?

6–9

TROUBLESHOOTING SERIES-PARALLEL CIRCUITS

The methods used for finding faults in series-parallel circuits are the same as for series circuits and parallel circuits. Short-circuit and open-circuit conditions are once again determined by using measuring instruments such as voltmeters and ohmmeters.

There are several causes for short circuits in series-parallel networks. Shorts may be caused by small pieces of conductive material, such as bits of solder or metal, bridged between the traces of a printed circuit board. Short circuits are also caused by electronic devices, such as resistors or semiconductors, melting as a result of excess current flow. When a short occurs in a series-parallel circuit, the amount of current flow in the circuit will increase due to the reduced resistance of the network.

Figure 6–33(a) shows a loaded voltage divider with a short across the load. This short circuit cancels out resistor R_2 since a 0 Ω path is now provided around this resistor. The only resistance left in the circuit is R_1, so the circuit is reduced to that of Figure 6–33(b). Resistor R_1 is often called a *series-dropping resistor* and is useful in preventing short-circuit current from flowing in the circuit. Unlike the parallel circuit, a short in a series-parallel network doesn't necessarily cause a direct short across the terminals of the supply. In Figure 6–33(a), both resistors R_1 and R_2 would have to fail in order for the voltage source to become short circuited.

The total voltage of a series-parallel circuit is generally not affected by a short-circuit condition. However, the voltage drops across individual components can vary drastically in the event of a short. For example, if the circuit of Figure 6–33(a) had identical voltage drops across R_1 and R_2 before a short occurred in R_L, then the voltage across R_1 would double and the voltage across R_2 would reduce to zero when a short occurs across R_L.

Providing overcurrent protection to series-parallel circuits is more difficult than a simple parallel network because a shorted device may not draw enough current to melt the

FIGURE 6–33 (a) Series-parallel circuit with short. (b) Equivalent circuit.

FIGURE 6–34 Loaded voltage divider.

fuse or trip the circuit breaker, although the excess current drawn by the short may damage any other components connected in series with the short. In loaded voltage divider circuits, such as Figure 6–34, a short across resistor R_{L1} would effectively short out all the resistors in the circuit. Consequently, short-circuit current would flow through only R_{L1}, the fuse would blow, and the other devices would not be affected. However, if R_{L3} became shorted, the current flowing through R_2 would increase and the excess heat dissipated by the increase in current could damage resistor R_2.

EXAMPLE 6–17

Determine which resistor in Figure 6–35 is short circuited. When the circuit is functioning properly, the three load voltages are $V_{RL1} = 15$ V, $V_{RL2} = 10$ V, and $V_{RL3} = 5$ V. The voltmeter readings shown are taken after the short has occurred. Assume a 10% bleeder current.

FIGURE 6–35 Circuit for Example 6–17.

Solution According to Figure 6–35, the load voltage V_{RL2} has fallen to 8.9 V and the load voltage V_{RL3} has risen to 8.9 V. Because V_{RL2} is now equal to V_{RL3}, resistor R_2 must be short circuited.

EXAMPLE 6–18

Solution If R_2 is shorted, the circuit is redrawn as shown in Figure 6–36. Because bleeder resistor R_3 normally draws 10% of the current through R_{L3}, the value of this resistor is found by Ohm's law.

In Figure 6–35, what is the total current if resistor R_2 is shorted?

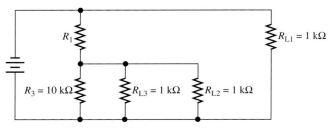

FIGURE 6–36   Redrawn circuit for Example 6–18.

$$R_3 = \frac{V_{R3}}{I_3} = \frac{5\text{ V}}{0.5\text{ mA}} = 10\text{ k}\Omega$$

The equivalent resistance of R_3, R_{L3}, and R_{L2} is determined as follows:

$$R_x = \frac{1}{1/R_3 + 1/R_{L3} + 1/R_{L2}}$$

$$= \frac{1}{1/10\text{ k}\Omega + 1/1\text{ k}\Omega + 1/1\text{ k}\Omega}$$

$$= 476.2\ \Omega$$

The voltage drop across the equivalent resistor R_x was measured at 8.9 V. Therefore, the current flowing in this branch is found by Ohm's law.

$$I_{Rx} = \frac{V_{Rx}}{R_x} = \frac{8.9\text{ V}}{476.2\ \Omega} = 18.69\text{ mA}$$

The current flowing through R_{L1} can also be found by Ohm's law, and the total current determined by Kirchhoff's current law.

$$I_{RL1} = \frac{V_{RL1}}{R_{L1}} = \frac{15\text{ V}}{1\text{ k}\Omega} = 15\text{ mA}$$

$$I_T = I_{Rx} + I_{RL1} = 18.69\text{ mA} + 15\text{ mA} = 33.69\text{ mA}$$

When an open-circuit condition occurs in a series-parallel circuit, the voltage drops around the circuit can vary substantially. Consider the circuit of Figure 6–37(a). In this circuit, the loaded voltage divider supplies 10 V to resistor R_L and drops 10 V across the series-dropping resistor R_1. If the load resistor opens, as in Figure 6–37(b), the total resistance increases and the voltage drops change in accordance with KVL.

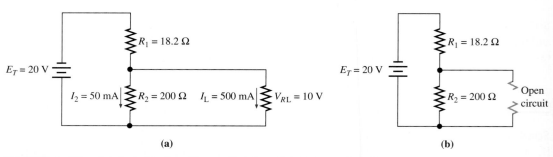

(a)   (b)

FIGURE 6–37   (a) Loaded voltage divider. (b) Open circuit.

EXAMPLE 6-19

Solution   The total resistance of the circuit with R_L opened consists of R_1 and R_2.

$$R_T = R_1 + R_2 = 18.2\ \Omega + 200\ \Omega$$

Calculate the voltages across R_1 and R_2 for the circuit of Figure 6–37 (b).

The voltage drops can now be found using the voltage divider rule.

$$V_{R1} = \left(\frac{R_1}{R_T}\right) E_T = \left(\frac{18.2\ \Omega}{218.2\ \Omega}\right) 20\ \text{V} = 1.67\ \text{V}$$

$$V_{R2} = \left(\frac{R_2}{R_T}\right) E_T = \left(\frac{200\ \Omega}{218.2\ \Omega}\right) 20\ \text{V} = 18.33\ \text{V}$$

Section Review

1. What happens to a series-parallel circuit when a short occurs in one of the components?
2. Why is it more difficult to provide overcurrent protection for a series-parallel circuit than a parallel circuit?
3. When an open-circuit condition occurs in a series-parallel circuit, the voltage drops in the circuit remain constant (True/False).

6-10

PRACTICAL APPLICATION SOLUTION

The design requirements for a dual-polarity power supply were outlined in the chapter opener. Your task was to design a power supply capable of supplying 200 mA for a +10 V and −10 V load. The following steps illustrate the method of solution for this practical application of series-parallel circuits.

STEP 1   Sketch a wiring diagram showing the reference point for the dual polarity supply (Figure 6–38). Three resistors are required for the circuit. Because the source is 24 V and the total voltage requirements are 20 V, a series-dropping resistor (R_1) must drop the difference (4 V). The other two resistors (R_2 and R_3) form a divider to provide the +10 V and −10 V values. The two load resistors are connected with R_2 and R_3 at the reference, or ground, point for the circuit.

FIGURE 6–38   *Circuit for practical application.*

STEP 2   Determine the ohmic value of the three resistors R_1, R_2, and R_3. In Figure 6–38, there is a total of 20 V between the −10 V and +10 V terminals. Therefore, the voltage dropped across resistor R_1 is 4 V.

$$V_{R1} = 4 \text{ V}, \quad V_{R2} = 10 \text{ V}, \quad V_{R3} = 10 \text{ V}$$

The current flowing through R_3 is 10% of I_{L2}, or 20 mA. The value of R_3 is found by Ohm's law.

$$R_3 = \frac{V_{R3}}{I_3} = \frac{10 \text{ V}}{20 \text{ mA}} = 500 \text{ }\Omega$$

The total current flowing in the circuit is the sum of bleeder current I_3 and the current through load R_{L2}.

$$I_T = I_3 + I_{L2} = 20 \text{ mA} + 200 \text{ mA} = 220 \text{ mA}$$

This value of current is the same as the current flowing through resistor R_1. Therefore, the resistance of R_1 can be determined by Ohm's law.

$$R_1 = \frac{V_{R1}}{I_T} = \frac{4 \text{ V}}{220 \text{ mA}} = 18.2 \text{ }\Omega$$

The current through R_2 is equal to the difference between I_T and I_{L1}.

$$I_2 = I_T - I_{L1} = 220 \text{ mA} - 200 \text{ mA} = 20 \text{ mA}$$

Therefore, $$R_2 = \frac{V_{R2}}{I_2} = \frac{10 \text{ V}}{20 \text{ mA}} = 500 \text{ }\Omega$$

STEP 3   Calculate the power dissipated by each resistor using the power equation.

$$P_{R1} = (I_{R1})^2 R_1 = (220 \text{ mA})^2 (18.2 \text{ }\Omega) = 0.88 \text{ W}$$

$$P_{R2} = (I_{R2})^2 R_2 = (20 \text{ mA})^2 (500 \text{ }\Omega) = 0.2 \text{ W}$$

$$P_{R3} = (I_{R3})^2 R_3 = (20 \text{ mA})^2 (500 \text{ }\Omega) = 0.2 \text{ W}$$

Summary

1. Circuits that contain both series and parallel combinations of resistances are called *series-parallel circuits.*
2. The flow of current in a series-parallel circuit must always obey Kirchhoff's current law.
3. The current divider rule is often used to solve for unknown currents in the branches of series-parallel circuits.
4. The current in the branch multiplied by the resistance equals the *IR* drop across any resistor in a series-parallel circuit.
5. The total power dissipated in a series-parallel circuit is equal to the sum of the power dissipated by each resistor.
6. Voltage dividers are used primarily to reduce the available voltage to a level below the value of the fixed supply.
7. The bleeder current in a loaded voltage divider represents a loss of 10% to 25% of the total load current.
8. The Wheatstone bridge is an instrument that measures resistance with a high degree of accuracy.

9. Series-dropping resistors in loaded voltage dividers help to prevent short-circuit conditions from occuring.
10. When an open-circuit condition occurs in a branch of a series-parallel circuit, the voltage drops around the circuit can vary substantially.

Answers to Section Reviews

Section 6–1

1. True
2. To reduce all parallel-connected resistors to series equivalent resistances, and to apply Ohm's law to components where two of the three values are known

Section 6–2

1. False
2. Which resistors form series circuits and which are parallel networks

Section 6–3

1. True
2. Resistor bank
3. True

Section 6–4

1. True
2. True

Section 6–5

1. $R_x = (10 \text{ k}\Omega \times 10 \text{ k}\Omega)/(10 \text{ k}\Omega + 10 \text{ k}\Omega) = 5 \text{ k}\Omega$
   $V_{Rx} = 5 \text{ k}\Omega/(30 \text{ k}\Omega + 5 \text{ k}\Omega) \times 40 \text{ V} = 5.71 \text{ V}$
2. $R_T = R_x + 30 \text{ k}\Omega = 5 \text{ k}\Omega + 30 \text{ k}\Omega = 35 \text{ k}\Omega$
   $I_T = 40 \text{ V}/35 \text{ k}\Omega = 1.14 \text{ mA}$

Section 6–6

1. True
2. By the product of the current squared, multiplied by the ohmic value of the resistance

Section 6–7

1. The output voltage would increase.
2. Bleeder current
3. A voltage divider capable of supplying both positive and negative voltages

Section 6–8

1. True
2. The Wheatstone bridge
3. A meter that measures small currents in either direction and is zero at center scale.

Section 6–9

1. The amount of current flow in the circuit will increase because of the reduced resistance of the network.
2. Because a shorted device may not draw enough current to melt the fuse or trip the circuit breaker
3. False

Review Questions

Multiple Choice Questions

6–1 A series-parallel circuit consists of
(a) Resistors connected in series
(b) Resistors connected in parallel
(c) Combinations of series and parallel resistors
(d) Voltage sources in series

6–2 In many cases, a complete series-parallel circuit can be replaced by an equivalent circuit consisting of
(a) A series circuit   **(b)** A parallel circuit
(c) A single resistor   **(d)** Two series-connected resistors

6–3 The circuit shown in Figure 6–39 consists of
(a) Two parallel networks connected in series
(b) Two series networks connected in parallel
(c) Two parallel circuits
(d) Two series circuits

FIGURE 6–39

6–4 The total resistance of the circuit of Figure 6–39 is
(a) 11.59 Ω   **(b)** 17.68 Ω   **(c)** 16.5 Ω   **(d)** 18.67 Ω

6–5 In a series-parallel circuit, the current, voltage, and resistance must obey
(a) Ohm's law   **(b)** Kirchhoff's current law
(c) Kirchhoff's voltage law   **(d)** All of the above

6–6 When solving for unknown currents in the branches of series-parallel circuits, if the total current is known _____ is often useful.
(a) Ohms law   **(b)** The voltage divider rule
(c) Kirchhoff's voltage law   **(d)** The current-divider rule

6–7 In Figure 6–40, the current I_A is equal to
(a) 107 mA   **(b)** 93 mA   **(c)** 87 mA   **(d)** 113 mA

FIGURE 6–40

6–8 When solving for unknown voltage drops in a series-parallel circuit, if the total voltage is known _____ is often useful.
(a) Ohm's law   **(b)** Kirchhoff's current law
(c) Kirchhoff's voltage law   **(d)** The voltage divider rule

6–9 The voltage drop across resistor R_1 in Figure 6–41 is equal to
(a) 5.4 V   **(b)** 4.5 V   **(c)** 7.5 V   **(d)** 6.6 V

FIGURE 6–41

$E_T = 12\text{ V}$

$R_1 = 1\text{ k}\Omega$

$R_2 = 10\text{ k}\Omega$ $R_L = 2\text{ k}\Omega$

6–10 Power in a series-parallel circuit can be found by
(a) IR **(b)** E^2R **(c)** I^2R **(d)** ER

6–11 The current that does not flow through any load in a loaded voltage divider is called
(a) Load current   **(b)** No-load current   **(c)** Bleeder current   **(d)** Total current

6–12 Bleeder current is typically _____ of the load current.
(a) 50%   **(b)** 10%   **(c)** 1%   **(d)** 100%

6–13 A dual-polarity voltage divider provides
(a) Positive and negative voltages   **(b)** DC and AC signals
(c) Negative ground   **(d)** Two voltage levels with the same polarity

6–14 The Wheatstone bridge was originally designed to
(a) Measure power   **(b)** Measure voltage
(c) Measure current   **(d)** Measure resistance

6–15 A galvanometer is a meter that measures
(a) Voltage   **(b)** Current   **(c)** Resistance   **(d)** Power

6–16 Short circuits in series-parallel networks are often caused by
(a) Blown fuses
(b) Electronic devices melting because of excess heat
(c) Open-circuited resistors
(d) None of the above

6–17 Short-circuit current can be prevented by
(a) Series-dropping resistors   **(b)** Bleeder resistors
(c) Solder bridges   **(d)** None of the above

Practice Problems

6–18 Determine the total resistance of the circuit shown in Figure 6–42.

FIGURE 6–42

$R_1 = 56\ \Omega$

$R_2 = 91\ \Omega$ $R_L = 85\ \Omega$

6–19   Calculate the total resistance for the circuit of Figure 6–43.

FIGURE 6–43

$R_1 = 1$ kΩ $R_3 = 3.3$ kΩ

$R_2 = 2.2$ kΩ $R_4 = 1.5$ kΩ

6–20   Determine the total resistance of the circuit shown in Figure 6–44.

$R_1 = 390\ \Omega$ $R_3 = 560\ \Omega$ **FIGURE 6–44**

$R_2 = 750\ \Omega$ $R_4 = 240\ \Omega$

6–21   Calculate the values of current flowing through resistor R_2 for the circuit of Figure 6–45.

FIGURE 6–45

$I_1 = 350$ mA R_1

R_2 $I_3 = 180$ mA R_3 R_4 $I_4 = 140$ mA

6–22   Find the values of currents I_A and I_B for the circuit of Figure 6–46.

$I_T = 200$ mA   **FIGURE 6–46**

I_A $R_1 = 3.9$ kΩ I_B $R_3 = 5.6$ kΩ

$R_2 = 2.7$ kΩ $R_4 = 1.8$ kΩ

6–23   Calculate the voltage drop across resistor R_1 in Figure 6–47.

FIGURE 6–47

$R_1 = 330\ \Omega$

$E_T = 15\ V$

$R_2 = 10\ k\Omega$ $R_3 = 1\ k\Omega$

6–24   Find the total current flowing in Figure 6–47.

6–25   Determine the power dissipated by resistor R_3 in Figure 6–47.

6–26   Calculate the values for all resistances in the loaded voltage divider circuit of Figure 6–48. Assume a 10% bleeder current.

FIGURE 6–48

R_1

$E_T = 20\ V$

R_2 $I_L = 2.5\ A$ $V_{RL} = 10\ V$

6–27   For the circuit of Figure 6–49, find the values of R_1, R_2, and R_3. Assume a 10% bleeder current.

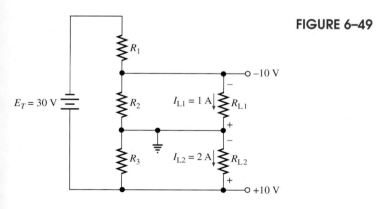

FIGURE 6–49

R_1

$-10\ V$

$E_T = 30\ V$

R_2 $I_{L1} = 1\ A$ R_{L1}

R_3 $I_{L2} = 2\ A$ R_{L2}

$+10\ V$

6–28   What is the power dissipated by R_3 in Figure 6–49?

6–29   For the Wheatstone bridge circuit of Figure 6–50. Find the value of R_2 if R_3 must be adjusted to 7.35 kΩ to balance the bridge.

FIGURE 6–50

6–30 Determine which resistor in Figure 6–51 is short circuited. When the circuit is functioning properly, the three load voltages are $V_{RL1} = 30$ V, $V_{RL2} = 20$ V, and $V_{RL3} = 10$ V. The voltmeter readings shown are taken after the short has occurred.

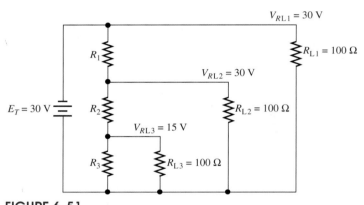

FIGURE 6–51

6–31 For the circuit of Figure 6–51, what is the total current if resistor R_1 is shorted?

6–32 What is the total current if the circuit of Figure 6–51 is functioning properly?

6–33 Calculate the voltages across R_1 and R_2 for the circuit of Figure 6–52.

FIGURE 6–52

Essay Questions

6–34 Define a series-parallel circuit.

6–35 List the five steps for analyzing series-parallel circuits.

6–36 Explain the relationship between currents in a series-parallel circuit.

6–37 What is a bank of resistors?

6–38 How is Kirchhoff's voltage law applied to series-parallel circuits?

6–39 Describe how power is calculated in a series-parallel network.

6–40 What is bleeder current?

6–41 How is a dual-polarity voltage divider constructed?

6–42 Explain the purpose of the Wheatstone bridge.

6–43 Describe two common faults when troubleshooting series-parallel circuits.

6–44 What is a series-dropping resistor?

6–45 Does a short in a series-parallel circuit always short out the supply voltage? Explain.

7

DC Measuring Instruments

Learning Objectives

Upon completion of this chapter you will be able to

- Explain the necessity of a shunt resistor in a DC ammeter circuit.
- Describe the effects of ammeter and voltmeter loading.
- Explain the basic operation of a multirange ammeter.
- Discuss the purpose of a multiplier resistor in a DC voltmeter.
- Define voltmeter sensitivity.
- Understand the operation of the ohmmeter.
- Discuss the basic principles of electronic and digital multimeters.

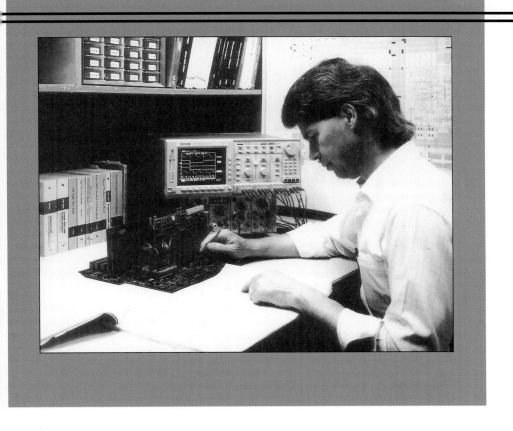

Practical Application

After successfully completing the practical application outlined in Chapter 6, your supervisor has now given you a field assignment where you are required to service a disabled computer in an office building. You have been provided with the service literature and schematic diagrams for the computer. The computer was damaged when a power surge occurred in the building and is now totally inoperative.

Upon arriving at the building, you switch on the computer and notice that the power light does not come on. You switch off the computer and remove the cover. A fuse inside the computer connected in series with the power supply and AC input is determined to be blown after being tested with an ohmmeter. After replacing the fuse, you switch on the computer and notice that the power light is now on but the computer is still inoperative. With the computer still turned on, you test the output of the power supply and record the following output voltages: 7.1 V, 7.1 V, and 20 V. According to the service literature, the output of the supply should be 5 V, 10 V, and 20 V. The schematic diagram of the power supply indicates that the three voltage levels are provided by a simple voltage divider. Your task is to determine what is wrong with the power supply.

To complete this practical application you will

1. Use an ohmmeter to check for a blown fuse.
2. Use a voltmeter to test the power supply.
3. Compare the voltage readings taken with data provided in the service literature.

7-1

INTRODUCTION

Measuring instruments or meters are often used in electronic circuits to determine quantities such as voltage, current, resistance, and power. There are two basic types of meters: analog and digital. Figure 7–1(a) shows an example of a **digital meter,** which provides a digital readout of a quantity being measured. An example of an **analog meter** is shown in Figure 7–1(b). An analog meter uses a movable pointer to indicate the value of a quantity being tested.

Although digital meters are becoming increasingly popular as test instruments, analog meters are still widely used in industry. Consider the speedometer in a typical automobile. Speedometers are now available either in the traditional analog form, with a "red line" type of pointer, or in digital form, with digital displays providing a digital readout of the automobile speed. Although digital speedometers are extremely accurate, they are difficult to read in bright sunlight and have a tendency to fluctuate because of the delay between the measurement taken and the reading displayed. If the speed of the automobile changes, an analog speedometer will track this change quite accurately with its pointer. However, a typical digital speedometer will often show rapidly fluctuating values that are not truly accurate until the speed remains constant for a fixed period of time.

In this chapter, we shall first examine the characteristics of analog meters and then study the various types of digital meters used in DC circuits. Analog, or moving coil meters, operate on a principle based on the reaction between two magnetic fields. One of these fields is obtained from a moving coil and the other is obtained from a permanent

FIGURE 7–1 *(a) Digital meter. (b) Analog meter.*

(a)   (b)

magnet. When current is fed through this moving coil, the resulting magnetic field reacts with the magnetic field of the permanent magnet and causes the coil to rotate. This is called the **D'Arsonval principle.** The D'Arsonval principle is the basis for practically all analog DC ammeters and voltmeters. The Weston-type instrument utilizes this principle.

The pointer of analog meters, which is very light and made of aluminum, is balanced by an adjustable counterweight. This helps to reduce the friction of the supportive pivots on the jewelled bearings. Because of the radial magnetic field, the deflection of the moving coil is virtually proportional to the current in the moving coil, allowing the scale divisions of the Weston-type instruments to be uniform throughout. The Weston-type instrument may be used as either an ammeter or a voltmeter, depending on whether the resistance of the coil is large or small. When used as a voltmeter, the moving coil has a resistance connected in series with it. When used as an ammeter, the coil is provided with a parallel resistor.

Section Review

1. Measuring instruments measure quantities such as voltage, current, and power (True/False).
2. Which meter provides a digital readout of quantities being measured?
3. Analog meters use a movable pointer to indicate the value of a quantity being measured (True/False).
4. What is the basic operating principle of analog meters?

7-2
DC AMMETER

An **ammeter** is an instrument used for measuring electric currents. The analog DC ammeter is almost always a Weston-type instrument. However, the moving coil of a Weston instrument cannot carry currents greater than 30 μA, so, to read larger values of current, a resistor is placed in parallel with the moving coil. The current going into the instrument will divide, with part of the current going through the coil and the rest through the resistor. This type of resistor is called a **shunt resistor.** A typical moving coil DC ammeter is shown in Figure 7–2.

FIGURE 7–2 *Typical DC ammeter.*

FIGURE 7–3 *Current division between an ammeter and a shunt resistor.*

The shunt is a low-value resistor, usually made of manganin strips, which are brazed to heavy copper blocks. This method ensures that the shunt has a low temperature coefficient of resistance. Manganin is an alloy consisting of copper, nickel, and ferromanganese. Instead of having a resistance rating, shunts are rated in terms of current and voltage. When an instrument is made with a built-in shunt, as shown in Figure 7–3, its scale is calibrated to read the total current directly.

The Weston-type ammeter is actually a voltmeter that measures the voltage drop across the shunt resistor. If the resistance of the shunt is constant, the voltage drop across R_{sh} is proportional to the current in the shunt. Therefore, the ammeter is just a divided circuit consisting of a millivoltmeter and a shunt, where the millivoltmeter measures the voltage drop across the shunt. Since Figure 7–3 is a divided circuit, the currents in the shunt and in the instrument vary inversely with their resistances. Expressed in equation form,

$$\frac{I_{sh}}{I_m} = \frac{R_m}{R_{sh}} \tag{7.1}$$

where R_{sh} = shunt resistance
   R_m = meter resistance
   I_{sh} = shunt current
   I_m = meter current

The internal resistance of an ammeter is very small. Ideally, the internal resistance should be zero. If the meter resistance were appreciable, its insertion would change the amount of current being read by the meter. This is referred to as the **loading effect** of an ammeter. Ammeter loading occurs mainly in low-voltage, low-resistance circuits.

EXAMPLE 7–1

What shunt resistance is required to extend the range of a 0 mA–10 mA movement having a meter resistance of 15 Ω, to read a total of 100 mA?

Solution

$$\frac{I_{sh}}{I_m} = \frac{R_m}{R_{sh}}$$

$$R_{sh} = \frac{R_m I_m}{I_{sh}}$$

$$= \frac{(15 \ \Omega)(10 \times 10^{-3} \ A)}{90 \times 10^{-3} \ A}$$

$$= 1.67 \ \Omega$$

A multirange ammeter can be constructed by using several values of shunt resistors and a rotary switch to select the desired range. The ammeter shown in Figure 7–4 has a switch, Sw, which selects a shunt resistor (R_1, R_2, or R_3) to be used across the coil of the meter. This curved type of switching mechanism is known as a *make-before-break switch*. The wide-ended moving contact connects to the next terminal to which it is being moved

FIGURE 7-4 Multi-range ammeter with individual shunts.

FIGURE 7-5 Ayrton shunt, or universal ammeter.

before it loses contact with the previous terminal. This is to ensure that, at all times, there is a shunt across the coil of the meter. Otherwise, an open circuit may occur and the full current would flow through the coil of the meter.

The Ayrton shunt, shown in Figure 7–5, is another method that is used to protect the ammeter coil from excessive current. When the moving contact of the rotary switch is connected to point A, the resistances of R_2 and R_3 are added to the coil resistance, while R_1 becomes the shunt for the meter. When the moving contact is switched to point B, the shunt becomes the sum of $R_1 + R_2$, and resistor R_3 is now in series with the meter. Finally, with the moving contact at point C, the resistance of the shunt in parallel with the instrument is the sum of all three resistors, $R_1 + R_2 + R_3$. Note that there is a shunt in parallel with the instrument at all times.

Section Review

1. The moving coil of a Weston-type instrument cannot carry currents greater than 30 mA (True/False).
2. What is a shunt resistor?
3. In what type of electronic circuit is ammeter loading most common?

_____ 7-3 __

DC VOLTMETER

When the Weston-type instrument is used as a **voltmeter,** it is performing essentially the same function as when it is used as an ammeter. Instead of a shunt resistor, the voltmeter has a high value of resistance placed in series with the moving coil. The size of the series resistor used depends on the current range of the movement and the voltage range desired. The resistance connected in series with the moving coil is called a **multiplier.** Figure 7–6 shows a typical moving coil DC voltmeter.

Because a voltmeter is placed across, or in parallel with, a circuit, it is desirable that it takes as little current as possible. Therefore, the moving coil of a voltmeter is usually wound with more turns of a finer wire than that of the ammeter. Less current is then required to develop the same torque for a full-scale deflection.

A basic voltmeter circuit is shown in Figure 7–7. It consists of a 0.5 mA meter movement with a multiplier resistor in series with it. To calculate the size of multiplier for an instrument, simply apply the rules of a series electric circuit.

FIGURE 7–6   DC voltmeter.

____ EXAMPLE 7–2 ____

The 0.5 mA instrument in Figure 7–7 has a moving coil resistance of 500 Ω. What size of multiplier is required to convert this to a 100 V voltmeter?

Solution

$$R_T = \frac{E_T}{I_T} = \frac{100 \text{ V}}{0.0005 \text{ A}} = 200,000 \text{ } \Omega$$

$$R_m \text{ (multiplier)} = 200,000 \text{ } \Omega - 500 \text{ } \Omega$$

$$\approx 200,000 \text{ } \Omega$$

Usually, one meter movement is arranged to read several different maximum voltage values by connecting resistors as shown in Figure 7–8(a) or 7–8(b). Unlike the ammeter, the rotary switch used with the voltmeter, as in Figure 7–8(a), should be a *break-before-make* type, which means the moving contact should disconnect from one terminal before connecting to the next terminal.

Voltmeter

FIGURE 7–7   Basic voltmeter circuit.

FIGURE 7–8
Switching configurations for a voltmeter.

(a)   (b)

Section Review

1. Voltmeters use shunt resistors to allow for large voltage readings (True/False).
2. What is a multiplier?
3. The moving coil of a voltmeter is wound with less turns of wire than an ammeter (True/False).

7–4

VOLTMETER SENSITIVITY

Because the voltmeter is connected in parallel with the circuit it measures, some of the circuit current will flow through the voltmeter. In order not to unbalance the circuit conditions, the meter current should be very small compared with the original circuit current. When voltage measurements are made in high-resistance circuits, it is necessary to use a high-resistance voltmeter to prevent this shunting action of the meter. The effect is less noticeable in low-resistance circuits because the shunting effect is less.

Voltmeter sensitivity is determined by dividing the resistance of the meter plus the multiplier by the full-scale deflection of the meter. Expressed in equation form,

$$\text{sensitivity} = \frac{R_m + R_s}{E} \qquad (7.2)$$

where R_m = meter resistance
R_s = series resistance (multiplier)
E = full-scale reading, in volts

Voltmeter sensitivity is expressed in ohms/volt. This rating can also be used to determine the current sensitivity of the meter. A voltmeter is considered more sensitive if it draws less current from the circuit. Therefore, the sensitivity of a voltmeter varies inversely with the current required for full-scale deflection. Expressed mathematically,

$$\text{sensitivity} = \frac{1}{I_{FS}} \qquad (7.3)$$

where I_{FS} is the current required for full-scale deflection of the meter movement. For example, the sensitivity of a 20 μA movement is the reciprocal of 0.00002 A, or 50,000 Ω/V.

_____ EXAMPLE 7–3 _____

What is the sensitivity of a
voltmeter having a total
internal resistance of
2.5 MΩ for a 100 V range?

Solution

$$\text{meter sensitivity} = \frac{2.5 \text{ M}\Omega}{100 \text{ V}} = 25 \text{ k}\Omega/\text{V}$$

_____ EXAMPLE 7–4 _____

What is the sensitivity of a
voltmeter that produces
full-scale deflection when
the meter current is
8.5 µA?

Solution

$$\text{meter sensitivity} = \frac{1}{8.5 \times 10^{-6}} = 117.6 \text{ k}\Omega/\text{V}$$

Section Review

1. The circuit current should be very large compared to the meter current (True/False).
2. How is voltmeter sensitivity determined?
3. Voltmeter sensitivity is expressed in volts/ohm (True/False).

— 7–5

VOLTMETER LOADING

When measuring voltage, it is necessary to connect the voltmeter *across* the circuit under test. Because a portion of the circuit current must flow through the voltmeter, the circuit behavior is modified somewhat. By connecting a voltmeter across two points in a highly resistive circuit, the meter acts as a shunt and, consequently, reduces the equivalent resistance in that portion of the network. This means that the voltmeter will actually produce a lower reading of the voltage drop than what existed before the meter was connected. This is referred to as the **loading effect** of a meter.

Often, the loading effect of a voltmeter can be ignored, especially if the meter has a high Ω/V, or sensitivity rating. However, if the meter has a low sensitivity, or the circuit under test has a high resistance, the loading effect of the meter must be taken into account. Consider the circuit shown in Figure 7–9(a). Two 10 kΩ resistors are connected in series to

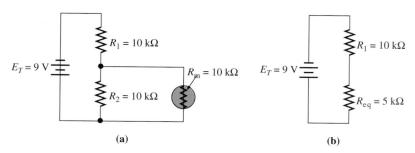

(a)   (b)

FIGURE 7–9 *Loading effect of a voltmeter. (a) Voltmeter with low sensitivity. (b) Equivalent circuit.*

a 9 V source, and a voltmeter is connected across R_2. Since the resistors are of equal value, each resistor will drop 1/2 of the applied voltage, or 4.5 V. Therefore, we would expect the voltmeter to read 4.5 V if it was connected across either resistor. If the voltmeter has a sensitivity of 1000 Ω/V, on the 0V–10 V range, it would have an internal resistance of 10,000 Ω. R_m is in parallel with R_2, so the equivalent resistance in that portion of the circuit is 5 kΩ, as shown in Figure 7–9(b). By connecting the voltmeter across R_2, the total resistance of the circuit has now been reduced to 15 kΩ, instead of the original 20 kΩ.

The voltage drop across $R_2 \| R_m$ is found by the voltage divider rule.

$$V_{R2} = \left(\frac{5 \text{ k}\Omega}{15 \text{ k}\Omega} \right) 9 \text{ V} = 3 \text{ V} \quad \text{(with meter connected)}$$

The percent error of a voltmeter is a ratio of the true voltage minus the voltage read by the meter, divided by the true voltage.

$$\% \text{ error} = \left(\frac{\text{true voltage} - \text{apparent voltage}}{\text{true voltage}} \right) 100\%$$

The meter used in the above example would have a percent error of

$$\% \text{ error} = \left(\frac{4.5 - 3}{4.5} \right) 100\% = 33.33\%$$

By increasing the sensitivity, the loading effect is minimized. For example, if a voltmeter with a sensitivity of 1 MΩ/V were to be used, the internal resistance on a 0 V–10 V scale would be 10 MΩ, as shown in Figure 7–10(a). The equivalent resistance of the 10 kΩ and 10 MΩ resistance is found by the product-over-sum rule.

$$R_{eq} = \frac{R_2 \, R_m}{R_2 + R_m} = \frac{(10 \times 10^3 \ \Omega)(10 \times 10^6 \ \Omega)}{10 \times 10^3 \ \Omega + 10 \times 10^6 \ \Omega} = 9990.01 \ \Omega$$

The equivalent resistance R_{eq}, shown in Figure 7–10(b) represents the resistance of R_2 and R_m combined. The voltage drop across this resistance is found by the voltage divider rule.

$$V_{R2} = \left(\frac{9,990.01 \ \Omega}{19,990.01 \ \Omega} \right) 9 \text{ V} = 4.498 \text{ V}$$

By increasing the sensitivity of the meter, the loading effect has been reduced to an inconsequential value. The percent error is now

$$\left(\frac{4.5 \text{ V} - 4.498 \text{ V}}{4.5 \text{ V}} \right) 100\% = 0.044\%$$

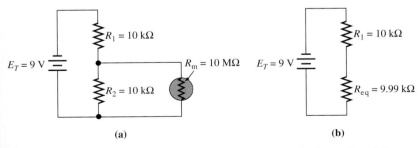

(a)   (b)

FIGURE 7–10   (a) Voltmeter with high value of internal resistance. (b) Equivalent resistance.

EXAMPLE 7–5

The voltage drop across the 100 kΩ resistor in the circuit of Figure 7–11 is to be measured. The voltmeter that is to be used has a sensitivity of 20,000 Ω/V, and is set to the 0 V–10 V range. Determine the

(a) Reading of the meter.
(b) Percent error.

FIGURE 7–11 *Circuit for Example 7–5.*

Solution

(a)　　$R_m = 10 \times 20{,}000 = 200 \text{ k}\Omega$

$$R_{eq} = \frac{R_m R_2}{R_m + R_2} = \frac{(200 \text{ k}\Omega)(100 \text{ k}\Omega)}{200 \text{ k}\Omega + 100 \text{ k}\Omega} = 66.667 \text{ k}\Omega$$

$$V_{R2} = \left(\frac{R_{eq}}{R_{eq} + R_1}\right) 24 \text{ V} = \left(\frac{66.667 \text{ k}\Omega}{66.667 \text{ k}\Omega + 150 \text{ k}\Omega}\right) 24 \text{ V} = 7.385 \text{ V}$$

(b)　　$V_{R2} \text{ (true voltage)} = \left(\frac{R_2}{R_1 + R_2}\right) 24 \text{ V} = \left(\frac{100 \text{ k}\Omega}{250 \text{ k}\Omega}\right) 24 \text{ V} = 9.6 \text{ V}$

$$\% \text{ error} = \left(\frac{\text{true voltage} - \text{measured voltage}}{\text{true voltage}}\right) 100\%$$

$$= \left(\frac{9.6 \text{ V} - 7.385 \text{ V}}{9.6 \text{ V}}\right) 100\% = 23.07\%$$

Section Review

1. What is the loading effect of a voltmeter?
2. The loading effect of a voltmeter can be ignored if the meter has a low-sensitivity rating (True/False).
3. What is the percent error of a voltmeter?

7–6
OHMMETER

An **ohmmeter** is an instrument that indicates the resistance of a circuit or part of a circuit directly in ohms without any need for calculation. The principle of the ohmmeter is based on supplying a fixed value of low voltage to a resistive circuit. A milliammeter is used to measure the current in the circuit. Because the voltage is a fixed value, the current varies inversely with the resistance, and the scale of the milliammeter is marked in the ohms required to give the corresponding value of current. Figure 7–12 shows a typical moving-coil ohmmeter.

If the external resistance connected to the meter is zero ohms, the pointer has a full-scale deflection. If the external resistance is infinity, the pointer has no deflection. The pointer will produce a mid-scale deflection when the external resistance being measured is equal to the internal resistance of the meter. The circuit of a simple series ohmmeter is

FIGURE 7–13
**Simplified
schematic diagram
of typical
ohmmeter circuit.**

FIGURE 7–12   *Ohmmeter.*

shown in Figure 7–13. The ohmmeter's pointer deflection is controlled by the amount of battery current passing through the moving coil.

When the test terminals of the ohmmeter are connected together, current will flow through the circuit. To obtain a full-scale deflection of the meter, the variable resistor is adjusted. This calibrating resistor is provided to compensate for reduction in terminal voltage in the battery due to different scale settings and aging.

After the ohmmeter is calibrated for zero reading, it is ready to be connected in a circuit to measure resistance. An ohmmeter of any type is *never* connected to an energized circuit. The additional voltage introduced by an energized circuit may be enough to establish a current through the moving coil much greater in magnitude than its current sensitivity. This could result in permanent damage to the ohmmeter.

The test leads of the ohmmeter are connected across the circuit to be measured. This causes the current produced by the meter's 3 V battery to flow through the circuit being tested. The greater the resistance of the circuit, the smaller the current flow will be, resulting in less deflection of the pointer. If no resistance is introduced, such as when the test terminals are short circuited, then the reading will be full scale.

Figure 7–14 shows that the scale of an ohmmeter is not divided into uniform divisions. The scale can be said to be nonlinear because of the unequal division of units. The most accurate readings are obtained around the center of the scale.

As the accuracy of an ohmmeter is greatest around the center of the scale, ohmmeters are designed with several different ranges. These devices are referred to as *multirange ohmmeters.* The circuit for a multirange ohmmeter is shown in Figure 7–15. The range switch of this ohmmeter has resistance ranges of R × 1, R × 10, R × 100, R × 1,000, and R × 10,000. When the range switch is on R × 100, the direct reading on the meter scale is multiplied by 100 for the true reading. For high-resistance readings, such as R × 10 k

FIGURE 7–14   *Ohmmeter with 1 mA movement.*

FIGURE 7–15 *Multirange ohmmeter circuit.*

on this meter, an additional 9 V battery is added. Generally, the maximum resistance that can be measured by an ohmmeter increases directly with the voltage of the battery used.

Section Review

1. The principle of the ohmmeter is based on supplying a varying value of low voltage to a circuit of low resistance (True/False).
2. Why should an ohmmeter never be connected to an energized circuit?
3. Why are ohmmeters designed with several different ranges?

___ 7–7 ___

MULTIMETERS

Because analog ammeters, ohmmeters, and voltmeters use the same meter movement, they are often combined into an instrument called a *volt-ohm-milliammeter* (**VOM**), or **multimeter.** The number of scales used in a particular meter will vary with the ranges of voltages and currents required. Figure 7–16 shows an example of a moving coil VOM.

The multimeter circuit shown in Figure 7–17 has a separate terminal provided for each scale. This method of completing the meter circuit for a test measurement requires a large number of terminals, but, when using this method, there is less chance of meter burn out. By using switches in place of the terminals, much greater convenience is obtained; however, if a switch is set incorrectly at the beginning of the measurement, or if it is carelessly moved without regard for the correct meter scale, the meter may be burned out.

Section Review

1. What is a VOM?
2. The number of scales used in a VOM will vary with the ranges of voltages and currents required (True/False).

(a)   (b)

FIGURE 7-16   *Typical multimeters. (a) Analog. (b) Digital.*

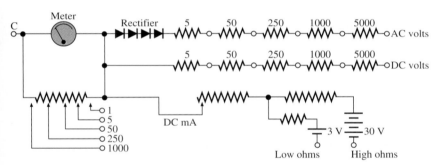

FIGURE 7-17   *Multimeter with terminals for various scale readings.*

_____ **7-8** ___

ELECTRONIC METERS

The electronic meter (EVOM) has several advantages over the passive type of multimeter. By electronically processing the incoming voltage or current, much greater sensitivity with less circuit loading can be obtained.

The following features of an electronic multimeter illustrate some basic differences between an EVOM and a VOM. For the electronic multimeter,

1. The zero control is used to electronically zero the meter on all ranges and functions.
2. The autopolarity function allows either test lead to be used on any point in a circuit, and the meter will indicate the relative polarity of the voltage at the + input of the meter. With this type of meter there is no danger of damaging the device by connecting it backward, or reversing polarity.

3. A low-ohm setting is available. This setting improves the accuracy of very low ohmic value readings.

Section Review

1. What is the purpose of zero control for an electronic meter (EVOM)?
2. Electronic VOMs have an autopolarity function (True/False).
3. What is the function of the low-ohm setting on an EVOM?

___ 7-9 ___

DIGITAL METERS

A **digital multimeter (DMM)** displays voltage, current, and resistance measurements as decimal numbers. Numerical readout is advantageous in many applications because it reduces human reading error and eliminates **parallax error.** Parallax is caused by looking at a meter from an angle, which will cause the pointer to appear left or right of the true position. Because the digital meter has no pointer, parallax is nonexistent. A typical digital meter is shown in Figure 7–18.

The output of a DMM is indicated by **seven-segment displays.** Each segment is individually controlled so that groups of segments can be formed to represent the decimal numbers 0 through 9. In the seven-segment format, each digit is composed of between two and seven segments. Figure 7–19(a) shows the ten possible numerical outputs for a typical seven-segment display. By energizing any combination of the seven segments, the numbers 0 through 9 can be illuminated, as well as a variety of letters from the alphabet, as shown in Figure 7–19(b).

The two types of seven-segment displays used by digital meters are **light emitting diode (LED)** and **liquid crystal display (LCD).** LEDs have two basic limitations: the semiconductor material is relatively expensive, and they need substantial amounts of power to become brightly illuminated. The most widely used power supply voltage for LED displays is 5 V. At standard brightness, each segment of a seven-segment display will typically require 16 mA. At high brightness, each segment requires 27 mA. To reduce the current requirements of a seven-segment display, the segments are *multiplexed.* This technique involves using a multiplexer to individually turn on and off segments very quickly. Each segment is energized in sequence, and the process is repeated many times per second so that the display appears to be continuously on. In addition to reducing the current demand of a seven-segment display, multiplexing also reduces the circuitry required to drive the segments of a display.

FIGURE 7-18   *Digital meter.*

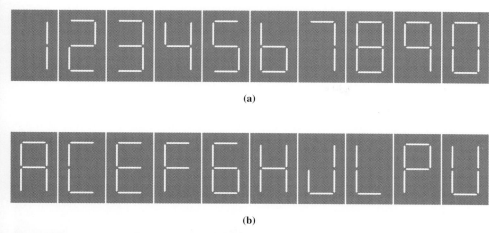

(a)

(b)

FIGURE 7–19 *(a) Seven-segment numeric display. (b) Limited alphabet.*

The LCD type of seven-segment display requires less power but is difficult to read in poor lighting. The LCD does not generate any light itself, but instead, it polarizes the light that passes through it. Basically, the LCD is a light-reflecting device that requires very low amounts of power to operate. Typical LCD seven-segment displays require 1 mW of power. LCDs are extremely popular in battery-powered DMMs because of their low power requirements. LCDs are available in a wide variety of sizes ranging from those used in digital watches to displays several inches high. Unlike LEDs, power consumption in an LCD does not increase significantly with size, making it well suited for meters with large displays.

Another popular type of display is the **dot matrix display.** Unlike the seven-segment display, the dot matrix format has the capability of displaying all 26 letters of the alphabet. Figure 7–20 shows a 5 × 7 format. Each display consists of a matrix of 5 × 7 dots that are individually energized according to a pattern of five horizontal and seven vertical conductors. Dot matrix displays can be either the LED or LCD type and are frequently used where both letters and symbols are required to be displayed. Multiplexing is even more important in the dot matrix display than in the seven-segment display because of the large number of dots. A 5 × 7 display consists of 35 dots that require a large number of interconnections without a multiplexer.

DMMs have many advantages over analog meters, including automatic overrange indication, overvoltage protection, and even automatic range selection. Some DMMs allow you to chose between manual and automatic ranging. A typical DMM will have four digits and a decimal point. DMMs are also available in five-digit configurations and have greater accuracy and resolution. Although the terms *accuracy* and *resolution* are

FIGURE 7–20 *Dot matrix display.*

often used interchangeably, they are related but do not mean the same thing. The resolution of a meter is its ability to display the difference between values. In other words, it is a ratio of the minimum value that can be displayed to the maximum value that can be displayed on a given range. The accuracy of a meter is an indication of the maximum error that can be expected between the actual signal being measured and the reading indicated by the meter.

In most DMMs, the left, or most significant digit, is known as a half digit because it can display only a 0 or 1. Consequently, a DMM with four digits is often called a *3 1/2-digit DMM*. A 3 1/2-digit meter has a total of 23 segments that can be arranged to display the digits. Figure 7–21 shows a 3 1/2-digit DMM with manual ranging. By

(a)   **(b)**

FIGURE 7–21 *Manual ranging digital multimeter.*

(a)

(b)

FIGURE 7–22 *Digital multimeters. (a) Hand-held DMMs. (b) Benchtop DMM.*

TABLE 7–1 *Comparison of analog and digital meters.*

	Digital	Analog
Accuracy	Extremely accurate for fixed signals. Less accurate for slow-changing and peak signals.	Meter reading is more accurate on right side of scale than on left side.
Reading errors	Unlikely due to digital display providing readout.	Errors in meter reading can result from parallax.
Range selection	Automatic or manual.	Manual.
Loading effect	Minimal on low-resistance circuits. Can be substantial on high-resistance circuits.	Readings can be severely affected.
Electromagnetic fields	Can adversely affect meter.	Have no effect on meter.

changing a range selector, the decimal point is moved so that the full-scale voltage can range from 1.999 V to 1999 V. In Figure 7–21(a), the range selector is set at 2 V. In this position, the meter is capable of reading voltages between 0.001 V and 1.999 V. In Figure 7–21(b), the selector is set at the 2000 V range and the meter can read voltages up to 1999 V. If the voltage exceeds the maximum selected by the range switch, the display will flash 1999 V, 199.9 V, 19.99 V, or 1.999 V, depending on the position of the range selector.

Figure 7–22 shows a DMM with autoranging. This meter also includes a test position for semiconductor diodes. When the selector switch is at the diode setting, the meter will indicate short circuit or open circuit conditions for the diode. A complete discussion of diode testing occurs in Chapter 19. Autoranging DMMs will automatically shift to a higher range when an overload is indicated.

When used as a voltmeter, a DMM has a very high internal resistance, typically 10 MΩ, which is one reason the percent error of this type of meter is so low. With such a high value of internal resistance, the loading effect caused by a DMM is almost negligible. Some DMM's are accurate to within ±0.005% of the true value. In DC current measurements, any ammeter will cause resistance to be inserted in the circuit under test. However, even the least expensive DMM will have a lower percent error than an analog meter. Table 7–1 lists the relative advantages and disadvantages of digital and analog meters.

Many DMMs also have high- and low-ohm settings that change the voltage at the test terminals of the meter. A typical voltage output on the low-ohm setting is 0.2 V, while the high-ohm setting will produce 2 V. The low-ohm setting allows for in-circuit testing of some components. This setting is not suitable for testing semiconductor devices such as diodes and transistors. A typical semiconductor must have at least 0.7 V applied across it in order to conduct. Consequently, the high-ohm setting must be used for testing these devices.

Section Review

1. What causes parallax error?
2. The output of a DMM is indicated by seven-segment displays (True/False).
3. What is a 3 1/2 digit DMM?

FIGURE 7–23 *Circuit for practical application.*

7–10

PRACTICAL APPLICATION SOLUTION

Your task in the practical application described in the chapter opener was to service an inoperative computer. The first test of the computer indicated that the power light did not turn on when the computer was switched on. After opening the cover, a test of the fuse revealed that it was blown. After replacing the fuse, the output of the power supply was found to be producing voltages below and above their rated values. The following steps illustrate how a multimeter is used in a practical application.

STEP 1   Check the fuse using an ohmmeter. If the fuse is good (not blown) the resistance of the fuse should be 0 Ω. A blown fuse should indicate infinity (∞) on an ohmmeter.

STEP 2   Connect the voltmeter across the three test points shown in the schematic diagram of Figure 7–23. Record the voltage readings taken at these points

STEP 3   Compare voltage readings taken in Step 2 with the data supplied by the manufacturer. From this comparison, it is apparent that resistor R_2 in the voltage divider for the power supply is shorted out. This conclusion is based on the fact that the voltage across R_3 and across $R_2 + R_3$ is the same.

Summary

1. Analog or moving coil meters operate on a principle based on the reaction between two magnetic fields.
2. An ammeter with a high internal resistance could produce a loading effect in low-voltage, low-resistance electronic circuits.
3. The sensitivity of a voltmeter is determined by dividing the resistance of the meter plus the multiplier by the full-scale deflection of the meter.
4. Voltmeter sensitivity is expressed in ohms per volt (Ω/V).
5. Analog voltmeters, ammeters, and ohmmeters are often combined into an instrument called a volt-ohm-milliammeter (VOM), or multimeter.
6. Ohmmeters should *never* be connected to energized circuits.
7. A digital multimeter (DMM) displays voltage, current, and resistance measurements as decimal numbers.
8. Numerical displays eliminate parallax error caused by misreading a pointer.
9. In most DMMs the left, or most significant digit, is known as a half digit because it can display only a zero or one.

Section 7–1

1. True
2. Digital meters
3. True
4. The reaction between two magnetic fields

Section 7–2

1. False
2. A resistor placed in parallel with a moving coil to allow larger values of current to be read
3. Low-voltage, low-resistance circuits

Section 7–3

1. False
2. The resistance connected in series with the moving coil of a voltmeter
3. False

Section 7–4

1. True
2. By dividing the resistance of the meter plus the multiplier by the full-scale deflection of the meter
3. False

Section 7–5

1. When the voltmeter produces a lower reading of the voltage drop than what existed before the meter was connected
2. False
3. A ratio of the true voltage minus the voltage read by the meter, divided by the true voltage

Section 7–6

1. False
2. Because the additional voltage introduced by an energized circuit may be enough to establish a current through the moving coil that is much greater in magnitude than its current sensitivity
3. To increase the accuracy of the meter reading

Section 7–7

1. An analog meter that combines an ohmmeter, voltmeter, and ammeter in the same meter movement
2. True

Section 7–8

1. To electronically zero the meter on all ranges and functions
2. True
3. To improve the accuracy of very low ohmic-value readings

Section 7–9

1. Looking at a meter from an angle, which will cause the pointer to appear left or right of the true position

2. True

3. A DMM in which the left, or most significant digit, is a half digit that can display only a 0 or 1

Review Questions

Multiple Choice Questions

7–1 All moving-coil instruments operate on a principle based on the reaction between
(a) Two electric fields **(b)** Two magnetic fields
(c) Two moving coils **(d)** Two permanent magnets

7–2 Weston-type meters are
(a) Digital meters **(b)** Electronic meters
(c) Based on the D'Arsonval principle **(d)** Never used as ammeters

7–3 To read large current values with an ammeter, a _____ resistor is used.
(a) Shunt **(b)** Load **(c)** Series-dropping **(d)** Bleeder

7–4 Ideally, the internal resistance of an ammeter is
(a) ∞ (infinity) **(b)** $0\,\Omega$ **(c)** $10\,\Omega$ **(d)** $0.01\,\Omega$

7–5 Ammeter *loading effect* occurs most often in _____ circuits.
(a) High-voltage, low-resistance **(b)** Low-voltage, high-resistance
(c) Low-voltage, low-resistance **(d)** High-voltage, high-resistance

7–6 The resistance connected in series with the moving coil of a voltmeter is called a _____ resistor.
(a) Shunt **(b)** Series-dropping **(c)** Bleeder **(d)** Multiplier

7–7 Ideally, the amount of current drawn by a voltmeter should be
(a) Equal to the current drawn by the load **(b)** 0 A
(c) Higher than the load **(d)** None of the above

7–8 Voltmeters are always connected in
(a) Parallel **(b)** Series **(c)** Series-parallel **(d)** All of the above

7–9 Voltmeter sensitivity is expressed in
(a) IR **(b)** V/Ω **(c)** IV **(d)** Ω/V

7–10 When a voltmeter produces a lower reading of a voltage drop than what existed before the meter was connected, it is called
(a) D'Arsonval principle **(b)** Multiplier **(c)** Loading effect **(d)** Sensitivity

7–11 If the sensitivity of a voltmeter is increased, the loading effect
(a) Increases **(b)** Decreases **(c)** Stays the same **(d)** Varies

7–12 Which instrument should *never* be used on energized circuits?
(a) Wattmeter **(b)** Voltmeter **(c)** Ammeter **(d)** Ohmmeter

7–13 When using a moving-coil ohmmeter, the accuracy is greatest at the _____ of the scale.
(a) Center **(b)** Left **(c)** Right **(d)** Bottom

7–14 Multimeters can measure
(a) Current, voltage, power **(b)** Voltage, resistance
(c) Current, voltage, resistance **(d)** Many voltage levels

7–15 Inaccurate meter readings result from
(a) Weston-type instruments **(b)** Wattmeters **(c)** VOM **(d)** Parallax

7–16   A digital meter is often referred to as a
(a) DMM   (b) VOM   (c) LED   (d) DMV

7–17   A DMM with 23 segments is called a
(a) 4-digit display   (b) 4 1/2-digit display
(c) 3 1/2-digit display   (d) 5-digit display

Practice Problems

7–18   An ammeter with an internal resistance of 0.75 Ω is connected in parallel with a 0.0015 Ω shunt. The line current flowing is 50 A. Find the value of meter current.

7–19   An ammeter has an internal resistance of 2.16 Ω and a shunt resistance is 0.004 Ω. What value of meter current flows when the line current is 100 A?

7–20   A meter movement has a full-scale deflection of 10 mA and a resistance of 20 Ω. Determine the value of the multiplier required to convert the movement to indicate 300 V full scale.

7–21   A meter movement with a full-scale deflection of 2500 μA and a coil resistance of 50 Ω is to be used as a voltmeter with the following ranges: 10 V, 50 V, 100 V. Determine the value of multiplier resistor for each range.

7–22   Determine the sensitivity of a voltmeter with a total internal resistance of 5 MΩ for a 50 V range.

7–23   If a voltmeter has an internal resistance of 8.5 MΩ when the 100 V range is selected, what is the sensitivity of the meter?

Essay Questions

7–24   What is the basic difference between an analog meter and a digital meter?

7–25   Describe the loading effect of an ammeter.

7–26   Explain the basic operation of a Weston-type voltmeter.

7–27   What is voltmeter sensitivity?

7–28   How does voltmeter loading affect a circuit being tested?

7–29   Describe how the loading effect of a voltmeter could be minimized.

7–30   Why should an ohmmeter never be used on an energized circuit?

7–31   What is a multimeter?

7–32   How does a digital meter eliminate parallax error?

7–33   List the two types of displays used in digital meters.

8

Network Theorems

Learning Objectives

Upon completion of this chapter you will be able to

- Apply loop analysis to DC circuits.
- Define nodal analysis.
- Explain Thévenin's theorem and its application to circuit analysis.
- Define Norton's theorem and understand how to use it to reduce a DC circuit to a simple equivalent.
- Convert voltage sources to current sources, and vice versa.
- Use Millman's theorem to reduce multiple voltage sources in parallel to a single equivalent voltage source.
- Apply superposition to a circuit with more than one voltage or current source.
- Define the maximum power transfer theorem.

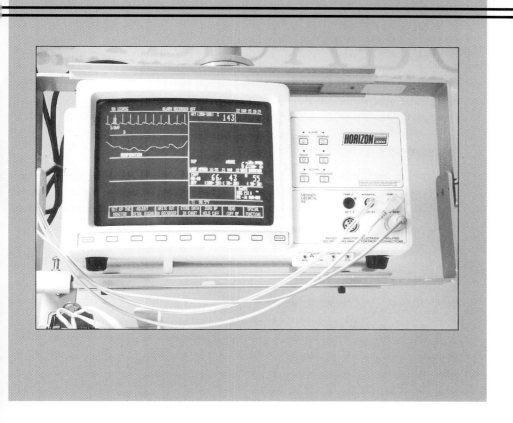

Practical Application

As a technician employed by a company manufacturing and servicing battery-powered emergency medical equipment, you are required to test the output voltage and current available from a battery-pack consisting of four 6 V batteries in parallel. Although the internal resistance of each battery is rated at 0.5 Ω, some of the resistances have increased because of aging. The internal resistance measurements of each battery are as follows: $R_1 = 2.7\ \Omega$, $R_2 = 2.4\ \Omega$, $R_3 = 3.2\ \Omega$, and $R_4 = 1.4\ \Omega$.

To complete this practical application, you will use Millman's theorem to determine the output voltage and current supplied by the battery pack to a 4 Ω load.

INTRODUCTION

When troubleshooting or designing electronic equipment, it is often necessary to analyze circuit behavior under certain operating conditions. The primary objective of circuit analysis is to determine the response of a circuit to a given voltage or current applied to the circuit. In most cases, circuit analysis techniques involve determining the performance of an isolated portion of a complex circuit. In these situations, it is necessary to replace the remainder of the circuit with a simplified equivalent network. Often, the equivalent circuit is reduced to a source and a resistor in series or parallel with the source. Thévenin's theorem and Norton's theorem enable us to do this.

A **complex circuit** is defined as any network that combines series and parallel elements in various interconnections. A systematic approach for setting up and solving several circuit equations simplifies solving for circuit parameters in complex networks. Two of the most common methods used are loop analysis and nodal analysis.

In this chapter, we will examine **linear circuits;** that is, circuits made up of resistors and driven by sources of constant voltage and current. An understanding of algebra is the only mathematical tool required for analyzing these circuits. The advantage of using algebraic methods of circuit analysis is that the same basic approach may be used on any circuit, regardless of the circuit's complexity.

Section Review

1. What is the main purpose of circuit analysis?
2. A complex circuit is any network that combines series and parallel elements in various interconnections (True/False).
3. What is a linear network?

LOOP ANALYSIS

In Chapter 6, Kirchhoff's voltage and current laws were applied as a method of solving series-parallel networks. These laws provide a means of solving a problem without reducing the circuit down to a simple series or parallel network. **Loop analysis** is a method of solving circuit problems by use of Kirchhoff's voltage law. Mathematically, this method of analysis results in a number of simultaneous equations in which the unknown quantities are the currents in various parts of the circuit. The three methods of solving simultaneous equations are substitution, determinants, and matrices.

As stated in Chapter 4, any closed path is called a *loop*. Usually, but not always, a current flows in a loop. This current is referred to as a **loop current.** An electric circuit may be analyzed by counting the unknown branch currents and unknown node voltages in the circuit. It is then necessary to write simultaneous equations to obtain a solution for the circuit. For a given network, the number of simultaneous equations to be solved may be determined by counting the minimum number of loops required.

The steps for solving a circuit problem by loop analysis are as follows:

STEP 1   Assign an arbitrary direction of current flow in each branch of the network, as shown in Figure 8–1.

FIGURE 8–1 *Loop analysis of circuit.*

FIGURE 8–2 *Properly labeled circuit for loop equations.*

STEP 2   Using the arrows as a guide, polarize each resistor for each current. Draw loops, as shown in Figure 8–2.

STEP 3   Write the Kirchhoff voltage law (KVL) equation for each loop. If the direction of the loop is from minus to plus across a component, *add* the voltage in the KVL equation. If the direction of the loop is from plus to minus, *subtract* the voltage in the KVL equation.

STEP 4   Solve the simultaneous equations for the unknown loop currents.

EXAMPLE 8–1

FIGURE 8–3   *Circuit for Example 8–1.*

For the circuit shown in Figure 8–3, solve for the currents in each of the three resistors.

Solution   Assign direction of current flow, indicate the polarity of the resistors, and draw loops as shown in Figure 8–4.

FIGURE 8–4   *Redrawn circuit for Example 8–1.*

Write the loop equations. Using point A as a reference, loops 1 and 2 are written as follows:

Loop 1:
$$2I_3 - 24 \text{ V} + 3I_1 = 0$$
$$3I_1 + 2I_3 = 24 \text{ V}$$

Loop 2:
$$2I_3 - 40 \text{ V} + 6I_2 = 0$$
$$6I_2 + 2I_3 = 40 \text{ V}$$

At this point, there are two equations in the three resistor currents. A third equation is derived by applying Kirchhoff's current law at node A.

$$I_3 = I_1 + I_2$$

Substituting the above equation into the two loop equations results in

$$3I_1 + 2(I_1 + I_2) = 24 \text{ V}$$
$$6I_2 + 2(I_1 + I_2) = 40 \text{ V}$$

which simplifies to

$$5I_1 + 2I_2 = 24 \text{ V}$$
$$2I_1 + 8I_2 = 40 \text{ V}$$

The loop equations are now solved by eliminating an unknown. This is done by multiplying one or more equations by an appropriate constant and adding or subtracting the resulting equations. In this example, the number 4 is chosen as the constant for the first equation.

$$20I_1 + 8I_2 = 96$$
$$\underline{2I_1 + 8I_2 = 40}$$
$$18I_1 = 56$$

$$I_1 = \frac{56}{18} = 3.11 \text{ A}$$

With I_1 known, it is substituted into either equation to obtain I_2.

$$5(3.11) + 2I_2 = 24 \text{ V}$$
$$15.55 + 2I_2 = 24 \text{ V}$$

$$I_2 = \frac{8.45}{2} = 4.23 \text{ A}$$

I_3 is now found by Kirchhoff's current law.

$$I_3 = I_1 + I_2$$
$$= 3.11 + 4.23 = 7.34 \text{ A}$$

_____ **EXAMPLE 8–2** _____

For the circuit of Figure 8–5, solve for the loop currents I_1, I_2, and I_3.

FIGURE 8–5 *Circuit for Example 8–2.*

Solution Identify the polarity of the resistors, and establish reference points, as shown in Figure 8–6.

FIGURE 8–6 *Redrawn circuit for Example 8–2.*

Using points A, B, and C as references for each loop, the following equations are derived:

Loop 1:
$$2I_1 + 3I_1 - 3I_2 - 10 \text{ V} = 0$$
$$5I_1 - 3I_2 = 10 \text{ V}$$

Loop 2:
$$4I_2 + 5I_2 - 5I_3 + 3I_2 - 3I_1 = 0$$
$$-3I_1 + 12I_2 - 5I_3 = 0$$

Loop 3:
$$6I_3 - 12 \text{ V} + 5I_3 - 5I_2 = 0$$
$$-5I_2 + 11I_3 = 12 \text{ V}$$

Combine loop 1 and loop 2. Multiply loop 1 by 3, and loop 2 by 5, and add the equations.

$$15I_1 - 9I_2 \qquad = 30$$
$$\underline{-15I_1 + 60I_2 - 25I_3 = 0}$$
$$51I_2 - 25I_3 = 30$$

Combine the result of loops 1 and 2 with loop 3. Multiply the last equation by 11 and loop 3 by 25. Add the equations.

$$561I_2 - 275I_3 = 330$$
$$\underline{-125I_2 + 275I_3 = 300}$$
$$436I_2 \qquad = 630$$

$$I_2 = \frac{630}{436} = 1.44 \text{ A}$$

I_2 is now substituted into loop 1.

$$5I_1 - 3(1.44) = 10$$
$$5I_1 = 14.32$$
$$I_1 = \frac{14.32}{5} = 2.86 \text{ A}$$

and into loop 3

$$-5(1.44) + 11I_3 = 12$$
$$11I_3 = 19.2$$
$$I_3 = \frac{19.2}{11} = 1.75 \text{ A}$$

Section Review

1. Loop analysis is a method of solving circuit problems using Kirchhoff's current law (True/False).
2. What is a loop current?
3. For a given network, the number of simultaneous equations to be solved may be determined by counting the minimum number of loops required (True/False).

NODAL ANALYSIS

Nodal analysis is a technique where equations based on Kirchhoff's current law are written and unknown voltages solved for. A *node* is a junction between two or more components. A **node voltage** is a voltage at a node with respect to a common reference point. The reference point is generally chosen as a node in the given circuit, although it can be referenced to a point outside the circuit. The number of node voltages in a circuit is always one less than the total number of nodes, if one node is chosen as ground.

When using nodal analysis to solve simultaneous equations, the **reference,** or ground, **node** is often chosen as the node with the largest number of components connected. Many electronic circuits are built on a metallic chassis, which is the practical choice for the ground node. In other circuits, such as electric power systems, the ground node is the earth. In either case, chassis ground or earth ground is assumed to be at zero volts, and node voltages will be at potentials above zero.

The steps to solve circuit problems by nodal analysis are as follows:

STEP 1 Identify the number of nodes. The circuit of Figure 8–7 contains four nodes. They are identified as points A, B, C, and D.

STEP 2 Use one node as a reference node. In this case, node B is chosen as the reference. Points C and D are called *dependent nodes* because the voltage sources determine the voltage. Node A is referred to as an *independent node* because the size of the resistor and circuit configuration determine the voltage. In other words, the voltage at this point is independent of the source voltages. The voltage between the independent node and the reference node is the unknown voltage, which is labeled V_x in Figure 8–8.

STEP 3 Identify and assign current direction at each independent node. The currents flowing into and out of node A are shown in Figure 8–9.

STEP 4 Use Kirchhoff's current law to write an equation for each independent node.

$$I_1 + I_2 + I_3 = 0$$

FIGURE 8–7 *Step 1 in nodal analysis.*

FIGURE 8–8 *Step 2 in nodal analysis.*

FIGURE 8–9 *Step 3 in nodal analysis.*

STEP 5   Use Ohm's law to express each current equation in terms of voltages.

$$I_1 = \frac{V_{R1}}{R_1} = \frac{V_x - V_C}{R_1}$$

$$I_2 = \frac{V_{R2}}{R_2} = \frac{V_x - V_D}{R_2}$$

$$I_3 = \frac{V_x}{R_3}$$

EXAMPLE 8–3

Use nodal analysis to find the voltage V_x for the circuit shown in Figure 8–10.

FIGURE 8–10   *Circuit for Example 8–3.*

Solution   Apply KCL at node A.

$$I_1 + I_2 + I_3 = 0$$

Apply Ohm's law.

$$I_1 = \frac{V_{R1}}{R_1} = \frac{V_x - V_C}{R_1} = \frac{V_x - 12}{3}$$

$$I_2 = \frac{V_{R2}}{R_2} = \frac{V_x - V_D}{R_2} = \frac{V_x - 24}{4}$$

$$I_3 = \frac{V_{R3}}{R_3} = \frac{V_x}{2}$$

Substitute values obtained in Ohm's law calculations into the KCL equation.

$$I_1 + I_2 + I_3 = 0$$

$$\frac{V_x - 12}{3} + \frac{V_x - 24}{4} + \frac{V_x}{2} = 0$$

The above equation is now solved in terms of V_x.

$$\left(\frac{1}{3} + \frac{1}{4} + \frac{1}{2}\right) V_x = \frac{12}{3} + \frac{24}{4}$$

$$\left(\frac{13}{12}\right) V_x = \frac{120}{12}$$

$$(1.083)V_x = 10$$

$$V_x = \frac{10}{1.083} = 9.23 \text{ V}$$

_____ **EXAMPLE 8–4** _____

Use nodal analysis to
solve for the unknown
voltages V_x and V_y in the
circuit of Figure 8–11.

FIGURE 8–11   *Circuit for Example 8–4.*

Solution   Redraw the circuit, assign current direction, and identify the independent nodes, as shown in Figure 8–12.

FIGURE 8–12   *Redrawn circuit for Example 8–4.*

Write the KCL equations at independent nodes.

Node A: $\qquad\qquad\qquad\qquad\qquad\qquad\qquad I_1 = I_2 + I_3$

Node B: $\qquad\qquad\qquad\qquad\qquad\qquad\qquad I_4 = I_3 + I_5$

Use Ohm's law to express each current in terms of voltage.

$$I_1 = \frac{V_{R1}}{R_1} = \frac{-40 - V_x}{10} \qquad I_4 = \frac{V_y}{R_4} = \frac{V_y}{10}$$

$$I_2 = \frac{V_x}{R_2} = \frac{V_x}{15} \qquad I_5 = \frac{V_{R5}}{R_5} = \frac{-60 - V_y}{15}$$

$$I_3 = \frac{V_{R3}}{R_3} = \frac{V_x - V_y}{30}$$

Substitute KCL equations with Ohm's law values.

Node A: $\qquad\qquad\qquad\qquad\qquad\qquad\qquad I_1 = I_2 + I_3$

$$\frac{-40 - V_x}{10} = \frac{V_x}{15} + \frac{V_x - V_y}{30}$$

$$-120 \text{ V} = 6V_x - V_y$$

Node B: $\qquad\qquad\qquad\qquad\qquad\qquad\qquad I_4 = I_3 + I_5$

$$\frac{V_y}{10} = \frac{V_x - V_y}{30} + \frac{-60 - V_y}{15}$$

$$-120 \text{ V} = -V_x + 6\,V_y$$

Multiply the equation for node A by 6 and add node B to node A.

$$36V_x - 6V_y = -720$$
$$\underline{-V_x + 6V_y = -120}$$
$$35V_x = -840$$

$$V_x = 24 \text{ V}$$

Substitute V_x into either node equation.

$$-6(-24) + V_y = -120$$
$$\underline{144 + V_y = 120}$$
$$V_y = -24 \text{ V}$$

Section Review

1. What is nodal analysis?
2. A node voltage is a voltage at a node with respect to a common reference point (True/False).
3. What is the first step when solving circuit problems by nodal analysis?

8–4

THÉVENIN'S THEOREM

In the study of electric circuits, it is often desirable to determine the effect that a change in a single resistance will have on the currents and voltages of a circuit, while all other resistances remain unchanged. Although this type of problem may be solved by loop equations and determinants, the solution by Thévenin's theorem is an easier method. It is particularly useful when studying the relationship between a load of some kind and the system of sources that supplies the load, as the theorem makes it possible to replace the entire supply system by one equivalent source in series with one resistance. This can be done regardless of the complexity of the supplying system.

Thévenin's theorem is stated as follows:

> Any two-terminal circuit, made up of fixed-value resistances and of voltage and current sources, can be replaced by a single voltage source in series with a single resistance, which will produce the same effects at the terminals.

Figure 8–13(a) shows a two-terminal linear circuit, which may consist of fixed resistances and any combination of constant current and voltage sources. A linear network is made up of components that have a directly proportional relationship between

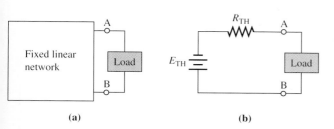

FIGURE 8–13 *(a) Fixed linear network. (b) Thévenin equivalent circuit.*

voltage and current. The **Thévenin equivalent circuit** for Figure 8–13(a) is shown in Figure 8–13(b). The Thévenin equivalent voltage, E_{TH}, is in series with the Thévenin equivalent resistance, R_{TH}.

The Thévenin voltage of Figure 8–13(a) could be measured by placing a high-resistance voltmeter between terminals A and B. It is possible to measure the equivalent resistance of Figure 8–13(a) by connecting an ohmmeter between points A–B. Any voltage sources in the circuit would have to be short circuited and any current sources open circuited for the ohmmeter reading to be accurate.

The steps in applying Thévenin's theorem are as follows:

STEP 1   Remove the portion of the circuit where the Thévenin equivalent circuit is to be determined.

STEP 2   Determine the Thévenin equivalent resistance, R_{TH}, by short circuiting the voltage sources and open circuiting the current sources.

STEP 3   Determine the Thévenin voltage, E_{TH}, across the portion of the circuit removed in Step 1. Return all sources to their original position.

STEP 4   Draw the Thévenin equivalent circuit. Place the resistance removed in Step 1 across the terminals of the Thévenin equivalent circuit. The current through this resistor can now be found by Ohm's law.

A very common problem in electronic circuits is to determine the value of current flowing in a loaded voltage divider network, such as the one shown in Figure 8–14. This type of circuit will often have a transistor connected where resistor R_3 is shown, and the value of current flowing into the transistor would be required. The following example illustrates how to solve this type of problem using Thévenin's theorem.

_____ **EXAMPLE 8–5** _____

For the circuit of Figure 8–14, determine the current through resistor R_3 using Thévenin's theorem.

FIGURE 8–14   Circuit for Example 8–5.

Solution   Disconnect resistor R_3, and short circuit the voltage source, as shown in Figure 8–15.

FIGURE 8–15   Resistor R₃ disconnected.

The Thévenin equivalent resistance is now determined.

$$R_{TH} = \frac{R_1 R_2}{R_1 + R_2} = \frac{(100 \times 10^3)(150 \times 10^3)}{(100 \times 10^3) + (150 \times 10^3)} = 60 \text{ k}\Omega$$

Insert the voltage source back into the circuit, as shown in Figure 8–16.

FIGURE 8–16 *Reinserted voltage source.*

Determine the current flow in the circuit with R_3 still disconnected.

$$I = \frac{E}{R_1 + R_2} = \frac{15 \text{ V}}{100 \text{ k}\Omega + 150 \text{ k}\Omega} = 60 \text{ μA}$$

The Thévenin equivalent voltage across terminals A–B is the voltage dropped across resistor R_3.

$$E_{TH} = V_{AB} = (150 \text{ k}\Omega)(60 \text{ μA}) = 9 \text{ V}$$

The Thévenin equivalent circuit can now be drawn, as shown in Figure 8–17.

FIGURE 8–17 *The Thévenin equivalent circuit for Example 8–5.*

The current through the 22 kΩ resistor is found by Ohm's law.

$$I_3 = \frac{E_{TH}}{R_{TH} + R_3} = \frac{9 \text{ V}}{60 \text{ k}\Omega + 22 \text{ k}\Omega} = 110 \text{ μA}$$

EXAMPLE 8–6

Use Thévenin's theorem to find the load current in Figure 8–18.

$R_1 = 3.3 \text{ k}\Omega$ $R_2 = 5.1 \text{ k}\Omega$

$E_T = 15 \text{ V}$ $R_3 = 10 \text{ k}\Omega$ $R_L = 2.4 \text{ k}\Omega$

FIGURE 8–18 *Circuit for Example 8–6.*

Solution   With R_L removed the voltage across the terminals will be

$$E_{TH} = \left(\frac{R_3}{R_1 + R_3} \right) E_T = \left(\frac{10 \text{ k}\Omega}{3.3 \text{ k}\Omega + 10 \text{ k}\Omega} \right) 15 \text{ V} = 11.28 \text{ V}$$

With the voltage source removed, the Thévenin resistance is now determined.

$$R_{TH} = \frac{R_1 R_3}{R_1 + R_3} + R_2 = \frac{(3.3 \text{ k}\Omega)(10 \text{ k}\Omega)}{3.3 \text{ k}\Omega + 10 \text{ k}\Omega} + 5.1 \text{ k}\Omega = 7.58 \text{ k}\Omega$$

The equivalent Thévenin circuit is shown in Figure 8–19. The load current is now solved using Ohm's law.

FIGURE 8–19   *The Thévenin equivalent circuit for Example 8–6.*

$$I_L = \frac{E_{TH}}{R_{TH} + R_L} = \frac{11.28 \text{ V}}{7.58 \text{ k}\Omega + 2.4 \text{ k}\Omega} = 1.13 \text{ mA}$$

___ **EXAMPLE 8–7** ___

For the bridge circuit shown in Figure 8–20, calculate the current through resistor R_L using Thévenin's theorem.

FIGURE 8–20   *Circuit for Example 8–7.*

Solution   Disconnect R_L and short circuit the voltage source, as in Figure 8–21.

FIGURE 8–21   *Resistor R_L disconnected.*

From Figure 8–21, it is apparent that resistors R_1 and R_2 are joined at point A, while their opposite ends are connected across the shorted voltage source. The same is true at point B for resistors R_3 and R_4. If the circuit is redrawn, as in Figure 8–22, the equivalent resistance of the circuit is found.

FIGURE 8–22 *Equivalent resistance for Example 8–7.*

For the circuit of Figure 8–22, the resistance between points A and B consists of the parallel-connected resistors R_1 and R_2 , which are in series with the parallel resistors R_3 and R_4. Therefore, the equivalent resistance is determined as follows:

$$R_{TH} = \frac{R_1 R_2}{R_1 + R_2} + \frac{R_3 R_4}{R_3 + R_4} = \frac{(20\ \Omega)\,(15\ \Omega)}{20\ \Omega + 15\ \Omega} + \frac{(30\ \Omega)\,(10\ \Omega)}{30\ \Omega + 10\ \Omega} = 16.07\ \Omega$$

Insert the voltage source back into the circuit of Figure 8–21, as in Figure 8–23.

FIGURE 8–23 *Reinserted voltage source.*

Use the voltage divider rule to find the voltages across resistors R_2 and R_4.

$$V_{R2} = \left(\frac{R_2}{R_1 + R_2} \right) E = \left(\frac{15\ \Omega}{20\ \Omega + 15\ \Omega} \right) 15\ V = 6.43\ V$$

$$V_{R4} = \left(\frac{R_4}{R_3 + R_4} \right) E = \left(\frac{10\ \Omega}{30\ \Omega + 10\ \Omega} \right) 15\ V = 3.75\ V$$

The potential difference between points A and B is the difference in voltage between V_{R2} and V_{R4}.

$$E_{TH} = V_{AB} = V_{R2} - V_{R4} = 6.43\ V - 3.75\ V = 2.68\ V$$

The Thévenin equivalent circuit can now be drawn, as shown in Figure 8–24.

FIGURE 8–24 *The Thévenin equivalent circuit for Example 8–7.*

The current through R_L is now solved by Ohm's law.

$$I_{RL} = \frac{E_{TH}}{R_{TH} + R_L} = \frac{2.68 \text{ V}}{16.07 \ \Omega + 12 \ \Omega} = 95.48 \text{ mA}$$

Section Review

1. State Thévenin's theorem.
2. Thévenin's theorem makes it possible to replace the entire supply system with one equivalent source in series with one resistance (True/False).
3. What is the first step in applying Thévenin's theorem?

8–5

NORTON'S THEOREM

Nortons's theorem is the dual of Thévenin's theorem. The use of Thevénin's theorem is limited to the identification of a voltage source in series with a resistance value. Norton's theorem is implemented to reduce a circuit to a current source and a parallel resistance value. **Norton's theorem** is stated as follows:

> Any two-terminal circuit can be replaced by an equivalent current source in parallel with an equivalent resistance. The current source determines the maximum possible current that would flow if the terminals of the original network were short circuited. The equivalent parallel resistance is the resistance value measured looking back into the original circuit.

The term *looking back* refers to the resistance of a circuit if voltage sources are removed and replaced by their internal resistance values and if all current sources are open circuited.

When applying Norton's theorem, the current and resistance must be determined to form a **Norton equivalent circuit,** as shown in Figure 8–25(b) The *Norton equivalent current, I_N,* is the short-circuit current between two points in a circuit. When the load terminals are short circuited, the *Norton equivalent resistance, R_N,* is calculated. Figure 8–25(a) consists of a fixed linear network. In Figure 8–25(b), the Norton equivalent current is found by shorting terminals A and B. The equivalent resistance is the resistance seen from terminals A–B when the independent sources of the fixed network are disconnected.

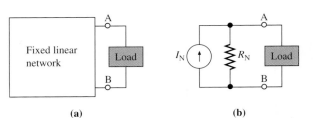

FIGURE 8–25 *Fixed linear network. (b) Norton equivalent circuit.*

(a)   (b)

EXAMPLE 8–8

FIGURE 8–26   *Circuit for Example 8–8.*

For the circuit shown in Figure 8–26, find the Norton equivalent circuit.

Solution   Short circuit the load, as shown in Figure 8–27. Resistors R_2 and R_3 are now in parallel.

FIGURE 8–27   *The Norton equivalent current for Example 8–8.*

The total resistance seen by the 40 V source is given by

$$R_T = \frac{R_2 R_3}{R_2 + R_3} + R_1 = \frac{(10\ \Omega)\,(15\ \Omega)}{10\ \Omega + 15\ \Omega} + 20\ \Omega = 26\ \Omega$$

The total current is found by Ohm's law.

$$I_T = \frac{E_T}{R_T} = \frac{40\ \text{V}}{26\ \Omega} = 1.54\ \text{A}$$

Short circuit the voltage source to find the Norton equivalent resistance of the circuit, as shown in Figure 8–28. R_1 and R_3 are in parallel; R_2 is in series with the parallel connection.

FIGURE 8–28   *The Norton equivalent resistance for Example 8–8.*

$$R_N = \frac{R_1 R_3}{R_1 + R_3} + R_2 = \frac{(20\ \Omega)\,(15\ \Omega)}{20\ \Omega + 15\ \Omega} + 10\ \Omega = 18.57\ \Omega$$

The Norton equivalent current is now found by using the current divider rule.

$$I_N = \left(\frac{R_3}{R_2 + R_3} \right) I_T = \left(\frac{15\,\Omega}{10\,\Omega + 15\,\Omega} \right) 1.54\,A = 0.924\,A$$

The Norton equivalent circuit is shown in Figure 8–29.

FIGURE 8–29 *The Norton equivalent circuit for Example 8–8.*

When current sources are in parallel, the total current is the sum of the individual current sources. The following example illustrates the use of Norton's theorem for a two-source network.

_____ **EXAMPLE 8–9** _____

Using Norton's theorem, solve for the current in resistor R_3 for the circuit of Figure 8–30.

FIGURE 8–30 *Circuit for Example 8–9.*

Solution   Short circuit resistor R_3, and draw arrows indicating current flow, as shown in Figure 8–31.

FIGURE 8–31 *Resistor R_3 shorted.*

Convert the voltage sources to current sources, as indicated in Figure 8–32.

FIGURE 8–32 *Voltage sources converted to current sources.*

$$I_{N1} = \frac{E_1}{R_1} = \frac{12 \text{ V}}{2 \text{ }\Omega} = 6 \text{ A}$$

$$I_{N2} = \frac{E_2}{R_2} = \frac{24 \text{ V}}{2 \text{ }\Omega} = 12 \text{ A}$$

The two current sources of Figure 8–32 can be combined into a single current source by adding the currents of the sources together. The equivalent resistance can be found by the product-over-sum rule. The Norton equivalent circuit is shown in Figure 8–33.

FIGURE 8–33 *The Norton equivalent circuit for Example 8–9.*

$$I_N = I_{N1} + I_{N2} = 6 \text{ A} + 12 \text{ A} = 18 \text{ A}$$

$$R_N = \frac{R_1 R_2}{R_1 + R_2} = \frac{(2 \text{ }\Omega)(2 \text{ }\Omega)}{2 \text{ }\Omega + 2 \text{ }\Omega} = 1 \text{ }\Omega$$

The current through R_3 is now determined by the current divider rule.

$$I_3 = \left(\frac{R_N}{R_N + R_3}\right) I_N = \left(\frac{1 \text{ }\Omega}{1 \text{ }\Omega + 5 \text{ }\Omega}\right) 18 \text{ A} = 3 \text{ A}$$

Section Review

1. Norton's theorem is the dual of Thévenin's theorem (True/False).
2. What does the term *looking back* refer to when applying Norton's theorem?
3. Norton's theorem is implemented to reduce a circuit to a current source in series with an equivalent resistance (True/False).

8–6

CONVERSION OF VOLTAGE AND CURRENT SOURCES

When using Thévenin's theorem or Norton's theorem for solving circuit problems, it is often advantageous to replace a curent source with a voltage source or a voltage source with a current source. This replacment of sources is called **source conversion.**

The terminal voltage of a practical voltage source is determined by the equation

$$V_T = E - IR_i$$

and, the terminal voltage of a practical current source is found by

$$V_T = IR_i - I_L R_i$$

A voltage source and a current source are *equivalent* if they produce identical currents in the same resistive load. Therefore, if the internal resistance of both sources are equal, the load current for a current source can be found by the current divider rule.

$$I_L = I\left(\frac{R_i}{R_L + R_i}\right)$$

The load current for a voltage source is found by Ohm's law.

$$I_L = \frac{E}{R_i + R_L}$$

In many practical applications, the internal resistance of a source is ignored. For this reason, many examples and problems discussed in this text do not have internal resistances shown with sources. However, it must be noted that to convert one type of source to another, the internal resistances must be identified.

_____ **EXAMPLE 8–10** _____

Transform the voltage source of Figure 8–34 to a current source, and determine the load current for each source.

FIGURE 8–34 *Circuit for Example 8–10.*

Solution As a voltage source, the load current is found by Ohm's law.

$$I_L = \frac{E}{R_i + R_L} = \frac{12\ \text{V}}{1.5\ \Omega + 8\ \Omega} = 1.26\ \text{A}$$

As a current source, the supply current is found by Ohm's law and the load current is found by the current divider rule.

$$I = \frac{E}{R_i} = \frac{12\ \text{V}}{1.5\ \Omega} = 8\ \text{A}$$

$$I_L = I\left(\frac{R_i}{R_L + R_i}\right) = 8\ \text{A}\left(\frac{1.5\ \Omega}{8\ \Omega + 1.5\ \Omega}\right) = 1.26\ \text{A}$$

The equivalent current source is shown in Figure 8–35.

FIGURE 8–35 *Equivalent current source for Example 8–10.*

_____ **EXAMPLE 8–11** _____

Determine the Thévenin equivalent circuit for the network to the left of terminals A–B in Figure 8–36.

FIGURE 8–36 *Circuit for Example 8–11.*

Solution   Disconnect the load, and replace the current source with an open circuit, as shown in Figure 8–37.

FIGURE 8–37
Redrawn circuit
for Example 8–11.

The resistance between points A and B is 2 Ω. This is the equivalent resistance of the circuit.

$$R_{TH} = R_i = 2\ \Omega$$

Calculate E_{TH} by returning the current source to its original position, as shown in Figure 8–38. With A–B open, the entire current of the circuit will flow through the internal resistance R_i. Therefore, the voltage at terminals A–B will be the product of the current through R_i and the ohmic value of R_i.

FIGURE 8–38   *Internal*
resistance R_i.

$$E_{TH} = IR_i = (20\ \text{A})\,(2\ \Omega) = 40\ \text{V}$$

The Thévenin equivalent circuit is shown in Figure 8–39.

FIGURE 8–39   *The Thévenin*
equivalent circuit for Example 8–11.

Section Review

1. What is source conversion?
2. In many practical applications, the internal resistance of a source is ignored (True/False).

MILLMAN'S THEOREM

Millman's theorem is used for circuits having more than one voltage or current source in parallel. Essentially, Millman's theorem combines the **source conversion theorem** with both the Thévenin and Norton theorems. Millman's theorem applies only to *sources connected directly in parallel*. If there are resistors between the sources, then this theorem can not be applied. To utilize Millman's theorem, all voltage sources must first be converted to current sources. Once the equivalent current source is determined, it can be transformed to a voltage source if desired. **Millman's theorem** is stated as follows:

> Any number of constant current sources in parallel may be combined into a single current source in which the equivalent current source is the algebraic sum of the individual source currents, and the resistance of the equivalent source is the equivalent resistance of the original parallel-connected resistors.

If all the sources in a circuit were constant voltage sources, the following equation would determine the Millman equivalent voltage source, E_M.

$$E_M = \frac{\pm G_1 E_1 \pm G_2 E_2 \pm G_3 E_3 \pm \cdots \pm G_n E_n}{G_1 + G_2 + G_3 + \cdots + G_n} \tag{8.1}$$

The plus or minus signs in the above equation are used to indicate whether the sources are supplying energy in the same direction. If the sources have the same polarity, they are added. Conversely, if the sources have opposite polarity, they are subtracted.

The Millman equivalent resistance R_M for parallel-connected voltage sources is found as

$$R_M = \frac{1}{G_1 + G_2 + G_3 + \cdots + G_n} \tag{8.2}$$

where $E_1, E_2, E_3, \ldots E_n =$ voltages of individual voltage sources
            $G_1, G_2, G_3, \ldots G_n =$ conductances of individual voltage sources

EXAMPLE 8-12

Use Millman's theorem to find the voltage across R_L and the current through R_L in the circuit of Figure 8–40.

FIGURE 8–40   *Circuit for Example 8–12.*

Solution   First, calculate the Millman equivalent voltage. Since the 15 V source is connected with its polarity opposite the 10 V and 12 V sources, the 15 V source is subtracted.

$$E_M = \frac{G_1 E_1 + G_2 E_2 - G_3 E_3}{G_1 + G_2 + G_3}$$

$$= \frac{(1/3)\,(10) + (1/2)\,(12) - (1/4)\,(15)}{1/3 + 1/2 + 1/4} = 5.15 \text{ V}$$

Next, determine the Millman equivalent resistance.

$$R_M = \frac{1}{G_1 + G_2 + G_3}$$

$$= \frac{1}{1/3 + 1/2 + 1/4} = 0.923 \ \Omega$$

The Millman equivalent circuit can now be drawn, as shown in Figure 8–41.

FIGURE 8–41 *The Millman equivalent circuit for Example 8–12.*

The voltage across R_L is found by the voltage divider rule, and the current through R_L is found by Ohm's law.

$$V_{RL} = \left(\frac{R_L}{R_M + R_L}\right) E_M = \left(\frac{8 \ \Omega}{0.92 \ \Omega + 8 \ \Omega}\right) 5.15 \ V = 4.62 \ V$$

$$I_{RL} = \frac{V_{RL}}{R_L} = \frac{4.63 \ V}{8 \ \Omega} = 0.58 \ A$$

Section Review

1. Millman's theorem combines the source transformation theorem with Thévenin's theorem and nodal analysis (True/False).
2. Can Millman's theorem be used on sources that are not connected directly in parallel?
3. State Millman's theorem.

SUPERPOSITION THEOREM

Circuits containing more than one source of emf may be solved by the **superposition theorem,** as well as by the Kirchhoff law method. The superposition theorem, as applied to DC circuits, may be stated as follows:

> The current that flows at any point in a circuit, or the potential difference between any two points in a circuit, resulting from more than one source of emf connected in the circuit, is the algebraic sum of the separate currents or voltages at these points. These values are the voltages and currents that would exist if each source of emf were considered separately and if each of the other sources were replaced by a unit of equivalent internal resistance.

To consider the effects of each source, the voltage sources must be short circuited (zero resistance), and the current sources open circuited (infinite resistance). When a voltage or current source contains an internal resistance, it must be taken into account when the source is removed.

_____ **EXAMPLE 8–13** _____

Use the superposition theorem to find the current through resistor R_2 in the circuit shown in Figure 8–42.

FIGURE 8–42 *Circuit for Example 8–13.*

Solution The first step is to remove the current source, as shown in Figure 8–43.

FIGURE 8–43 *Current source removed.*

The component current through resistor R_2 with the current source removed, I', can now be found by using Ohm's law.

$$I' = \frac{E}{R_1 + R_2} = \frac{12\ V}{10\ \Omega + 15\ \Omega} = 0.48\ A$$

The next step involves short circuiting the voltage source, as shown in Figure 8–44.

FIGURE 8–44 *Voltage source shorted.*

The component current through R_2 with the voltage source removed, I'', can now be determined by using the current divider rule.

$$I'' = \left(\frac{R_1}{R_1 + R_2}\right) I = \left(\frac{10\ \Omega}{10\ \Omega + 15\ \Omega}\right) 5\ A = 2\ A$$

Because the current I' is the same direction as the current I'', the current through resistor R_2 is the sum of these two currents.

$$I_{R2} = I' + I'' = 0.48 + 2 = 2.48\ A$$

EXAMPLE 8–14

FIGURE 8–45 *Circuit for Example 8–14.*

For the circuit shown in Figure 8–45, use superposition to determine currents I_1, I_2, and I_3. The values of the network are as follows: $R_1 = 5\ \Omega$, $R_2 = 15\ \Omega$, $R_3 = 20\ \Omega$, $E_1 = 60$ V, $E_2 = 100$ V.

Solution The circuit is redrawn, as shown in Figure 8–46, with voltage source E_1 short circuited.

FIGURE 8–46 *Voltage source E$_1$ shorted.*

$$R_1 \| R_3 = \frac{R_1 R_3}{R_1 + R_3} = \frac{(5\ \Omega)\,(20\ \Omega)}{5\ \Omega + 20\ \Omega} = 4\ \Omega$$

$$R_T = R_1 \| R_3 + R_2 = 4\ \Omega + 15\ \Omega = 19\ \Omega$$

$$I_2' = \frac{E_2}{R_T} = \frac{100\ \text{V}}{19\ \Omega} = 5.263\ \text{A}$$

$$E_1 \| E_3 = R_1 \| R_3 \times I_2' = (4\ \Omega)\,(5.263\ \text{A}) = 21.052\ \text{V}$$

$$I_1' = \frac{E_1 \| E_3}{R_1} = \frac{21.052\ \text{V}}{5\ \Omega} = -4.21\ \text{A}$$

$$I_3' = \frac{E_1 \| E_3}{R_3} = \frac{21.052\ \text{V}}{20\ \Omega} = 1.053\ \text{A}$$

The circuit is now redrawn, as shown in Figure 8–47, with source E_2 short circuited.

FIGURE 8–47 *Voltage source E$_2$ shorted.*

$$R_2\| E_3 = \frac{R_2 R_3}{R_2 + R_3} = \frac{(15\ \Omega)\,(20\ \Omega)}{15\ \Omega + 20\ \Omega} = 8.571\ \Omega$$

$$R_T = R_2\| R_3 + R_1 = 8.571\ \Omega + 5\ \Omega = 13.571\ \Omega$$

$$I_1'' = \frac{E_1}{R_T} = \frac{60\ \text{V}}{13.571\ \Omega} = 4.421\ \text{A}$$

$$E_2\| E_3 = R_2\| R_3 \times I_1'' = (8.571\ \Omega)\,(4.421\ \text{A}) = 37.892\ \text{V}$$

$$I_3'' = \frac{E_2\| E_3}{R_3} = \frac{37.892\ \text{V}}{20\ \Omega} = 1.895\ \text{A}$$

$$I_2'' = \frac{E_2\| E_3}{R_2} = \frac{37.892\ \text{V}}{15\ \Omega} = -2.526\ \text{A}$$

The total current is now found by superimposing the results.

$$I_1 = I_1' + I_1'' = (-4.21\ \text{A}) + 4.421\ \text{A} = 0.211\ \text{A}$$

$$I_2 = I_2' + I_2'' = 5.263\ \text{A} + (-2.526\ \text{A}) = 2.737\ \text{A}$$

$$I_3 = I_3' + I_3'' = 1.053\ \text{A} + 1.895\ \text{A} = 2.948\ \text{A}$$

EXAMPLE 8-15

Use the superposition theorem to find the current
through resistor R_3 in the
circuit shown in Figure
8–48. The values of the
network are $R_1 = 20\ \Omega$, R_2
$= 30\ \Omega$, $R_3 = 15\ \Omega$, $I_1 =$
5 A, $I_2 = 10$ A.

FIGURE 8–48 *Circuit for Example 8–15.*

Solution Redraw the circuit, as shown in Figure 8–49, with the current source I_2
open circuited.

FIGURE 8–49 *Current source I_2
open.*

$$\frac{1}{R_T} = \frac{1}{R_1} + \frac{1}{R_3} + \frac{1}{R_2} = 0.15\ \text{S}$$

$$R_T = 6.67\ \Omega$$

$$V' = I_1 R_T = (5\ \text{A})\,(6.67\ \Omega) = 33.33\ \text{V}$$

$$I_3' = \frac{V'}{R_3} = \frac{33.33\ \text{V}}{15\ \Omega} = 2.22\ \text{A}$$

The circuit is now redrawn, as in Figure 8–50, with current source I_1 disconnected.

FIGURE 8–50 *Current source I_1 open.*

$$V'' = I_2 R_T = (10 \text{ A}) (6.67 \text{ }\Omega) = 66.7 \text{ V}$$

$$I_3'' = \frac{V''}{R_3} = \frac{66.7 \text{ V}}{15 \text{ }\Omega} = 4.44 \text{ A}$$

The result is then found by superimposing the two currents for I_3.

$$I_3 = I_3' + I_3'' = 2.22 \text{ A} + 4.44 \text{ A} = 6.67 \text{ A}$$

Section Review

1. In addition to the Kirchhoff law method, circuits with more than one source of emf may be solved by the superposition theorem (True/False).
2. What must be done to the voltage sources and current sources in order to apply the superposition theorem?

MAXIMUM POWER TRANSFER THEOREM

Maximum power transfer is very important in electronic circuits such as antennas, power supplies, and audio amplifiers. In electronic communication systems, maximum power transfer is essential when analyzing antennas. Antenna circuits receive very low power signals that must be recovered completely for maximum signal strength. Consequently, the antenna circuit is adjusted to allow for the maximum transfer of power, regardless of the efficiency. When the load resistance and the source resistance are equal, the circuit is said to be *matched*, and maximum power transfer occurs between the antenna and load. The **maximum power transfer theorem** can be stated in the following manner:

> Maximum power is drawn from a source when the load resistance equals the internal resistance of the source.

Maximum power transfer also occurs in automobile starting circuits, with the 12 V battery acting as the source and the starter motor as the load. When it is extremely cold, the strength of the battery decreases and its internal resistance increases. As a result, larger values of current are required to supply the same load resistance.

The Thévenin resistance of a circuit is comparable to the internal resistance of the source, because it absorbs some of the available power from the source. The effect of the source resistance, R_i, on the power output of a DC source may be shown by an analysis of the circuit in Figure 8–51. The current for any value of load resistance, R_L, is found by Ohm's law.

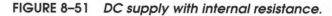

FIGURE 8–51 *DC supply with internal resistance.*

$$I = \frac{E}{R_i + R_L}$$

When the variable load resistor is set at the zero ohms position (equivalent to a short circuit), the current is limited only by the internal resistance of the source. The maximum current that may be drawn from the source would be found by

$$I = \frac{E}{R_i}$$

If the load resistance is increased, current drawn from the source will decrease. At the same time, the terminal voltage applied across the load will increase and approach a maximum value as the current approaches zero. To calculate the power delivered to the load, the following formula may be used:

$$P_L = I^2 R_L = \frac{E^2 R_L}{(R_i + R_L)^2} \tag{8.3}$$

The load power is dependent on both R_i and R_L; however, R_i is considered to be constant for any circuit. The ratio of output power to input power, or the power transfer, from the source to the load increases as the load resistance is increased. The efficiency of power transfer approaches 100% as the load resistance approaches a relatively large value compared with the source resistance. The efficiency of power transfer is expressed by the following equation

$$\eta = \frac{P_L}{P_T} = \frac{I^2 R_L}{I^2 R_L + I^2 R_i} \tag{8.4}$$

where P_L = the load power
P_T = the power developed by the source

The graph in Figure 8–52 shows the relationship between load power and efficiency with a load resistance that varies between 0 Ω and 200 Ω. When the load resistance is 100 Ω, the maximum load power of 100 W is obtained. This occurs when the resistance of the source is equal to the load resistance. However, the efficiency of power transfer is only 50% at the maximum power transfer resistance of 100 Ω and approaches zero efficiency at relatively low values of load resistance compared with that of the source.

The problem of obtaining high efficiency and maximum power transfer is resolved as a compromise between the low efficiency of maximum power transfer and the high efficiency of the high-resistance load. Where circuits deal with large values of power and efficiency is critical, the load resistance is made large in proportion to the source resistance. The source and load resistance are matched for maximum power, and the efficiency of the circuit is ignored.

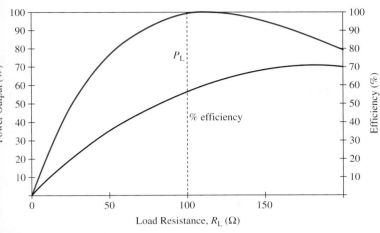

FIGURE 8–52 *Effect of load resistance on power output.*

EXAMPLE 8–16

Solution

(a) At full load the load voltage is found by dividing the rated power by the rated current. The difference between the no-load voltage and the full-load voltage is the internal voltage drop. The internal resistance is then found by Ohm's law.

$$V_L = \frac{P_L}{I_L} = \frac{80 \text{ W}}{3 \text{ A}} = 26.67 \text{ V}$$

$$V_i = V_{nl} - V_{fl} = 40 \text{ V} - 26.67 \text{ V} = 13.33 \text{ V}$$

$$R_i = \frac{V_i}{I_L} = \frac{13.33 \text{ V}}{3 \text{ A}} = 4.44 \ \Omega$$

(b)
$$R_L = \frac{V_L}{I_L} = \frac{26.67 \text{ V}}{3 \text{ A}} = 8.89 \ \Omega$$

(c) For maximum power transfer, $R_L = R_i = 4.44 \ \Omega$.

(d)
$$\text{efficiency} = \left(\frac{P_L}{P_T}\right) 100\% = \left(\frac{P_L}{V_{nl} \times I_L}\right) 100\%$$

$$= \left(\frac{80 \text{ W}}{40 \text{ V} \times 3 \text{ A}}\right) 100\%$$

$$= 66.67\%$$

A DC power supply has a no-load voltage of 40 V and delivers 80 W at a full-load current of 3 A. Find the

(a) Internal resistance of the supply.
(b) Value of load resistance when full-load current flows.
(c) Load resistance for maximum power transfer.
(d) Efficiency of the source at full load.

Section Review

1. State the maximum power transfer theorem.
2. When the load resistance and the source resistance are equal, the circuit is said to be matched (True/False).
3. How is the efficiency of power transfer determined?

PRACTICAL APPLICATION SOLUTION

In the chapter opener, your task was to determine the output voltage and current for four batteries with different internal resistances and the same output voltages. The following steps illustrate the method of solution using Millman's theorem.

STEP 1   Determine the Millman equivalent voltage.

$$E_M = \frac{G_1E_1 + G_2E_2 + G_3E_3 + G_4E_4}{G_1 + G_2 + G_3 + G_4}$$

$$= \frac{(1/2.7)(6) + (1/2.4)(6) + (1/3.2)(6) + (1/1.4)(6)}{1/2.7 + 1/2.4 + 1/3.2 + 1/1.4} = 6 \text{ V}$$

STEP 2   Determine the Millman equivalent resistance.

$$R_M = \frac{1}{1/2.7 + 1/2.4 + 1/3.2 + 1/1.4} = 0.55 \text{ } \Omega$$

STEP 3   Calculate the output voltages using the voltage divider rule.

$$V_{RL} = \left(\frac{R_L}{R_M + R_L}\right) E_M = \left(\frac{4 \text{ } \Omega}{0.55 \text{ } \Omega + 4 \text{ } \Omega}\right) 6 \text{ V} = 5.27 \text{ V}$$

STEP 4   Determine the load current supplied by the battery pack using Ohm's law.

$$I_{RL} = \frac{V_{RL}}{R_L} = \frac{5.27 \text{ V}}{4 \text{ } \Omega} = 1.3 \text{ A}$$

Summary

1. A complex circuit is any network that combines series- and parallel-circuit elements in various interconnections.
2. Linear networks are circuits made up of resistors driven by sources of constant voltage and current.
3. Any current that flows in a closed path is called a *loop current.*
4. Nodal analysis is a technique where equations based on Kirchhoff's current law are written and unknown voltages solved for.
5. A node voltage is a voltage at a node with respect to a common reference point.
6. The use of Thévenin's theorem is limited to the identification of a voltage source in series with a resistance value.
7. Norton's theorem is the dual of Thévenin's theorem.
8. In most practical applications, the internal resistance of a voltage or current source is ignored.
9. Millman's theorem combines the source conversion theorem with both Thévenin's and Norton's theorems.
10. The maximum power transfer theorem is based on the principle that maximum power is drawn from a source when the load resistance equals the internal resistance of the source.

Section 8-1

1. To determine the response of a circuit to a given voltage or current applied to the circuit
2. True
3. A circuit constructed using resistors and driven by sources of constant voltage and current

Section 8-2

1. False
2. Any current flowing in a closed path
3. True

Section 8-3

1. A technique where equations based on Kirchhoff's current law are written and unknown voltages solved for
2. True
3. To identify the number of nodes

Section 8-4

1. Any two-terminal circuit, made up of fixed-value resistances and of voltage and current sources, can be replaced by a single voltage source in series with a single resistance, which will produce the same effects at the terminals.
2. True
3. To remove the portion of the circuit where the Thévenin equivalent circuit is to be determined

Section 8-5

1. True
2. To the resistance of a circuit if voltage sources are removed and replaced by their internal resistance values and if all current sources are open circuited
3. False

Section 8-6

1. A method of replacing a voltage source with a current source and vice versa
2. True

Section 8-7

1. False
2. No
3. Any number of constant current sources in parallel may be combined into a single current source in which the equivalent current source is the algebraic sum of the individual source currents, and the resistance of the equivalent source is the equivalent resistance of the original parallel-connected resistors.

Section 8-8

1. True
2. The voltage sources must be short circuited and the current sources open circuited.

Section 8–9

1. Maximum power is drawn from a source when the load resistance equals the internal resistance of the source.
2. True
3. A ratio of the load power to the power developed by the source

Review Questions

Multiple Choice Questions

8–1 The primary objective of circuit analysis is to determine the response of a circuit to a given _____ applied to the circuit.
(a) Resistance or voltage   **(b)** Current or voltage
(c) Current or resistance   **(d)** Power or voltage

8–2 A complex circuit is defined as any network that combines
(a) Current and voltage   **(b)** Parallel circuits
(c) Unequal resistances   **(d)** Series and parallel devices

8–3 A linear network is a circuit made up of
(a) Inductors   **(b)** Capacitors   **(c)** Resistors   **(d)** Semiconductors

8–4 Loop analysis is a method of solving circuit problems by use of
(a) Kirchhoff's voltage law   **(b)** Ohm's law
(c) Kirchhoff's current law   **(d)** Resistors

8–5 If the direction of the loop is from the negative side to the positive,
(a) Add the current in the KCL equation
(b) Subtract the current in the KCL equation
(c) Add the voltage in the KVL equation
(d) Subtract the voltage in the KVL equation

8–6 Nodal analysis is a technique where equations based on _____ are written.
(a) Ohm's law   **(b)** Kirchhoff's voltage law
(c) Nodes   **(d)** Kirchhoff's current law

8–7 The number of node voltages is always _____ than the total number of nodes.
(a) 1 less   **(b)** 2 less   **(c)** 1 more   **(d)** 2 less

8–8 The node with the largest number of connections is often called the
(a) Node voltage   **(b)** Reference node
(c) Dependent node   **(d)** Independent node

8–9 Thévenin's theorem concerns the effect of change in _____ on the current and voltage of a circuit.
(a) Temperature   **(b)** Power   **(c)** Resistance   **(d)** Efficiency

8–10 A Thévenin equivalent circuit consists of a
(a) Voltage source in series with a resistor
(b) Current source in series with a resistor
(c) Voltage source in parallel with a resistor
(d) Current source in parallel with a resistor

8–11 Norton's theorem is the _____ of Thévenin's theorem.
(a) Dual   **(b)** Equivalent   **(c)** Opposite   **(d)** All of the above

8–12 The term *looking back* refers to the resistance of a circuit if voltage sources are removed and replaced by
(a) Current sources
(b) Short circuits
(c) Open circuits
(d) Their internal resistance

8–13 When current sources are in parallel, the total current is the _____ of the individual current sources.
(a) Product   (b) Equivalent   (c) Sum   (d) None of the above

8–14 Millman's theorem is used for circuits that have
(a) Unequal resistances
(b) Only one voltage source
(c) Sources connected in series
(d) More than one voltage or current source in parallel

8–15 When using the superposition theorem to consider the effect of each source in the circuit, the
(a) Voltage sources are shorted and the current sources are open circuited
(b) Voltage sources are opened and the current sources are short circuited
(c) Current sources are shorted and the voltage sources are shorted
(d) Current sources are open circuited and the voltage sources are opened

8–16 For a circuit to be matched for maximum power transfer, the load resistance must be _____ the source resistance.
(a) Less than   (b) Greater than   (c) Equal to   (d) Proportional to

8–17 Maximum power transfer occurs in circuits with
(a) High efficiency
(b) Small value load resistors
(c) 50% efficiency
(d) Low efficiency

Practice Problems

8–18 Calculate the equivalent resistance of the circuit shown in Figure 8–53.

FIGURE 8–53

$R_1 = 600\ \Omega$ $R_3 = 1\ k\Omega$

$R_2 = 800\ \Omega$ $R_4 = 2.2\ k\Omega$

8–19 If the applied voltage in Figure 8–53 is 24 V, what would be the current through each branch?

8–20 For the circuit shown in Figure 8–54, calculate the total resistance.

FIGURE 8–54

$R_1 = 3\ \Omega$

$R_3 = 8\ \Omega$ $R_4 = 4\ \Omega$

$R_2 = 6\ \Omega$

8–21 If the total current in the circuit shown in Figure 8–54 is 3.43 A, what value of current flows in resistor R_4?

8–22 In the circuit of Figure 8–54, find the current through R_3 if the applied voltage is 24 V.

8–23 Calculate the total resistance of the circuit shown in Figure 8–55.

FIGURE 8–55

8–24 In the circuit of Figure 8–55, find the current through R_5 if the applied voltage is 120 V.

8–25 If the total current flowing in Figure 8–55 is 400 mA, what is the voltage drop across resistor R_4?

8–26 What is the total power dissipated by the circuit of Figure 8–55 if the applied voltage is 24 V?

8–27 If the total current flowing in the circuit of Figure 8–55 is 1.8 A, how much power is dissipated by resistor R_6?

8–28 Calculate the total resistance of the circuit shown in Figure 8–56.

FIGURE 8–56

8–29 In the circuit of Figure 8–56, find the current through R_3 if the applied voltage is 6 V.

8–30 If the total current flowing in Figure 8–56 is 2 mA, find the voltage drop across R_3.

8–31 Find the total power dissipated by the circuit shown in Figure 8–56 if the applied voltage is 12 V.

8–32 Solve for currents I_1 and I_2 in the circuit shown in Figure 8–57.

FIGURE 8–57

8–33 Find the voltage drops across resistors R_1, R_2, and R_3 in Figure 8–57.

8–34 Use loop analysis to solve for the three currents flowing in the circuit shown in Figure 8–58.

FIGURE 8–58

8–35 Find the total power dissipated by the circuit of Figure 8–58.

8–36 Find the value of current flowing through resistor R_5 in Figure 8–59.

FIGURE 8–59

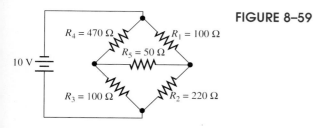

8–37 Solve for all currents in the circuit shown in Figure 8–60.

FIGURE 8–60

8–38 Use nodal analysis to find the voltage V_x for the circuit shown in Figure 8–61.

FIGURE 8–61

8–39 Solve for the five branch currents shown in Figure 8–62 using nodal analysis.

FIGURE 8–62

8–40 Solve for the currents I_1, I_2, and I_3 in the circuit of Figure 8–63.

FIGURE 8-63

8-41 Find the current through resistor R_3 in Figure 8-64 using Thévenin's theorem.

FIGURE 8-64

8-42 Find the voltage and current for load resistor R_L in the circuit shown in Figure 8-65.

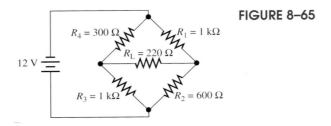

FIGURE 8-65

8-43 Using Thévenin's theorem, find the current flowing through resistor R_3 in the circuit shown in Figure 8-66.

FIGURE 8-66

8-44 For the circuit of Figure 8-67, find the current flowing through resistor R_3 by use of Thévenin's theorem.

FIGURE 8-67

8–45 Use Thévenin's theorem to find the equivalent voltage source for the circuit shown in Figure 8–68.

FIGURE 8–68

2 A ⊕   3 Ω

8–46 Use Norton's theorem to find the equivalent current source for the circuit shown in Figure 8–69.

FIGURE 8–69

4 Ω

12 V

8–47 For the circuit of Figure 8–70, calculate the load current and load voltage using Norton's theorem, if the load resistance is
(a) 47 kΩ
(b) 68 kΩ

FIGURE 8–70

$R_1 = 22$ kΩ $R_2 = 33$ kΩ

60 V $R_3 = 39$ kΩ R_L

8–48 Figure 8–71 shows a battery-generator charging circuit. Determine the load current and load voltage by using Norton's theorem.

FIGURE 8–71

$R_B = 0.5$ Ω $R_G = 0.3$ Ω

$E_B = 12.8$ V $R_L = 1.5$ Ω $E_G = 13.8$ V

8–49 Use the superposition theorem to determine the value of current flowing in resistor R_3 in the circuit shown in Figure 8–72.

FIGURE 8–72

$R_1 = 1$ kΩ $R_2 = 2.2$ kΩ

$E_1 = 9$ V $R_3 = 600$ Ω $E_2 = 12$ V

8–50 Find the values of the three currents flowing through the resistors shown in Figure 8–73 using the superposition theorem.

$R_1 = 3\ \Omega$ $R_2 = 5\ \Omega$ **FIGURE 8–73**

$E_1 = 35$ V $R_3 = 8\ \Omega$ $E_2 = 40$ V

8–51 Using superposition, determine the current through resistor R_2 in the circuit shown in Figure 8–74.

$R_1 = 100\ \Omega$ **FIGURE 8–74**

24 V $R_2 = 220\ \Omega$ 3 A

8–52 Find the values of the three currents in the circuit of Figure 8–75 using superposition. The values of the circuit components are as follows: $R_1 = 6\ \Omega$, $R_2 = 10\ \Omega$, $R_3 = 12\ \Omega$, $E_1 = 9$ V, $E_2 = 18$ V.

FIGURE 8–75

8–53 For the circuit shown in Figure 8–76, use the maximum power transfer theorem to find
(a) Ohmic value of R_L for maximum power
(b) Maximum power that can be delivered to R_L

$R_i = 2.8\ \Omega$ **FIGURE 8–76**

12 V R_L

8–54 For the circuit of Figure 8–76, determine the power for the following values of load resistance:
(a) 20 Ω
(b) 100 Ω
(c) 4 Ω

8–55   A DC power supply delivers 24 V to its output terminals when no-load is connected. When the rated full-load current of 5 A is drawn from the supply, it provides 100 W to the load. Calculate the following:

(a) The power supply's internal resistance

(b) The value of load resistance for full-load current

(c) The value of load resistance for maximum power transfer

(d) The full-load efficiency of the power supply

Essay Questions

8–56   What is circuit analysis?

8–57   Define a complex circuit.

8–58   Explain the advantage of using algebra to solve circuit problems.

8–59   What is loop analysis?

8–60   Describe the basic principle of nodal analysis.

8–61   Why is Thévenin's theorem useful when analyzing electronic circuits?

8–62   List the four steps in applying Thévenin's theorem.

8–63   What is Norton's theorem?

8–64   Explain how a voltage source is converted to a current source.

8–65   Under what circumstances would Millman's theorem be useful?

8–66   Briefly describe how superposition is applied to electronic circuits.

8–67   When is maximum power transfer essential?

9

Magnetism

Learning Objectives

Upon completion of this chapter you will be able to

- Explain Weber's theory.
- Define the term *domain*.
- Describe the principle of the magnetic field.
- List four characteristics of magnetic lines of force.
- List the three laws of magnetic attraction and repulsion.
- Name the three classifications of magnetic materials.
- Describe the field around a current-carrying conductor.
- Define the left-hand rule.
- List the three factors affecting the strength of an electromagnetic field.
- Explain how magnetic fields are used to store audio and video signals.
- Name two types of permanent magnets.
- Describe the Hall effect.

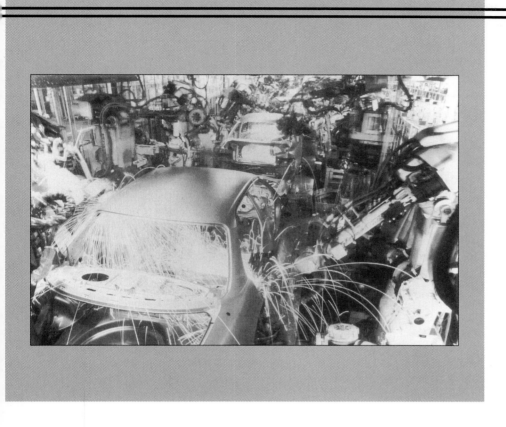

Practical Application

As a technician in a large manufacturing plant, you are required to design and install a low-voltage stop/start circuit to control a DC motor. The stop button is a momentary push button with a set of normally closed (NC) contacts, and the start button has a set of normally open (NO) contacts. Your supervisor has informed you that the stop/start station is to be controlled by a 12 V DC supply. The DC motor is to be connected to a separate 100 V supply. An electromagnetic relay with two sets of normally open contacts is to be used in this circuit. Because the start button is a momentary push button, a latching function is required to maintain current flow in the control circuit after the push button is released. The specifications for circuit operation are as follows:

1. When the start button is pressed, the control relay is energized and the two normally open contacts change state.

2. One contact provides the "seal-in", or latching, function to maintain current flow in the control circuit, and the other contact energizes the DC motor power circuit.

3. When the stop button is pressed, the normally open contacts return to their original state and power is interrupted to the DC motor.

To complete this practical application, you will draw a schematic diagram, design a latching-type relay operation, and use a low voltage to control a high voltage.

___ 9-1 _____
INTRODUCTION

The phenomenon called **magnetism** has been known and investigated for a very long period of time. At some point in the sixth century B.C. it was discovered that a lodestone would point itself in a particular direction if it were freely suspended.

If a lodestone is placed on a pile of iron filings, the filings will tend to move so that they are concentrated around two regions on the stone. These regions are called the **poles** of the stone. When the lodestone is suspended, a line drawn through the poles is found to point roughly north and south. The **magnetic pole** that points northward is called the **north pole,** and the other is referred to as the **south pole.**

Eventually, this magnetic effect was found to be caused by the presence of iron in the lodestone. More recent discoveries have shown that this property is shared by a group of materials, known as **ferromagnetic materials.** These include iron, cobalt, and nickel, as well as certain alloys. Magnets made of these materials can exhibit the effect of **ferromagnetism** to a much greater degree than the original lodestone.

Around A.D. 1600 William Gilbert undertook the first serious study of magnets. From his studies, Gilbert theorized that the earth is a huge magnet with its magnetic north pole near the geographical north pole, and the magnetic south pole located near the geographical south pole.

Section Review

1. The phenomenon called magnetism was first discovered in 1658 (True/False).
2. What was Gilbert's theory of magnetism?
3. What causes a magnetic effect in lodestone?

___ 9-2 _____
NATURE OF MAGNETISM

A popular theory of magnetism is known as **Weber's theory.** According to this theory, the molecules of magnetic material, such as iron, are tiny magnets, each with a north and south magnetic pole and with a surrounding magnetic field. All unmagnetized materials have the magnetic forces of their molecular magnets neutralized by adjacent molecular magnets, thereby eliminating any magnetic effect. In magnetic materials the molecular magnets align themselves so that their magnetic properties add to one another. They will be *lined up* so that the north pole of each molecule points in one direction and the south pole of each molecule points in the opposite direction. When all the molecules of a material are aligned in such a manner, it is said to have one effective north pole and one effective south pole. In nonmagnetic materials, the molecular magnets are in a random configuration and do not become aligned in the presence of a magnetic field.

If a magnet is split in half, Weber's theory appears to be proven correct, since each half will possess both a north and a south pole. If the magnet is again divided, there will also be one effective north pole and one south pole. The polarities of each subsequent division of the magnet will be in the same direction as the original magnet. Also, the magnetizing effect of a magnet can be nullified by any means that will disarrange the orderly array of the molecules, such as jarring or heating the magnet.

A more modern theory of magnetism is the modified version of Weber's molecular theory. Scientists now believe that a magnetic field is produced by a moving electric field. From the study of atomic structure it is known that the electrons of an atom rotate in concentric shells around the nucleus. Electrons are believed to spin on their axes, in the same way the earth turns on its axis, as they orbit around the nucleus. The phenomenon of magnetism seems to be associated with both the spinning and orbiting activities of electrons.

An electron carries a negative electrical charge. The spinning effect of the electron creates a magnetic field. The polarity of the magnetic field is determined by the direction the electron is spinning. The strength of the magnetic field of an atom depends on the number of electrons spinning in each direction. If an atom has equal numbers of electrons spinning in opposite directions, the magnetic fields surrounding the electrons cancel each other out, and the atom is unmagnetized. However, if more electrons spin in one direction than in the other, the magnetic fields do not cancel out completely, and the atom is said to be magnetized.

An atom such as iron has an atomic number of 26, meaning there are 26 protons in the nucleus and 26 revolving electrons arranged in four shells. If 13 electrons are spinning in a clockwise direction and 13 electrons are spinning in a counterclockwise direction, the opposing magnetic fields will be neutralized. However, in the case of iron, it has 14 electrons in its third shell. Of these 14 electrons, 9 spin in one direction and 5 in the other. The net result is an external magnetic field caused by the 4 electrons that do not have cancelled magnetic fields.

The atoms in a given magnetic material do not act independently but are bound together in groups. When a number of such atoms are grouped together, there is an interaction between the magnetic forces of various atoms. This interaction causes a number of neighboring atoms to line up parallel to each other in such a way that their magnetic fields aid each other. An electrostatic force, referred to as **exchange interaction,** maintains these neighboring groups in parallel even when thermally agitated. Exchange interaction is only effective against heat to a certain point. When heated above the Curie point, the atomic alignment breaks down and the material becomes nonmagnetic. For example, the Curie temperature of iron is 770°C, and when iron is heated above this temperature, it is no longer a magnetic material.

In solid materials, molecules and atoms form groups of regular geometric shapes, called **crystals.** The crystalline structures have magnetized regions, which are known as **domains.** A domain is a microscopic, needle-shaped crystal that contains a very large number of spinning electrons. The domain has a net magnetic effect oriented in a given direction and is similar to the molecular magnet postulated by Weber.

Ferromagnetic materials have a large number of domains. When the material is unmagnetized, the magnetization forces of all the domains display no overall directional characteristic. The domains in any material are always magnetized but are randomly oriented throughout the material. That is, the domains are pointing magnetically in all different directions. This random configuration results in the magnetic field of each domain being neutralized by opposing magnetic forces of other domains.

When an external field is applied to a magnetic material, the domains will line up with the external field. This occurs because increasing the external magnetic field alters the electron spin within the atoms in some of the domains. A material that has been magnetized will experience a decrease in the number of spinning electrons in one direction, while the electron spins in the other direction increase by the same amount. The magnetic strength of a magnetized material is determined by the number of domains aligned by the magnetizing force. When the magnetizing force is removed, the amount of magnetism remaining is dependent on the type of material being magnetized.

Section Review

1. What is the modern theory of magnetism?
2. Does a moving electron carry a negative electrical charge or a positive electrical charge?
3. A domain is a microscopic, needle-shaped crystal that contains a very large number of spinning electrons (True/False).

9-3

MAGNETIC FIELD

If a layer of iron filings is sprinkled over a piece of cardboard, and a magnetized strip, or bar, is laid upon it, the filings will arrange themselves in a definite pattern when they are moved by a slight tap on the cardboard. This pattern may be thought of as a map of the conditions surrounding the magnet, which are said to be because of the field of the magnet. Therefore, a **magnetic field** has the property of exerting a force on iron filings that tends to set them into a particular pattern.

Rather than use iron filings, the magnetic field surrounding a bar magnet may be investigated more accurately using a compass. A compass is actually a magnet that is mounted in such a way that it is free to rotate when affected by an outside force. If a small magnetic compass is placed near one end of a bar magnet, the force between the poles of the magnet and the poles of the compass-needle magnet will move the compass needle away from its usual north-south direction. As long as the compass needle is small in comparison with the bar magnet, the compass will point in the direction of force exerted by the bar magnet on the poles of the compass. More exactly, the compass will point in the direction of the magnetic field surrounding the bar magnet. Because the north pole of the compass will be repelled by the north pole of the bar magnet, the compass will point away from the north pole of the bar magnet. This implies that the direction of a magnetic field at any point is the direction of the force exerted on the north pole of a magnet placed at that point.

If the compass is moved steadily in the direction in which it points at any instant, it will follow a continuous path from the north pole to the south pole of the magnet. Any number of these paths can be traced out, depending on the starting point chosen for the plotting compass. The resulting pattern, which is a replica of the pattern formed by the iron filings, represents a map of the magnetic field of a magnet. Because the lines of this map represent the direction of the force exerted on the north pole of the plotting compass, they are called **lines of force.** The pattern formed by the magnetic lines of force is shown in Figure 9–1.

Some important properties of the lines of force can be deduced from the pattern of Figure 9–1. The lines are seen to emerge from the north pole of the magnet and to re-enter the magnet at the south pole. They are continuous in the sense that every line leaving the north pole will eventually arrive at the south pole. These lines of force do not begin at the

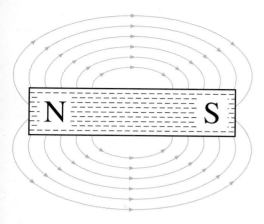

FIGURE 9–1 *Magnetic field around a bar magnet.*

north pole of a magnet and end at the south pole, but continue inside the magnet to form complete closed loops.

The characteristics of magnetic lines of force can be stated as follows:

1. Magnetic lines of force are continuous and will always form closed loops.
2. Magnetic lines of force cannot intersect each other.
3. Magnetic lines of force in a field around a magnet can be plotted by the use of iron filings.
4. Magnetic lines of force that are traveling in the same direction tend to repel each other. Parallel magnetic lines of force traveling in opposite directions tend to attract each other.

The total number of lines of force in a given region is called **magnetic flux.** Flux in a magnetic circuit corresponds to current in an electric circuit. The number of lines of force per unit area is called **flux density.** The unit of flux density is the **tesla, T.** The unit of magnetic flux is the **weber, Wb.** One weber is equal to one hundred million (10^8) lines of force. One tesla is equal to one weber per square meter (Wb/m^2). The relationship between flux density and lines of force is expressed in equation form as follows:

$$B = \frac{\Phi}{A} \tag{9.1}$$

where B = flux density, in teslas
Φ = total field flux, in webers
A = cross-sectional area, in square meters

EXAMPLE 9–1

An iron core with a cross-sectional area of $0.25\,m^2$ has a total flux of $750\,\mu Wb$. Calculate the flux density in the core material.

Solution

$$B = \frac{\Phi}{A} = \frac{750 \times 10^{-6}\ \text{Wb}}{0.25\ \text{m}^2} = 0.003\ \text{T}$$

Section Review

1. Magnetic lines of force are continuous but do not form closed loops (True/False).
2. What is the unit of flux density?
3. Express the relationship between flux density and lines of force as an equation.

9-4

MAGNETIC ATTRACTION AND REPULSION

When the north pole of one magnet is brought near the south pole of another magnet, a force of attraction is exerted between them. However, if the north pole of the magnet is brought near the north pole of another magnet, there is a force of repulsion between them. From this, the first two **laws of attraction and repulsion** are stated as

> 1. Like magnetic poles repel each other.
> 2. Unlike magnetic poles attract each other.

The third law of magnetic attraction and repulsion concerns the strength of the attraction and repulsion of the magnets based on their distance from each other. This third law is stated as

> 3. The attraction or repulsion between magnets varies directly with the product of their strengths and inversely with the square of the distance between them.

Section Review

1. Like magnetic poles attract each other (True/False).
2. State the third law of magnetic attraction and repulsion.
3. Unlike magnetic poles attract each other (True/False).

9-5

MAGNETIC MATERIALS

When a material is easy to magnetize, it is said to have a high **permeability.** Soft iron, being relatively easy to magnetize, has a high permeability, whereas steel is much harder to magnetize and has a lower permeability than iron. Materials that magnetize easily but lose this magnetism rapidly when the magnetizing influence is removed are called *temporary magnets.* When a material retains its magnetism for long periods of time, it is called a *permanent magnet.*

The ability of a material to retain its magnetism is known as *retentivity.* Soft iron has a low retentivity and hard steel has a high retentivity. The magnetism that is left in a material after the magnetizing influence has been removed is called **residual magnetism.**

Magnetic materials can be classified into one of three groups: **paramagnetic, diamagnetic,** and **ferromagnetic.**

Paramagnetic materials are those that become only slightly magnetized even though they are under the influence of a strong magnetic field. This slight magnetization is in the same direction as the magnetizing field. Materials of this type are aluminum, chromium, platinum, and air.

Diamagnetic materials exhibit a very slight opposition to magnetic lines of force. These materials are magnetized in a direction opposite to the external field that is being applied. In all known cases of materials that are diamagnetic, the diamagnetic effect is so small that very sensitive instruments are required to detect it. Examples of diamagnetic materials are copper, mercury, silicon, gold, and silver.

Materials possessing pronounced magnetic properties are said to be ferromagnetic. These are the materials used as permanent magnets and electromagnets. Ferromagnetic materials, such as iron, nickel and cobalt, are easy to magnetize and are considered to have high permeability.

Section Review

1. A material with high permeability is easy to magnetize (True/False).
2. What is retentivity?
3. Name the three classifications of magnetic materials.

9-6
FIELD AROUND A CURRENT-CARRYING CONDUCTOR

When an electric current is passed through a long straight conductor, a magnetic field is established in and around the conductor. In 1820, the Danish physicist Hans Christian Oersted discovered that when a compass is placed near a current-carrying conductor, the compass needle sets itself at right angles to the conductor. If the direction of current is reversed, the direction of the compass needle will also be reversed. If the current is switched off, the compass needle will return to its original north-south direction. If the compass needle is moved steadily in the direction in which it points at any instant, it will trace out a circular path around the conductor, with the current axis at the center of the circle. Any number of such circles can be traced out, depending on the starting radius of the plotting compass. This experiment demonstrates that the magnetic field exists in concentric circles around the conductor.

The direction of the magnetic field surrounding a conductor can be determined by what is known as the **left-hand rule.** This rule is illustrated in Figure 9–2 and is further specified by the following statement:

> If a current-carrying conductor is grasped in the left hand with the thumb pointing in the direction of current flow, the fingers will then point in the direction of the magnetic lines of force.

The symbol ⊙ is used in diagrams to denote a cross-sectional view of a conductor carrying current toward the reader, and the symbol ⊕ is used to indicate current flowing away from the reader. Figure 9–3 shows the use of these symbols.

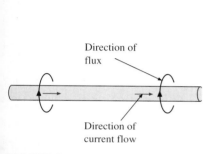

FIGURE 9–2   *Left-hand rule for determining direction of flux around a conductor.*

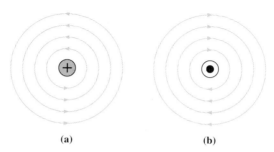

(a)   **(b)**

FIGURE 9–3   *Direction of magnetic field surrounding current in a conductor. (a) Current direction away from reader. (b) Current direction towards reader.*

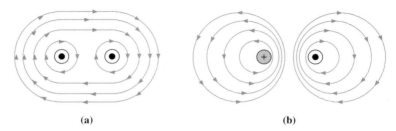

(a)   (b)

FIGURE 9–4   *Magnetic field around two parallel conductors. (a) Current flowing through two conductors in the same direction. (b) Current flowing in opposite directions.*

When two parallel conductors carry current in the same direction, the magnetic fields tend to encircle both conductors, drawing them together with a force of attraction, as shown in Figure 9–4 (a). Two parallel conductors carrying current in opposite directions are shown in Figure 9–4(b). The field around one conductor is opposite in direction to the field around the other conductor. The resulting lines of force are crowded together in the space between the conductors and tend to push the conductors apart. Therefore, two parallel, adjacent conductors carrying currents in the same direction attract each other and two parallel conductors carrying currents in opposite directions repel each other.

If the current-carrying conductor is formed into a loop, the magnetic lines of force will all pass through the center of the loop in the same direction. This is indicated approximately in Figure 9–5, where a few lines of force are sketched in as typical of those it would be possible to trace.

For any one segment of the loop, the surrounding magnetic field is much the same as it is for straight wire. This implies that the number of lines of force has not changed, even though the magnetic field is now concentrated in the smaller physical area of the conductor.

Figure 9–5 shows that the lines of force representing the magnetic field of the current in the loop leave the loop on one side and enter it at the other. If the current direction is assumed to be counterclockwise, the lines of force would appear to be coming toward the reader, when viewed from one side of the loop. If viewed from the other side of the loop, the current direction would be clockwise, and the lines of force would appear to be going away from the reader. In other words, the current loop is acting as a magnet, which has one side as a north pole and the other side as a south pole.

FIGURE 9–5   *Concentrating the magnetic field by forming the conductor into a single-turn loop.*

Direction of
current flow

Section Review

1. State the left-hand rule.
2. Draw the symbol used in diagrams to denote a cross-sectional view of a conductor carrying current towards you.
3. If a current-carrying conductor is formed into a loop, the magnetic lines of force will all pass through the center of the loop in the same direction (True/False).

9-7

MAGNETIC FIELD AROUND A COIL

The magnetic field around a current-carrying conductor exists at all points around its length. The magnetism associated with this conductor can be intensified by forming the conductor into a coil, or **solenoid.** If a current is passed through the coil, the field around each turn reinforces the field of adjacent turns. These lines of force produce a greatly strengthened magnetic field. The combined influence of all the turns produces a two-pole field similar to that of a simple bar magnet. One end of the coil will be a north pole and the other end will be a south pole, depending on current direction.

The polarity of any coil may be found by means of the left-hand rule for a coil, which may be stated in the following manner:

> Grasp the solenoid with the left hand so that the fingers follow the current direction around the circumference of the solenoid; the thumb then points in the direction of the magnetic lines of force through the solenoid.

Section Review

1. The magnetic field around a current-carrying conductor exists in all points around its length (True/False).
2. How can the magnetism associated with a conductor be intensified?
3. The polarity of a coil cannot be found by the left-hand rule (True/False).

9-8

ELECTROMAGNETS

An **electromagnet** is typically composed of a coil of wire wrapped around a an iron core. When an electric current is passed through the coil, a magnetic field is developed that is strengthened by the presence of the iron core. The magnetic field will have the same polarity regardless of whether the iron core is present or not. If the current direction is reversed through the coil, the polarity of both the coil and iron core are reversed.

The addition of the iron core accomplishes two things. First, the magnetic flux is greatly increased, because of the iron core being more permeable than air. Second, the flux is more highly concentrated than it would be if only air were used, because the iron bar has a smaller cross-sectional area than the air gap.

When a piece of soft iron is placed near the poles of an electromagnet, the iron is attracted to the magnet. This force of attraction between the magnet and iron is because of a property that tends to make flux contract and become as short as possible. Figure 9–6 shows a solenoid and an iron core placed near the coil. The lines of force are shown to extend through the soft iron and magnetize it. Since unlike poles attract, the piece of iron is drawn toward the coil.

FIGURE 9–6 *Solenoid with iron core.*

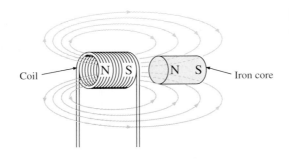

Coil — Iron core

If the bar is free to move, it will be drawn into the coil to a position near the center where the field is strongest. If a spring were attached to the iron bar, it would return to its original position when the current through the coil was interrupted. The solenoid-and-plunger type of magnet is employed extensively in electromechanical systems.

In practice, the solenoid is used for tripping circuit breakers, operating relays, and for operating contactors in automatic motor starters. In practically all cases a soft iron plunger, or armature, is necessary to obtain the tractive effort required of the solenoid.

Another popular type of electromagnetic device is the **relay.** A relay is essentially a switch that can be operated from a remote location. Relays use one or more sets of contacts to make or break control circuits. Figure 9–7 shows an example of a magnetically operated relay. When electric current passes through a coil, a magnetic north and south pole are produced across the gap separating the coil and armature. The relay is actuated, or *latched,* whenever a large enough value of current flows through the coil and the attractive force between the core and armature overcomes the spring tension, as shown in Figure 9–7(b). The relay remains latched as long as a specific value of current flows through the coil. When the current through the coil is reduced below this value, the core becomes unmagnetized and the armature is pulled up by spring action to its unactuated position.

Section Review

1. What is the basic construction of an electromagnet?
2. What three factors affect the strength of an electromagnetic field?
3. Relays use one or more sets of contacts to make or break control circuits (True/False).

FIGURE 9–7 *Electromagnetic relay. (a) De-energized. (b) Energized.*

As mentioned in Section 9–5, materials that remain magnetized for long periods of time are said to possess high retentivity. Permanent magnets are high-retentivity materials that require a strong external field to become magnetized. Once magnetized, the magnetic domains tend to stay locked in place. A common permanent magnet material is alnico. This alloy is lighter than steel and is capable of producing much stronger magnetic fields. Ceramic magnets are also classified as permanent magnets, they are lighter than metallic magnets and have resistivities as high as those of good insulators.

Permanent magnets are often used with electromagnets in audio and video devices for the recording and playback of signals. Magnetic recording tape is a type of permanent magnet consisting of a plastic film coated with ferrite oxide particles. The molecules of this oxide form domains, which comprise the smallest known permanent magnets. On an unmagnetized tape, these domains are randomly oriented over the entire surface of the tape. Each of the tiny magnetic particles on the surface of the tape produces a magnetic flux in the space surrounding the particle.

When a signal is recorded on magnetic recording tape, an electromagnet, called a *head,* is scanned along the tape. The magnetization from the record head orients the individual domains into varying degrees of direction. Once the tape has been recorded, the data on the tape can be stored indefinitely, provided the tape is not exposed to magnetic fields comparable in strength to those used in recording.

Permanent magnets are used extensively in audio loudspeakers. Figure 9–8 shows a typical loudspeaker. A field with high flux density is created in the loudspeaker by a permanent magnet made of a material such as alnico. A magnetic coil, called a *voice coil,* is held in place by an element called a *spider.* The spider is attached to the *speaker cone,* which is made of paper or cloth and is suspended from the main frame, or basket.

When current from an audio amplifier flows through the voice coil, the speaker cone vibrates and air pressure waves are created. As the current flows through the voice coil, a magnetic field with a changing polarity is developed. At one instant this field will aid the field of the permanent magnet and the cone will move inward. When the voice coil changes polarity and opposes the field of the permanent magnet, the cone will be pushed outward. This constant movement of the cone generates sound pressure waves at frequencies perceptible to the human ear.

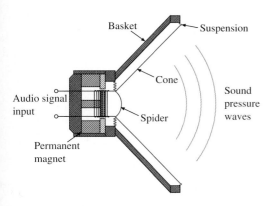

FIGURE 9–8 *Permanent magnet loudspeaker.*

Section Review

1. Permanent magnets are low-retentivity materials that require a strong external field to become magnetized (True/False).
2. What is a ceramic magnet?
3. Why are permanent magnets made from alnico instead of steel?

___ 9–10 ___

THE HALL EFFECT

In industrial electronic circuits, such as robotics, an extremely popular method for sensing angular position is the Hall effect sensor. This sensor is based on the *Hall effect,* first noted by Edwin H. Hall in 1879. Hall discovered that when a magnetic field was brought close to a gold strip carrying current, a voltage was produced. The voltage developed from this magnetic field is called a *Hall voltage.* The **Hall effect sensor,** named in his honor, detects the magnitude and polarity of a magnetic field. Semiconductor materials are now used in Hall devices instead of gold, and the Hall voltages obtained can be quite substantial.

Figure 9–9 shows the Hall effect on a semiconductor, and the relationships between the applied magnetic field, the current flowing through the semiconductor, and the voltage produced by the Hall effect. A magnetic field is applied at right angles to the semiconductor and causes charge carriers to be redistributed within the material. As a result, a Hall voltage is induced in a direction perpendicular to the current and magnetic field. The magnitude of the voltage is proportional to the product of the magnetic field and current. If the current is held constant, the Hall voltage will vary directly with the strength of the magnetic field that is perpendicular to the semiconductor material.

Section Review

1. What is Hall voltage?
2. Hall effect sensors detect the magnitude and polarity of a magnetic field (True/False).
3. The magnitude of the Hall voltage is inversely proportional to the product of the magnetic field and current (True/False).

Magnetic field

I_H I_H V_H

FIGURE 9–9 *Hall effect.*

___ 9–11 ___

PRACTICAL APPLICATION SOLUTION

In the chapter opener, your task was to design and install a low-voltage control circuit for a DC motor using two push buttons and a control relay with two sets of contacts. One contact was to provide a seal-in, or latching, function for the start button and the other contact

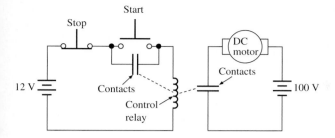

FIGURE 9–10 *Diagram of stop/start circuit with control relay.*

was to energize and de-energize the DC motor. The following steps demonstrate the method of solution for this practical application of magnetism.

STEP 1   Draw a diagram of how the circuit is to be connected. Figure 9–10 shows a diagram of the low-voltage control circuit and high-voltage power circuit. The dashed lines between the windings of the electromagnet and contacts indicate that the electromagnet controls these contacts. When the start button is pressed, the contact in parallel with the start button closes and the contact in series with the DC motor and 100 V supply also closes. The relay remains energized after the start button is released because a path for the current is provided through the parallel contact. Because the stop button is in series with the start button and holding contact, when the stop button is pressed, current flow is interrupted to the electromagnetic coil and the contacts return to their normally open state.

STEP 2   Test the circuit. Thoroughly check the wiring for the circuit and verify that the installation is complete before starting the motor.

Summary

1. A moving electron carries a negative electrical charge. The spinning effect of the electron creates a magnetic field.
2. The total number of lines of force in a given region is called *magnetic flux.*
3. The attraction or repulsion between magnets varies directly with the product of their strengths and inversely with the square of the distance between them.
4. When a material is easy to magnetize, it is said to have a *high permeability.*
5. When an electric current is passed through a long straight conductor, a magnetic field is established in and around the conductor.
6. The magnetic field around a current-carrying conductor can be intensified by forming the conductor into a coil, or solenoid.
7. Solenoids are electromagnetic devices with a moveable iron core.
8. Relays are electromagnetic devices that use one or more sets of contacts to make or break control circuits.
9. Permanent magnets are often used with electromagnets in audio and video devices for the recording and playback of signals.
10. The Hall effect is based on the Hall voltage, which is a voltage developed across a magnetic field.

Answers to Section Reviews

Section 9–1

1. False
2. That the earth is a huge magnet with its magnetic north pole near the geographical north pole and the magnetic south pole located near the geographical south pole
3. The presence of iron

Section 9–2

1. That it is produced by a moving electric field, and the phenomenon of magnetism is associated with both the spinning and orbiting activities of electrons.
2. Negative electrical charge
3. True

Section 9–3

1. False
2. The weber
3. $B = \Phi/A$

Section 9–4

1. False
2. A law that states the attraction or repulsion between magnets varies directly with the product of their strengths and inversely with the square of the distance between them
3. True

Section 9–5

1. True
2. The ability of a material to retain its magnetism
3. Paramagnetic, diamagnetic, and ferromagnetic

Section 9–6

1. If a current-carrying conductor is grasped in the left hand with the thumb pointing in the direction of current flow, the fingers will then point in the direction of the magnetic lines of force.
2. $\odot$
3. True

Section 9–7

1. True
2. By forming the conductor into a coil, or solenoid
3. False

Section 9–8

1. A coil of wire wrapped around an iron core
2. The amount of current in the coil, the number of turns of the coil, and the size and material of the core
3. True

Section 9–9

1. True
2. A type of permanent magnet that is lighter than metallic magnets and has a resistivity as high as a good insulator
3. Alnico is lighter than steel and is capable of producing much stronger magnetic fields.

Section 9–10

1. A voltage developed when a magnetic field is brought close to a current-carrying semiconductor
2. True
3. False

Review Questions

Multiple Choice Questions

9–1 The magnetic effect of lodestone is caused by
(a) Copper **(b)** Iron **(c)** Cobalt **(d)** Nickel

9–2 The first serious study of magnetism was undertaken in 1600 by
(a) Gilbert **(b)** Oersted **(c)** Weber **(d)** Tesla

9–3 According to Weber's theory, the molecules of magnetic material are
(a) Spinning electrons **(b)** Caused by exchange interaction
(c) Tiny magnets **(d)** Called domains

9–4 The spinning effect of an electron in a magnetic material creates a
(a) Negative electrical charge **(b)** Positive electrical charge
(c) Neutral charge **(d)** Magnetic field

9–5 In solid materials, molecules and atoms form groups of regular geometric shapes, called
(a) Domains **(b)** Crystals **(c)** Lines of force **(d)** Webers

9–6 Ferromagnetic materials have
(a) A small number of domains **(b)** Poor magnetizing ability
(c) A large number of domains **(d)** A small number of molecular magnets

9–7 The total number of lines of force is called
(a) Magnetic flux **(b)** Flux density **(c)** Domains **(d)** Tesla

9–8 The first two laws of attraction and repulsion are unlike poles
(a) Repel each other, like poles repel **(b)** Attract each other, like poles attract
(c) Repel each other, like poles attract **(d)** Attract each other, like poles repel

9–9 When a material is easy to magnetize, it is said to have a
(a) Low permeability **(b)** High permeability
(c) Low retentivity **(d)** High residual magnetism

9–10 Materials possessing pronounced magnetic properties are said to be
(a) Paramagnetic **(b)** Diamagnetic
(c) Ferromagnetic **(d)** Temporary magnets

9–11 When a material is difficult to magnetize, but retains the magnetism for long periods of time, it
(a) Has low retentivity **(b)** Is a temporary magnet
(c) Has low residual magnetism **(d)** Is a permanent magnet

9–12 Materials that are magnetized in a direction opposite to the applied external field are called
(a) Paramagnetic **(b)** Diamagnetic
(c) Ferromagnetic **(d)** Permanent magnets

9–13 In 1820, _____ discovered that when a compass is placed near a current-carrying conductor, the compass needle sets itself at right angles to the conductor.
(a) Oersted **(b)** Tesla **(c)** Gilbert **(d)** Hall

9–14 In electromagnetic circuits, iron cores are used to
(a) Decrease flux **(b)** Develop a specific polarity across the coil
(c) Maintain a constant polarity **(d)** Increase flux

9–15 Magnetic recording tape is a type of
(a) Temporary magnet **(b)** Permanent magnet
(c) Electromagnet **(d)** Paramagnetic material

9–16 The Hall effect sensors
(a) Detects magnitude and polarity of an electric field
(b) Measures flux density
(c) Detects magnitude and polarity of a magnetic field
(d) Measures total field flux

Practice Problems

9–17 An iron core with a cross-sectional area of 0.01 m^2, has a total flux of 300 μWb. Determine the flux density in the core material.

9–18 With a flux of 200×10^{-4} Wb through an area of 75 mm × 30 mm, what is the flux density?

9–19 Determine the amount of flux required to provide a flux density of 220×10^{-6} Wb/m^2 in a cross-sectional area measuring 0.0025 m × 0.00317 m.

9–20 If the flux density in an iron core is 3.35 T and the material has a cross-sectional area of 0.25 in.^{2}, find the flux through the core.

9–21 Calculate the flux density of a certain magnetic material having a cross-sectional area of 0.155 in.^{2} and a total flux of 275 μWb.

9–22 A core has a flux density of 7.2×10^3 Wb/m^2. If the flux is 26×10^2 Wb, find the cross-sectional area.

Essay Questions

9–23 What are ferromagnetic materials?

9–24 Explain Weber's theory.

9–25 Describe the basic difference between Weber's theory and the modern theory of magnetism.

9–26 What is a domain?

9–27 How is a compass used to demonstrate the magnetic field around a magnet?

9–28 Explain what is meant by the term *lines of force*.

9–29 List four characteristics of magnetic lines of force.

9–30 Describe the relationship between magnetic flux and flux density.

9–31 List the three laws of magnetic attraction and repulsion.

9–32 What is permeability?

9–33 How can it be proved that a magnetic field exists around a current-carrying conductor?

9–34 Explain the left-hand rule.

9–35 What is an electromagnet?

10

The Magnetic Circuit

Learning Objectives

Upon completion of this chapter you will be able to

- Define magnetomotive force.
- Express magnetic reluctance in terms of magnetomotive force and magnetic flux.
- Define field intensity.
- Understand the permeability curves of common magnetic materials.
- Describe the magnetic properties of some common materials.
- Define magnetic hysteresis and residual magnetism.
- Express Ampere's circuit law.
- Describe the effect of air gaps in a magnetic circuit.

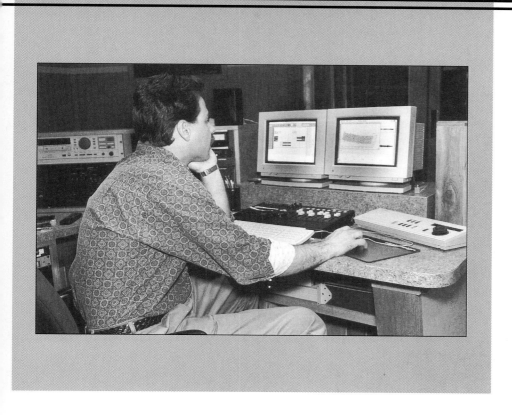

Practical Application

As a technician employed by a company involved in the manufacture of audio and video equipment, you are asked to assist in the design of a new audio-recording head. Your task is to determine the amount of current flowing in the windings to produce a flux of 0.3 mWb. The specifications for the recording head are as follows:

1. The core is made of sheet steel with a cross section of 2 cm × 2 cm and an average path length of 20 cm.
2. The air gap is 12.5 μm.
3. The magnetizing coil has 1200 turns.

INTRODUCTION

The magnetic circuit and the electric circuit have a great number of similarities. The electric circuit makes it possible to calculate voltage, resistance, and current, provided sufficient data are given. Likewise, with the magnetic circuit, it is possible to calculate magnetic quantities such as magnetomotive force, flux, and reluctance, provided sufficient data are given. For example, **magnetic flux** is equal to the magnetomotive force divided by the reluctance.

In the magnetic circuit, after the flux is established, no heat is produced in the circuit itself and, therefore, no energy is used in retaining this magnetism, provided the electric current in the coil is steady. All the power used in supplying current to the coil can be calculated as I^2R losses in the coil itself. Because no energy is used in the magnetic circuit, it indicates that there is no movement of the flux or flux lines.

Table 10–1 lists magnetic quantities and units with SI symbols. Although other systems of measurement, such as English and CGS, are still used occasionally, the SI system is considered the standard.

Section Review

1. Three quantities that can be calculated in a magnetic circuit are magnetomotive force, flux, and reluctance (True/False).
2. Without referring to Table 10–1, list four magnetic quantities.
3. Without referring to Table 10–1, name four SI magnetic units.

10-2

MAGNETOMOTIVE FORCE

The amount of flux developed in a solenoid is dependent on the current, I, and the number of turns, N. The product of I and N in a magnetic circuit is called **magnetomotive force (mmf)** and is a measure of the ability of a coil to produce flux. Mmf determines the degree to which electromagnetic effects are produced. Flux is the *current* of a **magnetic circuit,** and the magnetomotive force is the force that establishes it. Therefore, mmf is analogous to the emf of the electric circuit. Because the mmf developed is proportional to the current and to the number of turns in the coil, the unit of measurement is the ampere-turn (At) and is found by

TABLE 10–1 *SI units and symbols for magnetic quantities.*

Quantity	SI Symbol	SI Unit	SI Symbol
Flux	Φ	Weber	Wb
Flux density	B	Tesla	T, or Wb/m^2
MMF	$\mathscr{F}$	Ampere-turn	At
Reluctance	$\mathscr{R}$	Ampere-turns/Weber	At/Wb
Field Intensity	H	Ampere-turns/meter	At/m

$$\mathscr{F} = I \times N \qquad\qquad (10.1)$$

where $\mathscr{F}$ = magnetomotive force, in ampere-turns
  I = current through a coil, in amperes
  N = number of turns of a coil

EXAMPLE 10–1

Solution

Determine the magneto-motive force produced by a coil with 1500 turns and a 20 mA current.

$$\mathscr{F} = I \times N$$
$$= (20 \times 10^{-3})\,(1500)$$
$$= 30 \text{ At}$$

EXAMPLE 10–2

Solution

How much current is nec-essary for a 300-turn coil to produce a magnetizing force of 800 At?

$$I = \frac{\mathscr{F}}{N}$$
$$= \frac{800 \text{ At}}{300}$$
$$= 2.67 \text{ A}$$

Section Review

1. What magnetic quantity is a measure of the ability of a coil to produce flux?
2. What magnetomotive force is produced by a coil with 1000 turns and 50 mA of current?
3. Magnetomotive force is used to establish flux in a magnetic circuit (True/False).

10–3
RELUCTANCE

The opposition that a magnetic path offers to magnetic flux when a magnetomotive force is applied is called **reluctance.** The symbol for reluctance is the script letter R, and the SI unit of measurement is the ampere-turns/weber (At/Wb). The relationship between magneto-motive force, magnetic flux, and reluctance may be expressed by the following equation:

$$\mathscr{R} = \frac{\mathscr{F}}{\Phi} \qquad\qquad (10.2)$$

where $\mathscr{R}$ = reluctance, in ampere-turns/weber
  $\mathscr{F}$ = magnetomotive force, in ampere-turns
  Φ = magnetic flux, in webers

_____ EXAMPLE 10–3 _____

The magnetic flux produced by a 100-turn coil is 35 mWb. If a current of 300 mA flows in the coil, what is the reluctance?

Solution

$$\mathcal{R} = \frac{\mathcal{F}}{\Phi} = \frac{NI}{\Phi}$$

$$= \frac{(100)\,(300 \times 10^{-3})}{35 \times 10^{-3}}$$

$$= 857.14 \text{ At/Wb}$$

_____ EXAMPLE 10–4 _____

A coil has a magnetomotive force of 220 At and and reluctance of 8500 At/Wb. Determine the total flux of the magnetic circuit.

Solution

$$\Phi = \frac{\mathcal{F}}{\mathcal{R}}$$

$$= \frac{220 \text{ At}}{8500 \text{ At/Wb}}$$

$$= 25.88 \text{ mWb}$$

Section Review

1. What is reluctance in a magnetic circuit?
2. State the relationship between mmf, magnetic flux, and reluctance in equation form.
3. If the magnetic flux produced by a 500-turn coil is 15 mWb and a current of 10 mA flows in the coil, what is the reluctance?

_____ 10–4 _____
FIELD INTENSITY

Magnetizing force is an important consideration in a magnetic circuit, as it gives a measure of the magnetic stress imposed on a given core material for its total length. In the SI system, magnetizing force is in ampere-turns per meter (At/m). **Magnetic field intensity,** or magnetizing force, is a measurement of the mmf needed to establish a certain flux density in a unit length of the magnetic circuit. Therefore, the magnetomotive force per unit length is the **field intensity,** H, and is expressed in mathematical form as

$$H = \frac{\mathcal{F}}{l} \qquad\qquad (10.3)$$

where H = magnetic field strength, in ampere-turns/meter
$\mathcal{F}$ = applied mmf, in ampere-turns
l = average length of magnetic path, in meters

EXAMPLE 10–5

Solution

$$H = \frac{\mathscr{F}}{l} = \frac{IN}{l}$$

$$= \frac{(500 \times 10^{-3})\,(1000)}{12 \times 10^{-2}}$$

$$= 4166.67 \text{ At/m}$$

A current of 500 mA is flowing in a coil with 1000 turns. The coil is circular and has an average length of 12 cm. Determine the field intensity of the resulting magnetic field.

EXAMPLE 10–6

Solution

$$H = \frac{IN}{l}$$

$$Hl = IN$$

and

$$I = \frac{Hl}{N}$$

$$= \frac{(2000)\,(25 \times 10^{-2})}{300}$$

$$= 1.67 \text{ A}$$

A 300-turn coil is wound on a form. The core length is 25 cm. Calculate the amount of current required to produce a field intensity of 2000 At/m.

Section Review

1. What is magnetic field intensity?
2. State the relationship between field strength, mmf, and average length of magnetic path in equation form.
3. What is another term for the magnetic force per unit length of a magnetic circuit?

10–5

PERMEANCE

The **permeance** of a circuit is the reciprocal of the reluctance and may be defined as that property of the circuit that permits the passage of magnetic flux, or lines of induction. Consequently, in a magnetic circuit in which the paths of the lines of force are almost all through iron, the permeance of the circuit is relatively high compared with the permeance of air, or other nonmagnetic material. Permeance corresponds to conductance in the electric circuit. The SI unit for permeance is the **henry, H,** named after the American physicist Joseph Henry (1797–1878).

Section Review

1. Permeance is the reciprocal of reluctance (True/False).
2. What is the SI unit for permeance?
3. The permeance of iron is quite low compared to the permeance of air (True/False).

PERMEABILITY

The ability of a material to concentrate magnetic flux is called **permeability, μ.** The permeability of any material is a measure of the ease with which its atoms can be aligned or the ease with which it can establish lines of force. Permeability is measured in henrys per meter (H/m). Numerical values of μ for different materials are assigned by comparing their permeability with the permeability of air, or a vacuum. The μ of nonmagnetic materials such as air, copper, wood, and plastic is for all practical purposes equal to unity (1). Magnetic materials such as iron, nickel, steel, and their alloys have a value much greater than unity. For example, soft iron has a permeability several-hundred times that of air. Figure 10–1 shows a graph of four common magnetic materials and their permeability curves.

Space permeability (μ_0) is the permeability of air, or a vacuum, and is the standard reference value. It is also referred to as the absolute, or free space, permeability. The permeability of free space is $4\pi \times 10^{-7}$ H/m, or 1.26×10^{-6} H/m.

The permeability of any material can be stated as the ratio of the magnetic flux density to the magnetic field intensity of a material. Expressed as an equation,

$$\mu = \frac{B}{H}$$

(10.4)

where   μ = permeability of a material, in henrys/meter
H = magnetic field strength, in ampere-turns/meter
B = flux density, in teslas

The **relative permeability, μ_r,** of a material is the ratio of the absolute permeability μ to the permeability that would exist if the material were replaced with air, with the

FIGURE 10–1 *Permeability curves of common ferromagnetic materials.*

mmf on the material remaining unchanged. Relative permeability is defined by the equation

$$\mu_r = \frac{\text{number of lines produced with the material as a core}}{\text{number of lines produced with air as a core } (4\pi \times 10^{-7})} \qquad \textbf{(10.5)}$$

$$= \frac{\mu}{\mu_0}$$

EXAMPLE 10–7

Determine the absolute value of μ for a magnetic material whose μ_r is 750.

Solution

$$\mu_r = \frac{\mu}{\mu_0}$$

Therefore,

$$\mu = \mu_r \, \mu_0$$
$$= (750) \, (1.26 \times 10^{-6})$$
$$= 0.945 \times 10^{-3} \, \text{T/(At/m)}$$

EXAMPLE 10–8

Using the value μ obtained in Example 10–7, determine the flux density that results when the field intensity is 275 At/m.

Solution

$$\mu = 0.945 \times 10^{-3} \quad \text{(from Example 10–7)}$$

$$\mu = \frac{B}{H}$$

Therefore,

$$B = \mu H = (0.945 \times 10^{-3}) \, (275) = 0.26 \, \text{T}$$

Section Review

1. What is permeability?
2. What is permeability measured in?
3. What is the relative permeability of a nonmagnetic material such as wood?

10–7

MAGNETIC PROPERTIES OF MATERIALS

In Section 10–6 we stated that for nonmagnetic materials the permeability is unity and that for magnetic materials the permeability is dependent on the flux density. Figure 10–2 shows the typical **magnetization,** or **B–H, curves** for sheet steel, cast steel, and cast iron.

These curves, which must be determined experimentally, show the manner in which the flux density varies with the magnetizing force. In this *B–H* graph, the slope of the curve is the change in *B* divided by the corresponding change in *H*, which is equal to the permeability of the material. Eventually, the curve reaches a point where increasing the field intensity has virtually no effect on the flux density. At this point the material is said to be *saturated.* The term **magnetic saturation** means simply that to produce an increase in the flux density, an extremely large increase in magnetizing force is necessary.

Since permeability is defined as $\mu = B/H$, the permeability can only be constant if the magnetization curve is in a straight line. In other words, the steeper the slope of the magnetization curve, the greater the permeability.

FIGURE 10–3
Manner in which
the permeability of
a magnetic
material varies with
flux density.

FIGURE 10–2 *Magnetization and relative
permeability curves.*

For magnetic materials, the permeability varies with the flux density, as indicated in Figure 10–3. Starting from a value of $B = 0$, the permeability first peaks and then falls off fairly rapidly, as shown.

Section Review

1. What is a magnetization curve?
2. What is magnetic saturation?
3. For magnetic materials, the permeability varies with the flux density (True/False).

___ 10–8 ___

MAGNETIC HYSTERESIS, RESIDUAL MAGNETISM

In Figure 10–2, it was demonstrated that the flux density increases when the magnetizing force increases, even when the curve is carried to a point beyond saturation. When the magnetizing force is decreased, the flux density will not decrease along the same line as it increased. The reason for this shall be explained using the curves shown in Figure 10–4.

Experiment proves that iron, once magnetized, tends to retain its magnetism. Permanent magnets are all dependent on this property. If a ferromagnetic material is magnetized to a value represented in Figure 10–4 by OA, the core is said to become saturated along the normal magnetization curve. By the time the magnetizing force has fallen to zero, the flux density has only been reduced to AR, which is called the **residual magnetism,** or **residual flux density.** The residual flux density is the amount of flux density remaining in the material after the magnetizing force has been removed.

To reduce the flux to zero, the **coercive force,** $-H$, must be applied, tending to establish magnetic flux in the opposite direction. Coercive force is defined as the demagnetizing force necessary to remove the residual flux from the material.

If starting from the condition represented by the point Q, the magnetizing force is again gradually reversed, the magnetizing curve will be as shown by QUA. The curve for

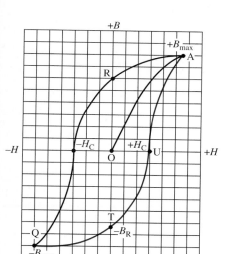

FIGURE 10-4 *Hysteresis curves.*

increasing magnetization always lies below that for decreasing magnetization. The flux density *B*, therefore, is said to lag behind the force *H*. This lagging is called **hysteresis.** Not only does magnetic hysteresis introduce a degree of uncertainty into the calculation of magnetic circuits but also, because of hysteresis, energy is wasted when the iron is subjected to alternating magnetism.

Section Review

1. What is residual flux density?
2. Coercive force is the demagnetizing force necessary to remove the residual flux from the material (True/False).
3. Hysteresis represents an energy loss in magnetic circuits (True/False).

_____ **10-9** ___

AMPERE'S CIRCUIT LAW

Ampere's circuit law states that the algebraic sum of the rises and drops of the mmf around a closed loop of a magnetic circuit is equal to zero. This law can also be expressed as the sum of the mmf rises are equal to the sum of the mmf drops.

When Ampere's circuit law is applied to magnetic circuits, sources of mmf are expressed by the following equations:

$$\mathscr{F} = NI$$

where $\mathscr{F}$ = mmf, in ampere-turns
N = the number of turns
I = the current, in amperes

$$\mathscr{F} = \Phi \mathscr{R}$$

where $\mathscr{F}$ = mmf, in ampere-turns,
Φ = the flux passing through a section of the magnetic circuit, in webers
$\mathscr{R}$ = the reluctance of that section, in ampere-turns/weber

$$\mathscr{F} = Hl$$

where $\mathscr{F}$ = mmf, in ampere-turns
$\quad\quad$ H = the magnetizing force on a section of a magnetic circuit, in ampere-turns/meter
$\quad\quad$ l = the length of the section, in meters

Section Review

1. State Ampere's circuit law.
2. Ampere's circuit law is used to find current in a magnetic circuit (True/False).

___ 10–10 ___
AIR GAPS

In a magnetic circuit, it is quite common for the magnetic core to not be continuous. This may be because of an **air gap** in the circuit. Air gaps are often deliberately placed in magnetic circuits to increase reluctance. By increasing the total reluctance, saturation of the iron core may be prevented, allowing a larger current flow through the coil wound on the core. In magnetic circuits the air gap may consist of a sheet of nonmagnetic material, such as fibreboard, which allows the length of the air gap to be accurately set.

When an air gap exists, as in Figure 10–5, the lines of flux passing through the gap are not limited to the projected area of the iron core. The spreading of the flux lines outside the area of the core for the air gap is known as **fringing.** The flux area increases due to fringing, resulting in a decrease in flux density. Therefore, the flux density in the air gap is slightly less than that in the iron sections of the magnetic circuit.

If the length of the air gap is small, the resulting fringing is also small. The correcting factor that takes fringing into account is to add the length of the gap to each cross-section dimension of the adjoining material. For example, if a core 7 cm × 3 cm has a 0.17 cm air gap, the air gap's effective area is A = (7.17) × (3.17) = 22.73 cm². For our purposes, we shall neglect fringing unless otherwise noted.

The flux density of the air gap in Figure 10–5 is defined by the following equation:

$$B_g = \frac{\Phi_g}{A_g} \tag{10.6}$$

where $\Phi_g = \Phi_{core}$
$\quad\quad$ $A_g = A_{core}$

FIGURE 10–5 *Fringing flux at an air gap.*

The magnetizing force of the air gap is determined in the following manner:

$$H_g = \frac{B_g}{\mu_0} \qquad (10.7)$$

where μ_0 is the permeability free space and has a value of $4\pi \times 10^{-7}$ H/m.

The magnetizing force of the air gap is equal to

$$H_g = (7.97 \times 10^5)\, B_g$$

A certain number of ampere-turns are required to send flux through a ferromagnetic material, and additional ampere-turns are required to send flux through the air gap of a magnetic circuit. Because the paths are in series, the total ampere-turns needed is the sum of the two.

Because flux density is the product of flux divided by area, flux density may also be expressed by the following equation:

$$B = \left(\frac{\mathscr{F}}{l}\right)\mu \qquad (10.8)$$

where B = flux density, in teslas
    $\mathscr{F}$ = mmf, in ampere-turns
    l = length of the material, in meters
    μ = permeability of the material

Solving for magnetomotive force, the equation may be rewritten as

$$\mathscr{F} = \frac{Bl}{\mu} \qquad (10.9)$$

Since the permeability for air is constant at $4\pi \times 10^{-7}$, the equation for ampere-turns in an air gap would be

$$\mathscr{F} = \frac{Bl}{4\pi \times 10^{-7}}$$

Consumer electronic products such as video cassette recorders (VCRs) and audio tape systems use air gaps for recording and playing back audio and video signals. Figure 10–6 shows a typical magnetic head with an air gap, or head gap, and windings. The air-gap length determines the frequency response of the recorded signal. Generally, the smaller the head gap, the higher the upper frequency limit. The gap depth determines the head efficiency and will influence the lifetime of the head.

FIGURE 10–6 *Magnetic head with air gap.*

Gap depth

Laminations

Windings

Gap length →| |←

FIGURE 10–7 *Magnetic tape recording head.*

Figure 10–7 shows a magnetic recording head used to store input signals on magnetic tape. The power supply provides current to the coils of wire that are wrapped around the head's magnetic pole pieces, resulting in a magnetic force that flows through the pole pieces and across the head gap. When a magnetic head is used for recording signals, the head gap causes a break in the magnetic flux and creates a reluctance to the magnetic force that has been set up. When this head comes in contact with magnetic tape, the tape's magnetic oxide offers a lower reluctance path to the flux than the air gap, and the flux path travels from one pole piece, through the tape, to the other pole piece.

In Figure 10–7, the stray flux that fringes the sides of the front gap tends to enlarge the gap cross-sectional area. Consequently, the reluctance of the gap is lowered and the head efficiency decreases.

Section Review

1. Why is it desirable to increase the total reluctance of a magnetic circuit?
2. What is fringing?
3. If the length of the air gap is small, the resulting fringing is large (True/False).
4. In audio and video recorders, what does the air gap length determine?

___ 10–11 ___
SERIES-MAGNETIC CIRCUIT

The method of calculating the magnetomotive force required to establish a given flux through a magnetic circuit consisting of a number of parts in series is very similar to the corresponding problem for a series-electric circuit. That is, the total reluctance of the path is simply the sum of the values for all the individual parts. However, the most convenient method of dealing with the problem is to calculate separately the actual magnetomotive force absorbed in establishing the flux in each part of the circuit and, then, add these values.

EXAMPLE 10–9

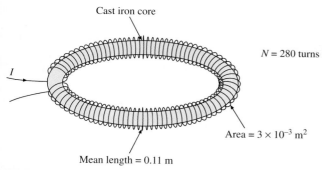

N = 280 turns

Cast iron core

I

Area = 3 × 10⁻³ m²

Mean length = 0.11 m

FIGURE 10–8 *Series-magnetic circuit for Example 10–9.*

For the series-magnetic circuit of Figure 10–8, find the

(a) Value of current required to develop a magnetic flux of $\Phi = 7 \times 10^{-4}$ Wb.
(b) Permeability of the material under these conditions.

Solution

(a) The flux density *B* is

$$B = \frac{\Phi}{A} = \frac{7 \times 10^{-4}}{3 \times 10^{-3}} = 0.233 \text{ T}$$

Using the *B–H* curves of Figure 10–2, we can determine the magnetizing force *H*.

$$H \text{ (cast iron)} = 600 \text{ At/m}$$

From Ampere's circuit laws, the following equation is derived.

$$NI = Hl$$

$$I = \frac{Hl}{N} = \frac{(600)\,(0.11)}{280} = 235.71 \text{ mA}$$

(b) The permeability of the material can be found in the following manner:

$$\mu = \frac{B}{H} = \frac{0.233}{600} = 3.88 \times 10^{-4} \text{ H/m}$$

A coil wound on a circular core, as in Example 10–9, is called a **toroid.** A toroidally wound coil has efficient magnetic characteristics. Because its core is endless and entirely within the coil winding, it concentrates the magnetic field to the maximum degree.

Section Review

1. Why are toroidal coils popular in certain applications?
2. The total reluctance of a series-magnetic circuit is the sum of the individual values of reluctance (True/False).

PRACTICAL APPLICATION SOLUTION

In the chapter opener your task was to determine the amount of current flow required to produce a flux of 0.3 mWb in a sheet steel core with 1200 turns and a 12.5 μm air gap. The following steps illustrate the method of solution for this practical application of the magnetic circuit.

STEP 1   Determine the magnetic flux of the air gap.

$$A_g = (20 \times 10^{-3})^2 = 0.4 \times 10^{-3} \text{ m}^2$$

$$B_g = \frac{\Phi}{A} = \frac{0.3 \times 10^{-3}}{0.4 \times 10^{-3}} = 0.75 \text{ T}$$

$$H = \frac{B_g}{\mu_0} = \frac{0.75}{4\pi \times 10^{-7}} = 596 \times 10^3 \text{ At/m}$$

$$(Hl)_g = (596 \times 10^3)(12.5 \times 10^{-6}) = 7.5 \text{ At}$$

STEP 2   Determine the flux density of the steel. Because the air gap is relatively small, the area of the air gap and the area of the steel are approximately equal. Thus,

$$B_g = B = \frac{\Phi}{A} = \frac{0.3 \times 10^{-3}}{0.4 \times 10^{-3}} = 0.75 \text{ T}$$

STEP 3   Use the graph of the B–H curves (Figure 10–2) to determine the magnetizing force of the sheet steel core and solve for the magnetomotive force (mmf).

$$H = 125 \text{ At/m}$$

$$Hl = (125)(0.2) = 25 \text{ At}$$

$$\mathscr{F} = (Hl)_g + (Hl) = 7.5 \text{ At} + 25 \text{ At} = 32.5 \text{ At}$$

Therefore,

$$I = \frac{\mathscr{F}}{N} = \frac{32.5}{1200} = 27 \text{ mA}$$

_ Summary

1. The amount of flux developed in a solenoid is dependent on the current and the number of turns.
2. The opposition that a magnetic path offers to magnetic flux when a magnetomotive force is applied is called *reluctance.*
3. The permeance of a circuit is the reciprocal of reluctance and represents the property of a circuit that permits the passage of magnetic flux.
4. The ability of a material to concentrate magnetic flux is called *permeability.*
5. Permeability of a material represents a ratio of the magnetic flux density to the magnetic field intensity of a material.
6. For magnetic materials, the permeability is dependent on the flux density.
7. The flux density that lags behind the magnetizing force is known as *hysteresis.*
8. The algebraic sum of the rises and drops of the mmf around a closed loop of a magnetic circuit is equal to zero.

9. Air gaps are often inserted in magnetic circuits to increase reluctance.
10. The spreading of the flux lines outside the area of the air gap core is called *fringing*.

<hr>

Answers to Section Reviews

Section 10–1

1. True
2. Flux, flux density, mmf, field intensity
3. Weber, tesla, ampere-turn, and ampere-turn/meter

Section 10–2

1. Magnetomotive force (mmf)
2. $\mathscr{F} = IN = (50 \text{ mA})(1000) = 50 \text{ At}$
3. True

Section 10–3

1. The opposition that a magnetic path offers to magnetic flux when a magnetomotive force is applied
2. $\mathscr{R} = \mathscr{F}/\Phi$
3. $\mathscr{R} = NI/\Phi = [(500)(10 \text{ mA})]/15 \text{ mWb} = 333.33 \text{ At/Wb}$

Section 10–4

1. A measurement of the magnetomotive force needed to establish a certain flux density in a unit length of the magnetic circuit
2. $H = \mathscr{F}/l$
3. Field intensity

Section 10–5

1. True
2. The henry (H)
3. False

Section 10–6

1. The ability of a material to concentrate magnetic flux
2. Permeability is measured in henrys per meter (H/m).
3. Unity (1)

Section 10–7

1. A curve showing the relationship between field intensity and flux density
2. The point where increasing the field intensity has virtually no effect on the flux density
3. True

Section 10–8

1. The amount of flux density remaining in the material after the magnetizing force has been removed
2. True
3. True

Section 10–9

1. The algebraic sum of the rises and drops of the mmf around a closed loop of a magnetic circuit is equal to zero.
2. False

Section 10–10

1. Because the saturation of the an iron core may be prevented, allowing a larger value of current to flow through the coil wound on the core
2. The spreading of the flux lines outside the area of the core for the air gap
3. False
4. The frequency response of the recorded signal

Section 10–11

1. Because the core is endless and entirely within the coil winding, so it concentrates the magnetic field to the maximum degree
2. True

Review Questions

Multiple Choice Questions

10–1   Magnetic flux can be measured in
(a) Ampere-turns   (b) Webers   (c) Teslas   (d) Ampere-turns/meter

10–2   The product of the current and number of turns is called
(a) mmf   (b) Field intensity   (c) Flux density   (d) Weber

10–3   The opposition that a magnetic path offers to magnetic flux when a magnetomotive force is applied is called
(a) Field intensity   (b) Flux density   (c) Reluctance   (d) Permeance

10–4   The magnetic force per unit length is defined as
(a) Reluctance   (b) Permeability   (c) Flux density   (d) Field intensity

10–5   The reciprocal of reluctance is
(a) Permeance   (b) Permeability   (c) Flux   (d) Hysteresis

10–6   The ability of a material to concentrate magnetic flux is called
(a) Permeance   (b) Flux density   (c) Reluctance   (d) Permeability

10–7   The flux density of a magnetic circuit increases when the
(a) Magnetizing force increases   (b) Magnetizing force decreases
(c) Coercive force decreases   (d) Hysteresis increases

10–8   Lagging flux density is also called
(a) Residual magnetism   (b) Permeance
(c) Hysteresis   (d) Coercive force

10–9   Air gaps are used in magnetic circuits to
(a) Increase hysteresis   (b) Increase reluctance
(c) Decrease reluctance   (d) Increase fringing

10–10   The spreading of the flux lines outside the area of the core for the air gap is called
(a) Saturation   (b) Flux density
(c) Hysteresis   (d) Fringing

10–11   The total reluctance of a series-magnetic circuit is equal to the
(a) Product of individual reluctances   (b) Sum of individual reluctances
(c) Square root of individual reluctances   (d) None of the above

10–12   A coil wound on a circular core is called
(a) An air gap   (b) A circular magnet
(c) A toroid   (d) A permanent magnet

10–13   Magnetic circuits are deliberately designed with air gaps in order to
(a) Increase reluctance   (b) Increase hysteresis
(c) Reduce leakage flux   (d) Increase leakage flux

10–14   When the length of an air gap increases, the
(a) Total reluctance increases   (b) Flux density decreases
(c) Fringing decreases   (d) Total reluctance decreases

10–15   When the length of an air gap decreases, the
(a) Total reluctance increases   (b) Flux density decreases
(c) Fringing decreases   (d) Fringing increases

Practice Problems

10–16   A toroidal coil consists of 300 turns and has a current of 200 mA flowing. If the length of the magnetic circuit is 8 cm, calculate the mmf and field intensity of the coil.

10–17   A solenoid has 50 turns and carries a current of 0.75 A. If the cross-sectional area of the solenoid's core is 0.035 m^2, and the flux density is 5.75 T, find the reluctance of the solenoid.

10–18   Determine the reluctance of a coil with an mmf of 72 At and a total magnetic flux of 20×10^{-6} Wb.

10–19   An iron core has a flux density of 2.15 T and a cross-sectional area of 0.13 in.^{2}. The coil around the core has 200 turns and carries a current of 0.35 A. Determine the reluctance.

10–20   If the reluctance of a magnetic path is 225×10^5 At/Wb, what value of mmf would be required for a flux of 120 μWb?

10–21   How much flux is established in the magnetic path of a toroidal coil with 20 turns carrying 2 A, if the reluctance of the material is 0.38×10^4 At/Wb?

10–22   In order to develop a flux of 0.033 Wb, how much current must flow through a coil with 300 turns in a magnetic circuit of 2.7×10^4 At/Wb?

10–23   A solenoid has a magnetomotive force of 150 At, its magnetic path is 0.21 m, and it has a permeability of 4.35×10^{-3} H/m. Find the flux density.

10–24   What value of flux density will be produced in a coil having a permeability of 246×10^{-5} H/m and a field intensity of 760 At/m?

10–25   A toroidal coil consists of 2500 turns and carries a current of 150 mA. The coil has a cross-sectional area of 6.25 cm^2, and the length of the magnetic circuit is 22 cm. The permeability of the material is 275×10^{-6} H/m. Calculate

(a) Magnetic field strength   (b) Flux density   (c) Total field flux

10–26   For the circuit shown in Figure 10–9, find the value of current required to develop a magnetic flux of 2000 μWb.

Essay Questions

10–27   Define magnetomotive force.

10–28   Describe the effect of reluctance on a magnetic circuit.

10–29   How is field intensity determined?

10–30   Explain the difference between permeance and permeability.

10–31   What is the relative permeability of a magnetic material?

FIGURE 10-9

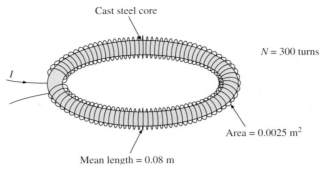

Cast steel core

N = 300 turns

Area = 0.0025 m²

Mean length = 0.08 m

10–32 When a magnetic material becomes saturated, what has occurred?

10–33 Define magnetic hysteresis.

10–34 Why are air gaps used in magnetic circuits?

10–35 Describe the effect of fringing due to air gaps.

10–36 Name two consumer electronic products that use electromagnets with air gaps.

10–37 What is a toroid?

11

Alternating Voltages and Currents

Learning Objectives

Upon completion of this chapter you will be able to

- Identify sine waves.
- Explain the instantaneous value of a sine wave.
- Convert radians to electrical degrees and vice versa.
- Define frequency, period, and wavelength.
- Determine the average and rms values of a sine wave.
- Explain the phase relationships between alternating current and voltage.
- Differentiate between a sinusoidal wave and a nonsinusoidal wave.
- Name three types of nonsinusoidal waves.
- Define harmonics.

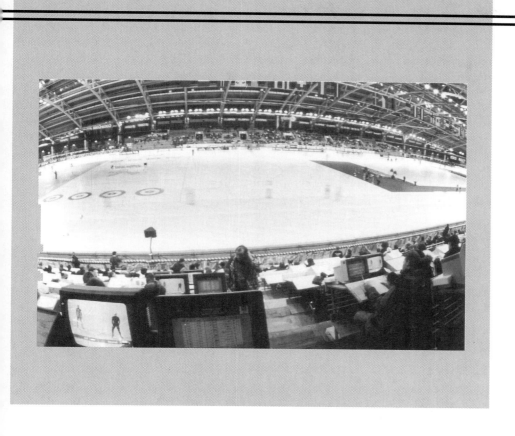

Practical Application

As a technician for a company specializing in large, arena-style lighting systems, you are asked to assist in the design of a new computerized lighting controller. The lighting system is divided into 64 channels, with each channel controlling four 1000 W lights connected in parallel to a 120 V_{rms} supply. The lighting intensity is varied for each channel by a high-current light dimmer. The brightness of the lamps is controlled by varying the point in the conduction cycle that the lamps are switched on. Because the lighting system must handle large amounts of current, your task is to determine the amount of instantaneous current drawn by each channel between 0° and 360° of the conducting cycle at 45° intervals.

INTRODUCTION

At the present time, over 95% of the electric energy used commercially is generated as alternating current. The main reason that alternating current is used is that alternating voltage may be easily raised or lowered in value. This is a tremendous advantage in electrical distribution systems, allowing AC power to be generated and distributed at a high voltage and then reduced to a more practical voltage at the load. High voltage is essential to transmit efficiently large amounts of power over long distances. In contrast, the voltage of direct current systems is limited to approximately 1500 volts per generator.

When using alternating current, the voltage may be raised and lowered economically by means of transformers. This provides an efficient method of transmitting power over long distances. The weight of a conductor required to transmit a given amount of power a given distance with a fixed loss varies inversely as the square of the transmission voltage. For example, when the transmission voltage is doubled, the weight of the conductor decreases by 25%. AC systems use transformers to step up voltage for economical power transmission and to step down to the lower voltages at which the power is utilized.

Section Review

1. What percentage of electric energy used commercially is generated as alternating current?
2. How is alternating current raised and lowered?
3. The weight of a conductor required to transmit a given amount of power a given distance with fixed loss varies inversely as the square of the transmission voltage (True/False).

THE SINE WAVE

The **sine wave** is a very common type of alternating current that is produced either by a rotating electrical machine, such as a generator, or by an electronic oscillator. Figure 11-1(a) shows a typical AC generator. This rotating machine converts mechanical energy into electrical energy and produces an electomotive force (emf). The emf produced by a generator is referred to as an *induced emf* because it is the result of mechanical motion between conductors and magnetic fields. An electronic signal generator is shown in Figure 11-1(b). **Signal generators** are electronic oscillators that are often used for testing electronic equipment such as audio amplifiers. These devices are used to alter the magnitude and frequency of a signal. Signal generators produce relatively low values of emf compared to rotating machines such as generators.

When a coil rotates in a uniform magnetic field, an emf is generated in that coil. This voltage is constantly varying in both direction and magnitude. Figure 11-2 shows an elementary alternating-current generator consisting of a one-turn coil, AB, which is free to rotate in a magnetic field. The end of the coil marked A is connected to slip-ring C, and slip-ring D is connected to the end of the coil marked B. The direction of coil rotation is assumed to be counter-clockwise. At the instant rotation begins, conductors A and B begin to move parallel to the magnetic field. Because the conductors do not cut the magnetic field, no voltage is induced. As the coil continues to rotate, the conductors begin to cut

FIGURE 11-1 *(a) Cutway view of an AC generator. (b) Signal generators.*

through the magnetic field and a voltage is induced in the coil. During the first 180° of rotation, the direction of the induced voltage is determined to be from A to B.

When the coil reaches 180°, the conductors once again move parallel to the flux, causing the induced voltage to fall to zero. As the counter-clockwise rotation continues, conductors A and B are moving across the magnetic field in opposite directions to their original motion. The reversal of conductors results in an emf of opposite polarity being developed during the second 180° of revolution. Therefore, as the coil passes through one revolution, the induced voltage rises from zero to a maximum value in one direction, decreases through zero to a maximum value in the opposite direction, and returns to zero.

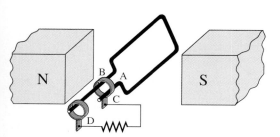

FIGURE 11-2 *Simple AC generator.*

FIGURE 11–3   *Waveshape of the coil in Figure 11–2.*

This type of voltage is referred to as **alternating emf.** The value of this emf at any given time is called the **instantaneous value** and is represented by the letter *e*. The greatest value that *e* can have is called the **maximum,** or **peak value,** E_p. The instantaneous value of current is represented by the letter *i* and the peak or maximum current by I_p. The **peak-to-peak value** *($E_{p\text{-}p}$)* of a sine wave is equal to $2E_p$.

If 360° of rotation are laid out along a horizontal axis, and the instantaneous voltages for various angular positions of the coil are plotted vertically, a waveshape similar to that of Figure 11–3 would be obtained.

The actual **waveform** of the alternating voltage induced in a single coil is determined by the pattern of flux distribution in the air gap. In commercial AC generators, several windings are connected in series and the alternating emf waveshape would more closely resemble the sinusoidal wave shown in Figure 11–4.

To determine the instantaneous value of emf at a given instant on a sine wave, it is necessary to understand some fundamental trigonometry. Trigonometric functions are based on the relationships among the lengths of the sides of a right-angle triangle as the size of one of its angles is varied. A basic right-angle triangle is shown in Figure 11–5. The sine of the angle θ is the length of the side opposite the angle divided by the length of the **hypotenuse,** which is the longest side and the one opposite to the right angle.

The sine waveform is the result of plotting the values of the sine against the values of angles to which each belongs. A convenient method of deriving sine values for plotting is by using a rotating **unit hypotenuse,** illustrated in Figure 11–6. As the unit hypotenuse rotates, the distance above or below the horizontal of its end point A represents the sine. This value can be projected directly to the right graph above the angle corresponding to θ that is represented by the position of line 0A.

Figure 11–6 shows the unit hypotenuse at 30° intervals during the first 180° of rotation. The fact that the rotating hypotenuse 0A is constructed to have one unit of length makes the length of the vertical lines AB equal to the sine of the angle. For example, triangle $0A_{30}\,B_{30}$ has a central angle 0 equal to 30°. Therefore, the length $A_{30}\,B_{30}$ is the sine 30° = 0.5. If a horizontal line is drawn from each A point toward the right, the line intersects with a scale to the right of the figure, indicating the value of the sine.

FIGURE 11–4
Sinusoidal waveform.

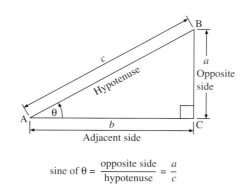

$$\text{sine of }\theta = \frac{\text{opposite side}}{\text{hypotenuse}} = \frac{a}{c}$$

FIGURE 11–5   *Definition of the sine function.*

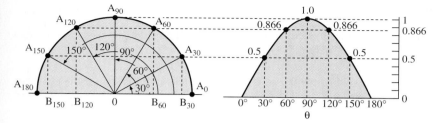

FIGURE 11-6 *Relationship between sine wave and rotating armature during first 180°.*

Section Review

1. Name two methods of producing sine waves.
2. What is the peak voltage of a sine wave?
3. A sine wave has 360 electrical degrees (True/False).

_____ 11-3 __

POLARITY OF A SINE WAVE

Essentially, a sine wave represents an equal combination of both positive and negative voltages. The polarity of this voltage changes when the signal crosses the zero axis. When the sine wave is below the zero line, it is considered as a negative polarity. When the wave is above the zero line it is said to be positive. Figure 11–7 shows an AC source connected across a resistance. When the sine wave is positive, or between 0° and 180°, the current flow is from the bottom of the source to the top. Consequently, the flow of current through the resistor is from point B to point A.

Figure 11–8 shows the direction of current in the AC circuit during the second 180°. The top of the AC source is now negative, and the current flow through the resistor is from point A to B. Therefore, the current flow through the resistor is reversing, or alternating, every 180°.

FIGURE 11-7 *Direction of current flow during positive alternation.*

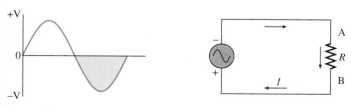

FIGURE 11-8 *Direction of current flow during negative alternation.*

Section Review

1. A sine wave represents an equal combination of both positive and negative voltages (True/False).
2. When does the polarity of a sine wave change?
3. When the sine wave is below the zero line, it is considered to be a positive polarity (True/False).

— 11–4

RADIAN MEASURE AND ANGULAR VELOCITY

In some situations, it is necessary to determine the relationship between the linear velocity of a conductor and the angular velocity of the conductor. The time rate of change in angular displacement is called the **angular velocity.** For example, the angular velocity of a wheel is a measure of how fast the wheel turns. If a wheel turns at 1000 revolutions per minute (rpm), it would have an angular velocity of 1000 rpm. Angular velocity is represented by the symbol ω (Greek letter omega).

Although angular velocity may be stated in revolutions per minute or revolutions per second, in electronic circuit problems radians per second is most commonly used. The **radian** is a convenient unit for calculating angular velocity. In AC circuits, angles are frequently measured in radians, rather than degrees. If the angle of rotation of the conductors is measured in radians, instead of degrees, one complete revolution represents 2π radians. The angle through which the conductors move in one second may be written as

$$\text{angular velocity} = \omega \ (\text{radians/second}) \tag{11.1}$$

The definition of a radian is illustrated in Figure 11–9. One radian is that angle that subtends a circular arc whose length is equal to the radius, r, of that arc. The arc subtended by 360° is a whole circle, which has a length of 2π times the radius. The central angle A0B in Figure 11–9 is equal to 1 radian, or 1 rad, because arc AB is equal to the radius 0A.

The circumference around a circle equals $2\pi r$. A complete circle, therefore, will have 2π radians, which is subtended by 360°. In other words, 2π radians = 360°, so the number of degrees in a radian can be found by dividing 360° by 2π.

$$1 \text{ radian} = \frac{360}{2\pi} = \frac{180}{\pi} = 57.3° \tag{11.2}$$

FIGURE 11–9 *Definition of a radian.*

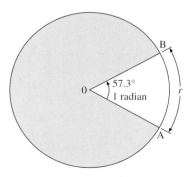

EXAMPLE 11–1

Solution

$$1 \text{ revolution} = 360° = 2\pi \text{ radians}$$

$$3 \text{ revolutions} = 6\pi \text{ radians} = 18.85 \text{ rad}$$

If a single loop of wire has completed three revolutions, how many radians has the coil traveled?

EXAMPLE 11–2

Solution

Change 2.5 radians into degrees.

$$\text{angle, in degrees} = \frac{\text{radians} \times 180}{\pi}$$

$$= \frac{2.5 \times 180}{\pi}$$

$$= 143.24°$$

EXAMPLE 11–3

Solution

A coil has rotated through 230°. What is this angle in radians?

$$\text{radians} = \frac{\text{degrees} \times \pi}{180} = \frac{230\pi}{180} = 4.01 \text{ rad}$$

Section Review

1. What is angular velocity?
2. What is the symbol for angular velocity?
3. One radian is equal to how many degrees?

11–5
FREQUENCY

The voltage wave generated by a conductor completing the passage of one north and one south magnetic pole is called a **cycle.** Therefore, one cycle of a voltage is generated every time an armature completes 360° of rotation. Anything that repeats itself periodically is said to occur in cycles. The number of cycles per second is defined as the **frequency,** f, of an AC voltage or current. The unit of frequency is the **hertz, Hz,** and was accepted as the SI unit for cycles per second in 1957 to honor the German physicist H.R. Hertz. One hertz is equivalent to one cycle per second (c/s).

Alternating current frequency is important because AC equipment requires a specific frequency as well as a specific voltage and current for proper operation. The standard commercial frequency used in North America is 60 Hz. Lower frequencies cause lights to flicker because of the lamp turning off each time the AC falls to zero. When 60 Hz is supplied to a lamp, the lamp turns on and off 120 times per second. Consequently, no flicker is noticeable because the human eye cannot react fast enough to see the light turn off and on.

In electronic circuits, frequency is also important in the study of audio, radio, and video systems. Audio frequencies are those frequencies that can be "heard" by the human ear. This range of frequencies is from approximately 20 Hz to 18,000 Hz. In the audible spectrum, the frequency is related to pitch. Essentially, pitch represents the frequency perceived by the listener. Frequencies with low pitch are like bass notes, and frequencies with high pitch represent the treble notes.

Radio and television transmitting stations convert sound waves and light waves to electrical impulses. These signals are then transmitted by use of high-frequency alternating currents. The alternating currents produce magnetic and electrical fields that radiate in all directions over long distances. The resulting magnetic and electrical fields are called **radio waves.** The frequencies of common radio waves are measured in hundreds of thousands or millions of cycles per second. Radio frequencies are typically between 30 kHz and 500 GHz. For convenience, the following engineering units are often associated with these frequencies:

$$1 \text{ kHz} = 1 \text{ kilohertz} = 1000 \text{ c/s}$$
$$1 \text{ MHz} = 1 \text{ megahertz} = 1 \times 10^6 \text{ c/s}$$
$$1 \text{ GHz} = 1 \text{ gigahertz} = 1 \times 10^9 \text{ c/s}$$
$$1 \text{ THz} = 1 \text{ terahertz} = 1 \times 10^{12} \text{ c/s}$$

When transmitting radio waves, the frequencies of the electromagnetic waves are much higher than the audio frequencies themselves. This higher frequency is called the *carrier frequency* of the transmitting station.

Section Review

1. What is a cycle?
2. The number of cycles per second is defined as the frequency of an AC voltage or current (True/False).
3. What is the unit of frequency?
4. What are radio waves?

— 11–6

PERIOD

The amount of time required for one cycle of change is referred to as the **time period** of the waveform. A **period** is the amount of time it takes a wave to pass through a complete cycle. The time period is represented by the letter T and is measured in seconds. A **periodic waveform** repeats over and over again in a fixed pattern. Figure 11–10 shows some typical periodic waveforms.

The period of a wave is inversely proportional to its frequency. Because the number of cycles per second is equal to the number of periods per second, the following equation may be derived:

$$f = \frac{1}{T}$$

(11.3)

where f = frequency, in hertz
          T = period, in seconds

(a)

(b)

(c)

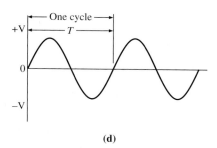

(d)

FIGURE 11–10 *Periodic waveforms. (a) Rectangular. (b) Triangular. (c) Sawtooth. (d) Sinusoidal.*

EXAMPLE 11–4

Solution

What is the time of one cycle when a signal has a frequency of 2650 kHz?

$$f = \frac{1}{T}$$

$$T = \frac{1}{f}$$

$$= \frac{1}{2.65 \times 10^6}$$

$$= 0.377\ \mu s$$

EXAMPLE 11–5

Solution

A signal has a time period of 140 μs. Calculate the frequency.

$$f = \frac{1}{T} = \frac{1}{140 \times 10^{-6}} = 7.14\ kHz$$

Section Review

1. What is the time period of a waveform?
2. What is a periodic waveform?
3. The period of a wave is directly proportional to its frequency (True/False).

____11-7____

WAVELENGTH

The distance that a radio wave travels in the time of one cycle is called its **wavelength.** The symbol for wavelength is the Greek letter λ (lambda). Wavelength represents a ratio of the velocity to the frequency. Because radio waves travel at the speed of light (300,000,000 m/s), if the frequency of a wave is known, the distance it will travel in the time required for one cycle can be found by the following equation:

$$\lambda = \frac{300,000,000}{f} \tag{11.4}$$

where   λ = wavelength, in meters
   f = frequency, in hertz

Because radio frequencies are often expressed in kilohertz (kHz) or megahertz (MHz) values, Equation 11.4 can be rewritten for convenience as

$$\lambda = \frac{300,000}{f \text{ (in kHz)}}$$

$$\lambda = \frac{300}{f \text{ (in MHz)}}$$

____ EXAMPLE 11-6 ____

What is the wavelength of an AM radio wave that has a frequency of 1400 kHz?

Solution

$$\lambda = \frac{300,000}{1400 \text{ kHz}} = 214.29 \text{ m}$$

____ EXAMPLE 11-7 ____

Determine the wavelength of a television video carrier wave with a frequency of 83.55 MHz.

Solution

$$\lambda = \frac{300}{83.55 \text{ MHz}} = 3.59 \text{ m}$$

The velocity of sound waves is very low compared to radio waves. Sound waves travel through air at approximately 340 m/s, depending on the temperature of the air.

$$v = \frac{340 \text{ m/s}}{f}$$

____ EXAMPLE 11-8 ____

What is the wavelength of the sound waves produced by a bass speaker at a frequency of 250 Hz?

Solution

$$\lambda = \frac{340 \text{ m/s}}{f} = \frac{340 \text{ m/s}}{250 \text{ Hz}} = 1.36 \text{ m}$$

Section Review

1. The distance that a radio wave travels in the time of one cycle is called its wavelength (True/False).
2. What is the symbol for wavelength?
3. The velocity of sound waves is very high compared to radio waves (True/False).

11-8

INSTANTANEOUS VALUES OF A SINE WAVE

In the sine wave, one complete cycle is represented by 360° or 2π radians. If the period of a sine wave is, for example, 0.3 s, then each degree of the cycle would be $0.3/360 = 0.833$ ms. The instantaneous value is determined from the value of sine θ at the particular angle. At any given point, the instantaneous value, e, of the sine wave is equal to the product of the peak value, E_p, of the sine wave and the sine of the angle corresponding to time. Expressed in equation form, the instantaneous value of voltage is

$$e = E_p \sin \theta \qquad (11.5)$$

To calculate the instantaneous value of current, the following equation may be used.

$$i = I_p \sin \theta \qquad (11.6)$$

In mathematics, the peak value is called the **amplitude,** and θ is called the **argument.** Amplitude is typically used in electronic circuits to specify the maximum value of voltage or current on the vertical axis.

EXAMPLE 11-9

Solution

$$i = I_p \sin \theta$$
$$= 2.2 \text{ A} \sin 25° = 0.93 \text{ A}$$

What is the instantaneous value of a sinusoidal current at 25° if $I_p = 2.2$ A?

Expressed in radian measure, the instantaneous value of voltage is calculated using the following equation:

$$e = E_p \sin \omega t \qquad (11.7)$$

where ω = angular velocity, in rads per second
       t = time, in seconds

Because there are 2π radians in a cycle, the instantaneous value of voltage may also be calculated as

$$e = E_p \sin 2\pi f t \qquad (11.8)$$

Because ω is expressed in radians per second, it is often necessary to convert radians to degrees when using the sine function.

_____ EXAMPLE 11-10 _____

An AC waveform with a frequency of 1.3 kHz has a maximum voltage of 100 V. Calculate the instantaneous value of voltage at 58 μs. (Assume $t = 0$ when the voltage is zero and increases in a positive direction.)

Solution

$$e = E_p \sin \omega t$$
$$= 100 \sin [2\pi(1.3 \times 10^3)(58 \times 10^{-6})]$$
$$= 100 \sin 0.47 \text{ rad V}$$

where 0.47 is the time angle in radians. Since 1 radian = 57.3°, e is calculated as follows:

$$e = 100 \sin (0.47 \times 57.3)$$
$$= 45.6 \text{ V}$$

_____ EXAMPLE 11-11 _____

An 850 Hz sine-wave signal has a maximum amplitude of 30 V. What is the instantaneous value of the signal 40 μs before it reaches its peak positive value?

Solution   First, determine the period of the signal.

$$T = 1/f = 1176.47 \text{ μs}$$

Next, determine the number of degrees per microsecond:

$$1176.47 \text{ μs} = 360°$$
$$1 \text{ μs} = 0.306°$$
$$40 \text{ μs} = 40 \times 0.306°$$
$$= 12.24°$$

Because the peak maximum positive value of a sine wave occurs at 90°, 12.24° before the peak value is equal to

$$90° - 12.24° = 77.76°$$

$$e = E_p \sin \omega t$$
$$= 30 \sin 77.76°$$
$$= 29.32 \text{ V}$$

Section Review

1. What is the instantaneous value of a sinusoidal current at 47° if the peak current is 60 mA?
2. In the sine wave, one complete cycle is represented by 360° or 4π radians (True/False).

_____ 11-9 _____

AVERAGE AND RMS VALUES OF A SINE WAVE

A sinusoidal waveform has an instantaneous value that is constantly changing and a maximum value that occurs only twice in each cycle. Because of the constant fluctuating nature of a sine wave, it is often desirable to know what the **average value** of the waveform is. An average is a value obtained by adding successive, equally spaced values and then dividing the sum by the number of values. By definition, the true average, or **mean,** value of a sine wave is zero, because the positive and negative half-cycles would be identical in

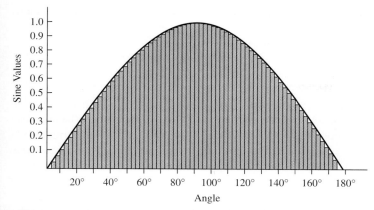

FIGURE 11–11 *Method of integrating area under curve to determine average value.*

shape and magnitude. However, when calculating the average value of a sine wave, only one-half of the cycle is used.

The principle of average value is applied to the sine wave of Figure 11–11. The positive half-cycle of the waveform has been divided into 3° intervals. Each 3° segment represents an equal time period. Therefore, the average value of the whole sine wave is the average of all the segment-center values. If the sine values for all angles up to 180° are added and then divided by the number of values, this average equals 0.637 and the peak value is 1.

$$\text{average value} = 0.637 \times \text{peak value} \qquad \textbf{(11.9)}$$

$$= \frac{2}{\pi} \times E_p$$

Figure 11–12(a) shows an alternating sine wave of current having a maximum instantaneous value of $\sqrt{2}$, or 1.414, amperes and an rms value of 1 ampere. If the wave is considered over one complete cycle, the average value is zero, since there is just as much negative current as there is positive. If a DC ammeter were connected to measure this current, it would read zero. This is because a DC meter reads the average values of voltage and current.

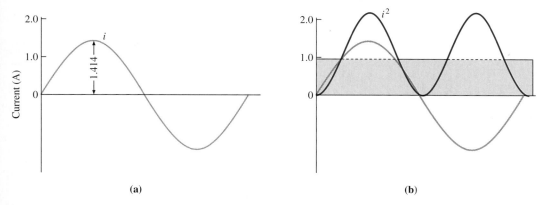

(a)

(b)

FIGURE 11–12 *(a) Maximum value of sine wave. (b) Current squared wave and proportion of heating effect.*

However, the rms value of an alternating current is based on its heating effect and is defined in the following manner:

> An AC ampere is the current that will produce heat in a pure resistance at the same rate as a DC ampere.

Because the heating effect varies as the square of the current (I^2R), the value in amperes of the current represented by the waveform in Figure 11–12(a) must be based on the squares of the instantaneous values of current. Figure 11–12(b) shows the current wave of Figure 11–12(a) plotted together with its squared values. The maximum value of this new waveform will be 1.414^2, because the maximum value of the original current wave is 1.414, or $\sqrt{2}$ amperes.

This squared wave of Figure 11–12(b) has a frequency that is double that of the original wave and a horizontal axis of symmetry that is at a distance of 1.0 ampere above the zero axis. The square root of this squared wave has an average value of 1.0 ampere, which is the average of the squares of the ordinates of the current wave. If this value of current were to flow through a resistance, its heating effect over one cycle would be proportional to the area of the shaded rectangle of Figure 11–12(b). This value of current is called the **effective** or **root-mean-square, rms, value** of current. Therefore, the sine wave of an AC ampere that produces heat at the same rate as a DC ampere has a maximum value of 1.414 amperes. The ratio of rms to maximum values is found by

$$E_{rms} = \left(\frac{1}{1.414}\right) E_p \qquad (11.10)$$

$$= 0.707\, E_p$$

$$= \frac{E_p}{\sqrt{2}}$$

$$I_{rms} = 0.707\, I_p \qquad (11.11)$$

When an alternating current or voltage is specified, it is always the effective, or rms, value that is implied, unless otherwise specified.

_____ **EXAMPLE 11–12** _____

An AC voltage has an rms value of 115 V. What is the peak voltage?

Solution

$$E_p = \frac{E_{rms}}{0.707} = \frac{115\ \text{V}}{0.707} = 162.6\ \text{V}$$

_____ **EXAMPLE 11–13** _____

A 240 V AC sine wave is connected to a 100 Ω resistor. Calculate the peak and rms values of current through the resistor.

Solution

$$E_p = \frac{E_{rms}}{0.707} = \frac{240\ \text{V}}{0.707} = 339.4\ \text{V}$$

$$I_p = \frac{E_p}{R} = \frac{339.4\ \text{V}}{100\ \Omega} = 3.4\ \text{A}$$

$$I_{rms} = \frac{E_{rms}}{R} = \frac{240\ \text{V}}{100\ \Omega} = 2.4\ \text{A}$$

EXAMPLE 11-14

Solution

Calculate the rms voltage of a signal that has a voltage of 105 V peak-to-peak.

$$E_{p\text{-}p} = 105 \text{ V}$$

$$E_p = \frac{1}{2} E_{p\text{-}p} = 52.5 \text{ V}$$

$$E_{rms} = 0.707 \, E_p = (0.707)(52.5) = 37.12 \text{ V}$$

In AC circuits, unless otherwise specified, power is calculated in terms of the effective values of current and voltage.

In some situations, it is convenient to base calculations initially using the average value of the emf over half a period, so that it becomes necessary to have some means of connecting this average value with the effective value. The correlation between these two values is called the **form factor** and is defined as

$$\text{form factor} = \frac{\text{effective value}}{\text{average value}} \qquad (11.12)$$

For sine waves the form factor is represented as

$$\frac{\text{effective value}}{\text{average value}} = \frac{E_p / \sqrt{2}}{(2/\pi)E_p} \qquad (11.13)$$

$$= \frac{\pi}{2\sqrt{2}}$$

$$= 1.11$$

Section Review

1. What is the true average, or mean, value of a sine wave?
2. An AC ampere is the current that will produce heat in a pure resistance at the same rate as a DC ampere (True/False).
3. State the equation for calculating rms voltage if the peak voltage is known.

11-10

PHASE RELATIONSHIPS

Whenever an alternating voltage causes a current to flow through a circuit, the current is said to be alternating at the same **fundamental frequency** as the applied voltage. Figure 11-13 shows a voltage and current waveform with different magnitudes but alternating at the same frequency. Under these conditions, when voltage and current pass through their zero values and increase to their maximum values at the same time and in the same direction, the two waves are considered to be **in phase** with each other.

If a sinusoidal voltage is applied to a purely resistive circuit, the resultant current will be as shown in Figure 11-13. However, in circuits containing combinations of resistance, capacitance, and inductance, the current and voltage waves do not pass through their maximum and minimum values at the same time. When this occurs, the current and voltage are said to be **out of phase** with each other.

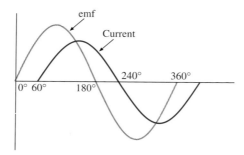

FIGURE 11–13   *Current and voltage waveforms in phase with each other.*

Figure 11–14 shows a current and a voltage waveform that have a phase displacement of 60°. Because the current waveform of Figure 11–14 starts a cycle 60° later than the voltage, the current is said to be **lagging** the voltage by 60°. Generally, a circuit that contains a predominant amount of inductance will have a phase displacement where the current lags the voltage. If a current starts before the voltage, it is said to be **leading** the voltage. A predominantly capacitive circuit will have a phase displacement where the current starts a cycle before, or leads, the voltage. The relationship between resistance, inductance, and capacitance is discussed in more detail in Chapter 16.

When the current and voltage of a circuit are in phase, the difference in time between the two waves is zero, since they begin and end their cycles at the same instant. When the current and voltage are out of phase, the phase displacement is usually measured in electrical degrees and is called the **phase angle.** In Figure 11–14, the current is lagging the voltage by 60°, implying that the phase angle is equal to 60°.

The sine and cosine values that are listed in a trigonometric table show that the cosine assumes the same values as the sine, except that, for the cosine, the angle is shifted 90° from the sine value. Therefore, a cosine wave is a sine wave shifted by 90°. Figure 11–15 illustrates the relationship between a sine wave and a cosine wave.

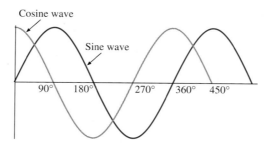

FIGURE 11–14   *Voltage and current waveforms with a phase displacement of 60°.*

FIGURE 11–15   *Displacement between sine and cosine wave.*

Section Review

1. What is meant by the term *in phase*.
2. If a sinusoidal voltage is applied to a purely resistive circuit, the current and voltage will be out of phase with each other (True/False).
3. What is phase angle?

NONSINUSOIDAL WAVEFORMS

The phase relationships of nonsinusoidal waveforms can be considerably different than that of sine waves. As mentioned in Section 11–6, waveforms such as the square wave, triangular wave, or sawtooth wave are called *periodic waveforms*. A periodic waveform is a repetitive waveform. Periodic by definition means recurring. A waveform does not have to be sinusoidal to be repetitive, although all nonsinusoidal periodic waveforms are composed of sine waves of different frequencies whose sum yields the original nonsinusoidal periodic waveform. The polarity of a periodic AC voltage reverses at regular time intervals, and the direction of a periodic AC current reverses at regular time intervals.

Nonsinusoidal waves include, triangular, sawtooth, pulse, and square waves. Signals used in electronic circuits are often nonsinusoidal. These waveforms are produced by television signals, music, digital data, and so forth. Nonsinusoidal waves are also used in electronic test equipment. Square waves are used to determine the frequency response of amplifiers and are also used as clock and signal sources in logic circuits. Triangular waves can be used to determine the overload, or clipping, point of an amplifier.

One of the most common nonsinusoidal waveforms is the pulse wave, shown in Figure 11–16. A **pulse** is a signal of relatively short duration. An ideal pulse signal is a perfect rectangle, where the signal instantly changes from one value to another. The difference between the original value and the new value is called the **pulse amplitude** and the length of time that the signal stays at the new value is called **pulse width.**

In practice, a real pulse takes a certain amount of time to reach a new value or return to the original value. The amount of time required for a signal to change from a low value to a higher value is called the *rise time*. The time required for the signal to go from a high value to a lower value is the *fall time*. In electronic circuits, rise time is defined as the amount of time required for the signal to rise to 90% of its maximum amplitude, and fall time is the time required to fall to 10% of its maximum amplitude.

The frequency at which the pulses occur is also called the **pulse repetition rate (PRR).** Because the pulse duration is often very short, time is measured in microseconds. The repetition rate is a measure of the number of pulses produced in one second. A group of consecutive pulses is called a **pulse train.** If all the pulses in a pulse train are identical in shape and evenly spaced, the signal is considered to be periodic. When the pulse width is multiplied by the repetition rate, a figure known as the **duty cycle** is obtained. The duty

FIGURE 11–16 *Pulse wave.*

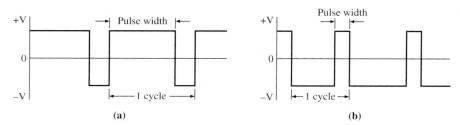

FIGURE 11–17   *Pulse wave. (a) High duty cycle. (b) Low duty cycle.*

cycle of a pulse waveform is a ratio of the pulse width to the period of one cycle. In equation form,

$$\% \text{ duty cycle} = \left(\frac{t_w}{T} \right) 100\% \qquad (11.14)$$

where t_w = pulse width
T = period of one cycle

Figure 11–17(a) represents a high duty cycle because the width of the positive-going pulse is a large percentage of the cycle (greater than 50%). Figure 11–17(b) shows a low duty cycle because the width of the positive-going pulse is a small percentage of the cycle (less than 50%).

_____ EXAMPLE 11–15 _____

For the nonsinusoidal wave shown in Figure 11–18, calculate the

(a) Period.
(b) Frequency.
(c) Duty cycle.

FIGURE 11–18   *Waveform for Example 11–15.*

Solution

(a)
$$T = 100 \text{ μs} + 400 \text{ μs} = 500 \text{ μs}$$

(b)
$$f = \frac{1}{T} = \frac{1}{500 \text{ μs}} = 2 \text{ kHz}$$

(c)
$$\% \text{ duty cycle} = \left(\frac{t_w}{T} \right) 100\%$$
$$= \left(\frac{100 \text{ μs}}{500 \text{ μs}} \right) 100\%$$
$$= 20\%$$

The average voltage of a pulse wave is found by multiplying the duty cycle by the peak-to-peak voltage and adding the baseline. The baseline represents the lowest value of the pulse voltage. Because pulse voltages can be either positive or negative, the baseline could be negative, positive, or zero. In equation form,

$$V_{avg} = \text{baseline} + (\text{duty cycle}) (E_{p-p})$$ (11.15)

EXAMPLE 11-16

Determine the average voltage of the pulse wave shown in Figure 11–19.

FIGURE 11-19 *Waveform for Example 11-16.*

Solution

$$\text{duty cycle} = \frac{t_w}{T} = \frac{250 \ \mu s}{1250 \ \mu s} = 0.2$$

$$E_{p-p} = 10 \ V$$

$$\begin{aligned} E_{avg} &= \text{baseline} + (\text{duty cycle}) (E_{p-p}) \\ &= -5 \ V + (0.2) (10 \ V) \\ &= -3 \ V \end{aligned}$$

If the duty cycle is exactly 50%, the pulse is considered to be a **square wave,** since the positive pulse is equal to one-half cycle. Figure 11–20 shows a typical square wave. Basically, a square wave represents a train of pulses that have identical pulse widths and identical time intervals between pulses. Because both the positive and negative sections of the wave are equal, the average value for one cycle is zero and there is no DC component.

Figure 11–21 shows a **triangular wave.** The triangular wave may be defined as a periodic ramp waveform where the voltage level changes from one level to another at a constant rate. The average value of the waveform over the entire period is zero. It should also be noted that the negative portion of the waveform is a mirror image of the positive portion. Therefore, the triangular wave is considered to be symmetrical.

The **sawtooth wave** shown in Figure 11–22 is another waveform that consists of two ramp voltages. This wave also has no DC component because the area under each half-cycle is equal, and the average value is zero. Sawtooth waves are used extensively in cathode-ray oscilloscopes, television receivers, and electronic triggering circuits.

FIGURE 11-20 *Square wave.*

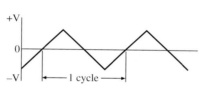

FIGURE 11-21 *Triangular wave.*

FIGURE 11–22 *Sawtooth wave.*

Complete mathematical analysis of a recurring nonsinusoidal waveform is made by applying principles that lead to the equation for the wave in a form called **Fourier series,** named after the French physicist and mathematician Baron Jean Fourier. According to Fourier's theorem, any vibration or oscillation that repeats itself indefinitely can be built up from a set of pure sinusoidal waves.

Periodic functions of time of period T can be represented with the Fourier series as an infinite sum of sinusoids having frequencies that are multiples of the fundamental frequency $f = 1/T$. By using a Fourier series, nonsinusoidal periodic waves that can be defined over a 2π interval can be expanded into a series of sine and cosine waves of different frequencies and a constant term called the *DC term.* The DC term is the average value of the wave over one full cycle.

Section Review

1. List three types of nonsinusoidal waveforms.
2. The amount of time required for a signal to change from a low value to a high value is called the rise time(True/False).
3. What is pulse repetition rate (PRR)?
4. What is a group of consecutive pulses called?

___ 11–12 ___

HARMONIC FREQUENCIES

The term **harmonics** in electronic circuits refers to components that have frequencies greater than the fundamental, or source, frequency. The fundamental sine wave is generally much greater in amplitude than the harmonics. When an electronic amplifier is used for music and speech, it must amplify the signal without affecting the signal's harmonic content. When the volume control on an amplifier is increased above its rated value, audible distortion results, called **harmonic distortion.** In most cases, this distortion is caused by the introduction of unwanted harmonics.

Nonsinusoidal waveforms, such as square waves, have not only a fundamental frequency but also a number of odd harmonic frequencies. When a sufficient number of these frequencies combine, they can form a square wave appearance. The resulting wave would be exactly square if an infinite number of odd harmonics are added. Generally speaking, the steeper the sides of the waveform, the more harmonics it contains.

Voltage and current harmonics are usually found by means of Fourier series analysis. A harmonic of a given wave is another wave having a frequency equal to an integral multiple of the frequency of the given wave. For example, a harmonic with a frequency double that of the fundamental frequency is called the **second harmonic.** If the source frequency is 60 Hz, then

Fundamental frequency = 60 hz
Second harmonic frequency = 120 hz
Third harmonic frequency = 180 hz
Fourth harmonic frequency = 240 hz

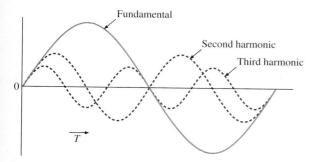

FIGURE 11–23 *Relationship between fundamental, second, and third harmonics.*

and so forth. The relation between a fundamental frequency and its second and third harmonics is shown in Figure 11–23.

There are two basic types of harmonics: those that are deliberately created, such as in waveshaping circuits, and those that are a cause of unwanted interference, such as distortion. In some electronic circuits, distortion is desirable in order to produce specific waveforms. However, other circuits are designed to operate with as little distortion as possible.

Since the magnitude of the harmonics decreases as the frequency increases, only the first few harmonics need to be considered when examining the effects of harmonics on electronic components and equipment. Harmonics generally are caused in electronic circuits by semiconductor devices such as thyristors switching on and off repetitively. Harmonics can damage electronic equipment by distorting the input sine wave and reducing its magnitude. Harmonics can also shift the zero crossing point of the load sine wave, resulting in a less efficient system with large values of heat loss.

Section Review

1. Harmonics refers to components that have frequencies below the fundamental frequency (True/False).
2. How are voltage and current harmonics usually found?
3. What are the two basic types of harmonics?

_____ **11–13** __

PRACTICAL APPLICATION SOLUTION

In the chapter opener, you were asked to assist in the design of a computerized lighting controller. The lighting system was described as consisting of 64 sets, or channels, each with four 1000 W lamps. Your task was to calculate the instantaneous values of current at 45° points in the conducting cycle. The following steps are necessary to successfully complete this practical application of alternating current.

STEP 1   Determine the peak current for each channel.

$$I_{rms} = \frac{P}{E} = \frac{4000 \text{ W}}{120 \text{ V}} = 33.33 \text{ A}$$

$$I_p = \frac{I_{rms}}{0.707} = \frac{33.33 \text{ A}}{0.707} = 47.15 \text{ A}$$

STEP 2   Solve for the instantaneous current at 45° intervals.

At 45°: $i = I_p \sin \theta = 47.15 \sin 45° = 33.34$ A

At 90°: $i = I_p \sin \theta = 47.15 \sin 90° = 47.15$ A

At 135°: $i = I_p \sin \theta = 47.15 \sin 135° = 33.34$ A

At 180°: $i = I_p \sin \theta = 47.15 \sin 180° = 0$ A

At 225°: $i = I_p \sin \theta = 47.15 \sin 225° = -33.34$ A

At 270°: $i = I_p \sin \theta = 47.15 \sin 270° = -47.15$ A

At 315°: $i = I_p \sin \theta = 47.15 \sin 315° = -33.34$ A

At 360°: $i = I_p \sin \theta = 47.15 \sin 360° = 0$ A

Summary

1. The sine wave is a common type of alternating current that is produced by either a rotating electrical machine or an electronic oscillator.
2. A sine wave represents an equal combination of positive and negative voltages.
3. The number of cycles per second is defined as the frequency of an AC voltage or current.
4. A periodic waveform repeats over and over again in a fixed pattern.
5. The distance that a radio wave travels in the time of one cycle is called its *wavelength.*
6. The term *amplitude* specifies the maximum value of voltage or current on the vertical axis.
7. The average value of a waveform is a value obtained by adding successive, equally spaced values and then dividing the sum by the number of values.
8. The effective, or rms, value of a current or voltage is the value that will produce the same amount of power as a direct current or voltage.
9. Nonsinusoidal waves include triangular, sawtooth, and pulse waves.
10. The term *harmonics* in electronic circuits refers to components that have frequencies greater than the fundamental frequency.

Answers to Section Reviews

Section 11–1

1. Over 95%
2. By using transformers
3. True

Section 11–2

1. Rotating electrical machines and electronic oscillators
2. The maximum instantaneous voltage that can occur during the sine wave
3. True

Section 11–3

1. True
2. When it crosses the zero axis
3. False

Section 11–4

1. The time rate of change in angular displacement
2. ω
3. $57.3°$

Section 11–5

1. The voltage wave generated by a conductor completing the passage of one north and one south magnetic pole
2. True
3. Hertz, Hz
4. High-frequency alternating currents that produce magnetic and electrical fields that radiate in all directions over long distances

Section 11–6

1. The amount of time required for one cycle of change
2. A waveform that repeats over and over again in a fixed pattern
3. False

Section 11–7

1. True
2. λ
3. False

Section 11–8

1. $i = 60 \text{ mA} \sin 47° = 43.88 \text{ mA}$
2. False

Section 11–9

1. Zero
2. True
3. $V_{rms} = 0.707 \, V_p$

Section 11–10

1. When voltage and current pass through their zero values and increase to their maximum values at the same time and in the same direction
2. False
3. The phase displacement between current and voltage that is measured in electrical degrees

Section 11–11

1. Triangular, sawtooth, and pulse waves
2. True
3. The frequency at which the pulses occur
4. A pulse train

Section 11–12

1. False
2. By Fourier series analysis
3. Those that are deliberately created, and those that are a cause of unwanted interference

Multiple Choice Questions

11–1   The main advantage of using alternating current for distribution to homes and factories is that the voltage
(a) Is sinusoidal   (b) Has frequency
(c) Can be as high as 1500 V   (d) Can be easily raised or lowered in value

11–2   When the voltage is doubled, the weight of the conductor
(a) Doubles   (b) Decreases by 25%   (c) Increases by 25%   (d) Triples

11–3   Sine waves are typically produced by
(a) Motors and generators   (b) Motors and oscillators
(c) Transformers   (d) Generators and oscillators

11–4   The peak-to-peak values of a sine wave is equal to
(a) $2 E_{rms}$ (b) $2 E_m$ (c) $2 E_{ave}$ (d) $2 I_{rms}$

11–5   The greatest value that an instantaneous signal can have is the
(a) Peak value   (b) Peak-to-peak value   (c) Rms value   (d) Average value

11–6   The time rate of change in angular displacement of a conductor is called
(a) Linear velocity   (b) Frequency   (c) Angular velocity   (d) Radian measure

11–7   Angular velocity is generally measured in
(a) Radians per second   (b) Seconds per cycle
(c) Revolutions per radian   (d) Radians per minute

11–8   Anything that repeats itself periodically is said to occur in
(a) Radians   (b) Seconds   (c) Frequencies   (d) Cycles

11–9   High-frequency alternating currents produce magnetic and electrical fields that are called
(a) Frequencies   (b) Cycles   (c) Radio waves   (d) Carrier waves

11–10   Radio frequencies are typically between
(a) 20 Hz and 20 kHz   (b) 60 Hz and 120 Hz
(c) 30 kHz and 50 GHz   (d) 30 kHz and 500 GHz

11–11   The amount of time required for one cycle of change is called the
(a) Time period   (b) Frequency   (c) Periodic waveform   (d) Wavelength

11–12   The distance that a wave travels in the time of one cycle is called the
(a) Period   (b) Wavelength   (c) Frequency   (d) Waveform

11–13   The maximum value of voltage or current is also called the
(a) Average   (b) Rms   (c) Amplitude   (d) Peak-to-peak

11–14   The mean value of a sine wave is also known as the
(a) Rms   (b) True average   (c) Peak   (d) Amplitude

11–15   In AC circuits, power is usually specified in terms of the _____ values of current and voltage.
(a) Average   (b) Peak   (c) Maximum   (d) Rms

11–16   The frequency at which pulses occur is known as
(a) Pulse amplitude rate   (b) Rise time
(c) Pulse repetition rate   (d) Fall time

11–17   A group of consecutive pulses is called a
(a) Pulse train   (b) Duty cycle   (c) Pulse inversion   (d) Period

11–18   The term *harmonics* in electronic circuits refers to components that have
(a) Low frequencies   **(b)** Frequencies above the fundamental
(c) Frequencies below the fundamental   **(d)** Nonsinusoidal waves

Practice Problems

11–19   How many radians has a single loop of wire traveled if it has completed 12 revolutions?

11–20   If a loop of wire has rotated 310°, how many radians has it traveled?

11–21   Convert 1.74 radians into degrees.

11–22   If a single loop of wire has traveled 25.13 radians, how many revolutions has the coil completed?

11–23   Calculate the time of one cycle if a signal has a frequency of 32.76 kHz.

11–24   Calculate the frequency of a signal with a time period of 15.7×10^{-6} seconds.

11–25   What is the wavelength of an AM radio wave that has a frequency of 1200 kHz?

11–26   Determine the instantaneous value of a sinusoidal voltage at 40° if the maximum voltage is 140 V.

11–27   An AC waveform with a frequency of 60 Hz has a maximum current of 10 A. Determine the instantaneous value of current at 0.7×10^{-3} seconds. Assume $t = 0$ when the current is zero and increases in a positive direction.

11–28   A 1500 Hz sinusoidal waveform has a maximum amplitude of 40 V. Determine the instantaneous voltage 30 µs before the wave reaches its peak positive value.

11–29   Determine the value of a sine wave voltage at 5 µs from the positive-going zero crossing, when the peak voltage is 12 V and the frequency is 25 kHz.

11–30   An AC voltage has an rms value of 208 V. Determine the peak voltage.

11–31   A 120 V AC sine wave is connected to an 80 Ω resistor. Determine the
(a) Peak value
(b) Rms value
of current through the resistor.

11–32   If an AC waveform has a peak-to-peak voltage of 187.11 V, find the rms value.

11–33   For the waveshape shown in Figure 11–24, calculate the period.

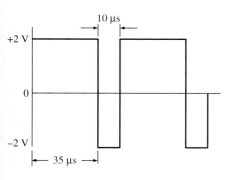

FIGURE 11–24

11–34   Find the frequency for the waveshape of Figure 11–24.

11–35   What is the % duty cycle of the waveform shown in Figure 11–24.

11–36   Determine the average voltage of the pulse wave shown in Figure 11–24.

Essay Questions

11–37 Explain the advantage of using alternating current for transmitting power.

11–38 What is meant by the term *peak-to-peak*?

11–39 Describe how the polarity of a sine wave affects the direction of current flow.

11–40 How is angular velocity calculated?

11–41 Define frequency.

11–42 Explain how radio waves are generated.

11–43 What is a carrier frequency?

11–44 List three examples of periodic waveforms.

11–45 Define wavelength.

11–46 How is the instantaneous value of a sine wave determined?

11–47 Explain the difference between average values and rms values.

11–48 What is fundamental frequency?

11–49 Describe a nonsinusoidal waveform.

11–50 How is the average voltage of a pulse wave determined?

11–51 Briefly explain Fourier's theorem.

11–52 Define harmonics.

12

AC Measuring Instruments and Troubleshooting Equipment

Learning Objectives

Upon completion of this chapter you will be able to

- Name two methods of frequency measurement.
- Describe the basic operating characteristics of an oscilloscope.
- Determine voltage and frequency values from oscilloscope displays.
- List two applications of signal generators.
- Define a function generator.

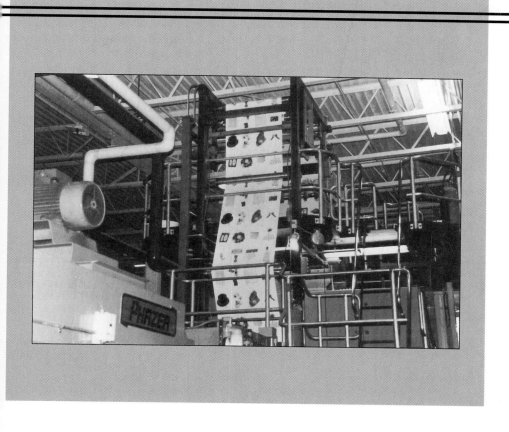

Practical Application

As a technician in a highly automated manufacturing plant, you are asked to test the output of a digital controller operating a printing press. The output of the controller consists of a series of voltage pulses that can be varied by a rotary dial on an operating console. According to the manufacturer's specifications for the controller, the average output voltage should be 5 V with a 0.5 duty cycle when the rotary dial is set to the mid-point position. A calibration potentiometer is provided inside the controller that allows the duty cycle to be increased or decreased as required.

Your task is to use an oscilloscope to determine the duty cycle and average output voltage of the controller. If any discrepancy exists between the measured values and the manufacturer's values, the calibration potentiometer is to be adjusted until the system is producing its rated values. With the dial set to the mid-point position, the oscilloscope is connected to the output of the controller and indicates a peak-to-peak voltage of 20 V, a pulse width of 40 ms, and a period of 100 ms.

12–1

INTRODUCTION

There are two basic types of AC test equipment: signal-measuring instruments and signal-generating instruments. Signal-measuring instruments include AC voltmeters, ammeters, oscilloscopes, and frequency counters. Signal-generating instruments include radio-frequency and audio-frequency generators, as well as function generators. When troubleshooting circuits, signal generators are used to *inject* a signal of a certain frequency, amplitude, and waveshape. Signal-measuring instruments provide a visual display of the signal. These AC meters can provide information regarding the frequency, magnitude, and waveshape of the component under test. When troubleshooting electronic equipment, it is often necessary to use both signal-generating and signal-testing equipment.

Section Review

1. Signal-measuring instruments and signal-generating instruments are the two basic types of AC test equipment (True/False).
2. Why are signal generators used for troubleshooting electronic circuits?
3. Signal-measuring instruments provide a visual display of the signal (True/False).

12–2

AC VOLTMETERS AND AMMETERS

AC meters often use a **rectifier** to convert the AC input signal to a DC value. Once the signal has been rectified, a D'Arsonval movement can be used in the measuring circuit. However, the rectifier-type D'Arsonval meter will not provide reliable readings for square waves, pulses, or irregular waveshapes. Figure 12–1 shows some typical waveforms found in electrical and electronic circuits.

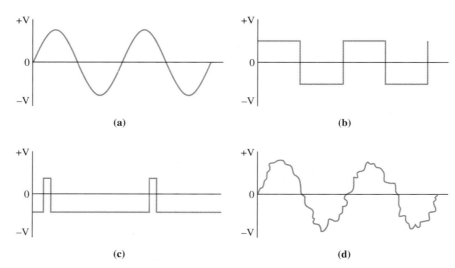

FIGURE 12–1 *(a) Sine wave. (b) Square wave. (c) Pulse wave. (d) Irregular wave.*

With the sine wave of Figure 12–1(a), an AC voltmeter will read 0.707 of the peak value. With the square wave of Figure 12–1(b), the meter will indicate a reading based on the form factor of the square wave. For the pulse waveform of Figure 12–1(c), the AC voltmeter reading is determined by the amplitude of the waveform and its duty cycle. The effective voltage increases for a high duty cycle and decreases for a low duty cycle. In the irregular waveform of Figure 12–1(d) a D'Arsonval meter would produce an incorrect reading, because this waveform requires a mathematical computation known as Fourier analysis to accurately determine its value.

The **electrodynamometer** is an analog meter that is capable of measuring both AC and DC voltages and currents. This meter provides very accurate measurements at low frequencies. It is similar to the D'Arsonval except that it has no permanent magnet and the magnetic field is produced by two field coils. In this meter, the stationary and moving coils establish flux in such a manner that like poles are adjacent to each other. The like poles repel and cause the moving coil to be deflected clockwise. Consequently, the pointer moves upscale. When the current flow is reversed, like poles are once again adjacent to each other and the pointer moves upscale from left to right.

Digital multimeters (DMMs) are capable of providing highly accurate AC measurements. These meters are becoming increasing popular because of their relatively low cost and high degree of accuracy. However, as mentioned in Chapter 7, DMMs can be severely affected by the presence of electromagnetic fields. Also, when a voltage signal is slowly changing, a DMM may show a confusing sequence of readings, where an analog meter would clearly display the signal fluctuations with its pointer. Consequently, many technicians still prefer to tune radio circuits with analog meters, adjusting the circuit until the pointer reaches a peak, or null, with the signal level.

Like its analog counterpart, a DMM is designed to read the rms value of an AC signal. Figure 12–2 shows a DMM connected to read the voltage at a 120 V AC outlet. Fixed voltages and frequencies, such as a 120 V, 60 Hz supply, are easily read by a DMM. Typical DMMs are limited to AC circuit measurements at frequencies between 45 Hz and 1000 Hz. For higher frequency voltage and current measurements, special metering equipment is required.

Section Review

1. AC meters often use a rectifier to convert the AC input signal to a DC value (True/False).
2. What is an electrodynamometer?
3. What are two disadvantages of using digital meters in AC measurements?

FIGURE 12–2 *DMM measuring AC voltage.*

OSCILLOSCOPES

An **oscilloscope** is an electronic measuring instrument consisting of a cathode-ray tube (CRT) and various associated circuit sections. The oscilloscope commonly is used to automatically plot a particular voltage variation versus time. Almost anything can be measured on the two-dimensional graph drawn by an oscilloscope. Parameters such as phase shift, rise time, decay time, peak-to-peak voltages, repetition rate, pulse duration, pulse delay time, period, and frequency can all be measured using this versatile instrument. A typical oscilloscope is shown in Figure 12–3.

The heart of the oscilloscope is the cathode-ray tube because it performs the basic functions to convert a signal into an image. The CRT is a vacuum tube similar in shape to a TV picture tube, as illustrated in Figure 12–4. A cathode-ray tube consists of an electron gun for supplying a concentrated beam of electrons, two pairs of deflection plates for changing the direction of the electron beam, and a screen coated with a substance that glows when struck by the electron beam. A transparent, ruled screen, called a **graticule,** generally is mounted in front of the fluorescent screen.

An electric current is passed through the heater of the electron gun in order to increase the temperature of the cathode to a point where electrons are emitted. The cathode is surrounded by a cylindrical cap that is at a negative potential. This cap, which has a small hole located along the longitudinal axis of the CRT, acts as the control grid. Because the control grid is at a negative potential, electrons are repelled away from the cylindrical walls and, consequently, stream through the hole where they move into the electric fields of the focusing and accelerating grids.

FIGURE 12–3 *Oscilloscope.*

FIGURE 12–4 *Basic components of a CRT*

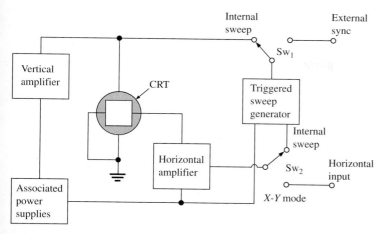

FIGURE 12–5 *Block diagram of a basic oscilloscope.*

Figure 12–5 shows a block diagram of a basic cathode-ray oscilloscope. The vertical input signal is applied to the vertical deflection plates via a multistage vertical amplifier. The vertical amplifier is the main factor in determining the sensitivity and bandwidth of an oscilloscope. The vertical sensitivity is a measure of how much the electron beam will deflect for a specified input signal.

The horizontal amplifier provides the deflection voltages required to deflect the beam across the *X*-axis of the CRT. With the switch set to **internal sync,** as it is for normal operation of the oscilloscope, the output of the vertical amplifier is applied to the **sweep generator.** The purpose of the sweep generator is to develop a voltage at the horizontal deflection plate that increases linearly with time. This linearly increasing voltage, called a **ramp voltage,** or **sawtooth waveform,** causes the beam to be deflected equal distances horizontally per unit of time. The rate at which the sawtooth wave rises establishes what is known as the *time base.*

Figure 12–6 illustrates a typical CRT faceplate with a graticule. Generally, graticules are laid out in an 8 × 10 pattern. Each of the 8 vertical and 10 horizontal lines divide the screen into 1 cm squares. The minor divisions on the vertical and horizontal center lines of the graticule represent increments of 0.2 cm. The volts/division switch and time/division switch on the front of an oscilloscope change the value of each major vertical and horizontal division on the graticule. For example, on the 5 V/division setting, each of the 8 major vertical divisions represents 5 V, so the entire screen can display up

Center graticule line

Center graticule line

FIGURE 12–6 *CRT faceplate with graticule.*

FIGURE 12-7 *Dual-trace oscilloscope.*

to 40 V from top to bottom. If the time/division switch is set at 10 ms, each of the 10 major horizontal divisions represents 10 ms, and the entire screen can display up to 100 ms from left to right.

Figure 12–7 shows the front panel of a dual-trace oscilloscope. The **intensity control** varies the brightness of the displayed waveform, and the **focus control** allows the focus of the wave to be adjusted.

The dual-trace oscilloscope allows two waveforms to be displayed on the screen at any given time. Two coaxial-type input terminals are used to connect the oscilloscope to the input signals. The selector switches beside the input connectors allow the operator to select AC or DC signals. The ground (GND) selector disconnects the input signals and grounds the input terminals. This allows each trace to be referenced to an appropriate position on the screen. By selecting MODE, input signal A, B, or signals A and B (dual) can be displayed on the screen.

The **vertical controls** for channels A and B allow an appropriate volts/division (VOLTS/CM) to be selected and the waveform to be moved up or down using the POSITION knob. The time/division (TIME/CM) switch selects the horizontal deflection sensitivity of the display. The horizontal position is adjusted by turning the position knob. The vernier knob in the center of both the TIME/CM and VOLTS/CM must be turned fully clockwise for these controls to be calibrated.

The **trigger control** is used to provide a stable waveform display. For the waveform to be stable, the display must commence at exactly the instant that the input waveform is at its zero position. The SOURCE selector switch allows the selection of INT (internal), or EXT (external) as triggering sources for the time base. When INT is selected, the time base is triggered by one of the input signals. The EXT trigger source allows the time base to be triggered from an external source connected to the EXT trigger terminal.

The most direct voltage measurement made with an oscilloscope is the peak-to-peak value. The peak-to-peak value of voltage is calculated as follows:

$$V_{\text{p-p}} = \left(\frac{\text{volts}}{\text{division}} \right) \left(\frac{\text{number of divisions}}{1} \right) \qquad (12.1)$$

The period and frequency of periodic signals are easily measured with an oscilloscope. The waveform must be displayed in such a manner that one complete cycle is displayed on the CRT screen. Accuracy is usually improved if the single cycle displayed fills as much of the horizontal distance across the screen as possible. The period is determined as

$$T = \left(\frac{\text{time}}{\text{division}} \right) \left(\frac{\text{number of divisions}}{\text{cycle}} \right) \qquad \textbf{(12.2)}$$

The frequency is then calculated as the reciprocal of the period, or

$$f = 1/T$$

_____ EXAMPLE 12–1 _____

Calculate the peak-to-peak voltage of the sine wave shown in Figure 12–8, if the volts/division switch is on 2 V/cm.

FIGURE 12–8 *Oscilloscope display for Example 12–1.*

Solution

$$V_{\text{p-p}} = \left(\frac{\text{volts}}{\text{division}} \right) \left(\frac{\text{number of divisions}}{1} \right)$$

$$= \left(\frac{2\ \text{V}}{\text{cm}} \right) \left(\frac{4}{1} \right)$$

$$= 8\ \text{V}$$

_____ EXAMPLE 12–2 _____

Solution

Determine the frequency of the sine wave shown in Figure 12–8 if the time/division switch is set at 1 ms/cm.

$$T = \left(\frac{\text{time}}{\text{division}} \right) \left(\frac{\text{number of divisions}}{\text{cycle}} \right)$$

$$= \left(\frac{1\ \text{ms}}{\text{cm}} \right) \left(\frac{3.2\ \text{cm}}{\text{cycle}} \right)$$

$$= 3.2\ \text{ms} \quad \text{or} \quad 0.0032\ \text{s}$$

$$f = \frac{1}{T}$$

$$= \frac{1}{0.0032\ \text{s}}$$

$$= 312.5\ \text{Hz}$$

_____ EXAMPLE 12–3 _____

For the square wave shown in Figure 12–9, determine the

(a) Peak-to-peak voltage.
(b) Frequency.

Assume the volts/division switch is set at 10 V/cm and the time/division switch is at 10 μs/cm

FIGURE 12–9 *Oscilloscope display for Example 12–3.*

Solution

(a)
$$V_{p-p} = \left(\frac{\text{volts}}{\text{division}} \right) \left(\frac{\text{number of divisions}}{\text{cycle}} \right)$$

$$= \left(\frac{10 \text{ V}}{\text{cm}} \right) \left(\frac{5.2}{1} \right) = 52 \text{ V}$$

(b)
$$T = \left(\frac{\text{time}}{\text{division}} \right) \left(\frac{\text{number of divisions}}{\text{cycle}} \right)$$

$$= \left(\frac{10 \text{ μs}}{\text{cm}} \right) \left(\frac{7.2 \text{ cm}}{\text{cycle}} \right)$$

$$= 72 \text{ μs} \quad \text{or} \quad 0.000072 \text{ s}$$

$$f = \frac{1}{T}$$

$$= \frac{1}{0.000072 \text{ s}} = 13.89 \text{ kHz}$$

_____ EXAMPLE 12–4 _____

Determine the average voltage of the pulse wave shown in Figure 12–10. Assume the volts/division switch is set at 5 V/cm and the time/division switch is set at 10μs/cm, and the baseline is 0 V.

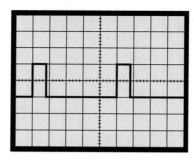

FIGURE 12–10 *Oscilloscope display for Example 12–4.*

Solution

$$V_{p-p} = \left(\frac{\text{volts}}{\text{division}} \right) \left(\frac{\text{number of divisions}}{\text{cycle}} \right)$$

$$= \left(\frac{5 \text{ V}}{\text{cm}} \right) \left(\frac{2}{1} \right) = 10 \text{ V}$$

$$\text{duty cycle} = \frac{t_w}{T}$$

$$= \frac{(0.8 \text{ cm}) (10 \text{ μs/cm})}{(5 \text{ cm}) (10 \text{ μs/cm})} = 0.16$$

$$E_{ave} = \text{baseline} + (\text{duty cycle}) (V_{p\text{-}p})$$
$$= 0 \text{ V} + (0.16) (10 \text{ V}) = 1.6 \text{ V}$$

Section Review

1. What is an oscilloscope?
2. What is a graticule?
3. A dual-trace oscilloscope allows two waveforms to be displayed on the screen at any given time (True/False).

12-4
FREQUENCY COUNTERS

In addition to oscilloscopes, frequency measurements can also be made with **frequency counters.** Although analog frequency counters are still available, **digital counters** are becoming increasingly popular. Digital electronic counters are used to measure the frequency of an unknown signal, the period of any signal, and the frequency ratio of two different signals. Frequency counters will typically provide a five-digit readout of the frequency being measured, with a range of between 2 Hz and 300 kHz. These counters are often equipped with plug-in capability that allows the frequency range to be extended beyond 12.5 GHz. An example of a digital frequency counter is shown in Figure 12–11.

Many frequency counters are also capable of counting in periods instead of cycles per second. Period measurements are quite useful when measuring unknown low-frequency signals accurately. Every frequency counter has an interval oscillator that serves as a reference, or time base, for both frequency and time measurements. An **oscillator** is a circuit that continuously generates a periodic, time-varying wave that can be either sinusoidal or nonsinusoidal in shape.

A simplified block diagram of a digital electronic counter is shown in Figure 12–12. Input signal A may be a sine wave that is passed through an amplifier and then through a **Schmitt trigger** circuit. A Schmitt trigger processes the input signal and is crucial in the

FIGURE 12–11 *Digital frequency counter.*

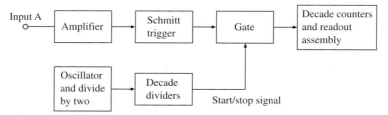

FIGURE 12–12 *Block diagram of digital frequency meter.*

operation of every electronic counter. The Schmitt trigger generates positive- and nega-tive-going spikes that correspond to the sine wave's zero crossings. A typical digital elec-tronic counter will use a 2 MHz oscillator that is divided by two to produce a stable 1 MHz clock. The 1 MHz clock signal is divided down in the decade dividers. The time elapsed between the start/stop signal depends on the setting of the frequency-time selector switch on the digital counter.

Figure 12–13 illustrates how the start/stop gate signal determines the actual fre-quency count. The length of time the start/stop gate is on determines the number of Schmitt trigger spikes allowed to pass through the gate and into the decade counters and digital readout assembly. For example, if the input signal is 1000 Hz and the frequency-time selector switch is set at 1 s, the start/stop gate will be 1 s long. This means that the 1 MHz clock must have been divided down through the decade dividers by 1 million times to gen-erate the 1 s start/stop gate. The decade counters and readout assembly receive 1000 spikes, corresponding to 1000 Hz. Therefore, the frequency displayed by a digital counter is simply the number of Schmitt trigger pulses per time period.

Section Review

1. Digital electronic counters are used to measure the frequency of an unknown signal, the period of any signal, and the frequency ratio of two different signals (True/False).
2. What is an interval oscillator?
3. Frequency counters will typically provide a range of between 0 Hz and 300 GHz (True/False).

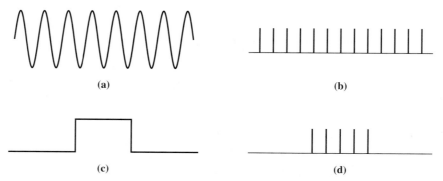

FIGURE 12–13 *Waveforms for frequency measurement. (a) Input.*
(b) Schmitt trigger. (c) Start/stop gate. (d) Frequency count.

TABLE 12–1 *Signal generator frequency-band limits.*

Abbreviation	Band	Approximate Range
AF	Audio frequencies	20 Hz–20 kHz
RF	Radio frequencies	Above 30 kHz
VLF	Very low frequencies	15 kHz–100 kHz
LF	Low frequencies	100 kHz–500 kHz
BDCST	Broadcast frequencies	0.5 kHz–1.5 Mhz
VIDEO	Video frequencies	1.5 MHz–5 MHz
HF	High frequencies	1.5 MHz–30 MHz
VHF	Very high frequencies	30 MHz–300 MHz
UHF	Ultrahigh frequencies	300 MHz–3000 MHz
MICROWAVE	Microwave frequencies	Above 3000 MHz

12–5

SIGNAL GENERATORS

Signal generators are designed to provide an alternating voltage at a certain frequency and amplitude. When troubleshooting electronic circuits, signal generators can be used to reproduce sinusoidal and nonsinusoidal waves of various frequencies. They are often used in testing and aligning radio transmitters, receivers, and amplifiers. The frequency bands covered by signal generators are shown in Table 12–1. A **band** is a specific range of frequencies.

Signal generators use oscillators to develop voltage waveforms of varying frequency. Most signal generators are voltage-output instruments and should not be expected to deliver large amounts of current.

There are two basic types of signal generators—audio frequency (AF) signal generators and radio frequency (RF) signal generators. Audio signal generators produce stable audio frequency signals for testing audio equipment. Radio frequency signal generators typically provide frequencies from 30 kHz to 3000 MHz. Figure 12–14 shows a typical

FIGURE 12–14 *AF signal generator.*

FIGURE 12–15 *Typical function generator.*

audio frequency generator with a frequency range of 20 Hz to 20 kHz. This frequency range is divided into separate, switch-selected ranges with continuous tuning within each band. Audio signal generators are used to inject signals into circuits, such as amplifiers, in order to measure the amplifier's output signal strength and frequency response.

Signal generators that are used for generating a wide variety of signals are called **function generators.** A typical function generator is capable of providing square, ramp, pulse, and sawtooth, as well as, sine waves. The repetition rate is generally selectable from 0.0005 Hz to 5 MHz. Although sine waves are useful for frequency-response tests, the square and pulse waves are used for testing transient response. Every function generator produces an output that has three important characteristics: amplitude, waveshape, and frequency. Function generators are also capable of supplying a DC offset, which adds a positive or negative DC component to any of the waveforms produced by the generator. An example of a function generator is shown in Figure 12–15.

Section Review

1. What is an audio frequency (AF) signal generator?
2. Radio frequency signal generators typically provide frequencies from 20 Hz to 20 kHz (True/False).
3. What is a function generator?

— 12–6

TROUBLESHOOTING WITH AC METERS

The techniques employed when troubleshooting with AC meters are similar to those used with DC meters. In both cases, troubleshooting represents a practical application of theory. It is a logical, systematic process of proceeding from effect to cause. Troubleshooting is also a process of elimination. In other words, narrow the problem by identifying the functioning parts of a circuit until all that remains are the nonfunctioning parts of the circuit. To properly troubleshoot any type of circuit, it is important to know "what should be there" when taking measurements. Adequate technical literature, knowledge of theory, and understanding the capabilities and limitations of test equipment form the basis of effective troubleshooting techniques.

When troubleshooting electronic equipment such as, TVs, VCRs, and audio receivers, the first test usually involves checking the output waveform at various points in the circuit using an oscilloscope. The waveform measurements at the output are made with the circuit energized and generally with some sort of signal generator connected as the input signal.

Function generators are often used for signal tracing when troubleshooting electronic circuits. The advantage of using a function generator is that the frequency, waveshape, and level of the signal can be set to specific, fixed values, allowing the signal to be easily identified at various points in the circuit. Consequently, any distortion or abnormal signal magnitudes can be pinpointed. In audio electronic circuits, the function generator can be used to replace the normal signal input, such as a microphone or tape head. The generator can also be used to inject a signal at some intermediate point in the circuit under test.

When troubleshooting with a signal generator, the amplitude and frequency of the generator must be correct. If at all possible, the actual waveforms with which the circuit normally operates are preferred for troubleshooting. Applying a signal to a circuit that is normally fed by some other point, or stage, in the circuit can create certain problems. For example, any DC in the circuit must be blocked or isolated from entering the signal generator. Otherwise, the correct amplitude for the injected signal may be difficult to estimate. Also, the generator output is essentially a short circuit when connected to DC voltage, so, if the generator is not isolated from the DC voltage present in the circuit under test, the generator or the circuit itself may be damaged.

Two very common tests performed when troubleshooting are voltage-gain tests and frequency-response tests. The voltage gain of an electronic circuit is determined as the ratio of the output voltage to the input voltage. The frequency response of a circuit is the manner in which the signal strength changes as the frequency changes. In some situations, the change in the phase shift between the input and output as the frequency varies is also an important characteristic of frequency response. When making frequency response measurements, the output signal from the circuit being tested should be monitored on an oscilloscope as the signal frequency is changed. If the signal appears distorted at some frequency, then the level of the function generator should be reduced as required.

Oscilloscopes are popular troubleshooting instruments because they reveal any distortion that may be present in the signal. Whenever the output waveshape is different from the input waveshape, the signal is said to be distorted. Except for a change in amplitude, the output signal and the input signal should always be identical. If there is a difference between the two signals, distortion is present. For example, in audio circuits, the presence of second or third harmonics can distort the fundamental signal. Although the oscilloscope is particularly useful for showing moderate to severe cases of audio frequency distortion, small amounts of distortion can be difficult to detect using the oscilloscope.

Section Review

1. Function generators are often used for signal tracing when troubleshooting electronic circuits (True/False).
2. What are two common tests performed when troubleshooting electronic circuits?

12–7
PRACTICAL APPLICATION SOLUTION

Your task was described in the chapter opener as using an oscilloscope to test the duty cycle and average output voltage of a digital controller. According to the manufacturer's specifications, with the rotary dial set at the mid-point position, the output voltage should be 5 V with a 0.5 duty cycle. An oscilloscope connected to the output of the controller indicates a peak-to-peak voltage of 20 V, a pulse width of 40 ms and a period of 100 ms. The following steps illustrate the method of solution for this practical application of AC measuring instruments and troubleshooting equipment.

STEP 1   Determine the duty cycle.

$$\text{duty cycle} = \frac{t_w}{T} = \frac{40 \text{ ms}}{100 \text{ ms}} = 0.4$$

STEP 2   Determine the average output voltage.

$$V_{\text{ave}} = (\text{duty cycle})(E_{\text{p-p}}) = (0.4)(20 \text{ V}) = 8 \text{ V}$$

STEP 3   Adjust the calibration potentiometer until the duty cycle and average output voltage correspond to manufacturer's values.

Summary

1. AC meters often use a rectifier to convert AC input signals to a DC value.
2. AC meters typically measure rms (effective) values of voltage and current.
3. An electrodynamometer is an analog meter that is capable of measuring both AC and DC voltages and currents.
4. An oscilloscope is an electronic measuring instrument consisting of a cathode-ray tube (CRT) and various associated circuit sections.
5. A dual-trace oscilloscope allows two waveforms to be displayed on the screen at any given time.
6. Digital frequency counters provide a numerical display of frequency measurements.
7. Signal generators are designed to provide an alternating voltage at a certain frequency and amplitude.
8. Signal generators that are used for generating a wide variety of waveshapes are called *function generators*.
9. Function generators are often used for signal tracing when troubleshooting electronic circuits.
10. Two common tests performed when troubleshooting with AC test equipment are voltage-gain tests and frequency-response tests.

Answers to Section Reviews

Section 12–1

1. True
2. To "inject" a signal of a certain frequency, amplitude, and waveshape
3. True

Section 12–2

1. True
2. An analog meter that is capable of measuring both AC and DC voltages and currents
3. Can be severely affected by the presence of electromagnetic fields and may produce erroneous readings for slow-changing signals

Section 12–3

1. An electronic measuring instrument consisting of a CRT and various associated circuit sections

2. A transparent, ruled screen mounted on the front of the CRT

3. True

Section 12–4

1. True

2. A device that serves as a reference, or time base, for both frequency and time measurements

3. True

Section 12–5

1. A generator that produces stable audio frequency signals for testing audio equipment

2. False

3. A generator that is capable of generating a wide variety of signals including square, pulse, and ramp waves

Section 12–6

1. True

2. Voltage-gain tests and frequency-response tests

Review Questions ___

Multiple Choice Questions

12–1   The two basic types of AC test equipment are
(a) Radio frequency and audio frequency generators
(b) Oscilloscopes and frequency counters
(c) Signal-measuring and signal-generating instruments
(d) Voltmeters and ammeters

12–2   D'Arsonval meters can measure AC when used with a
(a) Diode   (b) Bridge rectifier   (c) Electrodynamometer   (d) Signal generator

12–3   When measuring a square wave with no offset, the average voltage is equal to
(a) $0.637 \times$ peak   (b) Duty cycle $\times$ peak
(c) $0.707 \times$ peak   (d) The peak voltage

12–4   When reading a pulse wave with a voltmeter, the average voltage is equal to
(a) $0.637 \times$ peak + baseline   (b) Duty cycle $\times$ peak-to-peak + baseline
(c) $0.707 \times$ peak   (d) The peak voltage

12–5   An electrodynamometer is a meter that is capable of measuring
(a) Only AC voltages   (b) Only DC voltages
(c) Both AC and DC voltages   (d) AC voltages when used with a rectifier

12–6   The oscilloscope displays _____ on the vertical axis and _____ on the horizontal axis.
(a) Voltage, time   (b) Time, voltage
(c) Voltage, frequency   (d) Frequency, voltage

12–7   The transparent, ruled screen on the front of an oscilloscope is called the
(a) Beam finder   (b) CRT   (c) Vertical display   (d) Graticule

12–8   The purpose of a sweep generator in an oscilloscope is to
(a) Develop a voltage at the vertical deflection plate
(b) Control the brightness of the displayed wave
(c) Develop a voltage at the horizontal deflection plate
(d) Locate a waveform that has been shifted off the screen

12–9   The trigger control in an oscilloscope is used to
(a) Disconnect the input signals and ground the input terminals
(b) Provide a stable waveform display
(c) Allow the time base to be triggered by an external source
(d) Locate a waveform that has been shifted off the screen

12–10   In addition to oscilloscopes, frequency measurements can also be made with
(a) Voltmeters   **(b)** Signal generators   **(c)** Oscillators   **(d)** Frequency counters

12–11   A circuit that continuously generates a periodic, time-varying wave is called a(n)
(a) Schmitt trigger   **(b)** Oscilloscope   **(c)** Oscillator   **(d)** Frequency counter

12–12   Instruments designed to provide an alternating voltage at a certain frequency and amplitude are called
(a) Signal generators   **(b)** Oscilloscopes
(c) Oscillators   **(d)** Frequency counters

12–13   The two basic types of signal generators are called _____ frequency generators.
(a) Audio and video   **(b)** VHF and UHF
(c) Video and microwave   **(d)** Audio and radio

12–14   Signal generators that are used for generating a wide variety of signals are called
(a) Audio frequency (AF) generators   **(b)** AF and RF generators
(c) Function generators   **(d)** Oscillators

Practice Problems

12–15   A sinusoidal signal has a peak-to-peak display of six divisions on an oscilloscope. The vertical sensitivity is set at 5 V/division. Calculate
(a) $E_{p\text{-}p}$ **(b)** E_p **(c)** E_{rms}

12–16   A sinusoidal signal has a display of eight horizontal divisions on an oscilloscope for one cycle. The seconds/division switch is set at 50 μs/division. Calculate the frequency of the signal.

12–17   A sinusoidal signal has a peak-to-peak display of four divisions on an oscilloscope. The vertical sensitivity is set at 10 V/division. Calculate
(a) $E_{p\text{-}p}$ **(b)** E_p **(c)** E_{rms}

12–18   A sinusoidal signal has a display of 5 1/2 horizontal divisions on an oscilloscope for one cycle. The time/division switch is set at 100 μs/division. Calculate the frequency of the signal.

12–19   For the nonsinusoidal waveform shown in Figure 12–16, calculate the
(a) Positive and negative amplitudes of the signal
(b) Period and frequency of the signal
Assume the vertical and time-base controls are set for 10 V/cm and 5 ms/cm, respectively.

FIGURE 12–16

Essay Questions

12–20 What are the two basic types of AC test equipment?

12–21 List three types of signal generating instruments.

12–22 Describe the basic operation of a rectifier in an AC measuring instrument.

12–23 How can a D'Arsonval movement be used to measure AC signals?

12–24 Explain the fundamental difference between an electrodynamometer and a D'Arsonval meter.

12–25 What is an oscilloscope used for?

12–26 Describe the basic operating principle of a frequency counter.

12–27 Explain the purpose of using signal generators when troubleshooting electronic equipment.

12–28 List the two basic types of signal generators.

12–29 Describe the difference between a signal generator and a function generator.

12–30 What is distortion, and how is it identified when troubleshooting?

13

Capacitance and Capacitors

Learning Objectives

Upon completion of this chapter you will be able to

- Describe the electrostatic field between two charged surfaces.
- Determine the flux density of a capacitor.
- Define relative permittivity and dielectric strength.
- Express the capacitance of a device in terms of charge and potential difference.
- List three factors that determine the capacitance of a capacitor.
- Define the terms *leakage current* and *leakage resistance*.
- Describe various types of capacitors used in electronic circuits.
- Utilize the capacitor color code.
- Explain transients in *RC* circuits.
- Describe the universal time constant curve.
- Discuss the relationship between capacitors connected in series and in parallel.
- Define coupling capacitors and bypass capacitors.
- Troubleshoot capacitors.

Practical Application

As a service technician you are required to troubleshoot a power amplifier
that is producing an output voltage that is below its rated value. The amplifier
is at least 10-years old and uses electrolytic bypass and coupling capacitors.
The amplifier has three bypass capacitors, C_1, C_2, and C_3. The DC voltage
drops across the bypass capacitors are measured as 7.2 V, 5.5 V, and 8.3 V,
respectively. The AC voltage drops across the bypass capacitors are 0 V, 2.5
V, and 0 V, respectively.

13–1

INTRODUCTION

When two conducting surfaces are separated by an insulator, often called a **dielectric** and a difference of potential is applied between the conductors, a state of stress will be established in the dielectric. If the potential difference is sufficient, the stress will cause the insulation to break down, and a spark will pass between the conducting surfaces, puncturing the insulation. **Breakdown** is the point at which current begins to flow in an insulator.

In the dielectric, a field of electric force, or an electrostatic field, is established. Stored energy is contained in this electrostatic field. Figure 13–1 shows two parallel plates of a conducting material, separated by a dielectric and connected through a switch and a resistor to a battery.

If the parallel plates are initially uncharged and the switch is left open, no net positive or negative charge will exist on either plate. When the switch is closed, charges from the source will distribute themselves on the plates; that is, a current will flow. There will be a surge of current at first, limited in magnitude by the resistance present. As more charge is accumulated and more voltage is developed across the plates, the accumulated charge tends to oppose the further flow of charge. This action creates a net positive charge on the top plate. Electrons are being repelled by the negative terminal through the lower conductor to the bottom plate at the same rate they are being drawn to the positive terminal. Finally, when enough charge has been transferred from one plate to the other, a voltage equal to the applied emf will have been developed across the plates. The final result is a net positive charge on the top plate and a net negative charge on the bottom plate. Figure 13–2 shows the two plates as charged bodies and an electric field set up in the dielectric.

The element shown in Figure 13–2, constructed of two plates separated by an insulating material, is called a **capacitor. Capacitance** is a measure of a capacitor's ability to store charge on its plates. The total strength of the electric field is represented by the total number of lines of force, or **dielectric flux.** The Greek symbol ψ (psi) is used to represent this flux.

The unit of electric charge is the coulomb, C, named after Charles Augustine de Coulomb. Coulomb determined that

> The force of attraction or repulsion between two charged bodies is directly proportional to the square of the distance between them.

Deficiency of electrons results in positively charged plate

Direction of lines of force

Excess of electrons results in negatively charged plate

FIGURE 13–2
Electrostatic field between two charged surfaces.

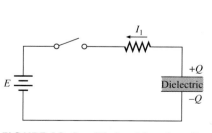

FIGURE 13–1   *Dielectric circuit.*

Mathematically, **Coulomb's law** can be expressed as

$$F = \frac{kQ_1 Q_2}{d^2}$$

(13.1)

where F = force between the bodies, in newtons
Q = charge of each body, in coulombs
k = a constant whose value depends on what material fills the volume of space in which the bodies are located (i.e., air, gas, water, etc.)
d = distance between the two charges, in meters

The proportionality constant, k, is called the **Coulomb constant,** and is often expressed as $k = 1/4\,\pi\epsilon_0$, where ϵ_0 is referred to as the **permittivity** of free space.

In SI units, the coulomb is defined as the unit of charge. A total of 6.24×10^{18} electrons are required to equal a charge of 1 C. Because the SI units of force, distance, and charge do not rely on Coulomb's law, the Coulomb constant, k, is determined as $k = 9.0 \times 10^9$ N $\cdot$ m^2/C^2.

_____ EXAMPLE 13–1 _____

Solution

$$F = \frac{kQ_1 Q_2}{d^2} = \frac{(9.0 \times 10^9)\,(4 \times 10^{-6})^2}{(0.06\ \text{m})^2} = 40\ \text{N}$$

Two charges, each having a magnitude of 4 μC are 6 cm apart. Determine the force of repulsion between the two charges.

Electric charge is measured in coulombs, and the coulomb is often used as the unit of electric flux. Therefore, a capacitor having a certain number of coulombs of charge will also have a similar amount of electric flux between its plates.

electric flux, $\psi = Q$

(13.2)

The larger the charge, Q, in coulombs, the greater the number of lines of flux extending or terminating per unit area. Twice the charge will produce twice the flux per unit area.

The charge-inducing capability of an electric field is called its **electric flux density, D.** Equation 13.2 implies that the electric flux density will be constant whenever the flux is evenly distributed, as it would be across the plates of a capacitor. The flux density that results from charge distribution across the surface of a capacitor is

$$D = \frac{\psi}{A} = \frac{Q}{A}$$

(13.3)

where A = area of plate, in square meters
D = flux density, in coulombs per square meter
Q = charge, in coulombs

If the distance between the plates shown in Figure 13–2 were decreased, the reaction of the energy field of each plate on the other is increased. This means that the flux increases without the applied voltage increasing. Therefore, the flux can be said to be inversely proportional to the distance between plates. **Elastance** can be defined as the opposition to the setting up of electric lines of force in an electric insulator or dielectric. The letter symbol for elastance is S. Elastance increases when the *distance* between plates increases. If the *area* of the plates were increased, the elastance would decrease.

Section Review

1. How is energy stored in a dielectric?
2. What is the symbol for electric flux?
3. What is electric flux density?

___ 13-2 _____

RELATIVE PERMITTIVITY (DIELECTRIC CONSTANT)

The elastance of any dielectric circuit depends on the material of the path. The **relative permittivity,** or **dielectric constant,** is a measure of how good a material is for the production of dielectric flux. The symbol for relative permittivity is the symbol ϵ_r.

The dielectric constant for a vacuum is taken as unity (1). For any other material, the constant will be more than 1, depending on how many more lines of force would be produced if the material were substituted for a vacuum as the path between the plates. Relative permittivity is dimensionless, since it is a ratio of the **absolute permittivity, ϵ,** of a material to the **absolute permittivity, ϵ_0,** of a vacuum. This ratio is expressed by the following equation:

$$\epsilon_r = \frac{\text{flux produced with a material as dielectric}}{\text{flux produced with a vacuum as dielectric}} \quad \text{(13.4)}$$

$$= \frac{\epsilon}{\epsilon_0}$$

The choice of ϵ_0 as the proportionality constant is the result of Gauss's law, which applies to any closed hypothetical surface (called a Gaussian surface). **Gauss's law** is stated as follows:

> The net number of electric lines of force crossing any closed surface in an outward direction is numerically equal to the net total charge within that surface.

The dielectric constant of a material will vary depending on the processing method in its manufacture. Some common values are given in Table 13–1.

TABLE 13–1 *Dielectric constants.*

Material	Typical Dielectric Constant
Vacuum	1.0
Air	1.0006
Ceramic (low loss)	5.0–570
Ceramic (high loss)	600–10,000
Glass	4.4–10.0
Mica (typical)	5.5
Mylar	3.0
Paper	4.0–6.0
Paraffin	2.1–2.5
Plastics	2.1–4.5
Porcelain	5.7
Rubber	2.5
Water	81.0
Wood	2.5–7.7

Section Review

1. What is relative permittivity?
2. State Gauss's law.
3. The dielectric constant of a material will vary depending on the processing method in its manufacture (True/False).

DIELECTRIC STRENGTH

Dielectric strength can be defined as the voltage per unit thickness at which breakdown occurs. The dielectric strength, therefore, corresponds to the field intensity required for breakdown. As the field intensity is increased, the polarization of the dielectric atoms becomes more pronounced. Finally, a value of field intensity may be reached at which so much force is exerted on the orbital electrons that they are torn free from their orbits. When breakdown occurs, the capacitor has characteristics very similar to those of a conductor. A typical example of breakdown is lightning, which occurs when the potential between the clouds and the earth is so high that charge can pass from one to the other through the atmosphere, which acts as a dielectric. The dielectric strengths of selected materials are listed in volts per mil in Table 13–2 (1 mil = 0.001 in.).

Capacitance exists between any two conductors separated by a dielectric. The conductors do not have to be plates. They may be wires, grids, or conductors of any shape. The dielectric may be air or any other material that is an insulator. In any two-wire cable, there will be capacitance between the two wires. Capacitance will also exist between circuit wiring and a metal chassis, between adjacent or opposite conductors on a printed circuit board, or between the collector, base, and emitter of a transistor. Capacitance resulting from these and other unwanted sources is referred to as **stray capacitance.** Stray capacitance is most easily minimized by keeping conductors as far apart as possible.

A capacitor has a capacitance of one **farad, F,** if one coulomb of charge is deposited on the plates by a potential difference of one volt across the plates. The farad is named after Michael Faraday, a nineteenth-century English chemist and physicist. Because the farad is, generally, too large a measure of capacitance for most practical

TABLE 13–2 *Dielectric strengths.*

Material	Strength (V/mil)
Air	76
Bakelite	150–500
Ebonite	30–100
Fiber	50
Glass (commercial)	760–3,800
Mica	760–5,600
Oil	100–500
Paper (kraft, dry)	250–600
Porcelain	100–250
Rubber	400–1,270
Vinyl (plastic)	15,800
Water (distilled)	380
Wood	25–75

applications, the microfarad is more commonly used. Expressed as an equation, capacitance is determined by

$$C = \frac{Q}{V} \tag{13.5}$$

where C = capacitance, in farads
Q = charge, in coulombs
V = potential difference, in volts

_____ **EXAMPLE 13-2** _____

What is the capacitance in which 200 V stores 3 coulombs?

Solution

$$C = \frac{Q}{V}$$

$$= \frac{3 \text{ C}}{200 \text{ V}}$$

$$= 0.015 \text{ F}$$

_____ **EXAMPLE 13-3** _____

A capacitance of 40 μF is connected for a considerable length of time across a 600 V source. How much charge is stored?

Solution

$$Q = CV$$

$$= (40 \times 10^{-6} \text{ F})(600 \text{ V})$$

$$= 0.024 \text{ C}$$

_____ **EXAMPLE 13-4** _____

A 1200 μF capacitor holds a charge of 0.016 C. What is the voltage across it?

Solution

$$V = \frac{Q}{C}$$

$$= \frac{0.016 \text{ C}}{1200 \times 10^{-6} \text{ F}}$$

$$= 13.33 \text{ V}$$

There are, essentially, three factors that determine the capacitance of a capacitor. They are

1. The effective area of the plates; the larger the area, the greater the capacitance
2. The distance between plates
3. The nature of the dielectric

A relationship can be derived from these three factors and expressed in the following equation:

$$C = \frac{\epsilon_0 \epsilon_r A}{d} \tag{13.6}$$

where C = capacitance, in farads
$\quad\quad \epsilon_r$ = dielectric constant (relative permittivity)
$\quad\quad \epsilon_0$ = permittivity of a vacuum
$\quad\quad d$ = distance between plates, in meters
$\quad\quad A$ = area of plates, in square meters

ϵ_0, the permittivity of a vacuum, is 8.85×10^{-12} farads/meter. For air, the value is only negligibly greater. The capacitance will be greater if the area of the plates is increased, or the distance between plates is decreased, or the dielectric is changed so that ϵ_r is increased.

What is the capacitance of the capacitor shown in Figure 13–3?

FIGURE 13–3   *Capacitor for Example 13–5.*

Solution

$$A = l \times w = 0.08 \times 0.4 = 0.032 \text{ m}^2$$

$$\epsilon_r \text{ (from Table 13–1)} = 5.5$$

$$C = \frac{(8.85 \times 10^{-12})\ (\epsilon_r)\ A}{d}$$

$$= \frac{(8.85 \times 10^{-12})\ (5.5)\ (0.032 \text{ m}^2)}{0.0002}$$

$$= 7.788 \times 10^{-9} \text{ F} = 7.788 \text{ nF}$$

A 500 pF capacitor is constructed using a poreclain dielectric between two plates, each having an area of 3×10^{-4} m². What is the thickness of the dielectric?

Solution   From Table 13–1, $\epsilon_r = 5.7$.

$$C = \frac{\epsilon_0 \epsilon_r A}{d}$$

$$d = \frac{\epsilon_0 \epsilon_r A}{C}$$

$$= \frac{(8.85 \times 10^{-12})\ (5.7)\ (3 \times 10^{-4} \text{ m}^2)}{500 \times 10^{-12} \text{ F}}$$

$$= 30.267 \times 10^{-6} \text{ m} = 30.267 \text{ μm}$$

Section 13–3 / Dielectric Strength   329

Section Review

1. What is dielectric strength?
2. What is stray capacitance?
3. State the equation for determining capacitance if charge and potential difference are known.
4. List the three factors that determine the capacitance of a capacitor.

— 13–4

LEAKAGE CURRENT

The **leakage current** in a capacitor is defined as the DC current that flows through the capacitor because of imperfections in the dielectric or that flows to surface paths from one plate to another. This very small value of current varies in inverse proportion to the insulation resistance of the dielectric.

Ideally, the dielectric insulating material has an infinite resistance. In reality, all dielectric materials have a finite resistance, called the **leakage resistance,** R_p. The leakage resistance will have a very large ohmic value and is shown in Figure 13–4 as being in parallel with the capacitor. The leakage resistor is the reason why it is impossible for a capacitor to maintain a charge indefinitely. If a charge is placed on a capacitor and not **refreshed,** it will eventually **leak off** through this leakage resistance. In most electric and electronic circuits, R_p is very large in comparison to other resistors in the circuit. Consequently, the effect of R_p on the circuit is considered to be negligible in most situations.

The leakage current of a capacitor multiplied by the applied voltage represents a power loss. A high value of leakage current will cause not only a rapid loss of charge, but will also result in the capacitor overheating. The **dissipation factor** of a capacitor is determined by the capacitor losses. These losses include both the power loss caused by leakage current as well as by **dielectric hysteresis.** Dielectric hysteresis is defined as an effect in a dielectric material caused by changes in orientation of electron orbits in the dielectric, such as the rapid reversal of the polarity of the line voltage.

If the losses are negligible and the capacitor returns the total charge to the circuit, it is considered to be a perfect capacitor. Therefore, the dissipation factor of a capacitor is a measurement of its efficiency. The dissipation factor of commercial capacitors varies between 0.0001 and 0.025.

Section Review

1. What is leakage current?
2. All dielectric materials have a finite resistance, called the leakage resistance (True/False).
3. How is the power loss of a capacitor calculated?

Leakage resistance

Capacitor

FIGURE 13–4 *Leakage resistance of a capacitor.*

Capacitors are available in many shapes and sizes. Generally, they are designated by their dielectric material. All capacitors can also be included under two headings: fixed and variable. The schematic symbols for fixed and variable capacitors are shown in Figure 13–5. The curved line of the capacitor symbol often represents the plate of the capacitor that is connected to the point of lower potential.

Fixed Capacitors

If its capacitance cannot be deliberately controlled, a capacitor is called *fixed.* Many types of fixed capacitors are available today. Some of the most common are paper, plastic, mica, ceramic, glass, and electrolytic.

The paper capacitor consists of aluminum foil and kraft paper dielectric rolled together and impregnated with wax or resin to exclude moisture. Paper capacitors commonly range from about 0.0005 µF to about 2 µF. A variation of the paper capacitor is the oil-filled capacitor. These capacitors are generally of a higher capacitance value (1.0 µF and up), with voltage ratings from 400 V to 5000 V. They are usually mounted in metal cases and, in the case of high-voltage ratings, the terminals are brought out through stand-off ceramic insulators.

The plastic film capacitor is very similar in construction to the paper capacitor. Plastics can withstand higher temperatures and are considerably more stable than paper. The disadvantage of plastic film capacitors is that for a given capacitance and voltage rating, plastic film capacitors are more costly than paper capacitors.

Mica capacitors are used mainly in the RF circuits of receivers and transmitters. Mica is one of the best natural insulators known. Radio transmitters use these capacitors because the voltage rating and current may go as high as 30 kV and 100 A. On account of its high cost, mica capacitors are seldom found with capacitance values greater than 0.05 µF. A variation of this kind of capacitor is the silver-mica capacitor. A thin layer of silver is deposited on the surface of the mica, and the resulting capacitor has excellent stability and tolerance. These silver-mica capacitors are used in frequency-selective (tuned) circuits and particularly for temperature (drift) compensation. Figure 13–6 shows some examples of mica capacitors.

Ceramic capacitors consist of a ceramic disk with silver plates attached to each flat surface. The leads are then attached through electrodes to the plates. An insulating coating of ceramic is then applied over the plates and dielectric. These capacitors have values ranging from a few picofarads to about 2 µF. Another type of ceramic capacitor is the multi-layer type, which consists of metal plates stacked alternately with ceramic dielectrics then molded into a single block. This construction is called monolithic. Some examples of ceramic capacitors are shown in Figure 13–7.

Glass was first used as a capacitor dielectric in the early 1950s. Capacitors of this type are characterized by extremely low losses and excellent stability and reliability. Glass

FIGURE 13–5 *Capacitor symbols. (a) Fixed. (b) Variable.*

(a)   (b)

FIGURE 13–6   *Mica capacitors.*

capacitors are available in capacitance values up to 0.01 μF and with voltage ratings up to 6000 V.

The electrolytic capacitor is used most commonly in situations where capacitances between 1 μF and several thousand microfarads are required. They are designed primarily for use in circuits where only DC voltages will be applied across the capacitor. An electrolytic capacitor consists of two plates separated by an electrolyte and a dielectric. One of the plates is oxidized and it is this oxide that forms the dielectric. These capacitors depend on chemical action within them to produce the dielectric. Current must flow through the capacitor to maintain this dielectric.

To maintain the dielectric film, electrolytic capacitors are connected into a circuit with the proper observation of polarity. For this reason, the terminals of this type of capacitor are marked positive and negative. The applied voltage must be connected positive to positive and negative to negative. If the capacitor is incorrectly connected, the current

FIGURE 13–7   *Ceramic capacitors.*

FIGURE 13–8 *Electrolytic capacitors.*

flowing through the device will be opposite in direction to the current that formed the dielectric, causing the dielectric oxide to be destroyed and short circuiting the capacitor. This can often cause the electrolytic capacitor to explode. Figure 13–8 shows some electrolytic capacitors.

Variable Capacitors

Variable capacitors generally use air as the dielectric and have one set of plates meshing between a second set of plates. The capacitance is altered by turning the shaft at one end to vary the common area of the movable and fixed plates. The greater the common area, the larger the capacitance. The fixed plate assembly is called the *stator*, and the movable portion is called the *rotor*. Because of a fringing effect, the minimum value of capacitance occurring when the plates are unmeshed is not zero but is a finite value. In high-voltage applications, the spacing between the plates of a variable capacitor is increased.

Another type of variable capacitor is the trimmer, or padding, capacitor. This capacitor consists of two or more plates separated by a dielectric of mica. A screw is mounted so that tightening the screw compresses the plates more tightly against the dielectric. By applying compression to the dielectric, the thickness of the dielectric is reduced and the capacitance increases. Some examples of trimmer capacitors are shown in Figure 13–9.

Section Review

1. What is the main application of mica capacitors?
2. Glass capacitors have extremely low losses and excellent stability and reliability (True/False).
3. What is the most common dielectric for variable capacitors?
4. How is the dielectric varied in a trimmer capacitor?

FIGURE 13–9 *Trimmer capacitors.*

TABLE 13–3 *Capacitor color code.*

Color	Significant Digit	Decimal Multiplier	Tolerance (%)	Voltage Rating
Black	0	10^0	±20	—
Brown	1	10^1	±1	100
Red	2	10^2	±2	200
Orange	3	10^3	±3	300
Yellow	4	10^4	±4	400
Green	5	10^5	±5	500
Blue	6	10^6	±6	600
Violet	7	10^7	±12.5	700
Gray	8	—	±30	800
White	9	—	±10	900
Gold	—	10^{-1}	±5	1000
Silver	—	10^{-2}	±10	2000
No color	—	—	±20	—

__ 13–6 __

CAPACITOR COLOR CODE

Some capacitors, such as mylar and molded mica, have color codes to indicate the voltage, tolerance, and value in picofarads of the capacitor. The method of determining the value of a capacitor by use of the color code is very similar to that of the resistor color code. The first two digits indicate the first and second significant digits, the third digit is the multiplier, the fourth digit is the tolerance, and the fifth digit, if any, is the rated voltage of the capacitor. Table 13–3 shows the chart for determining capacitance values based on color code.

____ EXAMPLE 13–7 ____

What is the value of a capacitor with the following color bands: brown, black, orange, red, gold?

Solution   The significant numbers are 10, the multiplier is 1000, the tolerance is ±2%, and the voltage rating is 1000 V. Therefore, the capacitor has a rating of 10,000 pF ±2% at 1000 V.

Section Review

1. Name two types of capacitors that use a capacitor color code.
2. The color code indicates voltage, tolerance, and value in microfarads (True/False).
3. What is the voltage rating of a capacitor with a fifth color band that is blue?

__ 13–7 __

TRANSIENTS IN *RC* CIRCUITS

Circuits that contain both resistance and capacitance are known as ***RC* circuits.** These circuits are used in a variety of applications including waveshaping and timing. When used in these types of applications, the resistor generally is connected in series with the capacitor to limit charging current.

A **transient** is the part of the change in a variable that disappears when going from one steady-state condition to another. In other words, a transient is a temporary, or short-lived occurrence. At **steady state,** the capacitor's charge and the voltage across it are constant and do not change with time. Because the potential of a capacitor cannot change instantaneously, a time period is required for the transition of the capacitor from the uncharged state to the charged state. The **transient state** of a capacitor would be the state of the capacitor between being fully charged and fully discharged. Although a device such as a switch would have a transient state, the term is usually applied to voltages and currents that increase or decrease in an exponential manner. Two typical exponential curves are shown in Figure 13–10, where x represents the horizontal component, such as time or frequency, and y represents the vertical component, such as current or voltage. In Figure 13–10(a), the exponential decay of the vertical component is given by the equation

$$y = e^{-x} \tag{13.7}$$

where $e = 2.71828$, the base of natural logarithms
  $x =$ a function of time, in seconds
  $y =$ vertical component, such as voltage or current

The exponential growth of the curve shown in Figure 13–10(b) is expressed by the following equation:

$$y = 1 - e^{-x}$$

Although both curves of Figure 13–10 begin at a definite point, they never reach a final value. However, in most cases the curve can be considered to rise to 100% and to fall to 0%.

Figure 13–11 shows a capacitor and a resistor connected through a switch to a DC power source. When the switch is in position A the capacitor will charge, and when it is in position B, the capacitor will discharge. If it is in the open position, there is no potential difference between the plates and no charge on the capacitor.

When the switch is at position A, the capacitor begins to charge. Initially (time = 0), the only limitation to the flow of current is the resistance (R), so the current will immediately rise to its maximum value. The instant the switch closes, electrons from the negative terminal of the power supply accumulate on the lower plate of the capacitor. The lower plate becomes negative, while the upper plate becomes positively charged from the positive terminal of the supply. A difference of potential now exists between the plates, which will oppose the applied voltage of the circuit. This opposition to the applied voltage is known as **counter emf.**

Although the net voltage of the circuit is the same, a slightly lower current continues to flow, which increases the counter emf and charge of the capacitor. This is a cumulative process; by increasing the charge on a capacitor the counter emf increases, and the charg-

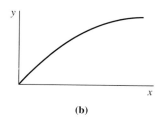

(a)   (b)

FIGURE 13–10 *(a) Exponential decay.*
(b) Exponential growth.

FIGURE 13–11 *RC circuit*
with switch.

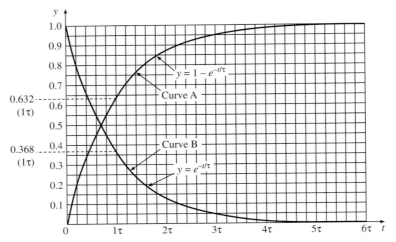

FIGURE 13–12 *Universal time constant curves.*

ing current decreases. When the counter emf is equal to the emf of the DC supply, the current ceases to flow.

The exponential curves, also called **charge curves,** of Figure 13–12 illustrate the values of the counter emf as the capacitor is charged (curve A) and as the capacitor is discharged (curve B). On curve A, the time required for the counter emf to reach approximately 63% of its maximum value is equal to one **time constant.** On curve B, the time required for the counter emf to fall to approximately 37% of its maximum value is also equal to one time constant. The value of one time constant is represented by the Greek letter τ (tau). From the chart of Figure 13–12 it can be seen that after five time constants a capacitor has been charged to within 1% of its maximum or discharged to within 1% of its minimum value. Since the graph represents an exponential curve, the curve never reaches 100% or 0%. Even after six time constants the capacitor is at 99.75% on the charging curve, 0.25% on the discharge curve.

The value of the time constant in seconds is equal to the product of the circuit resistance in ohms and its capacitance in farads. In equation form,

$$\tau = RC \tag{13.8}$$

where τ = time, in seconds
    C = capacitance, in farads
    R = resistance, in ohms

_____ **EXAMPLE 13–8** _____

What would be the value of one time constant in a series circuit containing a 0.001 µF capacitor and a 47 kΩ resistor?

Solution

$$\tau = RC$$
$$= (47 \times 10^3\ \Omega)\,(0.001 \times 10^{-6}\ \text{F})$$
$$= 47\ \mu s$$

Instantaneous voltage is the value of voltage at any specific instant in time. The **IEEE** (Institute of Electrical and Electronic Engineers) recommends the use of lowercase letter symbols to symbolize quantities that change with respect to time. The equation for the instantaneous voltage on a capacitor, when charging in a circuit containing resistance and capacitance, can be found using the following equation:

$$v_C = E(1 - e^{-t/\tau})$$ (13.9)

where v_C = capacitor voltage at time t
E = supply voltage
e = exponential constant = 2.71828
t = time from charge commencing, in seconds
τ = time constant, in seconds

EXAMPLE 13–9

FIGURE 13–13   Circuit for Example 13–9.

For the circuit shown in Figure 13–13, assume the capacitor is initially discharged. Determine the voltage across the capacitor after the switch has been closed for 81.6 ms.

Solution

$$\tau = RC = (1 \times 10^6 \ \Omega)(0.68 \times 10^{-6} \ F) = 0.68 \ s$$
$$v_C = E(1 - e^{-t/\tau})$$
$$= 110(1 - e^{-0.0816/0.68}) = 12.44 \ V$$

The following equation is used to calculate the length of time required for the voltage across a capacitor to rise to a certain value.

$$t = -\tau \ln\left(1 - \frac{v_C}{E}\right)$$ (13.10)

where E = initial value of voltage
v_C = voltage at any point on the charge curve for which time is desired
τ = time constant

EXAMPLE 13–10

Solution

$$\tau = RC = (1 \times 10^6 \ \Omega)(0.68 \times 10^{-6} \ F) = 0.68 \ s$$
$$t = -\tau \ln\left(1 - \frac{v_C}{E}\right)$$
$$= (-0.68 \ s) \ln\left(1 - \frac{83.5 \ V}{110 \ V}\right)$$
$$= 0.97 \ s$$

Given the circuit of Figure 13–13, where $R = 1 \ M\Omega$, $C = 0.68 \ \mu F$, and $E = 110 \ V$, determine the length of time required after the switch is closed for voltage across the capacitor to rise to 83.5 V.

In most practical applications, a capacitor is considered to be fully charged after five time constants. This is because

$$v_C = E(1 - e^{-5}) = E \times 0.993$$

which is considered to be close enough to 100% for the majority of electrical and electronic circuits.

EXAMPLE 13–11

A 56 kΩ resistor is connected in series with a 20 µF capacitor and a switch. If the circuit is connected across a 40 V supply and the capacitor is completely discharged initially, calculate the following when the switch is closed.

(a) Circuit time constant, τ

(b) Capacitor voltage at time $t = 2.5$ s

(c) Charge on the capacitor at time $t = 2.5$ s

(d) Time required for voltage to rise to 27.3 V

Solution

(a) $\tau = RC = (56 \times 10^3 \ \Omega)(20 \times 10^{-6} \ \text{F}) = 1.12$ s

(b) $v_C = E(1 - e^{-t/\tau}) = 40 \ \text{V} \ (1 - e^{-2.5/1.12}) = 35.71$ V

(c) $C = \dfrac{Q_C}{v_C}$

$Q_C = Cv_C = (20 \times 10^{-6} \ \text{F})(35.71 \ \text{V}) = 714.2 \ \mu\text{C}$

(d) $t = \tau \ \ln\left(1 - \dfrac{v_C}{E}\right) = 1.12 \ \ln\left(1 - \dfrac{27.3 \ \text{V}}{40 \ \text{V}}\right) = 1.285$ s

---

In Figure 13–14, when the switch is at position A, the capacitor is part of the load being supplied with energy from the source. When the switch is at position B, the capacitor *is* the source, and the resistor is its load. Essentially, the capacitor is functioning like a battery. When the switch is in position A, the battery is charging, and, when in position B, the battery is discharging.

Although the polarity of the capacitor remains the same during both the charging and discharging, the polarity of the resistor will reverse when the switch is changed from position A to position B. After five time constants, the capacitor is considered to be completely discharged. If the capacitor was fully charged and the switch changed to position B, the equation for the decaying voltage across the capacitor would be

$$v_C = Ee^{-t/\tau} \tag{13.11}$$

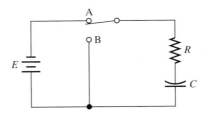

FIGURE 13–14 RC circuit with charge/discharge switch.

EXAMPLE 13–12

Solution

$$\tau = RC = (5.6 \times 10^3 \ \Omega)(100 \times 10^{-6} \ \text{F}) = 0.56 \ \text{s}$$

$$v_C = Ee^{-t/\tau} = 24e^{-1.05/0.56} = 3.68 \ \text{V}$$

The circuit shown in Figure 13–14 has a capacitance of 100 µF, a resistance of 5.6 kΩ, and a supply voltage of 24 V. Assume the capacitor is fully charged and the switch is changed to position B. Determine the voltage across the capacitor after the switch has been at position B for 1.05 s.

The length of time required for a voltage to decay to a certain value in an *RC* circuit is determined by the following equation:

$$t = \tau \ln\left(\frac{E}{v_C}\right) \qquad (13.12)$$

where E = initial value of voltage
v_C = voltage at any point on the decay curve for which time is desired

EXAMPLE 13–13

Solution

$$\tau = RC = (1 \times 10^3 \ \Omega)(10 \times 10^{-6} \ \text{F}) = 0.01 \ \text{s}$$

$$t = \tau \ln\left(\frac{E}{v_C}\right) = 0.01 \ \ln\left(\frac{21 \ \text{V}}{13 \ \text{V}}\right) = 4.8 \ \text{ms}$$

A 10 µF capacitor is charged to a value of 40 V and then is discharged into a 1 kΩ resistor. Determine the length of time required for the voltage to decrease from 21 V to 13 V.

Section Review

1. What is meant by the term *steady state?*
2. What is the transient state of a capacitor?
3. State the equation for determining the time constant of an *RC* network.
4. In most practical applications, a capacitor is considered to be fully charged after two time constants (True/False).

13–8

ENERGY STORED BY A CAPACITOR

When an ideal capacitor is connected to a voltage source, the source must expend energy to charge the capacitor. If the charging source is disconnected from the capacitor, energy will remain stored between its plates. The energy being stored in the capacitor is in the form of an electrostatic field.

As a capacitor is charged, the voltage between its plates at any given time will increase at an exponential rate. The total energy transferred to a capacitor during a given charging time will be determined by the average voltage. Since the voltage increases at a uniform rate with the charge current, the average voltage is one-half the final voltage. The energy stored by a capacitor can be found using the following equation:

$$W = \frac{CV_C^2}{2} \tag{13.13}$$

where W = energy stored, in joules
C = capacitance, in farads
V_C^2 = potential difference across capacitor, in volts

_____ EXAMPLE 13–14 _____

Calculate the energy stored by a 100 μF capacitor when 24 V is applied across the capacitor's plates.

Solution

$$W = \frac{CV_C^2}{2} = \frac{(100 \times 10^{-6} \text{ F}) (24 \text{ V})^2}{2} = 0.029 \text{ J}$$

Section Review

1. The energy being stored in the capacitor is in the form of an electromagnetic field (True/False).
2. State the equation for determining the energy stored by a capacitor.
3. What is the average voltage of a charging capacitor?

___ 13–9 ___

CAPACITOR IN SERIES AND PARALLEL

Capacitors may be connected in series or parallel to give resultant values. The resultant values may be either the sum of the individual values (in parallel) or a value less than that of the smallest capacitance (in series).

A circuit consisting of a number of capacitors in series is similar in some respects to a circuit containing several series-connected resistors. In a series capacitive circuit the same current flows through each part of the circuit and the applied voltage will divide across the individual capacitors.

When capacitors are connected in a series configuration, the magnitude of the charge on each plate must be the same. In Figure 13–15, the electrons that produce the negative

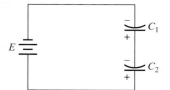

FIGURE 13–15 *Capacitors in series.*

charge on C_1 must come from the plate of C_2 and leave it with a positive charge. The battery maintains a potential difference between the positive plate of C_1 and the negative plate of C_2, transferring electrons from one to the other. The charge cannot pass between the plates of a capacitor. Therefore, the charge (Q) is the same in all parts of the circuit. In equation form,

$$Q = Q_1 = Q_2 = \cdots = Q_n$$

also if

$$C = \frac{Q}{V} \quad \text{and} \quad V = \frac{Q}{C}$$

then

$$V_T = \frac{Q}{C_T} = \frac{Q}{C_1} + \frac{Q}{C_2}$$

If we divide this last equation by Q, the result is

$$C_T = \frac{1}{1/C_1 + 1/C_2} \tag{13.14}$$

Equation 13–14 for series capacitors is similar to that for the total resistance of parallel resistors. For two capacitors in series, the total capacitance is found by the product-over-sum rule.

$$C_T = \frac{C_1 C_2}{C_1 + C_2} \tag{13.15}$$

_____ **EXAMPLE 13–15** _____

40 μF   60 μF   80 μF

100 V

FIGURE 13–16 *Circuit for Example 13–15.*

For the circuit of Figure 13–16, find the

(a) Total capacitance.
(b) Voltage across each capacitor.

Solution

(a) The capacitance is calculated from Equation 13.14.

$$\frac{1}{C_T} = \frac{1}{C_1} + \frac{1}{C_2} + \frac{1}{C_3}$$

$$= \frac{1}{40\ \mu F} + \frac{1}{60\ \mu F} + \frac{1}{80\ \mu F}$$

$$= 0.025 \times 10^6 + 0.0167 \times 10^6 + 0.0125 \times 10^6$$

$$= 54.17 \times 10^3$$

$$C = \frac{1}{54.17 \times 10^3}$$

$$= 18.46\ \mu F$$

(b)

$$Q_1 = Q_2 = Q_3 = Q = C_T V_T$$

$$= (100\ V)(18.46 \times 10^{-6}\ F)$$

$$= 18.46 \times 10^{-4}\ C$$

$$V_1 = \frac{Q_1}{C_1} = \frac{18.46 \times 10^{-4}\ C}{40 \times 10^{-6}\ F} = 46.15\ V$$

$$V_2 = \frac{Q_2}{C_2} = \frac{18.46 \times 10^{-4}\ C}{60 \times 10^{-6}\ F} = 30.77\ V$$

$$V_3 = \frac{Q_3}{C_3} = \frac{18.46 \times 10^{-4}\ C}{80 \times 10^{-6}\ F} = 23.08\ V$$

When capacitors are connected in parallel, one plate of each capacitor is connected directly to one terminal of the supply, and the other plate of each capacitor is connected to the other terminal of the supply. In Figure 13–17, because all the negative plates of the capacitors are connected together and all the positive plates are connected together, C appears as a capacitor with a plate area equal to the sum of all the individual plate areas. Connecting capacitors in parallel effectively increases plate area, consequently, the capacitance increases.

For capacitors connected in parallel, the total charge is the sum of all the individual charges.

$$Q_T = Q_1 + Q_2 + Q_3 + \cdots + Q_n$$

Since

$$Q_1 = C_1 V_1 \qquad Q_2 = C_2 V_2 \qquad Q_3 = C_3 V_3$$

where

$$V_T = V_1 = V_2 = V_3$$

the total charge is then

$$Q_T = C_T V_T = Q_1 + Q_2 + Q_3 = C_1 V_T + C_2 V_T + C_3 V_T$$

Dividing the last equation by V_T, we obtain an expression for the total parallel capacitance. This expression is

$$C_T = C_1 + C_2 + C_3 + \cdots + C_n \tag{13.16}$$

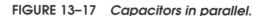

FIGURE 13–17 *Capacitors in parallel.*

_____ **EXAMPLE 13–16** _____

FIGURE 13–18 *Circuit for Example 13–16.*

In the circuit of Figure 13–18, solve for the

(a) Total capacitance.
(b) Charge on each capacitor.
(c) Voltage across each capacitor.

Solution

$$20 \ \mu F + 30 \ \mu F = 50 \ \mu F$$

(a)
$$C_T = \frac{(10 \ \mu F)(50 \ \mu F)}{10 \ \mu F + 50 \ \mu F} = \frac{500 \ \mu F}{60 \ \mu F} = 8.33 \ \mu F$$

(b)
$$Q_T = C_T V_T = (8.33 \times 10^{-6} \ F)(100 \ V) = 8.33 \times 10^{-4} \ C$$

This is the value of charge on C_1. To solve for C_2 and C_3, it is necessary to first find the capacitor voltages.

(c)
$$V_1 = \frac{Q_1}{C_1} = \frac{8.33 \times 10^{-4} \ C}{10 \times 10^{-6} \ F} = 83.3 \ V$$

$$V_2 = V_3 = V_T - V_1 = 100 \ V - 83.3 \ V = 16.7 \ V$$

The charge on capacitors C_2 and C_3 can now be found.

$$Q_2 = C_2 V_2 = (20 \times 10^{-6} \ F)(16.7 \ V) = 3.3 \times 10^{-4} \ C$$

$$Q_3 = C_3 V_3 = (30 \times 10^{-6} \ F)(16.7 \ V) = 5.0 \times 10^{-4} \ C$$

Section Review

1. State the relationship between total charge and the charge on each plate of a group of series-connected capacitors.
2. Connecting capacitors in parallel effectively increases the plate area (True/False).
3. When two capacitors are connected in parallel, what happens to the total capacitance?

COUPLING CAPACITORS AND BYPASS CAPACITORS

In electronic circuits, the term **coupling** refers to the transfer of energy from one circuit to another. The capacitor is popular in this application because it has the ability to block DC signals and allow AC signals to pass through. In AC amplifier circuits, DC voltages are often present in order for a transistor to operate properly. These DC voltages, called *biasing voltages,* are essential for the transistor's operation but are not intended to appear at the output signal of the amplifier circuit. By using **coupling capacitors,** it is possible to isolate the DC voltage in the circuit and still allow the AC signal to pass from the input to the output.

Figure 13–19(a) shows an example of a coupling capacitor in a circuit containing both AC and DC voltage sources connected across a load resistor. Figure 13–19(b) illustrates what an oscilloscope would display if connected across the terminals of the two supplies. The 30 V DC supply provides a solid DC voltage that represents the reference point for the AC source. During the first 180°, the AC input is positive going and *adds* to the DC voltage. Because the AC supply has a peak voltage of 10 V, the AC signal rises from the 30 V reference up to 40 V and falls back to 30 V at the 180° point. During the second 180°, the AC signal is now negative going and *opposes* the DC supply. Consequently, the AC voltage is now subtracted from the DC voltage and the voltage falls from 30 V to 20 V before returning to 30 V at the 360° point. In Figure 13–19(c), the coupling capacitor has blocked the DC voltage so that only the AC signal appears across the load. As a result, the load *sees* only the 10 V_p AC value.

Bypass capacitors are parallel-connected devices that are primarily used to provide a bypass around a component for AC signals, but still allow DC voltages to be present across the component. Figure 13–20(a) shows a bypass capacitor connected across a resis-

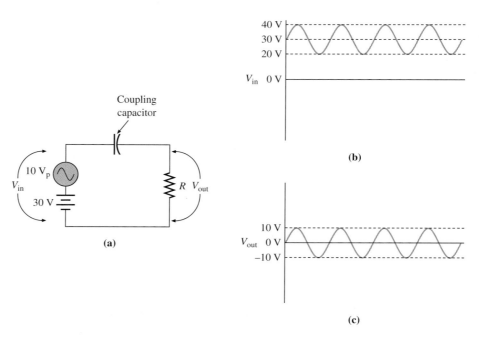

FIGURE 13–19 *(a) Coupling capacitor. (b) Input waveshape. (c) Output waveshape.*

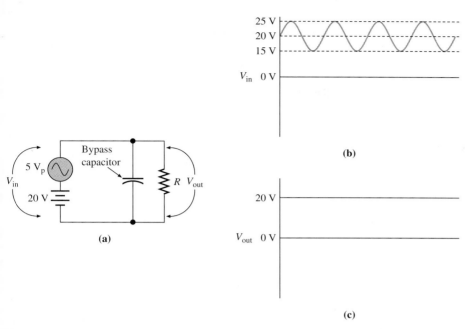

FIGURE 13–20 (a) Bypass capacitor. (b) Input waveshape. (c) Output waveshape.

tor. The waveshape across the resistor that results from the AC and DC supplies is shown in Figure 13–20(b). In Figure 13–20(c), if the voltage drop across the resistor begins to rise, the capacitor cannot charge instantly, and there is a delay before the voltage drop begins to increase. Consequently, if the voltage drop across the resistor begins to fall, the charge on the capacitor maintains an average voltage level. As a result, signal fluctuations across the resistor are virtually eliminated.

Section Review

1. In electronic circuits, what is coupling defined as?
2. What is the basic purpose of a coupling capacitor?
3. Bypass capacitors are series-connected devices (True/False).

_____ 13–11 __

TROUBLESHOOTING CAPACITORS

The test procedure for capacitors is significantly different from the procedure used to test resistors because of the energy-storage capability of capacitors. If a capacitor is to be tested using an ohmmeter, the charge stored in the capacitor is capable of severely damaging the ohmmeter. Consequently, a capacitor should always be discharged before attempting any kind of test on the device.

A very important consideration when troubleshooting capacitors is the dielectric leakage. As a capacitor ages, the dielectric resistance of the device may decrease, resulting in high values of leakage current. Because capacitors are often connected in series and parallel with other components, such as semiconductor devices, high leakage can have an adverse affect on these components. The operating characteristics of a circuit can be drastically altered by dielectric leakage, because the faulty capacitor is now providing a current path instead of having almost infinite resistance.

The ohmmeter is a popular test instrument for troubleshooting capacitors. Figures 13–21 through 13–23 show a standard test procedure for checking capacitors with an ohmmeter. In Figure 13–21 the capacitor is shorted out by using a jumper wire between the two terminals. Because capacitors are capable of storing a charge for long periods of time, it should never be assumed that a capacitor is discharged. Some capacitors are capable of retaining charge for days or weeks after power has been removed.

Figure 13–22 shows an ohmmeter connected across the terminals of the capacitor. The ohmmeter has an internal voltage source, so, when the ohmmeter is connected across the capacitor, the capacitor begins to charge because of the voltage applied from the ohmmeter battery. Initially, the ohmmeter pointer will swing towards zero, because a discharged capacitor acts as a short circuit. As the capacitor begins to charge, the meter read-

FIGURE 13–21   *Shorting out capacitor with jumper.*

FIGURE 13–22
Ohmmeter initially reads zero ohms.

FIGURE 13–23   *Pointer moves slowly towards infinity as capacitor charges.*

ing moves toward the infinity end of the scale, as shown in Figure 13–23. How quickly the pointer movement occurs depends on the size of the capacitor and the internal resistance of the ohmmeter. Capacitors with very small values (up to several hundred pF) will charge almost instantly, while capacitors with values of hundreds or thousands of farads can take several minutes to completely charge.

When the capacitor is fully charged, the pointer on the ohmmeter will stop moving. If the capacitor's dielectric resistance is very high, the pointer will indicate infinity. Anything less than infinity is an indication of the leakage resistance of the device. If a capacitor is acting as a short circuit, the ohmmeter will remain at zero ohms for the duration of the test. If the capacitor is open circuited, the ohmmeter will not start at zero ohms and gradually rise. Instead, the meter will start at infinity and remain there for the duration of the test.

When testing electrolytic capacitors with an ohmmeter, ensure that proper polarity is observed when connecting the terminals of the ohmmeter to the capacitor. The negative terminal of the meter must be connected to the negative terminal of the capacitor. If the terminals are reversed, the resistance readings will be false and, in the case of low-voltage capacitors, it is possible to destroy the capacitor by applying reverse-polarity voltage.

The principal faults that occur in variable capacitors are caused by age and the plates wearing out as a result of movement. Because trimmer capacitors use moving parts, wear is inevitable. In some situations, sliding contacts become noisy or intermittent. Often, a contact cleaner applied at such points eliminates the malfunction. In other cases, the plates of the variable capacitor may become bent because of mishandling. If the plates should touch, a short circuit will occur. Usually the fault can be located by using an ohmmeter connected across the capacitor. Rotating the shaft will indicate the position at which the short circuit is occurring.

Section Review

1. It is possible to test capacitors using an ohmmeter (True/False).
2. What effect does dielectric leakage have on the operating characteristics of a circuit?
3. What is the main cause of faults that occur in variable capacitors?

_13–12_
PRACTICAL APPLICATION SOLUTION

In the chapter opener your task was to troubleshoot a power amplifier that is producing an output voltage below its rated value. Preliminary voltmeter tests provided data regarding the voltage drops across the bypass capacitors. The amplifier was described as being relatively old and constructed with electrolytic bypasses and coupling capacitors. The following steps illustrate how this data can be used to arrive at a solution for this practical application of capacitance and capacitors.

STEP 1   Evaluate measured data. From the measurements given in the chapter opener and the discussion of bypass capacitors presented in Section 13–10, it is apparent that the AC voltage drop across a bypass capacitor should be 0 V. Capacitor C_2 has an AC voltage drop of 2.5 V, which implies that this capacitor is faulty. Because both AC and DC voltages are present across C_2, the capacitor has failed as an open circuit. This occurrence is relatively common in older electrolytic capacitors because of the dielectric drying out with age.

STEP 2   Replace the faulty capacitor and retest the amplifier. The output voltage should now be at its rated value.

___ Summary ___

1. Capacitance is a measure of a capacitor's ability to store charge on its plates.
2. The relative permittivity, or dielectric constant, is a measure of how good a material is for the production of dielectric flux.
3. A capacitor has a capacitance of one farad if one coulomb of charge is deposited on the plates.
4. Leakage current is the DC current that flows through a capacitor because of imperfections.
5. Capacitors are either fixed or variable.
6. A transient is the part of the change in a variable that disappears when going from one steady-state condition to another.
7. The value of one time constant, τ, is equal to the product of the circuit resistance and capacitance.
8. When capacitors are connected in a series configuration, the magnitude of the charge on each plate must be the same.
9. Connecting capacitors in parallel effectively increases plate area, which increases overall capacitance.
10. The principal faults that occur in variable capacitors are caused by age and the plates wearing out as a result of movement.

___ Answers to Section Reviews ___

Section 13–1

1. By an electrostatic field
2. ψ
3. The charge-inducing capability of an electric field

Section 13–2

1. A measure of how good a material is for the production of dielectric flux
2. The net number of electric lines of force crossing any closed surface in an outward direction is numerically equal to the net total charge within that surface.
3. True

Section 13–3

1. The voltage per unit thickness at which breakdown occurs
2. Capacitance resulting from unwanted sources such as conductors
3. $C = Q/V$
4. Effective area of plates, distance between plates, and nature of the dielectric

Section 13–4

1. The DC current that flows through the capacitor because of imperfections in the dielectric or that flows to surface paths from one plate to another
2. True
3. The leakage current of the capacitor multiplied by the applied voltage

Section 13–5

1. Used mainly in the RF circuits of receivers and transmitters
2. True

3. Air

4. By turning a screw to compress the dielectric

Section 13–6

1. Mylar and molded mica

2. False

3. 600 V

Section 13–7

1. When the capacitor's charge and voltage across it are constant and do not change with time

2. The state of the capacitor between being fully charged and fully discharged

3. $\tau = RC$

4. False

Section 13–8

1. False

2. $W = CV^2/2$

3. One-half the final voltage

Section 13–9

1. The magnitude of the charge on each plate of series-connected capacitors is the same in all parts of the circuit.

2. True

3. The total capacitance increases.

Section 13–10

1. Refers to the transfer of energy from one circuit to another

2. To block DC signals and allow AC signals to pass through

3. False

Section 13–11

1. True

2. High values of dielectric leakage can produce large amounts of leakage current with the capacitor providing a current path instead of having almost infinite resistance.

3. The plates become worn because of movement.

Review Questions ___

Multiple Choice Questions

13–1 The breakdown of dielectric material is caused by high
(a) Voltage   **(b)** Current   **(c)** Resistance   **(d)** Capacitance

13–2 Energy is stored in a capacitor in its
(a) Magnetic field   **(b)** Electrostatic field
(c) Magnetic and electrostatic field   **(d)** Dielectric flux

13–3 Capacitance is a measure of a capacitor's ability to store
(a) Magnetism   **(b)** Current   **(c)** Power   **(d)** Charge

13–4 The charge-inducing capability of an electric field is called its
(a) Dielectric flux   **(b)** Relative permittivity
(c) Electric flux density   **(d)** Elastance

13–5 If the charge on a capacitor is doubled, the flux per unit area
(a) Decreases by 50%   **(b)** Doubles   **(c)** Triples   **(d)** Decreases by 25%

13–6 The opposition to the setting up of electric lines of force is
(a) Elastance   **(b)** Permittivity   **(c)** Flux density   **(d)** Dielectric strength

13–7 Relative permittivity is also referred to as
(a) Elastance   **(b)** Dielectric flux   **(c)** Dielectric constant   **(d)** Flux density

13–8 All dielectric insulating materials have a finite resistance, called the
(a) Dielectric hysteresis   **(b)** Leakage resistance
(c) Flux density   **(d)** Elastance

13–9 The state of a capacitor between being fully charged and fully discharged is called the
(a) Instantaneous state   **(b)** Steady state
(c) Transient state   **(d)** Fluctuating state

13–10 The value of voltage at any specific instant in time is called
(a) Average voltage   **(b)** Effective voltage
(c) Peak voltage   **(d)** Instantaneous voltage

13–11 In electronic circuits, the transfer of energy from one point to another is called
(a) Bypass   **(b)** Transients   **(c)** Coupling   **(d)** Elastance

13–12 AC signals are shorted to ground by using
(a) Bypass capacitors   **(b)** Coupling capacitors
(c) Open-circuit capacitors   **(d)** RC circuits

13–13 A very important consideration when troubleshooting capacitors is the
(a) Voltage rating   **(b)** Dielectric leakage
(c) Hysteresis   **(d)** Dielectric flux

13–14 If a capacitor's dielectric resistance is very high, the pointer of an ohmmeter will read close to
(a) Zero ohms   **(b)** Infinity   **(c)** 20 Ω **(d)** Cannot be determined

Practice Problems

13–15 A 6.8 μF capacitor is connected to a 120 V supply. How much charge will it store?

13–16 What size capacitor is required to store a charge of 2.17 C from a 250 V supply?

13–17 A 200 μF capacitor holds a charge of 0.00325 C. What voltage is applied across the capacitor?

13–18 A capacitor is marked 0.0022 μF. What is its value in picofarads?

13–19 A computer-grade 0.39 nF capacitor has 12 V across its terminals. Determine the charge in coulombs.

13–20 Find the capacitance of a mica capacitor with a dielectric constant of 5.5, a plate area of 2.7 cm^2, and a parallel plate separation of 0.25 cm.

13–21 A capacitor has plates of 1.5 cm $\times$ 2.5 cm, separated by a ceramic dielectric ($\epsilon_r = 35$) 0.33 mm in thickness. Find the capacitance.

13–22 A variable capacitor with air dielectric ($\epsilon_r = 1$) has 8 stationary and 7 movable plates. When the plates are completely meshed and maximum capacitance is achieved, the area of each plate facing the dielectric is 0.00315 m^2 and the plate separation is 0.0015 m. Determine the maximum capacitance.

13–23 What is the capacitance in which 50 mV stores 100 μC?

13–24 How much charge is stored in a capacitor that has parallel plates with a plate area of 0.002 m², a plate separation of 0.01 m², a paper dielectric ($\epsilon_r = 4$), and an applied voltage across the plates of 12 V.

13–25 Find the capacitance of a capacitor made of two parallel plates each with areas of 3.09 cm² separated by an air gap of 0.225 cm.

13–26 A 0.0068 µF capacitor is constructed using a mylar dielectric ($\epsilon_r = 3$). If the distance between plates is 1 mm, what is the area of the plates?

13–27 Find the thickness of mica dielectric ($\epsilon_r = 5.5$) when used in a 0.0022 µF capacitor with parallel plates each having an area of 2.5×10^{-2} m².

13–28 Referring to Table 13–3, what is the value of a ceramic capacitor marked red, red, orange, brown, red?

13–29 What is the value in µF of a capacitor having the following color bands: blue, gray, yellow, white, brown?

13–30 What are the color bands of a 100 µF capacitor with a tolerance of 1% and a voltage rating of 300 V?

13–31 Find the value of one time constant in a series circuit containing a 0.022 µF capacitor and a 1 kΩ resistor.

13–32 What size resistor would have to be connected in series with a 470 µF capacitor to produce a time constant of 0.0031 s?

13–33 A 100 Ω resistor is connected in series with a capacitor and the resulting time constant is 0.022 s. Find the size of the capacitor.

13–34 The circuit shown in Figure 13–24 has components with the following values: $C = 100$ µF, $R = 33$ kΩ, $E = 24$ V. If the capacitor is initially discharged, determine the length of time required for v_C to rise to 11.5 V.

FIGURE 13–24

13–35 The capacitor shown in Figure 13–24 is fully charged. The component values are $E = 15$ V, $C = 22$ µF, $R = 100$ kΩ. How long will it take for the voltage to fall to 10 V if the switch is set at point B?

13–36 Three 40 µF capacitors are connected in series. Find the total capacitance.

13–37 Find the total capacitance of three 10 µF capacitors connected in parallel.

13–38 A 20 µF capacitor is connected in series with a 10 kΩ resistor and is separated from a 24 V source by a switch. Determine the
(a) Current flowing at the instant the switch is closed
(b) Value of current flowing after one time constant

13–39 A 50 µF capacitor is in series with a 30 kΩ resistor and is connected to a 12 V source. Determine the time required for the voltage to rise from 3 V to 9 V.

13–40 A 100 V source is applied across a 10 µF capacitor in series with a 200 kΩ resistor. What is the voltage across the capacitor after 2 s?

13–41 A 100 kΩ resistor is connected in series with a 0.2 μF capacitor. How much time does it take for the voltage across the capacitor to rise to 43 V after 120 V is applied to the circuit?

13–42 A 50 kΩ resistor and a 20 μF capacitor are connected in series to a switch and a 40 V source. If the capacitor is initially discharged, what is the voltage across the capacitor after the switch is closed for 2.75 s?

13–43 A fully charged 0.68 μF, 100 V, capacitor is in series with a 220 kΩ resistor and an open switch. How long will it take the capacitor to discharge to 20 V after the switch is closed?

13–44 Determine how long it will take for the voltage across a 10 μF capacitor to reach 15 V if it is connected in series with a 6 kΩ resistor and the applied voltage is 40 V. Assume the capacitor is initially discharged.

13–45 For the circuit shown in Figure 13–25, find the
(a) Total capacitance
(b) Voltage across each capacitor in the circuit

FIGURE 13–25

13–46 Determine the charge on each capacitor for the circuit shown in Figure 13–25.

Essay Questions

13–47 What is a capacitor?

13–48 Describe the relationship between flux density and the distribution of charge on the plates of a capacitor.

13–49 Define dielectric strength.

13–50 How is stray capacitance minimized?

13–51 List the three factors that determine the capacitance of a capacitor.

13–52 What is leakage resistance?

13–53 Name five types of capacitors.

13–54 Define transients in *RC* circuits.

13–55 How is energy stored in a capacitor?

13–56 Explain how electronic circuits are coupled using a capacitor.

13–57 What is a bypass capacitor?

13–58 How does the test procedure for capacitors differ from that used on resistors?

14

Inductance and Inductors

Learning Objectives

Upon completion of this chapter you will be able to

- Describe the principle of electromagnetic induction and flux linkages.
- List the four basic factors that determine the magnitude of an induced emf.
- Explain Lenz's law and the principle of counter emf.
- Define self-inductance and mutual inductance.
- List various types of inductors used in electrical and electronic circuits.
- Discuss the differences between inductors connected in series and in parallel.
- Explain inductive time constants and transients in *RL* circuits.
- Discuss energy stored in a magnetic field.
- Troubleshoot inductors.

Practical Application

The rapid increase in computers in factories and offices has created many problems such as harmonics and electromagnetic interference (EMI). As a technician employed by a company specializing in electronic power supply filters, you are required to select and install a choke coil to remove unwanted noise and interference from a communications cable in a manufacturing plant. The current flow through the communications cable is 105 mA. According to the design specifications, the coil must store 0.1 J of energy to provide the necessary filtering action. In this practical application of inductance and inductors, your task is to determine the correct value of inductance for the choke coil.

ELECTROMAGNETIC INDUCTION

In 1831, Michael Faraday discovered that if a conductor is moved through a magnetic field so that it cuts magnetic lines of flux, a voltage will be induced across the conductor. Figure 14–1 shows a moving conductor and a stationary magnetic field. If the speed with which the conductor passes through the field is increased, or if the strength of the magnetic field is increased, then the amount of induced voltage will also increase.

A magnetic field always exists at right angles to the current flowing in a conductor. Conversely, a changing magnetic field always has an emf associated with it that is at right angles to the direction of the lines of flux. The emf that is associated with a changing magnetic field is referred to as **induced emf.** Induced voltages that are developed by mechanical motion between conductors and magnetic fields are called **generated voltages.**

Electromagnetic induction is the process by which an electromotive force is induced, or generated, in an electric circuit when there is a change in the magnetic flux linking the circuit. Four basic factors determine the magnitude of an induced emf. They are

1. The number of turns in the conductor.
2. The strength of the magnetic field.
3. The relative speed between the coil and the magnetic field.
4. The angle at which the conductor passes through the magnetic field.

Section Review

1. What did Michael Faraday discover in 1831?
2. The emf associated with a changing magnetic field is referred to as induced emf (True/False).
3. What is electromagnetic induction?

FARADAY'S LAW

According to **Faraday's law,** when the magnetic flux linking a coil changes, a voltage proportional to the rate of flux change is induced in the coil. The magnitude of the voltage induced is proportional to the time rate of change of the flux linkage. **Flux linkage** exists when the flux and electric circuit are connected like two links of a chain. If a current flows

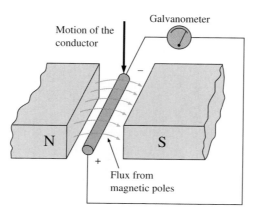

Motion of the conductor

Galvanometer

N

S

Flux from magnetic poles

FIGURE 14–1 *Conductors moving down through a magnetic field.*

FIGURE 14–2 *Examples of flux linkages. Dashed lines represent flux direction.*

in a conductor, a magnetic flux is set up around the conductor. The product of the turns of conductor and the number of lines of flux linking these turns is called the *linkage* of the circuit.

Several examples of flux linkages are shown in Figure 14–2. In Figure 14–2(a), the cross section of two current-carrying conductors illustrates magnetic linkage occurring from one conductor to the other. In Figure 14–2(b), which shows a portion of a magnetic circuit having two coils, the flux produced by the energized coil links each of the coils. The linkage of the coil shown in Figure 14–2(c) has its linkage flux produced externally. In Figure 14–2(d), a fixed coil is linked by the flux produced by a U-shaped permanent magnet.

Section Review

1. State Faraday's law.
2. The sum of the turns of a conductor and the number of lines of flux linking these turns is called *linkage* of the circuit (True/False).

_____ **14–3** ___

LENZ'S LAW

The induced emf in a circuit always tends to oppose any change in the amount of current in that circuit. This is commonly referred to as **Lenz's law,** which may also be stated as follows:

> In all cases of electromagnetic induction, current, which flows as a result of an induced emf, is in such a direction that the magnetic field established by the current reacts to stop the motion that generates the emf.

This law may be demonstrated by using a bar magnet, coil, and galvanometer, as shown in Figure 14–3. When the north pole of the bar magnet is inserted in the coil (Figure 14–3a), an emf is induced in the windings and the galvanometer will momentarily deflect. Because

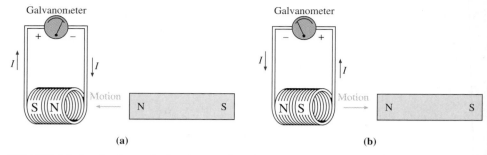

FIGURE 14–3   *Induced electromotive force. (a) North pole inserted in coil.*
(b) North pole withdrawn.

the electric circuit is completed by connecting the galvanometer between the terminals of the coil, current will flow in the circuit.

The direction of this current will establish an electromagnetic field with its north pole adjacent to the north pole of the bar magnet. Therefore, the direction of the electromagnetic field will oppose the north pole of the bar magnet entering the coil. If the magnet is withdrawn from the coil, as in Figure 14–3(b), the current in the coil reverses. The galvanometer will once again deflect, but the deflection is now opposite to that of Figure 14–3(a). The current in the coil has now reversed and the electromagnetic field is in the same direction as the field of the bar magnet. Because a north pole tends to move in the direction of the lines of force, this mmf would oppose the withdrawal of the bar magnet from the coil. The emf in each case is momentary and will cease when the change of flux linking the coil stops.

In Figure 14–4, a coil is connected to a battery with an open switch. If the switch is closed, current will flow in the coil. As the current increases in the coil, the flux increases, which increases the flux linking the coil. This action will induce an emf whose value is determined by the rate that the flux increases and by the number of turns in the coil. The induced voltage, often referred to as *counter emf*, initially opposes the increase in current and retards it, thereby preventing the current from reaching its maximum value immediately.

This emf is said to be *self-induced* because it is the changing current within the coil itself that induces the counter emf. Self-induction may, therefore, be defined as

> The emf that is generated in an electric circuit by a change of current within the circuit itself.

Circuits consisting of coils in which strong magnetic fields are set up have high self-inductance. Such circuits are generally constructed by winding a great many turns of wire on a continuous magnetic field core of high permeability.

FIGURE 14–4   *Principle of counter emf. Voltage induced in a coil tends to oppose the supply voltage.*

Section Review

1. The induced emf in a circuit always tends to oppose any change in the amount or direction of current in that circuit (True/False).
2. What is counter emf?
3. Define self-induction.

SELF-INDUCTANCE

Inductance is defined as the property of a circuit that opposes any change in the amount of current in that circuit. The ability of an inductor to induce a voltage into itself is referred to as **self-inductance.**

The unit of inductance is the **henry, H.** A circuit has a self-inductance of one henry when a current changing at the rate of one ampere per second induces an average of one volt. The symbol used to represent inductance is the capital letter L. When the length of a coil is much greater than its diameter and the permeability of the magnetic path is constant, the inductance of the coil can be found using the following equation:

$$L = \frac{\mu N^2 A}{l}$$

(14.1)

where μ = permeability of magnetic path
  N = total number of turns of coil
  A = area, in square meters
  l = length of coil, in meters

EXAMPLE 14-1

Solution

permeability of free space, $\mu_0 = 4\pi \times 10^{-7}$ H/m

$\mu = \mu_r\mu_0 = (1000)\,(4\pi \times 10^{-7})$

$L = \frac{\mu N^2 A}{l} = \frac{(1 \times 10^3)\,(4\pi \times 10^{-7})\,(1250)^2(0.015)^2}{0.175}$

$= 2.52$ H

A cast steel core is wound with 1250 turns of wire. The core has a cross-sectional area of 1.5 cm^2 and an average path length of 17.5 cm. The relative permeability at the rated current of the coil is 1000. Determine the inductance of the coil.

Section Review

1. The ability of an inductor to induce a voltage into itself is called *self-inductance* (True/False).
2. What is the symbol used to represent inductance?
3. A circuit has a self-inductance of one henry when a current changing at the rate of one ampere per second induces an average of one millivolt (True/False).

MUTUAL INDUCTANCE

When two windings are placed so that a change of current in one will cause its changing magnetic field to cut the turns of the other, an induced emf will be set up in the second coil. The two circuits are then said to possess mutual inductance.

An example of mutual inductance is shown in Figure 14–5(a). When the switch in series with the DC supply and coil is closed, a magnetic field is established in the iron core. This magnetic field links the iron core with the secondary winding. During the time the flux is increasing from a minimum to a maximum value, an emf is induced in the secondary coil. This emf is referred to as an emf of mutual induction. When the switch is closed, current will flow in the secondary winding in a direction that will establish an emf that opposes the increase in flux.

When the switch in the primary circuit is opened, the magnetic field will collapse. During this time interval, when the flux is decreasing from a maximum to a minimum value, an emf will also be developed in the secondary coil. This emf will cause current to flow in the direction illustrated by Figure 14–5(b).

The mutual inductance of two adjacent coils is dependent on such factors as the permeability of the core material, the number of turns on each coil, the physical dimensions of the coils, and the coefficient of coupling.

The **coefficient of coupling, *k*,** is defined as the degree of coupling that exists between two magnetic circuits. When maximum magnetic coupling exists between two coils, virtually all the magnetic flux created by one coil passes through the other. Under this condition the coefficient of coupling is said to be 1, or unity. This degree of closeness of the coils is known as **tight coupling.**

In practice, the coupling is never perfect. If two coils are wound on a common closed iron core, the coefficient of coupling can be considered as unity. However, if two coils are wound on separate iron cores, the coefficient of coupling is dependent on the distance between coils and the angle between the axes of the coils. **Loose coupling** produces a small value of coupling. For example, if 2% of the lines of force from one coil were cutting the second coil, then the coefficient of coupling would be 0.02. In many electronic circuits, two coils are wound on an air-core plastic form. The coefficient of coupling would be quite low (0.01 to 0.1) since the relative permeability of the core is unity.

(a)   (b)

FIGURE 14–5 *Direction of a current in a secondary winding as a result of the primary current increasing and decreasing. (a) Magnetic field increasing after switch is closed. (b) Magnetic field decreasing after switch is opened.*

Coil A

Coil B

FIGURE 14–6 *Effect of degree of coupling.*

Figure 14–6 shows two coils on separate iron cores. Coil B is placed at a right angle to coil A. The coils are centered so that equal parts of ΦA link equal numbers of turns on coil B, but in opposite directions. The emf induced in one part of coil B would be equal and opposite to that induced in the other part, and the net emf induced in coil B would be zero.

The mutual inductance between two coils, L_1 and L_2, may be expressed in terms of the inductance of each coil and the coefficient of coupling, k, as follows:

$$M = k\sqrt{L_1 L_2} \qquad\qquad (14.2)$$

where M = mutual inductance, in henrys
  k = coefficient of coupling
  L_1 = inductance of first coil, in henrys
  L_2 = inductance of second coil, in henrys

_____ **EXAMPLE 14–2** _____

Solution

$$M = k\sqrt{L_1 L_2}$$
$$= 0.91\sqrt{(7.5\ \text{mH})(4.9\ \text{mH})}$$
$$= 5.52\ \text{mH}$$

Two coils have inductances of 7.5 mH and 4.9 mH, respectively. If they are coupled with a coefficient of coupling of 0.91, what is the mutual inductance?

Section Review

1. What is mutual inductance?
2. When the coefficient of coupling is one, it is known as tight coupling (True/False).
3. Loose coupling produces a large value of coupling (True/False).

_____ **14–6** __

TYPES OF INDUCTORS

A device that introduces inductance into an electric circuit is called an inductor. Basically, all inductors are made by winding a length of conductor around a core. The two main categories of inductors are air core and iron core.

Air-core inductors contain no magnetic iron. They are usually wound on some tubular insulating material. This material may be made of treated cardboard, fiberglass,

FIGURE 14–7 *Air-core inductor.*

Bakelite, hard rubber, or plastic. The straight spiral coils of air-core inductors are said to be spiral wound. Large air-core inductors with high-power capabilities do not require an insulated core. These coils are wound with bare copper wire of large diameter and are sufficiently rigid to be self-supporting. The schematic symbol for an air-core inductor is illustrated in Figure 14–7.

Iron-core inductors have cores made of various alloys of iron and other materials to give the required characteristics for a particular application. Figure 14–8 shows some typical inductors.

The main cause of losses in an iron core inductor are **eddy currents.** An eddy current loss takes place in cores of magnetic material that are subjected to cycles of magnetization. The moving magnetic field induces an emf in the core, which in turn causes an electric current to flow. To reduce the magnitude of these currents and, thereby, minimize eddy-current loss, magnetic cores are constructed out of thin laminations. The surfaces of these laminations are treated so that high electrical resistance is offered to the flow of eddy currents from one lamination to another. The schematic symbol for an iron-core inductor is illustrated in Figure 14–9.

Hysteresis loss also occurs in an iron-core inductor. Hysteresis loss takes place in any magnetic material in which the magnetic field continually reverses direction. When the current in the inductor increases, a completely demagnetized iron core becomes magnetized in a nonlinear manner. If the inductor current is decreased to zero, the magnetization does not return to zero. Some residual magnetism is left in the core.

Both eddy current losses and hysteresis losses are considered to be **heat losses.** Hysteresis loss is a heat loss caused by the alternating movement of magnetic domains in the iron, and eddy current loss is a heat loss caused by an alternating current induced in the iron. Both losses are generated by the alternating magnetic field of the iron-core inductor.

(a) (b) (c)

FIGURE 14–8 *Typical inductors. (a) Fixed molded inductors. (b) Variable coil inductors. (c) Toroid inductor.*

(a)   (b)   (c)

FIGURE 14–10
Inductor symbols.
(a) Tapped coil.
(b) Moveable
core. (c) Slider.

FIGURE 14–9   *Iron-core inductor.*

Inductors can also be classified as fixed or variable. Fixed inductors are designed with set values of inductance that cannot be changed. Variable inductors have inductances that can be altered by using either a tapped coil, movable core, or slider.

Tapped coils are manufactured by placing taps at certain points in the winding. The value of inductance can be changed by using one fixed end of the winding and selecting any one of the taps.

Movable-core inductors are made so that the core can be moved into and out of the winding. As the length of the air gap increases, the value of reluctance also increases. Circuits that use movable-core inductors are often referred to as *permeability-tuned circuits.*

The slider inductor operates on the same principle as the slide wire bridge. The sliding contact varies the amount of inductance introduced into the circuit. The three types of variable inductor symbols are shown in Figure 14–10.

Section Review

1. What is an inductor?
2. What is the main cause of losses in an iron core?
3. Hysteresis losses are not classified as heat losses (True/False).
4. What is a permeability-tuned circuit?

14–7

INDUCTORS IN SERIES

When inductors are connected in series and are far enough apart, or shielded so that there is no coupling between them, the inductors may be added in the same manner in which resistors are added. Therefore, when inductors are connected in series and have no mutual inductance, the total inductance is equal to the sum of the individual inductances. In equation form,

$$L_T = L_1 + L_2 + L_3 + \cdots + L_n \tag{14.3}$$

When mutually coupled coils are connected in series, they can be connected either with their mmfs aiding or opposing. Figure 14–11 shows two series-connected inductors with aiding mmf. The inductance of two coils connected with their flux aiding each other can be calculated by adding the inductance of the first and second coils to the mutual inductance of the two coils. In equation form,

$$L_a = L_1 + L_2 + 2(k \sqrt{L_1 L_2}) \tag{14.4}$$

FIGURE 14–11 *Two series-connected coils with aiding flux.*

Φ

Coil 1   Coil 2

I

FIGURE 14–12 *Two series-connected coils with opposing flux.*

Φ

Coil 1   Coil 2

I

where L_a = total inductance of circuit
        L_1 = inductance of first coil
        L_2 = inductance of second coil
        k = coefficient of coupling

The two inductors shown in Figure 14–12 have their flux in opposite directions, which means they are connected series opposing. The inductance of two coils connected with their flux opposing each other can be calculated in the following manner:

$$L_o = L_1 + L_2 - 2(k \sqrt{L_1 L_2}) \tag{14.5}$$

If Equations 14.4 and 14.5 are combined, the result is as follows:

$$L_a - L_o = 4M$$

Therefore, in terms of series-aiding and series-opposing connections, the mutual inductance of two series-connected coils can be determined by the following equation:

$$M = \frac{L_a - L_o}{4} \tag{14.6}$$

___ **EXAMPLE 14–3** ___

A 50 mH inductor and a 30 mH inductor are connected in series aiding. The coefficient of coupling is 0.75. Calculate the inductance of the circuit.

Solution

$$L_a = L_1 + L_2 + 2(k \sqrt{L_1 L_2})$$
$$= 50 \text{ mH} + 30 \text{ mH} + 2\left[0.75 \sqrt{(50 \text{ mH}) (30 \text{ mH})}\right]$$
$$= 138.09 \text{ mH}$$

$M(+)$ $M(-)$

(a)   (b)

FIGURE 14–13 *Dot convention for coupled coils. (a) Additive polarity, mutual flux aiding. (b) Subtractive polarity, mutual flux opposing.*

_____ **EXAMPLE 14–4** _____

Solution

$$M = k \sqrt{L_1 L_2}$$

$$k = \frac{M}{\sqrt{L_1 L_2}} = \frac{12}{\sqrt{(17)(15)}} = 0.75$$

$$L = L_1 + L_2 - 2(k \sqrt{L_1 L_2})$$
$$= 17 \text{ H} + 15 \text{ H} - 2\left[0.75 \sqrt{(17 \text{ H})(15 \text{ H})}\right]$$
$$= 8.05 \text{ H}$$

A 17 H inductor and a 15 H inductor are connected series opposing. The mutual inductance of these two coils is 12 H. What is the coefficient of coupling and the inductance of the circuit?

A convenient method for indicating the instantaneous flux direction is to use the **dot convention.** The dot system allows us to represent the direction of flux produced by a coil without having to indicate the windings and the flux path. The dot represents the tip of the flux arrow. If the current through each of the mutually coupled coils enter the dot terminals, or terminals without dots, the mutually coupled coils will be positive, or series aiding. If the arrow indicating current direction enters a dot and the other current is shown leaving a dot, then the polarity of the mutually coupled coil is negative, or series opposing. Figure 14–13 illustrates the dot conventions for additively and subtractively coupled coils. Figures 14–11 and 14–12 would be represented by Figures 14–13(a) and 14–13(b), respectively.

Section Review

1. When inductors are connected in series, the total inductance is equal to the sum of the individual inductances (True/False).
2. What is the purpose of using the dot convention for inductors?

_____ **14–8** ___

INDUCTORS IN PARALLEL

If no mutual induction exists, inductors that are connected in parallel are added in the same manner as resistors in parallel. In equation form,

$$\frac{1}{L_T} = \frac{1}{L_1} + \frac{1}{L_2} + \frac{1}{L_3} + \cdots + \frac{1}{L_n} \qquad (14.7)$$

For two inductors in parallel, with no mutual inductance, the product-over-sum equation may be used.

$$L_T = \frac{L_1 L_2}{L_1 + L_2} \qquad (14.8)$$

When mutual inductance exists between two parallel-connected inductors, the following equation may be used to find the total inductance of the circuit.

$$L_T = \frac{L_1 L_2 (1 - k^2)}{L_1 + L_2 - 2(k \sqrt{L_1 L_2})}$$

(14.9)

_____ EXAMPLE 14–5 _____

A 35 H inductor and a 25 H inductor are connected in parallel. The coefficient of coupling is 0.6. Calculate the inductance of this circuit.

Solution

$$L_T = \frac{L_1 L_2 (1 - k^2)}{L_1 + L_2 - 2(k \sqrt{L_1 L_2})}$$

$$= \frac{(35)(25)[1 - (0.6)^2]}{35 + 25 - [2(0.6)\sqrt{(35)(25)}]}$$

$$= 22.85 \text{ H}$$

Section Review

1. Inductors in parallel are added in the same manner as resistors in series (True/False).
2. For two inductors in parallel with no mutual inductance, the product-over-sum equation may be used (True/False).
3. What is the inductance of a circuit with a 5 mH and a 10 mH inductor connected in parallel with a coefficient of coupling of 0.9?

___ 14–9 ___

INDUCTIVE TIME CONSTANT

When emf is first applied to an inductor, it takes a certain amount of time for current to build up and reach a constant value. A DC circuit containing inductance as well as resistance will have a gradual current change between zero and maximum. Because the buildup current in an inductive circuit starts at zero, it always must be changing as it approaches its stable value. During these changes, a relationship exists between the values of current reached and the time it takes to reach them. This relationship is expressed by a quantity called an **inductive time constant** and is defined as follows:

> The inductive time constant is the amount of time, in seconds, required for the current in an LR circuit to rise to 63.2% of its total value after an emf is initially applied across the series-connected inductance and resistance.

The value of the time constant is directly proportional to the inductance and inversely proportional to the resistance. As is the case with the time constant for a capacitive circuit, the time constant for an inductive circuit is represented by the symbol τ, which is the Greek lowercase letter tau.

If the values of resistance and inductance are known, the time constant can be calculated using the following equation:

$$\tau = \frac{L}{R}$$

(14.10)

where τ = time for current to reach 63.2% of its final value, in seconds
     L = inductance of the circuit, in henrys
     R = resistance of the circuit, in ohms

EXAMPLE 14–6

Solution

A 12 Ω resistor is connected in series with a 0.36 H inductor. What time is required for the circuit to build up to 63.2% of its final value?

$$\tau = \frac{L}{R}$$

$$= \frac{0.36 \text{ H}}{12 \text{ Ω}}$$

$$= 0.03 \text{ s}$$

For capacitive circuits, the product of $R \times C$ was the time constant. For inductive circuits the time constant is L/R. Figure 14–14 shows the universal time constant chart, which is identical to the chart shown in Figure 13–12. It is indeed a universal chart, with the charging and discharging paths applicable to both RC and RL circuits. Curve A would represent the charging time of an inductive circuit and curve B would represent the discharge time.

Section Review

1. The value of one inductive time constant is directly proportional to the resistance and inversely proportional to the inductance (True/False).
2. State the equation for determining the time constant of an RL circuit.
3. The time constant chart for charging and discharging inductors is the same as the chart for capacitors (True/False).

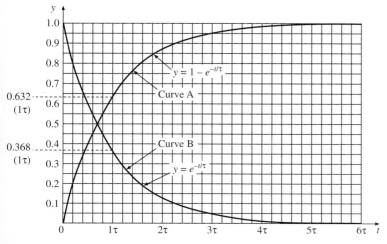

FIGURE 14–14 *Universal time constant curves.*

FIGURE 14–15 *The RL circuit.*

___ 14–10

TRANSIENTS IN *RL* CIRCUITS

Figure 14–15 shows an *RL* circuit with a switch and a source of emf. At any time after the switch is closed, the mathematical expression for current through the coil can be found by first applying Kirchhoff's voltage law:

$$E = v_R + v_L$$

where v_R and v_L are the instantaneous voltage drops across the resistor and coil, respectively.

By substituting v_R with iR, Kirchhoff's voltage law may now be expressed as

$$E = iR + v_L$$

Because the current, *i*, is the same for the resistor and the inductor, the equation now becomes

$$i_L = \left(\frac{E}{R}\right)\left[1 - e^{-t(R/L)}\right] \tag{14.11}$$

where i_L = instantaneous current, at time *t*, in amperes
E = supply voltage, in volts
R = series resistance, in ohms
e = exponential constant 2.71828
t = time from current commencement, in seconds
L = inductance of the inductor, in henrys

___ EXAMPLE 14–7 ___

The circuit of Figure 14–15 has an 8 H inductor and a 1 kΩ resistor connected in series to a 10 V supply. Calculate the current i_L at $t = 6$ ms when the switch is closed.

Solution

$$i_L = \left(\frac{E}{R}\right)\left[1 - e^{-t(R/L)}\right]$$

$$= \frac{10}{1 \times 10^3}\left[1 - e^{-0.006(1000/8)}\right]$$

$$= 5.28 \text{ mA}$$

In the *RC* circuit, the voltage v_C essentially reaches its final value in five time constants. The same is true for the current i_L in an *RL* circuit. If *R* is held constant and *L* is reduced, the rise time of five time constants decreases. When plotting a curve of time constants for an *RL* circuit, the ratio *L/R* always has the same numerical value. The larger the inductance, the larger the time constant, and the longer it will take i_L to reach its final value.

When a circuit that contains only resistance is opened, the current immediately drops to zero. In an inductive circuit, the current tends to drop to zero when the switch is opened; however, any change in current produces an emf that tends to keep the current flowing in the circuit. For this reason, the current through an inductor cannot change instantaneously. The equation for calculating a **discharge curve** for an *RL* circuit is

$$i_L = I(e^{-t/\tau}) \tag{14.12}$$

where I = the current flowing in the coil just prior to the start of the decay transient
        τ = time constant L/R

Since the current is a decaying exponential, the time constant is the time required for the current to decay to 36.8% of its original value.

_____ **EXAMPLE 14–8** _____

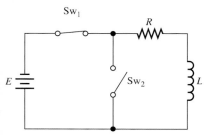

FIGURE 14–16 *Circuit for Example 14–8.*

The *RL* circuit of Figure 14–16 consists of a 43 Ω resistor, a 0.4 H coil, and an applied voltage of 12 V. Assuming switch Sw$_1$ has been closed for a very long period of time, calculate the current flowing 25 ms after switch Sw$_1$ is opened and Sw$_2$ is closed.

Solution   The current I is found by Ohm's law.

$$I = \frac{E}{R} = \frac{12\text{ V}}{43\text{ }\Omega} = 0.279\text{ A}$$

$$\tau = \frac{L}{R} = \frac{0.4\text{ H}}{43\text{ }\Omega} = 9.302 \times 10^{-3}\text{ s}$$

$$i_L = I(e^{-t/\tau}) = 0.279(e^{-0.025/0.009302}) = 18.98\text{ mA}$$

The voltage across the coil in an *RL* circuit can change instantaneously. Figure 14–17 shows the voltages across a series-connected inductance and resistance. At the instant the switch is closed, the current is zero because of the sudden rate of change of current causing the counter emf to equal the applied emf. Because there is no current, the voltage across the resistance is zero. Therefore, the entire applied voltage appears across the inductance.

Immediately, the voltage across the inductor begins to decline, and the voltage across the resistor increases. This continues until the total applied emf appears across R and none across L. The voltage across R will be at its maximum value when the current has reached its steady state, which means there is no counter emf in L. To calculate the voltage across an inductor in an *RL* circuit, the following equation is used.

$$v_L = E(e^{-t/\tau}) \tag{14.13}$$

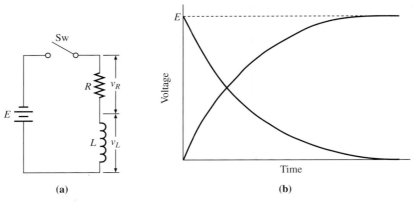

FIGURE 14–17 *(a) RL circuit. (b) Voltages varying with time immediately after emf is initially applied.*

EXAMPLE 14–9

The *RL* circuit of Figure 14–16 consists of a 30 Ω resistor and a 2 H coil. An emf of 20 V is applied to the circuit. What is the voltage across the inductor 0.02 s after switch Sw_1 is closed and Sw_2 opened?

Solution

$$\tau = \frac{L}{R} = \frac{2\text{ H}}{30\text{ }\Omega} = 66.67 \times 10^{-3}\text{ s}$$

$$v_L = E(e^{-t/\tau}) = 20(e^{-0.02/0.06667}) = 14.82\text{ V}$$

While the voltage across the inductor is decreasing with time, the voltage across the resistor is increasing exponentially. The voltage across the resistor increases at a rate determined by the current through the circuit. The equation for calculating the changing voltage drop across a resistor in an *RL* circuit is as follows:

$$v_R = E(1 - e^{-t/\tau}) \tag{14.14}$$

EXAMPLE 14–10

Using the values given in Example 14–9, solve for the voltage across the resistor. Verify by using Kirchhoff's voltage law.

Solution

$$\tau = \frac{L}{R} = \frac{2\text{ H}}{30\text{ }\Omega} = 66.67 \times 10^{-3}\text{ s}$$

$$v_R = E(1 - e^{-t/\tau}) = 20(1 - e^{-0.02/0.06667}) = 5.18\text{ V}$$

$$E = v_R + v_L$$

$$v_R = E - v_L = 20\text{ V} - 14.82\text{ V} = 5.18\text{ V}$$

The time, at any point on the curve during exponential decay of the voltage, is the same equation for both inductors and capacitors. That is,

$$t = \tau \ln\left(\frac{E}{v_L}\right)$$

EXAMPLE 14–11

Solution

In an *RL* circuit, the time constant is 7.5 ms. Determine the length of time required for the voltage to decay from an initial value of 20 V to 14.57 V.

$$t = \tau \ln\left(\frac{E}{v_L}\right)$$

$$= (7.5 \text{ ms}) \ln\left(\frac{20 \text{ V}}{14.57 \text{ V}}\right)$$

$$= 2.38 \text{ ms}$$

Section Review

1. In an *RL* circuit, the voltage across an inductor reaches its final value in three time constants (True/False).
2. Why is it impossible for the current through an inductor to change instantly?
3. In a series *RL* circuit, explain what happens to the voltage across the resistor if the voltage across the inductor is decreasing with time.

14–11
ENERGY STORED IN A MAGNETIC FIELD

During the period that a magnetic field is being established by a coil, a portion of the energy that is supplied by the source is stored in the field. The remainder of the energy supplied during the rise of the current is dissipated from the coil of the magnet as heat. After the magnetic field is established and the current has reached a steady state, all the energy input to the coil windings of the electromagnet is dissipated as heat, and no additional energy is required to maintain this magnetic field.

The amount of energy stored in the magnetic field for any value of current may be expressed using the following equation:

$$W = \frac{LI^2}{2} \tag{14.15}$$

where L = inductance, in henrys
   I = current, in amperes
   W = energy stored, in joules

EXAMPLE 14–12

Find the energy stored in a series-connected *RL* circuit having an inductance of 5 H and a resistance of 120 Ω. The supply voltage is 100 V.

Solution

$$I = \frac{E}{R} = \frac{100 \text{ V}}{120 \text{ }\Omega} = 0.833 \text{ A}$$

$$W = \frac{LI^2}{2}$$

$$= \frac{(5 \text{ H}) (0.833 \text{ A})^2}{2}$$

$$= 1.74 \text{ J}$$

Section Review

1. After an electric field has been established in an electromagnet, all the energy consumed by the magnet is dissipated as heat (True/False).
2. State the equation for calculating the energy stored in a magnetic field.

14–12

TROUBLESHOOTING INDUCTORS

Inductors almost always fail as either open-circuit or short-circuit conditions. Consequently, the ohmmeter is very useful when testing these devices. When testing the resistance of an inductor, the ohmic value should be low, but never zero ohms. The ohmic value of an inductor depends on the length of the coil and the wire size. Although inductors can be manufactured with resistances of several hundred ohms, in electronic circuits the resistance of an inductor is typically under 50 Ω.

Figures 14–18 and 14–19 show how an inductor is tested using an ohmmeter. Ideally, one end of the inductor should be isolated from the circuit to which it is connected. This will prevent false readings when the device is tested. For example, if a coil is open circuited but connected in parallel with a device having low resistance, then the coil may appear to have a low ohmic value when it is actually infinity.

A short circuit in an inductor may be the result of a total short, as indicated in Figure 14–18, or it may be caused by a partial short. In either case, excess current flow will result

FIGURE 14–18 *Testing inductor with an ohmmeter. Zero ohms indicates a short circuit.*

FIGURE 14–19 *Infinite resistance reading indicates an open circuit.*

because of the resistance of the coil being reduced. Partial short circuits in coils are caused by the insulation of the windings becoming damaged because of heat. When excess current flows through a coil, the coil heats up and the insulating material on the windings will eventually melt. The result is the short circuiting of adjacent turns of wire on the coil.

An open circuit, as indicated in Figure 14–19, can be caused by either the inductor itself, or the point where the inductor is connected to the circuit. A poorly soldered connection, or cold-soldered joint, will result in an open-circuit condition. This type of failure can be corrected by resoldering the connection between the inductor and the circuit board.

Inductors can also be checked in an energized circuit using a voltmeter. If the inductor is shorted, the voltage drop across its terminals will be zero. In a series circuit, an open-circuited inductor will have a voltage drop that is equal to the applied voltage of the circuit.

Section Review

1. What are the two most common types of inductor failures?
2. The resistance of a typical inductor is under 500 Ω (True/False).
3. It is possible to test an inductor using a voltmeter (True/False).

14–13

PRACTICAL APPLICATION SOLUTION

In the chapter opener, your task was to select and install a choke coil in a communications cable of a manufacturing plant to reduce unwanted noise. The current flow through the communications cable was given as 105 mA, and the coil was required to store 0.1 J of energy. The following steps illustrate the method of solution for this practical application of inductance and inductors.

STEP 1   Determine the inductance.

$$W = \frac{LI^2}{2}$$

$$L = \frac{2W}{I^2}$$

$$= \frac{2(0.1)}{(105 \text{ mA})^2}$$

$$= 18.15 \text{ H}$$

STEP 2   Select an inductor as close to the 18.15 H value as possible and install it in the circuit.

Summary

1. A changing magnetic field always has an emf associated with it that is at a right angle to the direction of the lines of flux.
2. When the magnetic flux linking a coil changes, a voltage proportional to the rate of flux change is induced in the coil.
3. The induced emf in a circuit always tends to oppose any change in the amount of current in the circuit.
4. A device that introduces inductance into an electronic circuit is called an inductor.
5. Inductors are added in series in the same manner as resistors.
6. The ability of an inductor to induce a voltage into itself is called *self-inductance*.
7. Mutual inductance is based on the principle of one coil inducing a voltage into another coil.
8. The main cause of losses in an iron core are eddy currents, which are currents produced by moving magnetic fields inside the magnetic material.
9. The inductive time constant is the amount of time required for the current in an *RL* circuit to rise to 63.2% of its total value.
10. Inductors almost always fail as either open-circuit or short-circuit conditions.

Answers to Section Reviews

Section 14–1
1. That, if a conductor is moved through a magnetic field so that it cuts magnetic lines of flux, a voltage will be induced across the conductor
2. True
3. The process by which an electromotive force is induced, or generated, in an electric circuit when there is a change in the magnetic flux linking the circuit

Section 14–2
1. When the magnetic flux linking a coil changes, a voltage proportional to the rate of flux change is induced in the coil.
2. False

Section 14–3
1. True
2. The induced voltage that opposes the increase in current and retards it, thereby preventing the current from reaching its maximum value immediately
3. The emf that is generated in an electric circuit by a change of current within the circuit itself

Section 14–4
1. True
2. *L*
3. False

Section 14–5

1. When two windings are placed so that a change of current in one will cause its changing magnetic field to cut the turns of the other, an induced emf will be set up in the second coil.
2. True
3. False

Section 14–6

1. A device that introduces inductance into an electronic circuit
2. Eddy currents
3. False
4. A circuit that uses a movable-core inductor

Section 14–7

1. True
2. To indicate the instantaneous flux direction produced by a coil

Section 14–8

1. False
2. True
3. $L_T = \dfrac{(5 \text{ mH})(10 \text{ mH})(1 - 0.9^2)}{5 \text{ mH} + 10 \text{ mH} - \left[2(0.9) \sqrt{(5 \text{ mH})(10 \text{ mH})}\right]} = 4.18 \text{ mH}$

Section 14–9

1. False
2. $\tau = L/R$
3. True

Section 14–10

1. False
2. Because any change in current produces an emf that tends to keep the current flowing in the circuit
3. The voltage across the resistor is increasing exponentially.

Section 14–11

1. True
2. $W = LI^2/2$

Section 14–12

1. Open circuits and short circuits
2. False
3. True

_____ *Review Questions* ___

14–1 Who discovered that if a conductor is moved through a magnetic field, a voltage will be induced across the conductor?
(a) Georg Simon Ohm   (b) Michael Faraday
(c) Charles Coulomb   (d) Joseph Henry

14–2 The emf that is associated with a changing magnetic field is called
(a) Instantaneous voltage   (b) Counter emf
(c) Induced emf   (d) All of the above

14–3 When the magnetic flux linking a coil changes, a voltage proportional to the rate of flux change is induced in the coil. This statement is known as
(a) Faraday's law   (b) Fleming's left-hand rule
(c) Lenz's law   (d) The principle of generator action

14–4 The product of the turns of a conductor and the number of lines of flux linking these turns is called
(a) Induced emf   (b) Flux linkage   (c) Generated voltage   (d) Flux direction

14–5 Circuits consisting of coils in which strong magnetic fields are set up have
(a) High self-inductance   (b) Low self-inductance
(c) No self-inductance   (d) No counter emf

14–6 The induced emf in a circuit always tends to oppose any change in the amount of current in that circuit. This is known as
(a) Faraday's law   (b) Lenz's law
(c) Fleming's left-hand rule (d) Self-inductance

14–7 The unit of inductance is the
(a) Ohm   (b) Farad   (c) Volt   (d) Henry

14–8 The ability of an inductor to induce a voltage into itself is referred to as
(a) Self-inductance   (b) Counter emf
(c) Mutual inductance (d) Faraday's law

14–9 The degree of coupling that exists between two magnetic circuits is called
(a) Tight coupling   (b) Loose coupling
(c) Coefficient of coupling (d) Mutual inductance

14–10 Circuits that use movable-core inductors are often referred to as
(a) Tapped coils   (b) Sliders   (c) Iron core   (d) Permeability tuned

14–11 The dot convention system is used to represent the direction of _____ produced by a coil.
(a) Voltage   (b) Flux   (c) Current   (d) Counter emf

14–12 The value of the inductive time constant is
(a) Directly proportional to the resistance and inversely proportional to the inductance
(b) Directly proportional to the resistance and inductance
(c) Directly proportional to the inductance and inversely proportional to the resistance
(d) Inversely proportional to the resistance and inductance

Practice Problems

14–13 An air-core solenoid is 10 cm long, has a diameter of 3 cm, and consists of 200 turns. Find the inductance of the solenoid.

14–14 A 250-turn coil has an air core, a length of 0.025 m, and a radius of 0.0035 m. Determine its inductance.

14–15 Two coils have inductances of 3.3 H and 5.4 H and are coupled with a coefficient of coupling of 0.83. Determine the mutual inductance.

14–16 Find the mutual inductance of 22 mH and 35 mH coils with a coefficient of coupling of 0.91.

14–17 Two coils having inductances of $L_1 = 350$ mH and $L_2 = 275$ mH are tightly wound on top of each other on a ferromagnetic core. Determine the maximum mutual inductance that is possible between the coils assuming unity coupling.

14–18 The mutual inductance between two coils is 0.39 H. The coils have self-inductances of 0.35 H and 0.79 H. Calculate the coefficient of coupling.

14–19 A 100 H coil and a 70 H coil are connected in a series-aiding configuration. The coefficient of coupling is 0.82. Determine the inductance of the circuit.

14–20 A 300 mH coil and a 200 mH coil are connected series aiding. The mutual inductances of these two coils is 150 mH. Determine the
(a) Coefficient of coupling
(b) Inductance of the circuit

14–21 A 20 H coil and a 15 H coil are connected series opposing. The coefficient of coupling is 0.77. Calculate the inductance of the circuit.

14–22 A 50 mH coil and a 35 mH coil are connected series opposing. The mutual inductance of these two coils is 28 mH. Determine the
(a) Coefficient of coupling
(b) Inductance of the circuit

14–23 Two series-connected coils, each having an inductance of 6 H, have a total inductance of 14 H when connected series aiding and 11 H when connected series opposing. Determine the
(a) Mutual inductance
(b) Coefficient of coupling of this circuit

14–24 Two 700 μH coils have a mutual inductance of 250 μH. Calculate the total inductance of the two coils when they are connected
(a) Series aiding
(b) Series opposing

14–25 A 400 mH coil and a 300 mH coil are connected in parallel. The coefficient of coupling is 0.85. Determine the inductance of the circuit.

14–26 A 300 Ω resistor is connected in series with a 20 H coil. Determine the length of time required for the circuit voltage to rise to 63.2% of its final value.

14–27 Determine the value of current flowing in the circuit of Figure 14–20 after the switch has been closed for 2.5 ms.

FIGURE 14–20

14–28 The *RL* circuit of Figure 14–21 consists of a 100 Ω resistor, a 250 mH coil, and an applied voltage of 24 V. Assuming switch Sw_1 has been closed for a very long period of time, calculate the current flowing 10 ms after switch Sw_1 is opened and Sw_2 is closed.

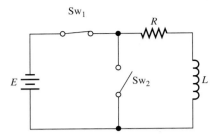

FIGURE 14–21

14–29 The *RL* circuit of Figure 14–21 has a 1 kΩ resistor and a 5 H coil. A 120 V emf is applied to the circuit. Determine the voltage across the inductor 10 ms after Sw_1 is closed and Sw_2 opened.

14–30 In an *RL* circuit, the time constant is 12.5 ms. Determine the length of time required for the voltage to decay from an initial value of 40 V to 28.65 V.

14–31 Determine the amount of energy stored in a series-connected *RL* circuit that has a 200 mH coil and a 5 Ω resistor. The supply voltage is 24 V.

14–32 Find the energy stored in a magnetic field when the inductance is 7.5 H and the current flowing is 1.5 A.

Essay Questions

14–33 Explain the principle of electromagnetic induction.

14–34 What are flux linkages?

14–35 List the four basic factors that determine the magnitude of an induced emf.

14–36 Describe Faraday's law.

14–37 Explain Lenz's law.

14–38 How is self-inductance different from mutual inductance?

14–39 Define tight coupling.

14–40 Describe the main cause of losses in an iron-core inductor.

14–41 What are the two main types of heat losses?

14–42 Name three types of variable inductors.

14–43 Why is the dot convention used?

14–44 How is energy stored in a magnetic field?

15

Transformers

Learning Objectives

Upon completion of this chapter you will be able to

- Explain the basic operating principles of the transformer.
- Draw the schematic symbols for iron- and air-core transformers.
- Explain the standard markings used to identify transformer windings.
- Understand the principles of reflected loads and impedance matching.
- List the various losses associated with transformers.
- Express the significance of transformer polarity.
- Differentiate between isolation transformers and autotransformers.
- Troubleshoot transformers.

Antenna circuits receive very low power signals that must be recovered completely for maximum signal strength. Transformers are used in antenna systems to transfer the maximum amount of power between the antenna and the load.

As a technician for a security system company, you are required to design a wireless burglar alarm. The alarm system consists of a wireless transmitter that generates a low-power signal when it detects any movement within a 50 ft radius. The antenna is connected to a receiver with a tunable resistor to allow for maximum power transfer between the antenna and receiver. The antenna has an impedance of 300 Ω and the receiver has a resistance of 75 Ω. Your task is to select an impedance-matching transformer that will provide maximum power transfer for the system.

INTRODUCTION

The **transformer** is a simple and reliable device that transfers energy from one circuit to another by electromagnetic induction. This transfer is usually, but not always, accompanied by a change of voltage. The transformer is basically two coils, often wound on an iron core, that forms a closed magnetic circuit. The only connection between the two coils is the common magnetic flux within the core of the transformer. When power is supplied to one coil at a specific frequency, power can be taken from the other coil at that same frequency.

Transformers are used for a variety of applications in electronic circuits. They are often used to step down the AC input voltage in power supply circuits and to couple circuits together. Transformers are also used to develop pulses for semiconductor devices, such as **silicon-controlled rectifiers (SCRs),** and to step up the voltage in circuits used for televisions.

The two windings are referred to as the *primary* and *secondary*. The **primary winding** receives energy from the supply circuit, and the **secondary winding** receives energy by induction from the primary. In most transformers the two windings are magnetically linked by a closed core of laminated iron or steel.

A transformer that receives energy at one voltage and delivers it at the same voltage is called a **one-to-one transformer.** When the transformer receives energy at one voltage and delivers it at a higher voltage, it is called a **step-up transformer.** If the energy received by the transformer is delivered at a lower voltage, it is referred to as a **step-down transformer.**

Section Review

1. How does the secondary winding of a transformer receive energy?
2. What is a step-up transformer?
3. A transformer that receives energy at one voltage and delivers it at another voltage is called a one-to-one transformer (True/False).

TRANSFORMER CONSTRUCTION

Transformers are divided into two general types: core type and shell type. In the core type, the winding surrounds the iron core. The shell type of transformer has the core built around the coils. Figure 15–1 shows the two types of transformers. The windings may be cylindrical in form and placed one inside the other, with the necessary insulation between them, or they may be built up in thin flat sections called *pancake coils.*

The magnetic circuit, or core, of the transformer is usually made of high-grade silicon steel, containing about 3% silicon. By using silicon, the magnetic characteristics of the core are improved and the iron losses are reduced. The steel core is in the shape of a hollow rectangle and is made up of laminations that are either rectangular or L-shaped. The sheet steel is laminated to reduce eddy current losses, which are generated by the alternating flux as it cuts through the iron core.

The primary and secondary windings may be placed on the same leg in either the core-type or shell-type transformer. The low-voltage winding and the high-voltage winding of a core-type transformer are wrapped around the core sides. In the shell-type trans-

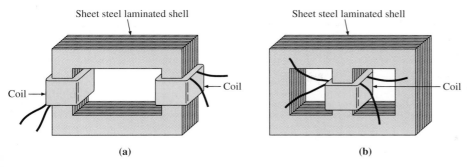

Sheet steel laminated shell

Coil

Coil

Sheet steel laminated shell

Coil

(a)

(b)

FIGURE 15–1 *(a) Core-type transformer. (b) Shell-type transformer.*

(a)

(b)

FIGURE 15–2 *Schematic symbol for transformers. (a) Iron or ferrite core. (b) Air core.*

former, the core has a middle leg on which both primary and secondary windings are placed. The shell-type transformer provides a much shorter magnetic path, but the average length per turn of wire is much greater. The entire flux passes through the central leg of the core. Once the flux passes through the central core it divides, with half of the flux going in each direction.

The schematic symbols used to represent transformers are shown in Figure 15–2. The symbol for an iron- or steel-core transformer is shown in Figure 15–2(a). A transformer with a nonmagnetic core material is represented by the air-core transformer symbol of Figure 15–2(b).

Section Review

1. What are the two general types of transformer construction?
2. Silicon is added to the core of a transformer to reduce iron losses (True/False).
3. Why is the steel core of a transformer laminated?

15–3

TRANSFORMER PRINCIPLES

Figure 15–3 illustrates a basic transformer, consisting of primary and secondary windings wrapped around a laminated iron or steel core. The primary winding is connected to a source of AC electric power, and the secondary winding is connected to the load. When a transformer is used to step up voltage, the low-voltage winding is the primary. When a transformer is used to step down voltage, the high-voltage winding is the primary. The primary winding is always connected to the source of power, regardless of the application.

Although any transformer may be used as a step-up or step-down transformer, it is very important that the windings be properly connected. For this reason, standard markings have been adopted for transformer terminals. The winding that is to be connected to a

FIGURE 15–3 *Transformer designed to step up voltage.*

high-voltage source or load is designated by H_1 and H_2. This winding is capable of handling high voltages and relatively low amounts of current, so its coil consists of a large number of turns of small-gauge wire. The winding that is to be connected to the low-voltage source or load is labeled X_1 and X_2. This winding consists of a small number of turns of large-sized copper, to carry the higher value of current. A step-up transformer would have its primary marked X_1 and X_2, and its secondary identified as H_1 and H_2. A step-down transformer would have its primary and secondary marked H_1, H_2, and X_1, X_2, respectively.

The transformer is based on the principle that energy may be efficiently transferred by magnetic induction from one winding to another by a varying magnetic flux. Therefore, when both windings are in the same magnetic circuit, mutual inductance exists between them. Because the primary is connected to an AC supply, the current and flux strength will rise to a maximum value and fall to zero twice in each cycle. During each cycle, the current and flux will flow in one direction for the first half-cycle and in the opposite direction for the second half-cycle.

In Chapter 14, we discussed that whenever changing flux links a coil of wire, an emf is developed in that coil. In a transformer, the magnetic lines of force from the primary coil link with the turns of the secondary coil and induce a voltage in the secondary winding. Voltage is induced in the secondary winding only if the current in the primary winding is alternating. A direct current in the primary winding has no effect on the secondary winding because the magnetic flux is stationary and does not cut, or link, the secondary turns.

The emf developed in the secondary coil is proportional to the rate of change of flux and is defined by the following equation:

$$E_{ave} = N \frac{\Phi_{pm}}{t} \tag{15.1}$$

where E_{ave} = average voltage induced into a coil, in volts
N = number of turns in the coil
Φ_{pm} = peak mutual flux, in webers
t = time required for the flux to rise from zero to its maximum value, in seconds

Equation 15.1 requires a steady value of magnetic flux. Because the winding does not move in a transformer, the flux changes sinusoidally and rises from zero to Φ_{pm} in one-quarter of a cycle.

$$E_{ave} = 4\Phi_{pm} fN$$

where   4 = multiplier that one full cycle of flux variation requires
f = frequency of the applied voltage, in hertz

The effective value of induced voltage becomes

$$E = \left(\frac{0.707}{0.637} \right) 4\Phi_{pm} fN$$

$$= 4.44\Phi_{pm} fN \qquad (15.2)$$

Equation 15.2 is called the **general transformer equation.** When the general transformer equation is applied to the primary winding of a transformer, the rms voltage of the primary winding is found by the following equation:

$$E_p = 4.44\Phi_{pm} fN_p \qquad (15.3)$$

where E_p = effective voltage of primary winding, in volts
      Φ_{pm} = peak mutual flux, in webers
      f = frequency of applied voltage, in hertz
      N_p = number of turns in primary winding

The flux developed by the primary's magnetomotive force links the secondary winding. In the ideal transformer, where no losses occur, all the flux developed by the primary will link the secondary. In this situation, a voltage will be induced in the secondary winding based on the following equation:

$$E_s = 4.44\Phi_{pm} fN_s \qquad (15.4)$$

EXAMPLE 15–1

Solution

The secondary winding of a 60 Hz transformer has 20 turns, and the flux in the core has a maximum value of 5.7×10^{-3} Wb. Determine the emf induced in the secondary.

$$E_s = 4.44\Phi_{pm} fN_s$$

$$= 4.44(5.7 \times 10^{-3} \text{ Wb})(60 \text{ Hz})(20)$$

$$= 30.37 \text{ V}$$

In an ideal transformer, the primary and secondary voltages are directly proportional to the number of turns in the windings. The ratio of primary winding turns to secondary winding turns is known as the **transformation ratio,** or **turns ratio,** and is given the letter symbol a. If the turns ratio is less than 1, the transformer would be called a step-up transformer, because E_s is greater than E_p. Conversely, if the transformation ratio is greater than 1, it would be referred to as a step-down transformer.

Because the flux shown in Figure 15–3 links each of the two windings, it follows that the same amount of emf per turn is induced in each winding, and the total induced emf in each winding must be proportional to the number of turns in that winding. To further illustrate this point, if we divide Equation 15.4 into Equation 15.3, the result is

$$\frac{E_p}{E_s} = \frac{N_p}{N_s} = a \qquad (15.5)$$

where a is the transformation ratio.

EXAMPLE 15–2

For the transformer shown in Figure 15–4, determine the

(a) Peak mutual flux.
(b) Number of primary turns.

$E_p = 120 \text{ V}$

$f = 60 \text{ Hz}$

$E_s = 24 \text{ V}$

$N_s = 80 \text{ turns}$

FIGURE 15–4   Circuit for Example 15–2.

Solution

(a)
$$E_s = 4.44\Phi_{pm}fN_s$$

$$\Phi_{pm} = \frac{E_s}{4.44fN_s}$$

$$= \frac{24 \text{ V}}{4.44\,(60 \text{ Hz})(80)} = 1.13 \text{ mWb}$$

(b)
$$\frac{E_p}{E_s} = \frac{N_p}{N_s}$$

$$N_p = \frac{E_pN_s}{E_s}$$

$$= \frac{(120 \text{ V})(80)}{24 \text{ V}} = 400 \text{ turns}$$

When a load is connected to the secondary winding of a transformer, a current will flow through the load because of the induced voltage E_s. The load current flowing through the secondary turns comprises a load component of magnetomotive force, which, according to Lenz's law, is in such a direction as to oppose the flux that is producing it. This opposition tends to reduce the transformer flux, but the attempted reduction in flux is accompanied by a reduction in the counter emf induced in the primary winding of the transformer, which keeps the flux constant. Because the internal resistance of the primary winding is low and the primary current is mainly limited by the counter emf in the winding, the primary current in the transformer increases.

In an ideal transformer, the current through the primary winding will continue to increase until the primary ampere-turns are equal to the secondary ampere-turns. The flux, by its rate of change, must balance the primary applied voltage, and because the primary counter emf varies only slightly from no-load to full-load current, the flux remains practically constant regardless of the load. If the no-load current is neglected in comparison with the total primary current, the primary and secondary ampere-turns are equal. Therefore,

$$N_pI_p = N_sI_s \quad \text{or} \quad \frac{N_p}{N_s} = \frac{I_s}{I_p} \tag{15.6}$$

By substituting Equation 15.5 into Equation 15.6, the following equation is derived.

$$\frac{E_p}{E_s} = \frac{N_p}{N_s} = \frac{I_s}{I_p} = a \qquad (15.7)$$

Therefore, in an ideal transformer $E_p I_p = E_s I_s$ and the power in, P_{in}, is equal to the power out, P_{out}.

EXAMPLE 15–3

Solution

$$\frac{E_p}{E_s} = \frac{I_s}{I_p}$$

$$I_p = \frac{E_s I_s}{E_p} = \frac{(12 \text{ V})(3 \text{ A})}{120 \text{ V}} = 300 \text{ mA}$$

A transformer delivers 3 A at 12 V. The primary voltage is 120 V. Determine the primary current.

Section Review

1. The primary winding is always connected to the source of power (True/False).
2. How are the windings that are connected to a high-voltage source or load identified?
3. State the general transformer equation.
4. In an ideal transformer, the primary and secondary voltages are directly proportional to the number of turns in the windings (True/False).

15–4

REFLECTED LOAD IN A TRANSFORMER

Any change in the load resistance connected to the transformer of Figure 15–4 will affect the current flowing through the secondary winding. Consequently, any variation in secondary current will cause the primary current to vary proportionately. This action is a result of Lenz's law. When a load is connected to the secondary, the current through the secondary develops a flux that is in opposition to the primary flux and cancels, or nullifies, part of the primary flux. The primary draws more current from the source to compensate for the flux canceled by the secondary. If we take the transformation ratio into account, the amount of current flow caused by a load resistor would be the same, regardless of whether it was connected to the primary or secondary of the transformer. This modified load resistance that appears as an input resistance at the primary is called the **reflected load** and is defined as follows:

> The value of a load resistance reflected from the secondary into the primary of a transformer is equal to the load resistance multiplied by the square of the transformation ratio.

In equation form,

$$R_p = a^2 R_L \qquad (15.8)$$

where R_p = reflected resistance at the primary, in ohms
  a = transformation ratio
  R_L = load resistance of secondary, in ohms

_____ EXAMPLE 15–4 _____

For the circuit of Figure 15–5, calculate the reflected resistance seen by the primary.

$R_L = 50\ \Omega$

FIGURE 15–5 *Circuit for Example 15–4.*

Solution

$$R_p = a^2 R_L = 6^2(50\ \Omega) = 1800\ \Omega$$

Section Review

1. The secondary draws more current from the source to compensate for the flux canceled by the primary (True/False).
2. Define reflected load in a transformer.
3. State the equation for reflected load.

___ 15–5 ___

MAXIMUM POWER TRANSFER

Reflected load calculations are generally used for simplifying transformer circuits when performing load-matching calculations to ensure maximum power transfer. The **maximum power transfer theorem** can be stated as follows:

> Maximum power is drawn from a source when the load resistance equals the internal resistance of the source.

The maximum power theorem also states that the greatest amount of power is transferred to the resistive load when the load resistance is equal to 50% of the total circuit resistance. In electronic communication systems, maximum power transfer is extremely important when analyzing signal sources such as antennas. When the load resistance and the source resistance of a circuit are equal, it is said to be a **matched circuit,** and maximum power transfer occurs between the antenna and load. The process of matching the output impedance of one circuit to the input impedance of another is also called **impedance matching.**

Transformers that match a source and load together in order to obtain maximum power transfer are called *impedance-matching transformers.* The term impedance will be discussed thoroughly in Chapter 16. At this point, we shall assume impedance to represent a general opposition to the flow of current. Impedance-matching transformers are popular in audio electronic circuits and telecommunication systems. In audio systems, matching transformers are used to connect the low speaker resistance to the power amplifier internal resistance. If the correct turns ratio is selected, a small load such as a 4 Ω speaker can be made to appear as large as the source resistance. When the resistance at the output terminals of an amplifier is the same as the resistance of the load, then maximum power will occur. The equation for determining the turns ratio is given by Equation 15.8 as

$$a = \sqrt{R_p/R_L}$$

EXAMPLE 15–5

Solution

$$a = \sqrt{R_p/R_L} = \sqrt{128\ \Omega/\ 8\ \Omega} = 4$$

The transformer must have 4 turns in the primary winding for every turn in the secondary in order for maximum power transfer to occur.

An impedance-matching transformer is required to match the 8 Ω impedance of a loudspeaker to the 128 Ω output impedance of a power amplifier. Calculate the required turns ratio of the audio transformer.

Section Review

1. State the maximum power transfer theorem.
2. What is an impedance-matching transformer?
3. State the equation for determining the turns ratio for maximum power transfer.

15–6

TRANSFORMER LOSSES

Losses in transformers are a result of eddy currents and hysteresis in the core and are known as *iron,* or *core, losses.* Losses that occur because of the resistance of the windings are referred to as *copper losses.* Because the magnetic flux in a transformer is changing direction many times a second, heat is developed because of the hysteresis of the magnetic material. Lower hysteresis losses are obtained by using **grain-oriented steel.** That is, the sheets of steel are cut so that the magnetic flux flows in the direction of the structural grain of the material. By using grain-oriented steel, the molecular friction is reduced and the hysteresis of the magnetic material is lowered.

The main purpose of laminating the core is to reduce eddy currents. Eddy-current loss is an I^2R loss that is set up by emfs induced in the core by the changing magnetic flux. These emfs are in a direction perpendicular to the flux path, so the layers of the laminations are laid parallel to the direction of the flux. After the sheets of steel are annealed, a coat of oxide is formed on the surface of the sheets, which increases the resistance in the path of the eddy currents. To insulate one sheet from another more effectively, a coat of insulating varnish is put on the steel sheets.

Even when no load is connected to the transformer, there is some loss of energy owing to the magnetizing current flowing in the primary winding. This I^2R loss also appears as heat generated in the winding and must be dissipated to prevent too great a temperature rise. When a load is connected to a transformer and current is flowing in both the secondary and primary windings, further losses of electrical energy occur. These losses are called **copper losses,** Cu, and are also considered to be I^2R losses because the loss in each winding is proportional to the resistance of the winding and the square of the current flowing in it. Copper losses may be calculated if the primary and secondary resistances are known. If R_H and R_X are the resistances of the high- and low-voltage windings, respectively, then the copper loss is calculated using the following equation:

$$\text{copper loss} = I_H^2 R_H + I_X^2 R_X \qquad (15.9)$$

From the above equation, it is obvious that copper losses vary with the load, being practically negligible at no load and at a maximum at full load. Transformers with high voltage and current ratings require conductors of large cross section to minimize copper loss. Unfortunately, a transformer requires a large number of turns to produce a high percentage of flux linkage, so a compromise between the size of conductors and the number of turns must be made.

Section Review

1. What are losses that occur because of the resistance of the windings known as?
2. How are low hysteresis losses obtained?
3. Copper losses represent I^2R losses (True/False).

__ 15–7

EFFICIENCY

The efficiency of a transformer is a ratio of the output power to the input power. In equation form,

$$\text{efficiency} = \frac{\text{output}}{\text{input}} \times 100\% = \eta$$

Because the output power is less than the input power, owing to the losses that occur in the transformer, it is often necessary to use these losses in the efficiency calculation. The sum of these losses is always very small, so the efficiency of a transformer is usually above 95%. As long as the supply frequency remains constant, the core losses are virtually constant for all conditions of loading, as well as at no load. The copper losses vary as the square of the volt-ampere (VA) output of the transformer and are considered variable for that reason. When expressed in terms of losses, the equation for the efficiency of a transformer becomes

$$\eta = \frac{P_{out}}{P_{out} + P_{losses}}$$

As the transformer goes from full load towards no load, the efficiency reaches a maximum value and proceeds to drop off. This maximum value, which is known as **maximum efficiency,** will occur when the copper losses are equal to the core losses.

Section Review

1. How is the efficiency of a transformer determined?
2. When does maximum efficiency of a transformer occur?
3. The efficiency of a transformer is usually above 95% (True/False).

__ 15–8

AUTOTRANSFORMER

The **autotransformer** is a device that accomplishes the desired transformer action within one coil, as compared to two or more coils of a standard transformer. Autotransformers are also known as *variacs*. In a standard transformer, the primary and secondary coils are magnetically coupled with no physical connection between the primary and secondary. In an autotransformer, the coils are magnetically coupled as well as physically connected,

FIGURE 15-6 *The autotransformer.*

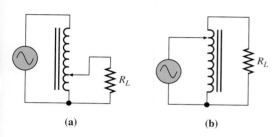

FIGURE 15-7 *Autotransformer.*
(a) Step-down configuration.
(b) Step-up configuration.

(a)   (b)

because one of the coils is common to both the primary and the secondary circuits. This coil is referred to as the **common winding.** Figure 15–6 shows a typical autotransformer.

Figure 15–7(a) shows an autotransformer used in a step-down configuration. The AC input is connected across the full winding, but only a small portion is common to both the input and output sides. Consequently, the primary voltage is higher than the secondary voltage. In Figure 15–7(b), the input supply is now connected across a portion of the total winding. The load is connected across the entire winding, so the secondary has more turns than the primary. This results in a higher voltage at the output than at the input.

Section Review

1. Autotransformers are also known as variacs (True/False).
2. What is the main difference between an autotransformer and a standard transformer?

15-9
ISOLATION TRANSFORMER

In AC power supplies that use a two-prong plug, the ground conductor, or **neutral,** is occasionally connected directly to the metal chassis of an electronic system. The current-carrying conductor, or *hot* as it is sometimes called, goes directly to the circuit that converts the AC to DC. If the connecting plug is reversed, the hot conductor would be connected to the ground of the chassis, and the chassis would be *energized.* This could result in electric shock, or electrocution, for someone who touches the chassis.

Figure 15–8(a) shows a properly connected AC plug and chassis. The neutral is connected to the chassis, and the AC supply is connected to the electronic circuit. Fig-

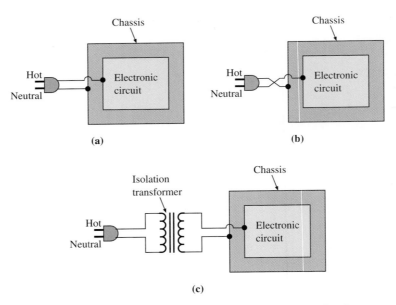

FIGURE 15–8 *(a) AC source connected to a load using a two-prong plug. (b) Connections reversed. (c) Isolation transformer connected between source and load.*

ure 15–8(b) shows the effect of reversing the plug. The chassis is now energized, and the electronic circuit is grounded. If a transformer were used between the AC supply and the electronic equipment, as in Figure 15–8(c), the equipment would now be *isolated* from the supply. These transformers, called **isolation transformers,** have a transformation ratio of 1 because they neither step up nor step down the voltage.

The entire purpose of an isolation transformer is to prevent equipment connected on the secondary side from being connected directly to the supply on the primary side. The secondary winding of an isolation transformer is said to be *floating* because it is no longer referenced to the ground on the primary side. The risk of electric shock has now been eliminated because there is no physical connection between the secondary and the supply.

Section Review

1. Name one hazard of using two-prong electrical plugs.
2. What is the purpose of an isolation transformer?
3. The primary winding of an isolation transformer is said to *float* (True/False).

___ 15–10 ___

CENTER-TAP TRANSFORMER

A **center-tap transformer** is designed to provide two separate secondary voltages that are 180° out of phase with each other. Figure 15–9(a) shows a typical center-tap transformer. The schematic diagram shown in Figure 15–9(b) illustrates the tapped lead in the exact center of the secondary winding. The center-tap connection provides a common point for two equal but opposite secondary voltages. If the center tap is grounded, the output from one terminal of the secondary is positive going (with respect to ground) while the voltage

FIGURE 15–9 *(a) Center-tap transformer. (b) Schematic diagram.*

(a)　　　　　　(b)

at the other secondary terminal is negative going. The total output voltage for the secondary winding is the sum of the two secondary-winding voltages.

Section Review

1. What is a center-tap transformer?
2. If the center-tap is grounded, the output from one terminal of the secondary is positive going and the other is negative going (True/False).

MULTIPLE-WINDING TRANSFORMER

In electronic circuits, one transformer is often used to supply a variety of voltages. These different voltage connections are called *taps*. Multiwinding transformers can provide either step-up or step-down voltages, or a combination of both, depending on the application. An example of a multiple-winding transformer is shown in Figure 15–10. This configuration represents a typical connection for a television receiver. The picture tube for the television set requires high voltage levels, and the control circuit components, such as integrated circuits (ICs), require very low voltages.

Figure 15–11 shows an example of a multiple-winding transformer supplying a number of different secondary voltage levels. This configuration is popular in power supply circuits, where a variety of lower-voltage levels are required for different components in the electronic circuitry. The various taps of the transformer are usually color coded as a means of referencing the transformer leads to schematic diagrams and other documentation when

FIGURE 15–10
Multiwinding transformer for a TV receiver.

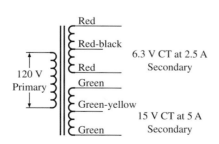

FIGURE 15–11
Multiwinding step-down transformer.

troubleshooting. All primary wires are generally black or black with a colored stripe. The secondary wires in Figure 15–11 are either green, yellow, red, black, or combinations of the four colors.

Section Review

1. What is a multiple-winding transformer?
2. How are the primary wires of a multiple-winding transformer identified?
3. The secondary voltage terminals of a multiple-winding transformer are called taps (True/False).

____ 15–12 _____

TRANSFORMER POLARITY

The primaries and secondaries of transformers usually consist of two or more windings instead of a single winding. This allows the windings on each side of the transformer to be connected in either series or parallel and several voltage and current ratings to be obtained from one transformer. Single-phase transformers are also interconnected to supply three-phase power. The winding connections must be made so that the correct phase relations exist among the winding voltages. **Transformer polarity** is defined as the relative polarity of the secondary voltage with respect to the primary voltage.

There are two methods used for identifying the winding terminals of a transformer. One method, which was described in Section 15–3, uses a set of letters and numbers established by the American Standards Association. The high-voltage terminals are labeled H_1 and H_2, and the low-voltage terminals are marked X_1 and X_2. The subscripts 1 and 2 denote the instantaneous polarity of the terminals with respect to each other. The terminals H_1 and X_1 have the same instantaneous polarity, and the terminals H_2 and X_2 have the same instantaneous polarity. When transformers have two coils on each side, labels H_3 and H_4 would be added on the high-voltage side, and X_3 and X_4 on the low-voltage side. Terminals H_3 and X_3 would have the same instantaneous polarity, as would terminals H_4 and X_4.

The other method of identifying the terminals of a transformer is called the **dot convention.** When this type of marking is used, terminals with the same instantaneous polarity are identified by dots and the other terminals are left blank. An example of the dot convention is shown in Figure 15–12.

The primary windings in Figure 15–13 are interconnected so that the direction of the magnetizing current in both coils at any instant tends to establish a flux in the same direction in the core. Because the induced voltages in both windings are caused by the same sinusoidal flux, the direction of the current in both windings will be the same. Therefore, the windings shown in Figure 15–13 are connected in **additive polarity.**

When the primary coils are connected so that the mmfs are set up in opposite directions in the core, the resulting flux is zero. The coils are then said to be connected in **subtractive polarity.**

Whether the external polarity is additive or subtractive is determined entirely by the manner in which the leads from the windings are connected to the external terminals. Because the windings in a transformer are sealed and not visible, it is necessary to rely on the manufacturer's external polarity markings for proper *phasing* of the windings. How-

FIGURE 15–12 *Transformer terminals identified using the dot convention.*

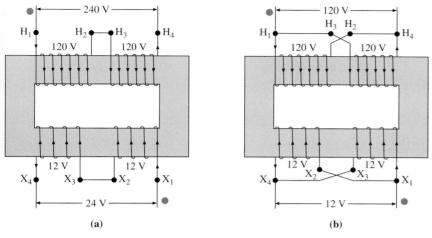

FIGURE 15–13 *Additive polarity transformer with (a) primary and secondary in series and (b) primary and secondary in parallel.*

ever, if a transformer has markings that are difficult to decipher, or nonexistent, a simple voltage test can be performed to determine the external polarity. Although any safe voltage is suitable for this test, Figure 15–14 shows 120 V being used, and the voltmeter should have a range of about twice the testing voltage. Assuming no markings are on the transformer, the right- and left-hand terminals, as viewed from the high-voltage side of the transformer, may be marked H_1 and H_2, respectively. A connection is made between H_1 and the low-voltage terminal directly opposite H_1. A voltmeter is then connected between H_2 and the low-voltage terminal directly opposite H_2.

If the voltmeter reading is greater than the input voltage, as in Figure 15–14(a), then the transformer is said to have additive polarity, and the low-voltage terminal directly opposite H_1 is marked X_2. The other low-voltage terminal is marked X_1. If the voltmeter reads below the input value, then the transformer is considered to be subtractive, and its low-voltage terminals are marked as shown in Figure 15–14(b).

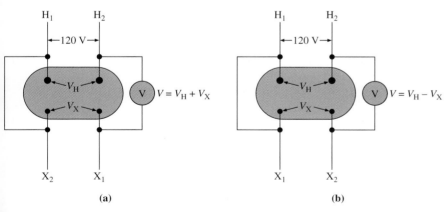

FIGURE 15–14 *Method of testing polarity of transformer using a voltmeter. (a) Voltage reading over 120 V indicates additive polarity. (b) Voltage reading under 120 V indicates subtractive polarity.*

Section Review

1. What is transformer polarity?
2. The winding terminals of a transformer can be identified using a standard set of markings or by the dot convention (True/False).
3. What is subtractive polarity?

___ 15–13 _____

PULSE TRANSFORMER

The pulse transformer is widely used in electronic circuits. Pulse transformer technique was originally developed for use in radar systems. However, the increased use of square-wave voltages has led to extending pulse transformer theory to a variety of applications including video signals and inverters. Figure 15–15 shows the output of a pulse transformer for a rectangular input pulse. The pulse width is shown as a value at approximately one-half of the output amplitude. The **pulse width** of the input signal is defined as the time between the start of the rise and the start of the fall of the input pulse. The pulse width of the input pulse and the output pulse are not necessarily equal. The rise and fall times of the pulse are often determined by the time elapsed between 10% and 90% of the pulse amplitude.

Pulse transformers are often used to couple a trigger pulse to a thyristor, such as an SCR, to obtain electrical isolation between two circuits. Pulse transformers used for thyristor control are generally either a 1:1 ratio (two windings) or a 1:1:1 ratio (three windings). Figure 15–16 shows an SCR connected to a 1:1 ratio pulse transformer. Diode D_1 is connected in series with the gate of the SCR to prevent reversal of the pulse transformer's output voltage. SCRs are described in more detail in Chapter 20.

Figure 15–17 shows how a 1:1:1 pulse transformer is used to drive an inverse-parallel pair of SCRs. In this circuit, the two SCRs and the trigger pulse generator are fully isolated.

Section Review

1. Define the term *pulse width*.
2. Pulse transformers are used in video signal applications and to trigger thyristors (True/False).

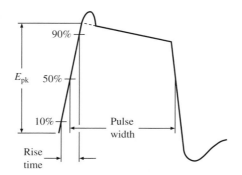

FIGURE 15–15 *Output of pulse transformer.*

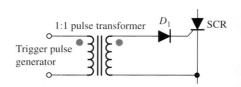

FIGURE 15–16 *Basic pulse transformer coupling.*

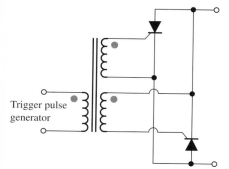

FIGURE 15-17 *1:1:1 pulse transformer controlling two SCRs.*

Trigger pulse generator

15-14

TROUBLESHOOTING TRANSFORMERS

The procedure used for troubleshooting transformers is similar to the method used when testing inductors. Because a transformer is essentially a coil, faults are typically caused by either open circuits or short circuits. Open circuits can occur at any point in either the primary or secondary winding, although the open circuit is most likely at the point where the wire of the winding is connected to the transformer terminal. This type of break is often located by visual inspection and repaired by resoldering the terminal.

Short-circuit conditions inside the transformer are often caused by excess current flowing through the windings. As the current flow through the windings increases, the temperature of the windings increases. If this temperature rises above the transformer's rated value, the insulation of the windings melts and the turns of wire are shorted together. The insulation used on the windings is a type of varnish that will produce a distinct smell when burnt. There is no practical method of repairing a shorted transformer. Consequently, these devices must be replaced when a short-circuit condition has occurred.

Opens and shorts are tested easily by using an ohmmeter. Recall from previous chapters that a short circuit will produce zero ohms at the terminals of the device being tested, and a reading of infinity implies an open circuit. When a transformer is being tested, a short circuit can produce a reading greater than zero ohms, but below the rated value. This is caused by a partial short, where some of the windings are shorted together, but are still in series with other windings that are undamaged. A partial short will change the transformation ratio of the transformer. Figure 15–18(a) shows a properly functioning step-down transformer with a transformation ratio of 10. This particular transformer has 500 turns on the primary and 50 turns on the secondary. Figure 15–18(b) shows how the transformer is affected if one-half the primary windings are shorted out. The transformation ratio is now 5, because there are now only 250 turns being used on the primary. This would cause an increase in the output voltage of the secondary.

Voltmeters are well suited for testing transformers that are energized. Because transformers operate on the principle of magnetic induction, an open-circuit condition across the primary terminals will result in a reading of 0 V across the secondary terminals, as

FIGURE 15-18 *(a) Transformer with a turns ratio of 10. (b) Effect of shorting out half of primary windings.*

FIGURE 15–19 *(a) Primary with open circuit. (b) Secondary with open circuit.*

shown in Figure 15–19(a). The voltmeter connected across the terminals of the primary winding would produce a reading equal to that of the applied voltage. If the open circuit occurs in the secondary winding, the voltage at the secondary is again 0 V, as shown in Figure 15–19(b). Once again this reading is because there is no induced voltage in the secondary. The voltage at the primary will be equal to the voltage drop across the primary windings.

Section Review

1. Short-circuit conditions inside a transformer are often caused by excess current flowing through the winding (True/False).
2. A partial short will change the transformation ratio (True/False).

15–15

PRACTICAL APPLICATION SOLUTION

In the chapter opener, a wireless burglar alarm system was described. The system uses a wireless transmitter, antenna, and receiver. Your task was to select an impedance-matching transformer that will provide maximum power transfer for the wireless system. The impedance of the antenna was given as 300 Ω and the impedance of the receiver was 75 Ω. The following steps illustrate the method of solution for this practical application of transformers.

STEP 1   Determine the turns ratio of the impedance-matching transformer.

$$a = \sqrt{300/75} = 2$$

STEP 2   Select and install an impedance matching transformer with a turns, or transformation, ratio of 2.

Summary

1. Transformers transfer energy from one circuit to another by electromagnetic induction.
2. The two transformer windings are called primary and secondary.
3. In an ideal transformer, all the flux developed by the primary will link the secondary.
4. The primary and secondary voltages of a transformer are directly proportional to the number of turns in the windings.
5. Transformers used for maximum power transfer are called impedance-matching transformers.
6. The efficiency of a transformer is a ratio of the output power to the input power.
7. Autotransformers use one coil instead of two to achieve transformer action.
8. A center-tap transformer is designed to provide two separate secondary voltages.

9. Transformers can be connected in additive or subtractive polarity.
10. An open-circuit condition across the primary terminals will result in a reading of 0 V across the secondary terminals.

Answers to Section Reviews

Section 15–1

1. By induction from the primary winding
2. A transformer that receives energy at one voltage and delivers it at a higher voltage
3. False

Section 15–2

1. Core type and shell type
2. True
3. To reduce eddy current losses

Section 15–3

1. True
2. H_1 and H_2
3. $E_p = 4.44\Phi_{pm} fN_p$
4. True

Section 15–4

1. False
2. The value of a load resistance reflected from the secondary into the primary of a transformer is equal to the load resistance multiplied by the square of the transformation ratio.
3. $R_p = a^2 R_L$

Section 15–5

1. Maximum power is drawn from a source when the load resistance equals the internal resistance of the source.
2. A transformer that matches a source and load together to obtain maximum power transfer
3. $\sqrt{a} = R_p/R_L$

Section 15–6

1. Copper losses
2. By using grain oriented steel, or laminated plates
3. True

Section 15–7

1. As a ratio of the output power to the input power
2. When the copper losses are equal to the core losses
3. True

Section 15–8

1. True
2. The autotransformer requires only one coil compared with a standard transformer that requires two coils.

Section 15–9

1. If the connecting plug is reversed, the current-carrying conductor may be connected to the metal chassis of an electronic system.
2. To prevent equipment connected on the secondary side from being connected directly to the supply on the primary side
3. False

Section 15–10

1. A transformer that is designed to provide two separate secondary voltages that are 180° out of phase with each other
2. True

Section 15–11

1. A transformer capable of supplying a number of different secondary voltage levels
2. All primary wires are generally black or black with a colored stripe.
3. True

Section 15–12

1. The relative polarity of the secondary voltage with respect to the primary voltage
2. True
3. When the coils are connected so that the mmfs are set up in opposite directions in the core and the resulting flux is reduced

Section 15–13

1. The time between the start of the rise and the start of the fall of the input pulse
2. True

Section 15–14

1. True
2. True

____ Review Questions ____

Multiple Choice Questions

15–1 When power is supplied to one coil of a transformer at a specific frequency, power can be taken from the other coil at
(a) A lower frequency   (b) A higher frequency
(c) The same frequency   (d) Varying frequencies

15–2 A transformer that receives energy at one voltage and delivers it at the same voltage is called a
(a) Step-up transformer   (b) Step-down transformer
(c) Voltage transformer   (d) One-to-one transformer

15–3 Transformers are divided into two general types
(a) Primary and secondary   (b) Core type and shell type
(c) Step-up and step-out   (d) Air transformers and autotransformers

15–4 Silicon is used in the magnetic circuit of a transformer to
(a) Reduce iron losses   (b) Reduce eddy currents
(c) Increase eddy currents   (d) Reduce hysteresis

15–5 When a transformer is used to step-down voltage,
(a) The high-voltage winding is the secondary

(b) The low-voltage winding is the primary
(c) The high-voltage winding is the primary
(d) The frequency is also stepped down

15–6   The winding that is to be connected to a high-voltage source or load is designated by
(a) X_1 and X_2 **(b)** H_1 and H_2 **(c)** X_1 and H_1 **(d)** X_2 and H_2

15–7   In an ideal transformer, the primary and secondary voltages are _____ the number of turns in the windings.
(a) Equal to   **(b)** Inversely proportional to
(c) Directly proportional to   **(d)** Greater than

15–8   The transformation ratio is also known as the
(a) Voltage ratio   **(b)** Current ratio   **(c)** Power ratio   **(d)** Turns ratio

15–9   Maximum power is drawn from a source when the
(a) Load resistance equals the internal resistance of the source
(b) Load resistance equals 1/4 of the circuit resistance
(c) Load resistance equals 1/2 of the internal resistance of the source
(d) Load resistance is double the circuit resistance

15–10   Transformers that connect a source and load together to obtain maximum power transfer are called
(a) One-to-one transformers   **(b)** Pulse transformers
(c) Impedance-matching transformers   **(d)** Autotransformers

15–11   The main purpose of laminating the core of a transformer is to reduce
(a) Hysteresis   **(b)** Copper losses
(c) Losses in the windings   **(d)** Eddy currents

15–12   The efficiency of a transformer is a ratio of the
(a) Output power to the input power   **(b)** Output voltage to the input voltage
(c) Output current to the input current   **(d)** All of the above

15–13   Autotransformers are also known as
(a) Isolation transformers   **(b)** Multiple-winding transformers
(c) Variacs   **(d)** Matching transformers

15–14   The wire brought out from the center of two secondary windings in a multiple-winding transformer is called
(a) Primary   **(b)** Floating conductor
(c) High voltage   **(d)** Center tap

15–15   The rise and fall times of pulses produced by a pulse transformer are often determined by the time elapsed between
(a) 10% and 90% of the pulse amplitude   **(b)** 20% and 80% of the pulse amplitude
(c) 10% and 80% of the pulse amplitude   **(d)** 20% and 90% of the pulse amplitude

Practice Problems

15–16   An ideal transformer with a turns ratio of 20 has an AC voltage of 120 V applied to its primary winding. Determine the secondary voltage.

15–17   The primary and secondary currents of a transformer were measured at 2 A and 20 A, respectively. Determine the primary voltage if the secondary is 12 V.

15–18   The 240 V primary of a 60 Hz transformer has 96 turns. Calculate the peak mutual flux developed.

15–19   A 12/120 V, 60 Hz step-up transformer has 60 turns on the secondary winding. Determine the peak mutual flux of the transformer.

15–20 A 120/24 V, 60 Hz step-down transformer has 80 turns on the primary winding. Determine the peak mutual flux of the transformer.

15–21 An ideal transformer has a turns ratio of 20 and a load resistance of 8 Ω connected to the secondary. Use reflected resistance to calculate the resistance "looking into" the primary winding.

15–22 An ideal transformer is supplying a secondary current of 8 A. The transformation ratio is 6, and the impedance at the primary is 100 Ω. Determine the

(a) Primary voltage

(b) Secondary voltage

(c) Resistance connected to the secondary

15–23 An impedance-matching transformer is to be used in a circuit to allow for maximum power transfer. The load connected to the secondary has a value of 600 Ω and the line resistance is 75 Ω. Determine the transformation ratio that will allow for maximum power transfer.

15–24 The output resistance of an audio power amplifier is measured as 2500 Ω. An impedance-matching transformer with 1000 turns on its primary is to be used to match the amplifier output to an 8 Ω loudspeaker. Determine the number of turns on the secondary winding of the impedance-matching transformer.

15–25 A 120/24 V transformer has a primary current of 2.2 A and a secondary current of 9.63 A. Determine the efficiency of the transformer.

Essay Questions

15–26 Describe the basic operating principle of a transformer.

15–27 List three applications for transformers in electronic circuits.

15–28 What are the two windings of a transformer called?

15–29 Why is silicon added to the core of a transformer?

15–30 How are the windings of a transformer marked for identification?

15–31 Explain the meaning of transformation ratio.

15–32 Define reflected load.

15–33 What is maximum power transfer?

15–34 List the two main types of losses in a transformer.

15–35 Describe the difference between an autotransformer and an isolation transformer.

15–36 Why are multiple-winding transformers used in electronic circuits?

15–37 What are pulse transformers?

16

Alternating Current Circuits

Learning Objectives

Upon completion of this chapter you will be able to

- Understand the difference between vectors and phasors.
- Describe the phase relationship between voltage and current in an AC circuit.
- Explain the effects of inductive reactance and capacitive reactance on an AC circuit.
- Define impedance.
- Utilize the voltage divider rule in AC calculations.
- Explain admittance and susceptance in AC circuits.
- Discuss power in AC circuits.

As a quality control electronic technician in an automobile manufacturing plant, one of your duties involves testing the current drawn by the output of a voltage-controlled oscillator (VCO) in a solid-state voltage regulator. A VCO is a circuit that converts a variable voltage into a variable frequency. For the test, the output voltage of the VCO is set at 2 V_{rms} and the frequency is 1 kHz. The VCO is connected to a load simulation board, which is a simple parallel *RLC* circuit with the following values: $R = 600\ \Omega$, $C = 0.68\ \mu F$, and $L = 50$ mH. Your task is to determine the output current drawn by the oscillator and compare this value to the current measured using an ammeter.

16–1

INTRODUCTION

In this chapter we will examine the effects of resistance, inductance, and capacitance when connected to an AC supply. In an AC circuit that contains only resistance, the current and voltage are said to be in phase with each other. If the circuit contains only inductance, the current lags the voltage by 90°. The opposition to current flow in this type of circuit is called the **inductive reactance** and is measured in ohms. When an AC circuit contains only capacitance, the current leads the applied voltage by 90°, and the ratio of voltage to current is equal to the **capacitive reactance** of the circuit in ohms. These three types of circuits are solved easily because Ohm's law may be directly applied to solve for current, voltage, resistance, or reactance.

In an AC circuit that contains both resistance and reactance, the opposition to current flow is the combined phasor sum of these quantities. This total opposition to the flow of alternating current is called **impedance.** The impedance of a circuit is expressed in ohms and its symbol is the letter Z.

Section Review

1. What is the relationship between current and voltage in a purely inductive circuit?
2. What symbol is used to represent impedance?
3. Impedance is the opposition to current flow offered by a circuit (True/False).

16–2

PHASORS

When solving alternating-current circuit problems, it is often convenient to use phasors. A **phasor** is a complex number that is used in AC circuits to represent quantities such as voltage or current. Although the terms phasor and vector are sometimes used interchangeably, there is a fundamental difference between these two measurements. A **vector** is defined as a quantity that has both magnitude and direction. Vector quantities are represented by an arrow, whose length represents the magnitude and whose direction represents the angle at which the force acts. Physical forces, such as those illustrated in Figure 16–1, are conveniently represented by vectors. If a force of 200 newtons (N) is applied at right angles to a force of 300 N, a resultant force would be established as indicated. This force is at an angle theta (θ) with respect to the first force.

200 N   **FIGURE 16–1**   *Representation of vector forces.*

θ

Resultant

300 N

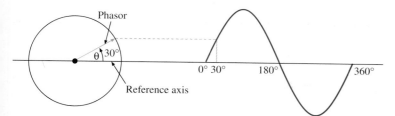

FIGURE 16–2 *Instantaneous values of current from a rotating phasor.*

A similar technique can be utilized in AC circuits to represent currents and voltages. Strictly speaking, AC voltages and currents are *not* vectors because they cannot be defined in terms of magnitude and direction alone. These quantities also have a *time* relationship to each other and must be solved by using complex algebra. This type of algebra is used for solving numbers that exist in the two-dimensional complex plane. To avoid possible complications between vectors using linear algebra and AC quantities in the time domain, these current and voltage arrows have been renamed phasors.

Phasors do not normally represent instantaneous values, because they are of fixed length in any given situation. They are, essentially, used to represent effective, or rms, values, although they can also represent an instantaneous value for one specific instant of time. The phasor corresponds to the entire cycle of current or voltage, but is shown at only one angle because the complete cycle is known to be a sine wave.

A **phasor diagram** consists of a number of phasors, one for each quantity represented. Each phasor is drawn from a common origin. The phasor in Figure 16–2 is stopped in one of its many possible positions. The horizontal axis is referred to as the *reference axis*. A phasor lying on the reference axis is considered to be the **reference phasor.** When a phasor is rotated in a counterclockwise direction from the reference axis, the resulting phase angle, theta, is taken to be positive. When the phasor is rotated in a clockwise direction, theta assumes a negative value.

Earlier in this chapter, we mentioned that, when a sinusoidal voltage is applied to a pure resistance, the resulting current is sinusoidal and is in phase with the voltage. Figure 16–3 illustrates how this information would be indicated on a phasor diagram. Note the capital letters **E** and **I**. When phasor quantities are shown as effective, or rms, values on a phasor diagram, they are usually shown in capital letters using boldface type. Instantaneous values of voltage and current are usually represented by lowercase letters. An instan-

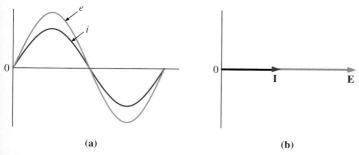

FIGURE 16–3 *(a) Current and voltage relationship for a purely resistive circuit. (b) Phasor diagram for a purely resistive circuit.*

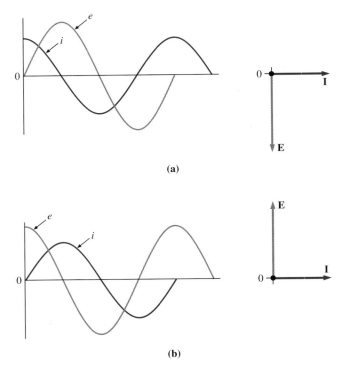

FIGURE 16–4 *(a) Current and voltage relationship in a purely capacitive circuit. Phasor diagram shows voltage lagging the current. (b) Current and voltage relationship in a purely inductive circuit. Phasor diagram shows voltage leading the current.*

teous voltage drop is represented by the letter v, and voltage source by e, and an instantaneous current by lowercase i.

When labeling a phasor quantity, the sinusoidal expression is not used. For example,

$$e = 141.4 \sin(\omega t + 30°)$$

becomes
$$\mathbf{E} = 100 \angle 30° \text{ V}$$

where $\mathbf{E}$ = rms value
  $\angle$ = phase angle

Figure 16–4(a) shows an example of a circuit where the current leads the voltage by 90°. A phasor diagram for a circuit where the current lags the voltage by 90° would be represented by the phasor diagram shown in Figure 16–4(b). In both examples, the phasor representing current is taken as the reference phasor. The choice of current or voltage as a reference phasor is arbitrary and is left up to the individual. However, when applying phasor diagrams to series circuits, it is customary to use current as a reference phasor because current is common to all parts of a series circuit. Consequently, a parallel circuit is usually represented in phasor diagrams with voltage as the reference phasor.

Section Review

1. What is a phasor?
2. A vector has magnitude and direction (True/False).
3. Why are AC voltages and currents not classified as vectors?
4. What is a reference phasor?

16–3

RESISTANCE IN AC CIRCUITS

The simplest AC circuit is one that contains resistance only. The relationship between voltage and current is given by Ohm's law $(E = IR)$. In such a circuit, the only voltages involved are the applied emf and the voltage drop across the resistor. The voltage and current magnitudes have a constant ratio, which is equal to the resistance and is not a function of time.

A sinusoidal voltage is applied to a resistance as shown in Figure 16–5. The equation for solving for the instantaneous current is given as

$$i = I_m \sin \omega t = I_m \sin 2\pi ft \qquad \textbf{(16.1)}$$

where f = frequency, in hertz
I_m = maximum, or peak, current

The resulting time variations in the current and voltage waveforms are shown in Figure 16–6. Because the applied emf must be entirely utilized in forcing the current through the resistor, R, the applied voltage must be equal to the voltage drop across the resistor. Therefore, the circuit voltage, e, at any instant is equal to the IR drop.

$$e = IR = I_m \times R \sin \omega t \qquad \textbf{(16.2)}$$

From the waveforms shown in Figure 16–6, it is apparent that the voltage wave and the current wave are both sinusoidal and have the same frequency. At the 90° point, sin $\omega t = 1$, and both the current and the voltage are simultaneously at their maximum positive values.

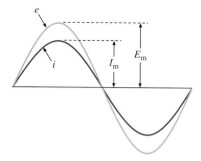

FIGURE 16–5
Resistive load connected across an AC source.

FIGURE 16–6
Instantaneous and phasor values of current and voltage in a purely resistive circuit.

(a)   (b)

FIGURE 16–7 *Phasor diagram of (a) maximum and (b) effective values of voltage and current.*

Because both current and voltage waveforms are in phase, the phasors that display their values must also be in phase with each other. In Figure 16–7(a), the current phasor and voltage phasor are shown at their maximum values. In Figure 16–7(b), the phasors are shown at their effective, or rms, values.

EXAMPLE 16–1

The applied emf of the circuit shown in Figure 16–5 is $e = 200 \sin (377t)$. Calculate the

(a) Instantaneous value of voltage at $t = 0.002$ s.
(b) Effective value of current for a resistance of 1200 Ω.

Solution

(a)
$$e = 200 \sin [(377)(0.002)(57.3)] = 136.92 \text{ V}$$

(b)
$$I_{rms} = \frac{E_{rms}}{R} = \frac{(0.707)(200 \text{ V})}{1200 \text{ Ω}} = 117.83 \text{ mA}$$

Section Review

1. In a purely resistive circuit, the current and voltage waveforms are in phase with each other (True/False).
2. In which AC circuit do the voltage and current magnitudes have a constant ratio that is not a function of time?

— 16–4

INDUCTIVE REACTANCE

When inductance was discussed in Chapter 14, the opposition that an inductance offers to a changing current was called *self-induced voltage,* or *counter emf,* and was measured in volts. Because a coil reacts to a current change by generating a counter emf, a coil is said to be *reactive.* Therefore, the opposition to current flow offered by a coil is called **inductive reactance,** which is measured in ohms and represented by the symbol X_L. The equation for determining inductive reactance is

$$X_L = 2\pi f L \tag{16.3}$$

where X_L = inductive reactance, in ohms
f = frequency, in hertz
L = inductance, in henrys

EXAMPLE 16–2

Solution

$$X_L = 2\pi f L = 2\pi(60 \text{ Hz})(12 \text{ H}) = 4523.89 \ \Omega$$

Calculate the inductive reactance of a 12 H coil at 60 Hz.

EXAMPLE 16–3

Solution

$$X_L = \frac{E_{\text{rms}}}{I_{\text{rms}}} = \frac{50 \text{ V}}{0.2 \text{ A}} = 250 \ \Omega$$

$$L = \frac{X_L}{2\pi f} = \frac{250 \ \Omega}{2\pi(60)} = 0.66 \text{ H}$$

The effective voltage across an inductor is 50 V when the effective current is 200 mA. The frequency of the circuit is 60 Hz. Determine the inductance.

EXAMPLE 16–4

Solution

$$X_L = 2\pi f L = \omega L = 150(5) = 750 \ \Omega$$

$$I_{\text{m}} = \frac{E_{\text{m}}}{X_L} = \frac{30 \text{ V}}{750 \ \Omega} = 0.04 \text{ A}$$

In a purely inductive circuit, *i* lags *e* by 90°.

$$i = I_{\text{m}}\sin(\omega t - 90°)$$
$$= 0.04 \sin[150(0.025)(57.3) - 90°]$$
$$= 32.82 \text{ mA}$$

The instantaneous voltage across a 5 H inductor is *e* = 30 sin 150*t*. Calculate the instantaneous current when *t* = 25 ms.

To calculate series and parallel combinations of inductive reactance, the same method is used as for solving resistance combinations. To find the total reactance of two or more series connected inductors,

$$X_{LT} = X_{L1} + X_{L2} + \cdots + X_{Ln} \tag{16.4}$$

For two parallel-connected inductors, the product-over-sum equation may be used.

$$X_{LT} = \frac{X_{L1}X_{L2}}{X_{L1} + X_{L2}} \tag{16.5}$$

To solve for two or more parallel-connected inductors,

$$\frac{1}{X_{LT}} = \frac{1}{X_{L1}} + \frac{1}{X_{L2}} + \frac{1}{X_{L3}} + \cdots + \frac{1}{X_{Ln}} \tag{16.6}$$

EXAMPLE 16–5

For the circuit shown in Figure 16–8, calculate the values of X_{LT}, I_T, V_{L1}, and V_{L2}.

FIGURE 16–8 Circuit for Example 16–5.

Solution

$$X_{LT} = X_{L1} + X_{L2} = 800\ \Omega + 1500\ \Omega = 2300\ \Omega$$

$$I_T = \frac{E_T}{X_{LT}} = \frac{120\ \text{V}}{2300\ \Omega} = 52.17\ \text{mA}$$

$$V_{L1} = I_T X_{L1} = (52.17\ \text{mA})(800\ \Omega) = 41.74\ \text{V}$$

$$V_{L2} = I_T X_{L2} = (52.17\ \text{mA})(1500\ \Omega) = 78.26\ \text{V}$$

EXAMPLE 16–6

For the circuit shown in Figure 16–9, calculate the values of X_{LT}, I_{L1}, I_{L2}, and I_T.

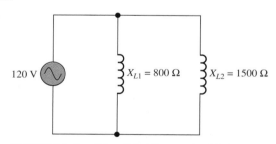

FIGURE 16–9 Circuit for Example 16–6.

Solution

$$X_{LT} = \frac{X_{L1}X_{L2}}{X_{L1} + X_{L2}} = \frac{(800\ \Omega)(1500\ \Omega)}{800\ \Omega + 1500\ \Omega} = 521.74\ \Omega$$

$$I_{L1} = \frac{120\ \text{V}}{800\ \Omega} = 150\ \text{mA}$$

$$I_{L2} = \frac{120\ \text{V}}{1500\ \Omega} = 80\ \text{mA}$$

$$I_T = I_{L1} + I_{L2} = 150\ \text{mA} + 80\ \text{mA} = 230\ \text{mA}$$

Section Review

1. What is the symbol for inductive reactance?
2. State the equation for finding inductive reactance.
3. What is the inductive reactance of a 100 mH coil at 500 Hz?

16–5

CAPACITIVE REACTANCE

Capacitance was defined in Chapter 13 as that quality of a circuit that enables energy to be stored in an electric field. A capacitor is a device that possesses the quality of capacitance.

In the discussion of DC circuits, it was shown that a capacitor is charged when a voltage is applied to its terminals and discharged when it is short circuited. If an alternating voltage is applied to a capacitor, it will be alternately charged and discharged, and an alternating current of the same frequency will flow in the circuit.

When an alternating emf is applied to the capacitor shown in Figure 16–10(a), the charge on the plate at any instant is proportional to the voltage E and will, therefore, be in phase with this voltage. Because the voltage is alternating, the capacitor is charged first in one direction and then in the other, and an alternating current will flow. During the first 90° of the sine wave, the voltage is increasing as the capacitor is being charged.

At 90° of the sine wave, the rate of change in voltage is zero, and zero current flows, even though the charge in the capacitor is at a maximum. Between 90° and 180° the voltage is decreasing, and, consequently, the charge on the capacitor is decreasing. This means that the capacitor must also be discharging at this point in the cycle. For the capacitor to discharge, the current must flow in the opposite direction to the applied voltage. This reverse polarity of current is shown below the horizontal line of Figure 16–10(b). Between 180° and 270°, the capacitor is being charged in the opposite direction, but the current is once again considered to have the same direction as the applied emf. The final 90° point of the cycle shows the current in the opposite direction to the voltage as the capacitor discharges. For the purely capacitive circuit of Figure 16–10(a), the current leads the voltage by 90°. This relationship is shown in the phasor diagram of Figure 16–10(c).

As in the case of a pure inductor, a pure capacitor has no resistance component. The ratio of effective voltage across a capacitor to the effective current is called the **capacitive reactance** and is represented by the symbol X_C. Capacitive reactance is measured in ohms and is defined as the opposition offered by a capacitor or by any capacitive circuit to the flow of current. The reactance of a capacitor varies inversely with its capacitance. The rate by which the voltage changes is determined by the frequency of the applied voltage.

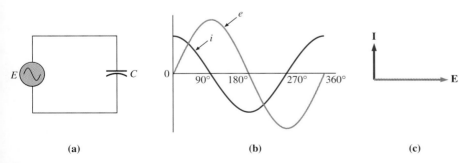

(a)   (b)   (c)

FIGURE 16–10   *Effect of capacitance in an AC circuit.*

Because the amount of current flowing in a capacitive circuit is determined by the rate at which the voltage is changing, a lower frequency results in a slower rate of change and less current will flow. The rate of change of voltage in an AC circuit is determined by the **angular velocity** of the applied voltage.

The reactance of a capacitor varies inversely with capacitance and with angular velocity. Capacitive reactance is determined by the following equation:

$$X_C = \frac{1}{2\pi f C} \tag{16.7}$$

where X_C = capacitive reactance, in ohms
f = frequency, in hertz
C = capacitance, in farads

_____ **EXAMPLE 16–7** _____

A 100 μF capacitor is supplied from a 120 V AC supply with a frequency of 60 Hz. Determine the

(a) Capacitive reactance.
(b) Current that flows in the circuit.

Solution

(a)

$$X_C = \frac{1}{2\pi f C}$$

$$= \frac{1}{2\pi(60 \text{ Hz})(100 \times 10^{-6} \text{ F})}$$

$$= 26.53 \ \Omega$$

(b)

$$I_{rms} = \frac{E_{rms}}{X_C}$$

$$= \frac{120 \text{ V}}{26.53 \ \Omega}$$

$$= 4.52 \text{ A}$$

_____ **EXAMPLE 16–8** _____

A capacitor has a reactance of 250 Ω at 550 Hz. What is its reactance at 60 Hz?

Solution If the reactance at one frequency is known, the reactance at the another frequency can be found by multiplying the known reactance by the ratio of the two frequencies.

$$\frac{X_{C1}}{X_{C2}} = \frac{f_1}{f_2}$$

$$X_{C2} = X_{C1} \frac{f_1}{f_2}$$

$$= 250 \ \Omega \ \frac{550 \text{ Hz}}{60 \text{ Hz}}$$

$$= 2291.67 \ \Omega$$

To calculate series and parallel combinations of capacitive reactance, the same method is used as for solving inductive reactance combinations. To find the total reactance of two or more series-connected capacitors,

$$X_{CT} = X_{C1} + X_{C2} + X_{C3} + \cdots + X_{Cn} \tag{16.8}$$

To solve for two or more parallel-connected capacitors,

$$\frac{1}{X_{CT}} = \frac{1}{X_{C1}} + \frac{1}{X_{C2}} + \frac{1}{X_{C3}} + \cdots + \frac{1}{X_{Cn}}$$ (16.9

EXAMPLE 16–9

FIGURE 16–11 *Circuit for Example 16–9.*

For the circuit shown in Figure 16–11, determine the values of X_{CT}, I_T, V_{C1}, and V_{C2}.

Solution

$$X_{CT} = X_{C1} + X_{C2} = 500\ \Omega + 1000\ \Omega = 1500\ \Omega$$

$$I_T = \frac{E}{X_{CT}} = \frac{120\ \text{V}}{1500\ \Omega} = 80\ \text{mA}$$

$$V_{C1} = I_T X_{C1} = (80\ \text{mA})(500\ \Omega) = 40\ \text{V}$$

$$V_{C2} = I_T X_{C2} = (80\ \text{mA})(1000\ \Omega) = 80\ \text{V}$$

EXAMPLE 16–10

FIGURE 16–12 *Circuit for Example 16–10.*

For the circuit shown in Figure 16–12, calculate the values of X_{CT}, I_{C1}, I_{C2}, and I_T.

Solution Because only two capacitors are given, the product-over-sum equation may be used.

$$X = \frac{X_{C1}X_{C2}}{X_{C1} + X_{C2}} = \frac{(500\ \Omega)(1000\ \Omega)}{500\ \Omega + 1000\ \Omega} = 333.33\ \Omega$$

$$I_{C1} = \frac{120\ \text{V}}{500\ \Omega} = 0.24\ \text{A}$$

$$I_{C2} = \frac{120\ \text{V}}{1000\ \Omega} = 0.12\ \text{A}$$

$$I_T = I_{C1} + I_{C2} = 0.24\ \text{A} + 0.12\ \text{A} = 0.36\ \text{A}$$

1. Define capacitive reactance.
2. The reactance of a capacitor varies directly with its capacitance (True/False).
3. What is the symbol for capacitive reactance?
4. The rate of change of voltage in an AC circuit is determined by the angular velocity of the applied voltage (True/False).

___ 16–6 ___

IMPEDANCE

Impedance is defined as the total opposition to current flow in an AC circuit. Impedance may be pure resistance or pure reactance, but usually it is a combination of resistance and reactance. The symbol for impedance is the letter Z, and it is measured in ohms. The current in an AC circuit is directly proportional to the voltage across the circuit and inversely proportional to the impedance of the circuit. In equation form,

$$Z = \frac{E}{I}$$ (16.10)

Because the amount of impedance is the ratio between voltage and current, the impedance angle is the difference in phase angle between the voltage and current.

___ **EXAMPLE 16–11** ___

The voltage and current for an AC load are 120 $\angle 45°$ V and 2.5 $\angle 30°$ A, respectively. Determine the impedance of the circuit.

Solution

$$Z = \frac{E}{I}$$

$$= \frac{120 \angle 45° \text{ V}}{2.5 \angle 30° \text{ A}} = 48 \angle 15° \ \Omega$$

When an AC circuit contains resistance and reactance, the total voltage drop, IZ, is equal to the phasor sum of the resistance, IR, and the reactance, IX, drops. By the **Pythagorean theorem,**

$$IZ = \sqrt{(IR)^2 + (IX)^2}$$ (16.11)

If the current, I, is canceled, the equation becomes

$$Z = \sqrt{R^2 + X^2}$$ (16.12)

Because impedance, resistance, and reactance are fixed quantities that usually do not vary with time, they may be represented by a **vector diagram.** Vector diagrams are usually drawn in rectangular form to distinguish them from phasor diagrams. Vector diagrams containing R, X, and Z are referred to as **impedance triangles** and can prove to be very useful in problem solving. Two examples of impedance triangles are illustrated in Figure 16–13. The impedance triangle shown in Figure 16–13(a) represents an inductance and resistance connected in series, while the impedance triangle of Figure 16–13(b) represents a capacitance and resistance connected in series. The reactive component in both circuits is on the vertical axis, so if a circuit contained both capacitance and inductance, the two values would be *subtracted* from each other.

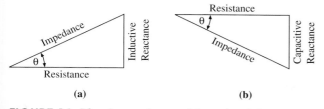

(a)   (b)

FIGURE 16–13 *Impedance triangle. (a) Resistance and inductive reactance in series. (b) Resistance and capacitive reactance in series.*

The phase angle θ in an impedance triangle is solved for in the same manner as for current and voltage relationships.

$$\theta = \tan^{-1} \frac{X}{R} \qquad\qquad (16.13)$$

Section Review

1. State the relationship between impedance, voltage, and current in equation form.
2. What is the impedance angle of a circuit?
3. Impedance, resistance, and reactance may be represented by a vector diagram (True/False).

THE SERIES *RL* CIRCUIT

A series circuit containing resistance, *R*, and inductance, *L*, is shown in Figure 16–14(a). The phasor diagram for this circuit is illustrated in Figure 16–14(b). Because current is the same through all parts of a series circuit, the current phasor is drawn as the reference phasor. The voltage, V_R, across resistance *R*, is always in phase with the current through the resistance. Because the voltage across an inductance leads the current by 90°, the phasor V_L is drawn 90° ahead of the reference phasor. The applied voltage, **E,** is the resultant of

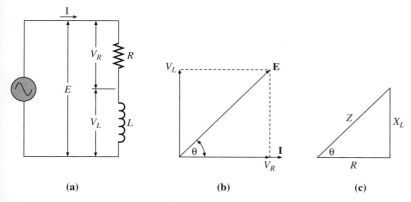

(a)   (b)   (c)

FIGURE 16–14 *(a) Series RL circuit. (b) Phasor diagram. (c) Impedance triangle.*

the two component voltages V_R and V_L. The impedance triangle is shown in Figure 16–14(c).

The triangle formed by Z, X_L, and R is similar to that formed by E, V_L, and V_R in the phasor diagram of Figure 16–14(b). It is evident that the impedance of this type of circuit equals the vector sum of the resistance and the inductive reactance. Also, it is evident from the impedance triangle that

$$\theta = \sin^{-1} \frac{X_L}{Z} = \cos^{-1} \frac{R}{Z} = \tan^{-1} \frac{X_L}{R} \tag{16.14}$$

$$X_L = Z \sin \theta \tag{16.15}$$

$$R = Z \cos \theta \tag{16.16}$$

$$X_L = R \tan \theta \tag{16.17}$$

EXAMPLE 16–12

A series RL circuit contains a 100 Ω resistor and a 0.35 H coil connected to a 170 V_p, 60 Hz, supply. Determine the voltages across the resistor and inductor.

Solution

$$X_L = 2\pi f L = 2\pi (60 \text{ Hz})(0.35 \text{ H}) = 131.95 \ \Omega$$

$$Z = \sqrt{R^2 + X_L^2} = \sqrt{(100 \ \Omega)^2 + (131.95 \ \Omega)^2} = 165.56 \ \Omega$$

$$\theta = \tan^{-1} \frac{X}{R} = \tan^{-1} \frac{131.95 \ \Omega}{100 \ \Omega} = 52.8°$$

The impedance angle of 52.8° is the same angle that the voltage leads the current.

$$V_R = 170 \cos 52.8° = 102.78 \text{ V}$$

$$V_L = 170 \sin 52.8° = 135.41 \text{ V}$$

EXAMPLE 16–13

A 2 H coil and a 500 Ω resistor are connected in series to a 240 V, 60 Hz supply. Determine the magnitude and phase of the current with respect to the applied voltage of the circuit.

Solution

$$X_L = 2\pi f L = 2\pi (60 \text{ Hz})(2 \text{ H}) = 754 \ \Omega$$

$$\theta = \tan^{-1} \frac{X_L}{R} = \tan^{-1} \frac{754 \ \Omega}{500 \ \Omega} = 56.5°$$

$$Z = \sqrt{R^2 + X^2} = \sqrt{(500 \ \Omega)^2 + (754 \ \Omega)^2} = 904.71 \ \Omega$$

$$I = \frac{E}{Z} = \frac{240 \ \angle 0° \text{ V}}{904.71 \ \angle 56.5° \ \Omega} = 265.28 \ \angle{-56.5°} \text{ mA}$$

Because this circuit contains inductance, the current *lags* the voltage by 56.5°.

$$I = 265.28 \ \angle{-56.5°} \text{ mA}$$

Section Review

1. In a series RL circuit, why is the current phasor used as the reference phasor?
2. In a series RL circuit, the current leads the voltage (True/False).

FIGURE 16–15   *Parallel RL circuit.*

THE PARALLEL *RL* CIRCUIT

Resistors and inductors connected in parallel are similar to any parallel circuit. For example, the voltage is the same across each branch, and the current divides in inverse proportion to the ohmic value of each component. Figure 16–15 shows a typical parallel *RL* circuit.

The impedance of the circuit shown in Figure 16–15 is found using the following equation:

$$Z = \frac{RX_L}{\sqrt{R^2 + X_L^2}} \qquad (16.18)$$

The phase angle between the AC supply voltage and the total current is

$$\theta = \tan^{-1} \frac{R}{X_L}$$

EXAMPLE 16–14

$R = 220 \, \Omega$ $X_L = 100 \, \Omega$

For the circuit of Figure 16–16, determine the impedance and phase angle.

FIGURE 16–16   *Circuit for Example 16–14.*

Solution

$$Z = \frac{RX_L}{\sqrt{R^2 + X_L^2}} = \frac{(220 \, \Omega)(100 \, \Omega)}{\sqrt{(220 \, \Omega)^2 + (100 \, \Omega)^2}} = 91.04 \, \Omega$$

$$\theta = \tan^{-1} \frac{R}{X_L} = \tan^{-1} \frac{220 \, \Omega}{100 \, \Omega} = 65.55°$$

In a pure inductor, the current lags the voltage by 90°, and the current flowing through a resistor is in phase with the voltage. Figure 16–17 shows the phasor diagram for a resistor and inductor connected in parallel. In this diagram, the applied voltage and the current through the resistor are shown as being in phase with each other. The current flowing through the inductor, I_L, is shown lagging the current through the resistor by 90°.

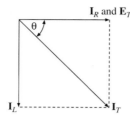

FIGURE 16–17 *Phasor diagram for current in a parallel RL circuit.*

The total current represents the phasor sum of I_R and I_L. If I_R and I_L are equal, the total current would be at an angle of $-45°$ from the applied voltage.

Section Review

1. Resistors and inductors in parallel have similar characteristics to any parallel circuit (True/False).
2. In a parallel *RL* circuit, if the inductor current and resistor current are equal, what would be the phase angle between the total current and the applied voltage?

—— 16–9

SERIES-PARALLEL *RL* CIRCUITS

In this section we will examine the effects of resistors and inductors connected in series-parallel configurations. To determine values of current, voltage, and impedance in series-parallel *RL* circuits, it is necessary to apply the rules pertaining to series *RL* circuits and parallel *RL* networks.

When solving series-parallel AC circuit problems, it is sometimes desirable to substitute an equivalent series circuit for a parallel combination, or an equivalent parallel circuit for a series combination. A series impedance, which draws current of the same magnitude and phase angle as a parallel circuit when connected across the same applied voltage, is called the *equivalent series circuit* of the parallel combination. Figure 16–18(a) shows a resistance and reactance connected in parallel. When these components are connected in series, as in Figure 16–18(b), a certain resistance value R_s and reactance value X_s will produce the same effect as the original resistance and reactance connected in parallel. In equation form,

$$R_s = \frac{R_p X_p^2}{R_p^2 + X_p^2} \qquad (16.19)$$

$$X_s = \frac{X_p R_p^2}{R_p^2 + X_p^2} \qquad (16.20)$$

(a)

FIGURE 16–18
(a) Equivalent parallel circuit.
(b) Equivalent series circuit.

(b)

EXAMPLE 16–15

Solution

$$R_s = \frac{R_p X_p^2}{R_p^2 + X_p^2} = \frac{(60\ \Omega)(80\ \Omega)^2}{(60\ \Omega)^2 + (80\ \Omega)^2} = 38.4\ \Omega$$

$$X_s = \frac{X_p R_p^2}{R_p^2 + X_p^2} = \frac{(80\ \Omega)(60\ \Omega)^2}{(60\ \Omega)^2 + (80\ \Omega)^2} = 28.8\ \Omega$$

Calculate the resistance and reactance values of a series equivalent circuit that may be substituted for a parallel circuit containing a resistance of 60 Ω and a reactance of 80 Ω.

To match line current and phase angle between a parallel circuit and an equivalent series circuit, the total impedance Z and the circuit phase angle θ must be the same for both circuits. Therefore,

$$R_s = Z_T \cos \theta$$

$$X_s = Z_T \sin \theta$$

EXAMPLE 16–16

FIGURE 16–19   *Circuit for Example 16–16.*

For the circuit of Figure 16–19, find the

(a) Total resistance.
(b) Total inductive reactance.
(c) Total impedance.

Solution   This circuit can be solved by converting the parallel-connected components into equivalent series values.

(a) The inductive reactances are determined first:

$$X_{L1} = 2\pi f L_1 = 2\pi(2000\ \text{Hz})(10 \times 10^{-3}\ \text{H}) = 125.66\ \Omega$$

$$X_{L2} = 2\pi f L_2 = 2\pi(2000\ \text{Hz})(5 \times 10^{-3}\ \text{H}) = 62.83\ \Omega$$

The equivalent resistance and inductive reactance for R_2 and X_{L2} are found using Equations 16.19 and 16.20.

$$R_{eq} = \frac{R_2 X_{L2}^2}{R_2^2 + X_{L2}^2} = \frac{(220\ \Omega)(62.83\ \Omega)^2}{(220\ \Omega)^2 + (62.83\ \Omega)^2} = 16.59\ \Omega$$

The total resistance is equal to the sum of R_1 and R_{eq}.

$$R_T = R_1 + R_{eq} = 120\ \Omega + 16.59\ \Omega = 136.59\ \Omega$$

(b) The equivalent series-connected inductive reactance for the parallel portion of Figure 16–19 is found using Equation 16.20.

$$X_{eq} = \frac{X_{L2} R^2}{R_2^2 + X_{L2}^2} = \frac{(62.83\ \Omega)(220\ \Omega)^2}{(220\ \Omega)^2 + (62.83\ \Omega)^2} = 58.09\ \Omega$$

Two series-connected inductances have a total value equal to the sum of the individual values. Therefore,

$$X_{LT} = X_{L1} + X_{eq} = 125.66 \ \Omega + 58.09 \ \Omega = 183.75 \ \Omega$$

(c) The total impedance is now determined using the Pythagorean theorem.

$$Z_T = \sqrt{R_T^2 + X_{LT}^2} = \sqrt{(136.59 \ \Omega)^2 + (183.75 \ \Omega)^2} = 228.96 \ \Omega$$

Section Review

1. In series-parallel *RL* circuits, what is an equivalent series circuit?
2. To match line current and phase angle between a parallel circuit and an equivalent series circuit, what must be the same for both circuits?

— 16-10

THE SERIES *RC* CIRCUIT

In an AC series circuit that contains resistance and capacitance, the applied voltage can be resolved into two components: V_R, which is in phase with the current, and V_C, which lags the current by 90°. The phasor diagram of Figure 16–20(b) shows this relationship. Current is used as a reference and is drawn on the horizontal axis. The phasor sum of V_R and V_C gives the applied voltage E. The value of the phase angle between the applied voltage and the current depends on the ratio of V_C to V_R. The magnitude of resultant voltage is found using the Pythagorean theorem:

$$E = \sqrt{V_R^2 + V_C^2} \tag{16.21}$$

The impedance triangle of Figure 16–20(c) is similar to the triangle formed by **E**, V_R, and V_C in the phasor diagram of Figure 16–20(b).

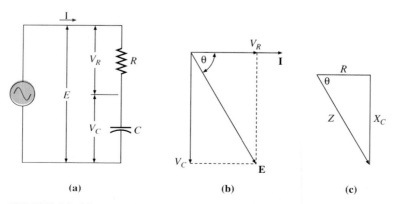

(a)   (b)   (c)

FIGURE 16–20 *(a) Series RC circuit. (b) Phasor diagram. (c) Impedance triangle.*

It is evident from the impedance triangle that the impedance of this circuit equals the vector sum of the resistance and the capacitive reactance. Also, it is evident from the impedance triangle that

$$-\theta = \sin^{-1}\frac{X_C}{Z} = \cos^{-1}\frac{R}{Z} = \tan^{-1}\frac{X_C}{R} \qquad (16.22)$$

$$X_C = Z\sin\theta \qquad (16.23)$$

$$R = Z\cos\theta \qquad (16.24)$$

$$X_C = R\tan\theta \qquad (16.25)$$

_____ EXAMPLE 16–17 _____

A resistance of 60 Ω is connected in series with a 50 μF capacitance across a 220 V, 60 Hz supply. Determine the

(a) Impedance.
(b) Current.

Solution

(a)
$$X_C = \frac{1}{2\pi fC} = \frac{1}{2\pi(60\text{ Hz})(50\times10^{-6}\text{ F})} = 53.05\ \Omega$$

$$Z = \sqrt{(60\ \Omega)^2 + (53.05\ \Omega)^2} = 80.09\ \Omega$$

(b)
$$I = \frac{E}{Z} = \frac{220\text{ V}}{80.09\ \Omega} = 2.75\text{ A}$$

_____ EXAMPLE 16–18 _____

Determine the value of a capacitor required to be connected in series with a 1200 Ω resistor to limit the current in a circuit to 25 mA. The supply voltage is 120 V, 60 Hz.

Solution

$$Z = \frac{E}{I} = \frac{120\text{ V}}{25\text{ mA}} = 4800\ \Omega$$

$$X_C = \sqrt{Z^2 - R^2} = \sqrt{(4{,}800\ \Omega)^2 - (1200\ \Omega)^2} = 4647.58\ \Omega$$

$$X_C = \frac{1}{2\pi fC}$$

$$C = \frac{1}{2\pi fX_C} = \frac{1}{2\pi(60\text{ Hz})(4647.58\ \Omega)} = 0.57\ \mu\text{F}$$

Section Review

1. State the Pythagorean theorem in equation form for finding the resultant voltage in a series RC circuit.
2. What is the impedance of a 2.7 kΩ resistor connected in series with a capacitive reactance of 1.65 kΩ?

_____ 16–11 _____

THE PARALLEL _RC_ CIRCUIT

When resistance and capacitive reactance are connected in parallel, the current through the resistance is in phase with the applied voltage, and the current through the capacitance leads the applied voltage by 90°.

The effect of varying either the frequency or capacitance of a parallel RC circuit is of particular interest in the field of electronics. For example, the bypass capacitor is based on the principle that a very small reactance in parallel with a vary large resistance effectively will eliminate the current variations passing through the resistor. Because capacitive reactance and frequency are inversely proportional to each other, capacitive reactance is low at high frequencies but approaches infinity when the frequency is near zero. Bypass capacitors in practical applications are chosen so that their reactance at the lowest frequency is one-tenth of the ohmic value of the resistor being bypassed.

The impedance and phase angle of a parallel RC circuit is determined using the following equations:

$$Z = \frac{RX_C}{\sqrt{R^2 + X_C^2}}$$
(16.26)

$$-\theta = \tan^{-1} \frac{R}{X_C}$$
(16.27)

___ EXAMPLE 16–19 ___

For the circuit of Figure 16–21, determine the impedance and phase angle.

$R = 220\ \Omega$ $X_C = 100\ \Omega$

FIGURE 16–21 *Circuit for Example 16–19.*

Solution

$$Z = \frac{RX_C}{\sqrt{R^2 + X_C^2}} = \frac{(220\ \Omega)(100\ \Omega)}{\sqrt{(220\ \Omega)^2 + (100\ \Omega)^2}} = 91.04\ \Omega$$

$$\theta = \tan^{-1} \frac{R}{X_C} = \tan^{-1} \frac{220\ \Omega}{100\ \Omega} = -65.6°$$

As with any parallel circuit, the voltage across each component remains constant and the total current is the sum of the individual currents. Figure 16–22 shows a phasor diagram that represents the currents in a parallel RC circuit. The resistive current, I_R is in phase with the applied voltage. The capacitive current, I_C leads the voltage by 90°. The total current, I_T, is the phasor sum of the resistive and capacitive currents.

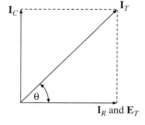

FIGURE 16–22 *Phasor diagram of currents in a parallel RC circuit.*

The total current in a parallel RC circuit is found by the Pythagorean theorem.

$$I_T = \sqrt{I_R^2 + I_C^2}$$ **(16.28)**

The phase angle is derived using the following equation:

$$\theta = \tan^{-1} \frac{I_C}{I_R}$$ **(16.29)**

EXAMPLE 16–20

Find the value of current through each branch and the total current and phase angle for the circuit of Figure 16–23.

$E_T = 10$ V
$f = 5$ kHz
$R = 1.5$ kΩ
$C = 0.01$ μF

FIGURE 16–23 *Circuit for Example 16–20.*

Solution

$$X_C = \frac{1}{2\pi f C} = \frac{1}{2\pi (5000 \text{ Hz})(0.01 \text{ μF})} = 3.18 \text{ kΩ}$$

$$I_R = \frac{E_T}{R} = \frac{10 \text{ V}}{1.5 \text{ kΩ}} = 6.67 \text{ mA}$$

$$I_C = \frac{E_T}{X_C} = \frac{10 \text{ V}}{3.18 \text{ kΩ}} = 3.15 \text{ mA}$$

$$I_T = \sqrt{I_R^2 + I_C^2} = \sqrt{(6.67 \text{ mA})^2 + (3.15 \text{ mA})^2} = 7.38 \text{ mA}$$

$$\theta = \tan^{-1} \frac{I_C}{I_R} = \tan^{-1} \frac{3.15 \text{ mA}}{6.67 \text{ mA}} = 25.28°$$

Section Review

1. Capacitive reactance and frequency are directly proportional to each other (True/False).
2. What is the total impedance of a 300 Ω resistance connected in parallel with a 200 Ω capacitive reactance?

16–12

SERIES-PARALLEL *RC* CIRCUITS

The rules for solving problems in series-parallel RC circuits are the same as those applied to any type of series-parallel circuit. In many situations, it is convenient to convert the parallel-connected components into equivalent series values, or vice versa. The equations used for finding equivalent resistance and reactance in Section 16–9 (Equations 16.19 and 16.20) can also be used for determining equivalent values for parallel RC circuits.

EXAMPLE 16–21

For the series-parallel *RC* circuit shown in Figure 16–24, determine the total capacitive reactance, total resistance, and total impedance.

FIGURE 16–24 *Circuit for Example 16–21.*

Solution   First, convert the parallel components, R_2 and X_{C2} into an equivalent series circuit.

$$R_{eq} = \frac{R_2 X_{C2}^2}{R_2^2 + X_{C2}^2} = \frac{(270\ \Omega)(450\ \Omega)^2}{(270\ \Omega)^2 + (450\ \Omega)^2} = 198.53\ \Omega$$

$$X_{eq} = \frac{X_{C2} R_2^2}{R_2^2 + X_{C2}^2} = \frac{(450\ \Omega)(270\ \Omega)^2}{(270\ \Omega)^2 + (450\ \Omega)^2} = 119.12\ \Omega$$

The circuit is now redrawn as a series circuit (Figure 16–25) and the total resistance and reactance is found by adding the individual values.

FIGURE 16–25 *Equivalent series circuit for Example 16–21.*

$$X_{CT} = X_{C1} + X_{eq} = 200\ \Omega + 119.12\ \Omega = 319.12\ \Omega$$

$$R_T = R_1 + R_{eq} = 330\ \Omega + 198.53\ \Omega = 528.53\ \Omega$$

$$Z_T = \sqrt{R_T^2 + X_{CT}^2} = \sqrt{(528.53\ \Omega)^2 + (319.12\ \Omega)^2} = 617.4\ \Omega$$

Section Review

1. The rules for solving problems in series-parallel *RC* circuits are the same as those applied to any type of series-parallel circuit (True/False).

When an AC circuit consists of resistance, inductance, and capacitance connected in series, as shown in Figure 16–26(a), the components of the applied voltage E are the voltage drop across the resistor V_R and the total reactance voltage V_X. The total reactance voltage is the difference between the capacitive reactance voltage, V_C, and the inductive reactance voltage, V_L. The applied voltage and phase angle are expressed using the following equations:

$$E = \sqrt{V_R^2 + (V_L - V_C)^2} \tag{16.30}$$

$$\theta = \tan^{-1} \frac{(V_L - V_C)}{V_R} \tag{16.31}$$

The phasor diagram of Figure 16–26(b) is for a circuit in which V_L is greater than V_C. The current **I** is plotted on the reference axis, V_R is in phase with **I**, and V_X is plotted 90° ahead of **I**. Because V_L is greater than V_C, the total reactive voltage V_X is considered as an inductive reactance voltage.

Figure 16–26(c) is the phasor diagram for a circuit in which V_L is less than V_C. When this condition exists, V_X lags 90° behind **I** and is treated as a capacitive reactance voltage. The impedance triangle of Figure 16–27(a) is for a series circuit in which $X_L > X_C$. Figure 16–27(b) is the impedance triangle for a series circuit in which $X_L < X_C$. The total, or equivalent, reactance is expressed by the equation

$$X = X_L - X_C \tag{16.32}$$

The impedance of a series *RLC* circuit is the vector sum of R and X regardless of whether X is inductive or capacitive in its effect. The value of this impedance is found by the Pythagorean theorem,

$$Z = \sqrt{R^2 + X^2} \tag{16.33}$$

$$\theta = \tan^{-1} \frac{X}{R}$$

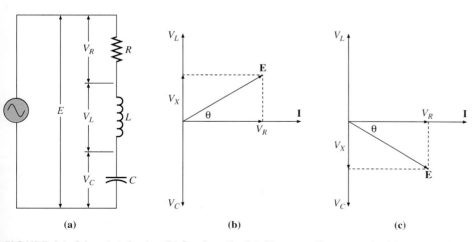

(a) (b) (c)

FIGURE 16–26 *(a) Series RLC circuit. (b) Phasor diagram for $X_L > X_C$. (c) Phasor diagram for $X_C > X_L$.*

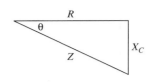

_____ **EXAMPLE 16–22** _____

For the circuit of Figure 16–28, determine the

(a) Impedance and phase angle of the circuit.
(b) Total current.
(c) Values of V_R, V_L, and V_C.

FIGURE 16–28 *Circuit for Example 16–22.*

Solution

(a)

$$X = X_L - X_C = 50\ \Omega - 40\ \Omega = 10\ \Omega$$

$$Z = \sqrt{R^2 + X^2} = \sqrt{(15\ \Omega)^2 + (10\ \Omega)^2} = 18.03\ \Omega$$

$$\theta = \tan^{-1}\frac{X}{R} = \tan^{-1}\frac{10\ \Omega}{15\ \Omega} = 33.7°$$

(b)

$$I_T = \frac{E_T}{Z} = \frac{12\ V}{18.03\ \Omega} = 665.56\ mA$$

(c)

$$V_R = IR = (665.56\ mA)(15\ \Omega) = 9.98\ V$$

$$V_L = IX_L = (665.56\ mA)(50\ \Omega) = 33.28\ V$$

$$V_C = IX_C = (665.56\ mA)(40\ \Omega) = 26.62\ V$$

Although the voltages determined in Example 16–22 appear to defy Kirchhoff's voltage law, it is important to remember that these AC voltages are phasor values and they not only have magnitude but also a phase angle. If the phasor sum of the three voltage drops was determined, it would indeed equal the applied voltage.

EXAMPLE 16–23

Solution

$$X_L = 2\pi fL = 2\pi(500 \text{ Hz})(0.2 \text{ H}) = 628.32 \ \Omega$$

$$X_C = \frac{1}{2\pi fC} = \frac{1}{2\pi(500 \text{ Hz})(0.5 \times 10^{-6} \text{ F})} = 636.62 \ \Omega$$

$$X = X_L - X_C = 628.32 \ \Omega - 636.62 \ \Omega = -8.3 \ \Omega$$

$$Z = \sqrt{R^2 + X^2} = \sqrt{(10 \ \Omega)^2 + (-8.3 \ \Omega)^2} = 13 \ \Omega$$

$$\theta = \tan^{-1} \frac{X}{R} = \tan^{-1} \frac{-8.3 \ \Omega}{10 \ \Omega} = -39.7°$$

Calculate the total impedance and phase angle of a
500 Hz series circuit containing a 0.2 H coil,
0.5 µF capacitor, and a
10 Ω resistor.

Section Review

1. A series *RLC* circuit has the following values: $V_L = 87$ V, $V_C = 54$ V, and $V_R = 75$ V. What is the total circuit voltage?
2. Determine the phase angle for a series *RLC* circuit with the following values: $V_L = 35$ V, $V_C = 15$ V, and $V_R = 24$ V.
3. The impedance of a series *RLC* circuit is the vector sum of R and X regardless of whether X is inductive or capacitive in its effect (True/False).

16–14
ADMITTANCE AND SUSCEPTANCE

In the discussion of resistance in DC circuits in Chapter 2, we demonstrated how the reciprocal of resistance, or conductance, was utilized in problem solving. In equation form,
conductance was expressed as

$$G = \frac{1}{R}$$

where G = conductance, in siemens
        R = resistance, in ohms

In AC circuit applications, the ease with which current flows in a given component,
or circuit, is called **admittance,** which is measured in siemens and represented by the letter Y. Admittance is defined as the reciprocal of impedance and is considered to be a vectoral quantity.

$$Y = \frac{1}{Z} \qquad\qquad (16.34)$$

where Y = admittance, in siemens
        Z = impedance, in ohms

As admittance is a vectoral quantity, it must have two rectangular components. For
impedance circuits these components are resistance and reactance. For admittance circuits

these components are **conductance** and **susceptance.** Susceptance is the reciprocal of reactance and is represented by the letter B.

$$Y = \sqrt{G^2 \pm B^2} \qquad (16.35)$$

where Y = admittance, in siemens
G = conductance, in siemens
B = susceptance, in siemens

The phase angle of an admittance circuit is found by the following equation:

$$\theta = \tan^{-1} \frac{B}{G} \qquad (16.36)$$

The sum of conductance and susceptance in a circuit is found by phasor addition. The ± symbol in Equation 16.35 is used to stipulate the type of susceptance in the circuit. The + symbol is used for capacitive susceptance, B_C, and a − symbol represents inductive susceptance, B_L.

When resistors are connected in parallel, the total conductance is calculated by the following equation:

$$G_T = G_1 + G_2 + G_3 + \cdots + G_n \qquad (16.37)$$

The total susceptance in a parallel circuit is determined by

$$B_T = B_1 + B_2 + B_3 + \cdots + B_n \qquad (16.38)$$

When the admittance of each parallel branch is known, the total admittance is found by the following equation:

$$Y_T = Y_1 + Y_2 + Y_3 + \cdots + Y_n \qquad (16.39)$$

To solve AC circuit problems that contain impedance, Ohm's law can be used to find the current, $I = E/Z$. The equivalent equation for calculating current in a circuit containing admittance is

$$I = EY \qquad (16.40)$$

When solving parallel AC circuit problems, it is possible to utilize an **admittance triangle,** which is based on the same principles as its counterpart, the impedance triangle. Figure 16–29(a) shows an admittance triangle containing conductance and capacitive susceptance. The admittance triangle shown in Figure 16–29(b) contains conductance and inductive susceptance.

(a)

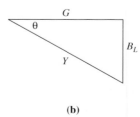

(b)

FIGURE 16–29
Admittance triangles. (a) Capacitive susceptance. (b) Inductive susceptance.

EXAMPLE 16–24

FIGURE 16–30 *Circuit for Example 16–24.*

Determine the total admittance and phase angle for the parallel circuit of Figure 16–30.

Solution

$$G = \frac{1}{R} = \frac{1}{10\ \Omega} = 0.1\ \text{S}$$

$$B_C = \frac{1}{X_C} = \frac{1}{15\ \Omega} = 0.067\ \text{S}$$

$$Y_T = \sqrt{G^2 + B^2} = \sqrt{(0.1\ \text{S})^2 + (0.067\ \text{S})^2} = 0.12\ \text{S}$$

$$\theta = \tan^{-1}\frac{B}{G} = \tan^{-1}\frac{0.067\ \text{S}}{0.1\ \text{S}} = 33.82°$$

Section Review

1. State the SI unit and symbol for conductance.
2. Define admittance.
3. State the Ohm's law equation for calculating current in a circuit that contains admittance.

THE PARALLEL *RLC* CIRCUIT

The parallel circuit shown in Figure 16–31(a) has three load branches, and the supply voltage E is common to all components. The phasor diagram may be determined in the same manner as for a series *RLC* circuit, except that the diagram now consists of current phasors. The phasor diagram of Figure 16–31(b) shows the relationship between the currents in the three components of the circuit. The reference phasor, E, is drawn with the current through the pure resistor, I_R, in phase with it. Because the inductive current lags the applied voltage, the I_L phasor is drawn 90° behind E. The capacitive current leads E by 90°, so it is drawn 90° ahead of the reference phasor.

The various factors of a parallel circuit can be determined with the least difficulty by observing the following order:

1. Find the impedance of each branch.
2. Find the current of each branch.
3. Draw a phasor diagram of the currents.
4. Find the total in-phase, or resistance, current.
5. Find the total reactance current.
6. Determine total current. Once the individual branch currents have been found, the total current is obtained by using the Pythagorean theorem.

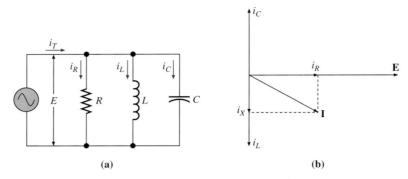

(a)   **(b)**

FIGURE 16–31   *(a) Parallel RLC circuit. (b) Phasor diagram in which $I_L > I_C$*

7. Find the impedance of the circuit. The total impedance is obtained by applying Ohm's law to the total current and the applied voltage of the circuit.

In parallel circuits containing both capacitive and inductive reactance, it is often advantageous to determine the admittance, conductance, and susceptance of the circuit. To determine the total current and phase angle in a circuit when the individual branch currents are known, the following equations may be used:

$$I_T = \sqrt{I_R^2 + I_X^2} \tag{16.41}$$

$$\theta = \tan^{-1} \frac{I_X}{I_R} \tag{16.42}$$

___ **EXAMPLE 16–25** ___

For the circuit shown in Figure 16–32, calculate I_R, I_L, I_C, I_T, and the phase angle between the applied voltage and the total current.

FIGURE 16–32 *Circuit for Example 16–25.*

Solution

$$X_L = 2\pi f L = 2\pi(1000 \text{ Hz})(5 \times 10^{-3} \text{ H}) = 31.42 \ \Omega$$

$$X_C = \frac{1}{2\pi f C} = \frac{1}{2\pi(1000 \text{ Hz})(10 \times 10^{-6} \text{ F})} = 15.92 \ \Omega$$

$$I_R = \frac{E}{R} = \frac{24 \text{ V}}{50 \ \Omega} = 0.48 \text{ A}$$

$$I_L = \frac{E}{X_L} = \frac{24 \text{ V}}{31.42 \ \Omega} = 0.76 \text{ A}$$

$$I_C = \frac{E}{X_C} = \frac{24 \text{ V}}{15.92 \ \Omega} = 1.51 \text{ A}$$

Because I_L and I_C are 180° out of phase with each other, the total reactive current, I_X, represents the difference between I_L and I_C.

$$I_X = I_C - I_L = 1.51 \text{ A} - 0.76 \text{ A} = 0.75 \text{ A}$$

$$I_T = \sqrt{I_R^2 + I_X^2} = \sqrt{(0.48 \text{ A})^2 + (0.75 \text{ A})^2} = 0.89 \text{ A}$$

The phase angle of the total current is found by using Equation 16.42.

$$\theta = \tan^{-1} \frac{I_X}{I_R} = \tan^{-1} \frac{0.75 \text{ A}}{0.48 \text{ A}} = 57.38°$$

_____ EXAMPLE 16–26 _____

FIGURE 16–33 *Circuit for Example 16–26.*

The circuit shown in Figure 16–33 contains the following values: $R = 800 \ \Omega$, $X_C = 1200 \ \Omega$, and $X_L = 1500 \ \Omega$. Determine the

(a) Susceptance.
(b) Conductance.
(c) Admittance.
(d) Total current.

Solution

(a)
$$G = \frac{1}{R} = \frac{1}{800 \ \Omega} = 1.25 \text{ mS}$$

(b)
$$B_C = \frac{1}{X_C} = \frac{1}{1200 \ \Omega} = 0.833 \text{ mS}$$

$$B_L = \frac{1}{X_L} = \frac{1}{1500 \ \Omega} = 0.667 \text{ mS}$$

$$B_T = B_C - B_L = 0.833 \text{ mS} - 0.667 \text{ mS} = 0.166 \text{ mS}$$

(c)
$$Y = \sqrt{G^2 + B^2} = \sqrt{(1.25 \text{ mS})^2 + (0.166 \text{ mS})^2} = 1.26 \text{ mS}$$

(d)
$$I = EY = (120 \text{ V})(1.26 \text{ mS}) = 151.2 \text{ mA}$$

Section Review

1. What is the first step when analyzing AC circuit problems?
2. Find the phase angle of a parallel *RLC* circuit with a reactive current of 275 mA and a resistive current of 600 mA.
3. What is the total current for a parallel *RLC* circuit with a reactive current of 9.53 A and a resistive current of 7.75 A?

POWER IN AC CIRCUITS

In the study of DC circuits, power was defined as the product of voltage and current. In AC circuits this will be true only when the current is in phase with the voltage, such as circuits containing only resistance. In cases where the current is not in phase with the voltage, the product of the voltage and current will not be equal to the power actually consumed by the circuit. In an AC circuit that contains reactance, the current through the circuit either leads or lags the applied voltage. This implies that there is a phase angle between the two components that must be taken into account when calculating the power consumed by the circuit.

To find the corresponding value of power in resistive and reactive circuits, the product of voltage and current is multiplied by the cosine of the phase angle. The **average power,** or **real power,** as it is sometimes called, is the power delivered to and dissipated by the load. This is the power indicated when measured by a **wattmeter.**

The equation for determining average power in an AC circuit is given as

$$P = E_{rms} \times I_{rms} \times \cos \theta \quad \text{or} \quad \frac{E_m \times I_m}{2} \times \cos \theta \qquad (16.43)$$

Figure 16–34(a) shows a resistor connected across an AC source. Figure 16–34(b) indicates how voltage, current, and the power dissipated in the resistor change with time. The power varies between zero and maximum and has a frequency that is *double* that of the source. The power developed by a resistive component is also referred to as the **true power.** The dotted horizontal line represents the average power, which is half of the maximum instantaneous power. The average power is considered to be the DC equivalent of the AC power supplied by the source.

In a purely resistive circuit, because voltage and current are in phase, $\theta = 0°$. Therefore,

$$P = E_{rms} \times I_{rms} \times \cos 0° \qquad (16.44)$$
$$= E \times I \times 1$$

Figure 16–35(a) shows an AC voltage source applied across a pure inductor. In Figure 16–35(b), the product of the voltage and current crosses the zero axis. For one-half of the cycle, energy is being supplied to the inductor. During the other half of the cycle, energy is being returned by the inductor to the source. Consequently, the total power dissi-

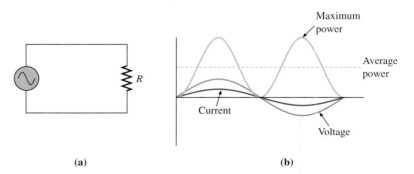

(a)   (b)

FIGURE 16–34 *(a) Resistive AC circuit. (b) Phase relationships of current, voltage, and power.*

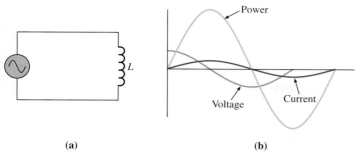

(a)   **(b)**

FIGURE 16–35 *(a) Inductive AC circuit. (b) Phase relationships.*

pated in a cycle by an ideal inductor is *zero*. A practical inductor will contain a small amount of internal resistance, so there will be some heat dissipation during the cycle. An inductor is often referred to as a **storage element,** which receives electrical energy, stores it in the form of a magnetic field, and returns the energy to the circuit at a later point in time.

In a purely inductive circuit, voltage leads current by 90°. The average power is calculated as

$$P = E_{rms} \times I_{rms} \times \cos 90° \qquad\qquad (16.45)$$
$$= E \times I \times 0$$
$$= 0$$

A capacitor is shown connected across an AC voltage source in the circuit of Figure 16–36(a). Figure 16–36(b) shows the waveshapes for the current, voltage, and instantaneous power for the circuit of Figure 16–36(a). A capacitor, like an inductor, is an energy storage element that stores energy during one half-cycle and returns energy during the other half-cycle.

In a purely capacitive circuit, voltage lags current by 90°. The average power is determined by the following equation:

$$P = E_{rms} \times I_{rms} \times \cos -90° \qquad\qquad (16.46)$$
$$= E \times I \times 0$$
$$= 0$$

The average power in a purely capacitive circuit is also zero.

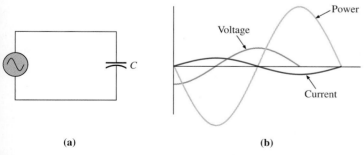

(a)   **(b)**

FIGURE 16–36 *(a) Capacitive AC circuit. (b) Phase relationships.*

EXAMPLE 16–27

Calculate the average power dissipated by the circuit shown in Figure 16–37.

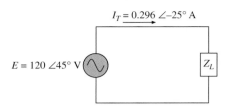

FIGURE 16–37 *Circuit for Example 16–27.*

Solution The phasor diagram of Figure 16–38 shows the angular displacement between the applied voltage and the current flowing through the circuit. From this diagram it is apparent that the applied voltage and load current are 70° out of phase with each other.

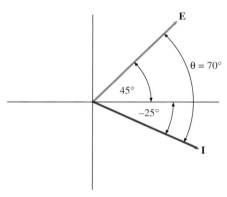

FIGURE 16–38 *Phasor diagram for Example 16–27.*

$$\theta = 45° - (-25°) = 70°$$

$$P = EI \cos \theta = (120 \text{ V})(0.296 \text{ A}) \cos 70° = 12.15 \text{ W}$$

EXAMPLE 16–28

Calculate the average power dissipated by a circuit having an instantaneous voltage of $e = 70 \sin(t + 60°)$ V, and an instantaneous current of $i = 15 \sin(t + 80°)$ A.

Solution

$$E_m = 70 \angle 60° \text{ V}$$

$$I_m = 15 \angle 80° \text{ A}$$

$$\theta = 60° - 80° = -20°$$

$$P = \frac{E_m \times I_m}{2} \cos \theta = \frac{70 \text{ V} \times 15 \text{ A}}{2} \cos 20° = 493.34 \text{ W}$$

The product of voltage and current is sometimes referred to as the **apparent power.** Because it is not actually a true power measurement, it is more commonly known as **volt-amperes, VA,** and is represented by the letter S. In equation form,

$$S = EI \qquad \text{(16.47)}$$

EXAMPLE 16–29

Solution

(a) $P = EI \cos \theta = (48 \text{ V})(800 \text{ mA}) \cos 30° = 33.26 \text{ W}$

(b) $S = EI = (48 \text{ V})(800 \text{ mA}) = 38.4 \text{ VA}$

An AC circuit has a load current of 800 mA, and an effective voltage of 48 V. The phase angle is 30°. Calculate the

(a) True power.
(b) Apparent power.

The ratio of true power to apparent power is called the **power factor, PF** of the circuit.

$$PF = \frac{\text{true power}}{\text{apparent power}} = \frac{E_{rms} I_{rms} \cos \theta}{E_{rms} I_{rms}} = \cos \theta \qquad \text{(16.48)}$$

Equation 16.48 shows that the power factor is equal to $\cos \theta$. For this reason, the phase angle is often referred to as the *power factor angle.* In an inductive circuit, where the current lags the voltage, the power factor is said to be a *lagging power factor.* If the current leads the voltage, the power factor is called a *leading power factor.* The value of the power factor depends on how much the current and voltage are out of phase with each other. Therefore, the power factor of a circuit is determined by the relative amounts of resistance and reactance in the circuit. If the impedance is known, the power factor is also known, because it is also the ratio of resistance to impedance.

$$PF = \frac{R}{Z} \qquad \text{(16.49)}$$

In a purely reactive circuit, the voltage and current are 90° out of phase with each other, and the power dissipated is zero. However, reactance does draw current from the supply, and the product of this current and the applied voltage has units of joules/second, which is the same unit as the watt. Because capacitance and inductance cannot dissipate power, the reactive component alternately stores and releases power as its magnetic field alternately builds up and collapses. The product of the voltage and current of the reactive component is directly proportional to the amount of energy stored and returned to the reactive component each time the current changes direction. This product is referred to as **reactive power** and is represented by the letter Q.

Reactive power is received from the supply during the first 180° and returned to the supply during the second 180° of each cycle. For this reason, reactive power is also called *wattless power.* To distinguish between true power and reactive power, true power is measured in watts, and reactive power is measured in **volt-ampere-reactive, or VARs.**

The reactive power can be calculated as

$$Q = I^2 X \qquad \text{(16.50)}$$

or

$$Q = E_{\text{rms}} \times I_{\text{rms}} \sin \theta \qquad \text{(16.51)}$$

_____ **EXAMPLE 16–30** _____

Calculate the reactive power of a 15 μF capacitor drawing 3.51 A from a 400 Hz supply.

Solution

$$X_C = \frac{1}{2\pi f C} = \frac{1}{2\pi (400 \text{ Hz})(15 \times 10^{-6} \text{ F})} = 26.53 \ \Omega$$

$$Q = I^2 X_C = (3.51 \text{ A})^2 (26.53 \ \Omega) = 326.85 \text{ VARs}$$

The ratio of real power to apparent power can be drawn as the sides of a right angle triangle, where real power is the side adjacent to angle θ and apparent power is the hypotenuse. Because real power is an *in phase,* or resistive, component, it would lie on the horizontal axis. Reactive power is drawn on the vertical axis. This type of triangle is called a **power triangle** and is shown in Figure 16–39. The two power triangles illustrated are based on current as the reference phasor, so they are assumed to be the triangles of a series circuit. Power triangles that are based on parallel circuits with the voltage as the reference phasor would have their reactive components drawn in the opposite direction to that shown in Figure 16–39. The power triangle of a series circuit corresponds to an impedance triangle, whereas the power triangle of a parallel circuit corresponds to admittance triangle.

The power triangle for a series-connected resistor and inductor is shown in Figure 16–39(a), and the triangle for a resistor and capacitor is shown in Figure 16–39(b). If a circuit contains both inductive and capacitive power, the reactive component of the power triangle will be determined by the difference between the reactive power delivered to each. In the right angle triangles of Figure 16–39, the Pythagorean theorem proves that

$$S = \sqrt{P^2 + Q^2} \qquad \text{(16.52)}$$

where   S = apparent power, in volt-amperes
P = true power, in watts
Q = reactive power, in VARs

Because θ represents the power factor of the circuit, power triangles are very useful in solving problems related to the power factor.

The total apparent power of an AC circuit must be determined by using the Pythagorean theorem for the total real and reactive powers. The total apparent power cannot be determined from the apparent power of each individual branch. Also, when solving for series or parallel circuits, the method of calculating total real, reactive, or apparent power is done by simply finding the sum of each branch. For these circuits, the total real

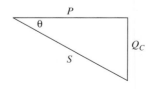

FIGURE 16–39 *Power triangles. (a) Resistor and inductor in series. (b) Resistor and capacitor in series.*

and reactive power is found by taking the rectangular form of the apparent power of each branch and adding each branch to find the sum of the real and imaginary values.

Equipment that is installed to improve the power factor is referred to as **power factor correction.** Power factor correction for an inductive load is accomplished by connecting capacitance in parallel with the load. The leading current of the capacitor branch supplies a lagging component of current to the inductive load and tends to reduce the line current accordingly.

Section Review

1. What is average power in an AC circuit?
2. What is another name for the power developed by a resistive component in an AC circuit?
3. The product of voltage and current in an AC circuit is also known as real power (True/False).
4. What is the power factor of a circuit?

16–17

EFFECTIVE RESISTANCE

When alternating current flows in a conductor, there is a larger number of flux linkages affecting the portion near the center of the conductor than near the surface. This phenomenon, known as *skin effect,* causes a conductor to offer a higher resistance to an alternating current than it offers to a direct current. Skin effect is one factor that contributes to the **AC resistance,** or **effective resistance** of a conductor. The effective resistance of an AC circuit is directly proportional to the frequency. Consequently, high-frequency electronic circuits are especially susceptible to skin effect. At frequencies above 10,000 MHz, a plating of gold or silver on the surface of a thin-walled hollow conductor is often utilized. Power transmission companies generally use aluminum cable with a steel core for strength. The current in the steel core is negligible in comparison with the aluminum because of skin effect and the higher conductivity of aluminum.

The following factors comprise the effective resistance of an electrical or electronic circuit:

1. Dielectric hysteresis loss
2. Dielectric leakage loss
3. Eddy current loss
4. Magnetic hysteresis loss
5. Ohmic (DC) resistance
6. Radiation loss
7. Skin effect
8. Temperature coefficient

Losses such as eddy current and hysteresis also increase as frequency increases. This is why inductors with laminated iron cores are rarely used for radio-frequency applications. **Radiation loss** is a phenomenon that is quite pronounced in antenna circuits. Antennas use power by radiating it away in the form of a radio wave. The resistance of an antenna has two components: radiation resistance and ohmic (DC) resistance. Radiation resistance is defined as the ratio of the power radiated by the antenna to the square of the current at the feed point. The radiation resistance of an antenna may be several hundred ohms, while the DC resistance is less than 0.1 Ω.

Effective resistance in high frequency circuits can be reduced by using fine, multi-strand conductors, called *Litz wire*. Because radio-frequency currents tend to flow near the surface of a conductor, Litz wire reduces AC resistance by insulating each strand of wire, allowing more current flow per circular mil.

Section Review

1. What is skin effect?
2. How is effective resistance reduced in high-frequency circuits?
3. Radio-frequency currents tend to flow near the surface of a conductor (True/False).

__ 16–18

PRACTICAL APPLICATION SOLUTION

In the chapter opener, your task was described as testing the output of a voltage-controlled oscillator (VCO) in a solid-state voltage regulator. The VCO was connected to a test board consisting of an *RLC* parallel circuit with $R = 600\ \Omega$, $C = 0.68\ \mu F$, and $L = 50\ mH$. The output of the VCO is set at 2 V_{rms} and 1 kHz for the test. The following steps illustrate the method of solution for this practical application.

STEP 1   Measure and record the current reading indicated by ammeter. Draw a sketch of the circuit (Figure 16–40).

STEP 2   Calculate total current.

$$X_L = 2\pi fL = 2\pi(1\ kHz)(0.05\ H) = 314.16\ \Omega$$

$$X_C = \frac{1}{2\pi fC} = \frac{1}{2\pi(1\ kHz)(0.68\ \mu F)} = 234.05\ \Omega$$

$$I_R = \frac{E}{R} = \frac{2\ V}{600\ \Omega} = 3.33\ mA$$

$$I_L = \frac{E}{X_L} = \frac{2\ V}{314.16\ \Omega} = 6.37\ mA$$

$$I_C = \frac{E}{X_C} = \frac{2\ V}{234.05\ \Omega} = 8.55\ mA$$

$$I_X = I_C - I_L = 8.55\ mA - 6.37\ mA = 2.18\ mA$$

$$I_T = \sqrt{I_R^2 + I_X^2} = \sqrt{(3.33\ mA)^2 + (2.18\ mA)^2} = 3.98\ mA$$

STEP 3   Compare the calculated value from Step 2 with the measured value. These two values should be approximately equal.

FIGURE 16–40
Diagram of VCO and test board.

Summary

1. In a purely resistive AC circuit, the current and voltage waveforms are in phase with each other.
2. The opposition to current flow offered by a coil is called *inductive reactance.*
3. The ratio of effective voltage across a capacitor to the effective current is called the *capacitive reactance.*
4. Impedance is the total opposition to current flow in an AC circuit.
5. In a series *RL* circuit, the voltage across an inductor leads the current by 90°.
6. In a series *RC* circuit, the voltage across an inductor lags the current by 90°.
7. The average, or real, power is the power delivered to and dissipated by the load.
8. Susceptance is the reciprocal of reactance, and conductance is the reciprocal of resistance.
9. Admittance is the phasor sum of susceptance and conductance.
10. Skin effect causes a conductor to offer higher resistance to an AC voltage than a DC voltage.

Answers to Section Reviews

Section 16–1

1. The current lags the voltage by 90°.
2. Z
3. True

Section 16–2

1. A complex number that is used in AC circuits to represent quantities such as voltage or current
2. True
3. AC currents and voltages cannot be defined in terms of magnitude and direction alone, they also have a time relationship to each other.
4. A phasor lying on the reference axis

Section 16–3

1. True
2. Purely resistive circuits

Section 16–4

1. X_L
2. $X_L = 2\pi fL$
3. $X_L = 2\pi(500 \text{ Hz})(100 \text{ mH}) = 314.16 \ \Omega$

Section 16–5

1. The opposition offered by a capacitor or by any capacitive circuit to the flow of current
2. False
3. X_C
4. True

Section 16–6

1. $Z = E/I$
2. The difference in phase angle between the voltage and current
3. True

Section 16–7

1. Because current is the same in all parts of the circuit
2. False

Section 16–8

1. True
2. $-45°$

Section 16–9

1. A series impedance that draws current of the same magnitude and phase angle as a parallel circuit when connected across the same applied voltage
2. The total impedance and the circuit phase angle

Section 16–10

1. $E = \sqrt{V_R^2 + V_C^2}$
2. $Z = \sqrt{(2.7 \text{ k}\Omega)^2 + (1.65 \text{ k}\Omega)^2} = 3.16 \text{ k}\Omega$

Section 16–11

1. False
2. $Z = \dfrac{(300 \ \Omega)(200 \ \Omega)}{\sqrt{(300 \ \Omega)^2 + (200 \ \Omega)^2}} = 166.4 \ \Omega$

Section 16–12

1. True

Section 16–13

1. $E = \sqrt{(75 \text{ V})^2 + (87 \text{ V} - 54 \text{ V})^2} = 81.94 \text{ V}$
2. $\theta = 39.8°$
3. True

Section 16–14

1. Siemens, S
2. The ease with which current flows in an AC circuit
3. $I = EY$

Section 16–15

1. Find the impedance of each branch.
2. $\theta = \tan^{-1} 275 \text{ mA}/600 \text{ mA} = 24.6°$
3. $I_T = \sqrt{(7.75 \text{ A})^2 + (9.53 \text{ A})^2} = 12.28 \text{ A}$

Section 16–16

1. The power delivered to and dissipated by the load
2. True power
3. False
4. The ratio of true power to apparent power

Section 16–17

1. When alternating current flows in a conductor, there is a larger number of flux linkages affecting the portion near the center of the conductor than near the surface.
2. By using fine multistrand conductors, called Litz wire
3. True

Multiple Choice Questions

16–1 The total opposition to the flow of alternating current is called
(a) Resistance   (b) Inductive reactance
(c) Capacitive reactance   (d) Impedance

16–2 A vector quantity
(a) Has both time and direction   (b) Has time and magnitude
(c) Has magnitude and direction   (d) Is the same as a phasor quantity

16–3 The opposition to current flow offered by a coil is called
(a) Inductive reactance   (b) Capacitive reactance
(c) Impedance   (d) Resistance

16–4 Inductive reactance is measured in
(a) Henrys   (b) Farads   (c) Hertz   (d) Ohms

16–5 The opposition to current flow offered by a capacitor is called
(a) Inductive reactance   (b) Capacitive reactance
(c) Impedance   (d) Resistance

16–6 For a purely capacitive circuit, the current
(a) Leads the voltage by 90°   (b) Lags the voltage by 90°
(c) Is in phase with the voltage   (d) Leads the voltage by 180°

16–7 Capacitive reactance is measured in
(a) Farads   (b) Hertz   (c) Coulombs   (d) Ohms

16–8 In a pure inductor, the current
(a) Leads the voltage by 90°   (b) Lags the voltage by 90°
(c) Is in phase with the voltage   (d) Leads the voltage by 180°

16–9 Conductance is the reciprocal of
(a) Resistance   (b) Reactance   (c) Impedance   (d) Capacitance

16–10 The ease with which current flows in a given AC circuit is called
(a) Impedance   (b) Admittance   (c) Conductance   (d) Susceptance

16–11 The power developed by a resistive component is often called the
(a) Apparent power   (b) Reactive power   (c) Wattless power   (d) Real power

16–12 The power indicated when measured by a wattmeter is the
(a) True power   (b) Apparent power   (c) Reactive power   (d) VAR

16–13 The product of voltage and current is sometimes referred to as the
(a) Real power   (b) True power   (c) Wattless power   (d) Apparent power

16–14 The power factor angle is a ratio between the
(a) Reactive power and true power   (b) Reactive power and apparent power
(c) True power and apparent power   (d) Wattless power and true power

16–15 The effective resistance of an AC circuit is
(a) Inversely proportional to frequency   (b) Directly proportional to frequency
(c) 1/2 of the frequency   (d) 1/4 of the frequency

Practice Problems

16–16 A 2.2 kΩ resistor is connected to an AC source. The applied emf of the circuit is e = 170 sin (377t). Determine the
(a) Rms value of current through the resistor
(b) Instantaneous value of voltage at $t = 0.0025$ s

16–17 If a 2.2 kΩ resistor is connected to an applied emf with $e = 100 \sin(100\pi t)$, determine the instantaneous voltage at $t = 0.008$ s.

16–18 Determine the inductive reactance of a 5.5 mH choke coil at 25 kHz.

16–19 The reactance of a coil at 60 Hz is 3350.1 Ω. Determine the inductance.

16–20 When an inductor is connected to a 24 V, 60 Hz supply, an rms current of 40 mA flows. Determine the inductance of the coil.

16–21 A voltage of 120 ∠45° V is applied across a purely inductive load with a reactance of 8 Ω. Determine the resulting current.

16–22 The instantaneous voltage across a 25 mH choke is $e = 22 \sin(377t)$. Determine the instantaneous current when $t = 10$ ms.

16–23 The equation for a sinusoidal voltage is $e = 200 \sin(5250t + 90°)$. Determine the instantaneous voltage at $t = 100$ μs.

16–24 Two inductors, $X_{L1} = 75$ Ω and $X_{L2} = 100$ Ω, are connected in series to a 24 V supply. Determine the
(a) Total inductive reactance   (b) Total current
(c) Voltage drop across L_1 (d) Voltage drop across L_2

16–25 Two parallel-connected inductors, $X_{L1} = 350$ Ω and $X_{L2} = 500$ Ω, are connected to a 208 V supply. Calculate the
(a) Total inductive reactance   (b) Current through L_1
(c) Current through L_2 (d) Total current in the circuit

16–26 Determine the capacitive reactance of a 0.68 μF capacitor when connected to a 1200 Hz supply.

16–27 A capacitor connected to a 60 Hz supply has a capacitive reactance of 2285 Ω. Determine its capacitance.

16–28 A capacitor has a reactance of 600 Ω at 1 kHz. Find its reactance when connected to a 60 Hz supply.

16–29 Two capacitors are connected in series to a 24 V, 60 Hz supply. C_1 has a capacitance of 10 μF and $C_2 = 15$ μF. Determine the
(a) Total capacitive reactance of the circuit   (b) Total current
(c) Voltage drop across C_1 and C_2

16–30 Two capacitors, $C_1 = 220$ μF and $C_2 = 300$ μF, are connected in parallel across a 120 V, 60 Hz source. Find the
(a) Total capacitive reactance   (b) Current through each parallel path
(c) Total current

16–31 Determine the impedance of an AC load having a load voltage of 12 ∠60° V and a load current of 30 ∠25° mA.

16–32 Calculate the impedance of an AC load having an instantaneous load voltage of $e = 64 \sin(\omega t + 45°)$ V and an instantaneous current of $i = 85 \sin(\omega t + 30°)$ mA.

16–33 A series *RL* circuit consists of a 1200 Ω resistor and a 5 H coil connected to a 208 V, 60 Hz supply. Determine the
(a) Circuit impedance   (b) Voltage drop across the resistor and inductor

16–34 A series *RC* circuit consisting of a 1.2 kΩ resistor and a 6 μF capacitor are connected to a 12∠0° V, 60 Hz power supply. Determine the
(a) Impedance   (b) Current that will flow through the circuit

16–35 In order to cause a current of 100 mA to flow in a series *RC* circuit, what size capacitor must be connected to a 200 Ω resistor? The supply voltage is 120 V, 60 Hz.

16–36 An AC *RLC* series circuit has the following values: $R = 10$ Ω, $X_C = 6$ Ω, $X_L = 12$ Ω. The supply voltage is 24 V, 60 Hz. Determine the
(a) Circuit impedance   **(b)** Circuit current   **(c)** Value of V_R, V_L, and V_C

16–37 Determine the total impedance of a 1 kHz *RLC* series circuit with the following components: $R = 2.2$ kΩ, $L = 0.1$ H, $C = 0.68$ μF.

16–38 A 12 Ω resistor is connected in parallel with a 100 μF capacitor. If the applied voltage is 120∠0° V, 60 Hz, determine the total current.

16–39 Calculate the resistance and reactance values of a series equivalent circuit that may be substituted for a parallel circuit containing a resistance of 12 Ω and reactance of 18 Ω.

16–40 An AC circuit has an impedance of 200∠60° Ω and rms voltage of 50∠−25° V. Determine the
(a) True power   **(b)** Apparent power of the circuit

16–41 Determine the reactive power of a 2 H inductor drawing 400 mA from a 60 Hz power supply.

Essay Questions

16–42 Describe the difference between a phasor and a vector.

16–43 What is a phasor diagram?

16–44 Define inductive reactance.

16–45 How is capacitive reactance determined?

16–46 What is impedance?

16–47 Describe the relationship between current and voltage in a
(a) Purely capacitive circuit   **(b)** Purely inductive circuit

16–48 What is admittance and susceptance?

16–49 Explain the difference between true power and apparent power.

16–50 Define reactive power.

16–51 Why is an inductor also called a storage element?

16–52 How is the power factor angle determined?

16–53 What is effective resistance?

16–54 List eight factors affecting effective resistance.

17

Resonance

Learning Objectives

Upon completion of this chapter you will be able to

- Define resonance.
- Explain the *Q* factor of an AC circuit.
- Discuss bandwidth of resonant circuits.
- Describe the basic operation of a tank circuit.
- Name the three resonant conditions of a parallel *RLC* circuit.
- Understand the purpose of damping resistors.
- List the three basic functions performed by a tuning circuit.
- Explain why the decibel is used when discussing cutoff frequencies in resonant circuits.

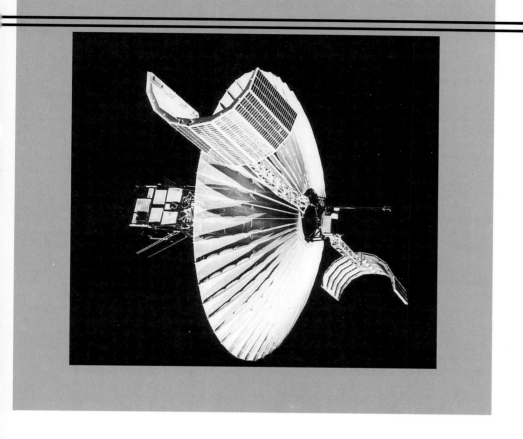

As a technician in the aerospace industry, you are working with a design team on the development of a satellite antenna system. One of the preliminary design considerations involves calculation of a tuned, series-resonant circuit that is part of a down-converter/receiver. In satellite systems, down-converters typically receive a 500 MHz block of signals from a low-noise amplifier. The low-noise amplifier boosts signals received from the satellite dish. You are asked to determine the amount of capacitance that must be connected in series with a 0.1 μH inductor to produce a resonant frequency of 4060 MHz in the down-converter/receiver.

Typical satellites transmit signals at an output power of only 5 W to 24 W, and, by the time the signal reaches earth, it has lost almost 200 dB of the original signal. In addition to finding the capacitance required to produce a resonant frequency of 4060 MHz, you are also asked to determine the dB loss for the system if the satellite is transmitting 22 W and receiving 0.00015 μW.

INTRODUCTION

Resonance is a physical condition that exists when the frequency of an external force applied to a structure is equal to the natural frequency, or vibration, of the structure. When the frequency of the external force is the same as the natural frequency, the energy and amplitude of the vibration may increase substantially. For example, a suspension bridge will tend to oscillate when a strong wind is blowing. Suspension bridges have been known to collapse as a result of sympathetic vibrations set up by strong gusts of wind. The oscillations that occur are a result of a kinetic energy being converted to potential energy and vice versa. When an object loses energy as a result of vibration, the object is said to be *damped,* and the magnitude of the vibration decreases.

In electronic circuits, resonance is caused by the flow of energy stored in magnetic fields (kinetic) and energy stored in electric fields (potential). For a given value of inductance and capacitance, resonance occurs at only one frequency, known as the **resonant frequency.** Resonance exists because the current or voltage phasors for both the inductor and capacitor are equal in magnitude but are in opposing directions. Circuits that contain both inductance and capacitance are also referred to as **tuned circuits.** The frequency that allows the maximum value of current flow represents the tuned frequency in a series AC circuit.

Resonance can be obtained in either series or parallel circuits containing resistance, capacitance, and inductance. When the circuit has equal values of inductive reactance and capacitive reactance, it tends to reject signals having frequencies removed from the resonant frequency. In other words, it will reject signals that are either above or below the frequency that produces resonance. Therefore, in resonant circuits, certain signals are selected to pass through while others are rejected, or blocked. This selection characteristic is known as **selectivity** and refers to the degree that the circuit accepts desirable signal frequencies and rejects undesirable signal frequencies.

Section Review

1. When an object gains energy as a result of vibration, the object is said to be damped (True/False).
2. What is a tuned circuit?
3. What is selectivity?

RESONANCE IN SERIES AC CIRCUITS

An AC series circuit is considered to be in resonance when the inductive reactance, X_L, equals the capacitive reactance, X_C. Under this condition, the total reactance, X, equals zero and the impedance is equal to the resistance of the circuit. The phase angle between the current and voltage in a resonant circuit is zero. Therefore, the current in an AC series circuit reaches its greatest magnitude when the network is in resonance. The condition of resonance exists for a given circuit *at only one frequency.* At all other frequencies, the circuit is either inductive or capacitive.

A series-resonant circuit offers low impedance to the flow of current at a particular frequency. The circuit is said to be *resonant* when the frequency of the applied voltage is adjusted to produce maximum current while the magnitude of the voltage is held constant.

FIGURE 17–1 *Series RLC circuit.*

The frequency of this voltage and current is called the *resonant frequency* and is defined in general terms as the frequency at which a given system or object will respond with maximum amplitude. In the circuit of Figure 17–1, the series *RLC* network has an impedance defined by the equation

$$Z = \sqrt{R^2 + (X_L - X_C)^2}$$

The impedance of an AC circuit varies with frequency. This is because inductive reactance varies directly with frequency, and capacitance varies inversely with frequency. The relationship between frequency and reactance are expressed by the following equations:

$$X_L = L = 2\pi f L$$

$$X_C = \frac{1}{\omega C} = \frac{1}{2\pi f C}$$

At resonant frequency, the inductive and capacitive reactances cancel each other out, leaving only the resistance to oppose the flow of current. Because X_L has a linear relationship with frequency, its graph is shown in Figure 17–2 as having a straight line. As X_C is inversely related to frequency, the resulting graph of the quantity is curved. The total reactance is shown as a dashed line. The point where the total reactance intersects the reference axis is the resonant frequency, f_r.

At resonant frequency, $X_L = X_C$. By using the reactance equations, the following formula is derived.

$$2\pi f_r L = \frac{1}{2\pi f_r C}$$

$$f_r^2 = \frac{1}{(2\pi)^2 LC}$$

$$f_r = \frac{1}{2\pi \sqrt{LC}} \qquad\qquad (17.1)$$

The graphs shown in Figures 17–3(a) and 17–3(b) are referred to as **response curves.** Figure 17–3(a) is the amplitude-frequency response curve for a series circuit, which illustrates the rise of current amplitude to a maximum value at the resonant frequency. Below f_r the impedance of the circuit is mainly capacitive, above f_r the impedance is essentially inductive. As the frequency increases, the circuit impedance decreases until $Z = R$. Because current is inversely proportional to impedance, when the impedance is at its minimum, the maximum amount of current flows in the circuit.

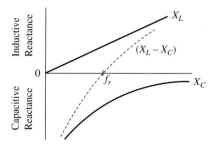

FIGURE 17–2 *Resonant frequency curve.*

(a)

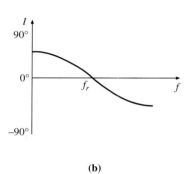

(b)

FIGURE 17–3 *Response curves for series circuit. (a) Amplitude-frequency response. (b) Phase-frequency response.*

Figure 17–3(b) is a phase-frequency response curve for a series circuit. At frequencies below f_r, the current is *leading* the supply voltage. At frequencies above f_r, the current is *lagging* the supply voltage. At the resonant frequency the circuit has resistance but no effective reactance, so the current is in phase with the applied voltage, and the phase angle is zero.

Because maximum current flows when $Z = R$, the resonant current is calculated by Ohm's law.

$$I_r = \frac{E}{R}$$

A series-resonant circuit will, therefore, have a low impedance when the resonant frequency signal is applied, and a higher impedance value for signal frequencies either above or below resonance. The high values of impedance that occur on either side of the resonant frequency offer much greater opposition to current flow. As a result, the flow of current is restricted and the signal is said to be rejected.

_____ **EXAMPLE 17–1** _____

A 100 V series *RLC* circuit has a resistance of 1200 Ω, an inductance of 0.2 H, and a capacitance of 0.057 μF. Calculate the frequency at which the circuit resonates.

Solution

$$f_r = \frac{1}{2\pi \sqrt{LC}}$$

$$= \frac{1}{2\pi \sqrt{(0.2 \text{ H}) (0.057 \times 10^{-6} \text{ F})}}$$

$$= 1490.62 \text{ Hz}$$

EXAMPLE 17–2

Solution

(a)
$$f_r = \frac{1}{2\pi \sqrt{LC}}$$

Therefore,
$$C = \frac{1}{4\pi^2 f_r^2 L}$$

$$= \frac{1}{4\pi^2 \,(400 \text{ Hz})^2 \,(0.01 \text{ H})} = 15.83 \text{ μF}$$

(b)
$$I_r = \frac{E}{R} = \frac{24 \text{ V}}{2.5 \text{ Ω}} = 9.6 \text{ A}$$

A coil with an inductance of 10 mH and a resistance of 2.5 Ω is connected in series with a capacitor and a 24 V, 400 Hz supply. Determine the

(a) Value of capacitance that will cause the circuit to be in resonance.

(b) Current at resonant frequency.

The energy stored in a resonant circuit is constant; although the energy stored in the electric field of a capacitor varies from zero to a maximum and back to zero each half-cycle, and the energy stored in the magnetic field of a coil varies in a similar manner, the total energy stored is not changing with time. The resonant frequency is the frequency at which the coil releases energy at the same rate as the capacitor requires it during one quarter-cycle, and absorbs energy at the same rate the capacitor releases it during the next quarter-cycle. Consequently, the external circuit is not required to supply energy to either the coil or the capacitor. Instead, the external circuit supplies the losses that result from resistance in the resonant circuit.

Section Review

1. Resonance occurs at only one frequency (True/False).
2. When does maximum current flow occur in a series *RLC* circuit?
3. At resonant frequency the capacitive reactance and resistance cancel each other out (True/False).

17–3
THE Q Of A SERIES CIRCUIT

The ratio between the reactive power of the capacitance or inductance at resonance and the real power of a resonant circuit is referred to as the **Q factor.** The Q factor is a measure of the quality of a resonance circuit. A resonance circuit is considered to be of good quality if it is capable of storing a required amount of energy with a minimum of energy loss. The Q factor indicates the amount of energy stored in a circuit compared to the amount of energy dissipated by a circuit. The smaller the level of dissipation, the higher the Q factor.

$$Q = 2\pi \,\frac{\text{maximum energy stored}}{\text{energy dissipated per cycle}}$$

The Q factor is also known as the *voltage magnification factor.* The voltage across the entire circuit is magnified so that it appears across each reactance multiplied by the value of Q. Whenever Q is greater than 1, the voltage across the coil and/or the capacitor in the circuit exceeds the supply voltage.

FIGURE 17–4 *Voltage relations in a series RLC resonant circuit.*

Figure 17–4 shows the voltage drops across a series-connected capacitor, inductor, and resistor. As the supply frequency varies, the voltages V_C, V_L, and V_R also change. At resonant frequency V_R is at its maximum value and is equal to the supply voltage. V_L, and V_C are greater than V_R at resonance but are not at their maximum values. This condition does not violate Kirchhoff's voltage law because V_L and V_C are 180° out of phase with each other. In this example, V_C is at its maximum value at a frequency below f_r and V_L is at its maximum at a value above f_r.

The Q factor of a circuit is also known as the *selectivity* and represents the ability of a tuned circuit to select a particular group of frequencies while rejecting other frequencies above and below this group, or band, of frequencies in terms of inductances. In a series-resonant circuit, the Q factor is calculated using the following equation:

$$Q = \frac{X_L}{R_s} = \frac{1}{\omega_s L R_s} \qquad (17.2)$$

where Q = the figure of merit
   X_L = inductive reactance at resonant frequency, in ohms
   R_s = resistance in series with X_L, in ohms

Q is a numerical factor that is not expressed in units. The Q factor for a series-resonant circuit in terms of capacitance is determined using the following equation:

$$Q = \frac{X_C}{R_s} = \frac{1}{\omega_s C R_s} \qquad (17.3)$$

The Q factor in a series-resonant circuit in terms of resistance, inductance, and capacitance is found by the following equation:

$$Q = \frac{1}{R} \sqrt{L/C} \qquad (17.4)$$

_____ **EXAMPLE 17–3** _____

A series-resonant circuit consists of a 750 µH inductance with an internal resistance of 9 Ω, and a 0.035 µF capacitor. Determine the Q factor of the circuit.

Solution

$$Q = \frac{1}{R} \sqrt{L/C}$$

$$= \frac{1}{9\ \Omega} \sqrt{(750 \times 10^{-6}\ \text{H})/(0.035 \times 10^{-6}\ \text{F})} = 16.3$$

Another factor that affects the Q of a circuit is the resistance encountered at very high frequencies. This high-frequency resistance, known as the *skin effect,* was defined in Chapter 16 as an AC resistance in which the effective resistance is directly proportional to the frequency of a circuit. Skin effect can change the circuit Q by contributing resistance in radio-frequency circuits such as antenna coils and radio-frequency transformers. Litz wire is used to reduce skin effect because of its low resistance to high-frequency current.

Section Review

1. What is the Q factor of a resonant-series RLC circuit?
2. What is another name for Q factor?
3. What effect does skin effect have on the Q factor?

_____ 17-4 __

BANDWIDTH OF RESONANT CIRCUITS

The **bandwidth, BW** of a resonant circuit is defined as the total number of cycles below and above the resonant frequency for which the current is equal to or greater than 70.7% of its resonant value. The width of this band of frequencies is also referred to as the **bandpass** of the circuit.

The frequency-response curve of a series-resonant circuit is shown in Figure 17–5. This curve has a maximum current at resonance. When the current drops to 0.707 of the peak value, the power in the circuit is reduced to 50% of the maximum value. The two frequencies in the curve that are at 0.707 of the maximum current are called **band, or half-power, frequencies.** These frequencies are identified on the curve as f_1 and f_2, and are often referred to as the **critical frequencies, or cutoff frequencies,** of a resonant circuit. f_1 is the lower half-power frequency; f_2 is the upper half-power frequency. The half-power relationship is derived in the following manner:

$$P_m = (I_m)^2 R$$

$$P_{f1} = (0.707 \times I_m)^2 R \qquad (17.5)$$
$$= 0.5\,(I_m)^2 R$$
$$= 0.5\,P_m$$

The resonant frequency can be determined from the critical frequencies by the following equation:

$$f_r = \sqrt{f_1 f_2} \qquad (17.6)$$

Because bandwidth is considered to be the difference between the two conditions where the impedance of the circuit is minimal, bandwidth can be expressed mathematically as

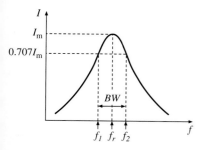

FIGURE 17–5 *Bandwidth of a resonant circuit.*

$$BW = f_2 - f_1 \qquad (17.7)$$

The bandwidth can also be given in terms of resonant frequency and quality factor using the following formula:

$$BW = \frac{f_r}{Q} \qquad (17.8)$$

_____ **EXAMPLE 17–4** _____

For the circuit shown in Figure 17–6, calculate the bandwidth at resonant frequency.

FIGURE 17–6 *Circuit for Example 17–4.*

$R = 250\ \Omega$

$L = 10\ \mu H$

$C = 5\ pF$

Solution

$$f_r = \frac{1}{2\pi\ \sqrt{LC}} = \frac{1}{2\pi\ \sqrt{(10 \times 10^{-6}\ \text{H})\ (5 \times 10^{-12}\ \text{F})}} = 22.51\ \text{MHz}$$

$$X_L = 2\pi f_r L = 2\pi(22.51 \times 10^6\ \text{Hz})\ (10 \times 10^{-6}\ \text{H}) = 1414\ \Omega$$

$$Q = \frac{X_L}{R} = \frac{1414\ \Omega}{250\ \Omega} = 5.66$$

$$BW = \frac{f_r}{Q} = \frac{22.51 \times 10^6\ \text{Hz}}{5.66} = 3.98\ \text{MHz}$$

_____ **EXAMPLE 17–5** _____

If the applied voltage for the circuit of Figure 17–6 is 24 V, determine the power dissipated in the circuit at the half-power frequencies.

Solution

$$Z_T = R = 250\ \Omega$$

$$I = \frac{E}{Z_T} = \frac{24\ \text{V}}{250\ \Omega} = 0.096\ \text{A}$$

$$P_{f1} = \frac{1}{2}\ P_m$$

$$= \frac{1}{2}\ I_m^2 R$$

$$= \left(\frac{1}{2}\right)(0.096\ \text{A})^2(250\ \Omega)$$

$$= 1.15\ \text{W}$$

EXAMPLE 17–6

Solution

(a)

$$BW = f_2 - f_1 = 125 \text{ kHz} - 15 \text{ kHz} = 110 \text{ kHz}$$

(b)

$$f_r = \sqrt{f_1 f_2} = \sqrt{(125 \text{ kHz})(15 \text{ kHz})} = 43.3 \text{ kHz}$$

A circuit with a frequency response curve similar to that shown in Figure 17–5 is said to *favor* signals at, or near, the resonant frequency. The ability of a resonant circuit to select one particular frequency and reject other frequencies is called the *selectivity of the circuit*. In other words, selectivity represents the degree of how well a circuit responds to frequency changes. The smaller the bandwidth, the higher the selectivity. Highly selective circuits have small bandwidths and are, therefore, able to sense small changes in frequency. Generally, a value of Q that is less than 10 is considered to be a wide band frequency, while a Q value greater than 10 is said to be a narrow band.

The selectivity and bandwidth are largely dependent on the resistance of a resonant circuit. As the circuit resistance increases, the bandwidth also increases. Therefore, a circuit with low values of resistance will produce narrower bandwidths, be more selective, and have sharper tuning characteristics.

Section Review

1. What is the bandwidth of a resonant circuit?
2. Half-power frequencies are the frequencies where the current drops to one-half of its maximum power (True/Fale).
3. State the equation for finding bandwidth when the Q and resonant frequency of the circuit are known.

17–5

RESONANCE IN PARALLEL AC CIRCUITS

Series-resonant circuits are used whenever maximum current is required for a specific frequency or band of frequencies. **Parallel-resonant circuits** are used whenever the signal strength of any one frequency or band of frequencies is to be reduced to a minimum.

The parallel *RLC* circuit has three resonant conditions:

1. When $X_L = X_C$
2. When the current is at its minimum
3. When the opposition to current flow is purely resistive ($\theta = 0$)

The impedance of a parallel-resonant circuit will be at its highest possible value when the inductive reactance and capacitive reactance are equal. This is the opposite condition to that encountered in the study of series-resonant circuits. The high impedance of a parallel-resonant circuit decreases the flow of current through the circuit. When the frequency is above or below resonance in a parallel circuit, the impedance is low.

At resonance, a parallel-resonant circuit acts as a storage device. If the resistance of the circuit is negligible, the energy from the signal is exchanged between the inductance and the capacitance at a rate corresponding to the frequency of the applied signal. This exchange of energy would continue even if the signal source were removed, because the capacitor and inductor would constantly charge and discharge each other. The oscillations of charge and energy that occur at the natural *LC* resonant frequency is known as the **flywheel effect.**

The parallel-resonant circuit is often connected as shown in Figure 17–7. At low frequencies, the capacitive reactance will be higher than the inductive reactance, and more current will flow through the inductor. At high frequencies, the inductive reactance has a higher ohmic value, and more current will flow through the capacitor. At resonant frequency, both reactances will be equal, and, as a result, equal values of current flow through *L* and *C*. If the currents have equal magnitude and opposite direction, they will cancel each other out.

The circuit of Figure 17–7 is often referred to as a **tank circuit** because of the storage of energy in the inductor and capacitor. In a tank circuit, the resonant frequency is influenced by how much resistance there is in the circuit. Parallel-resonant circuits are designed so that they have as little loss as possible, which means a *Q* of very high value is preferable.

If the resistor shown in Figure 17–7 were removed, there would be a pure inductor connected in parallel with a pure capacitor. In this situation, the current at resonance would be zero, because the current flowing through the two reactances would be of equal and opposite value. The series-equivalent impedance of this circuit becomes infinite at resonance, because of the line current being zero and $Z = E/I$. The equation for the impedance of a parallel-resonant circuit is expressed as follows:

$$Z_r = \frac{X_L \times X_C}{R} = \frac{L}{RC} \tag{17.9}$$

where Z_r is the input impedance at resonance.

If the impedance of the tank circuit shown in Figure 17–7 is plotted to a logarithmic frequency base, the impedance will peak at the resonant frequency. The magnitude of impedance drops off sharply with frequencies either above or below resonance. At resonance, a tank circuit with a low value of resistance will contain a high value of impedance and possess a narrow bandwidth. Because the condition for parallel resonance is that $I_L = I_C$, which implies that $X_L = X_C$, the resonant frequency for a parallel-resonant circuit is calculated in the same manner as for a series-resonant circuit. That is,

$$f_r = \frac{1}{2\pi \sqrt{LC}}$$

FIGURE 17–7   *Parallel-resonant circuit.*

EXAMPLE 17-7

FIGURE 17-8   Circuit for Example 17-7.

For the circuit shown in Figure 17-8, calculate the

(a) Resonant frequency.
(b) Total impedance.
(c) Line current at resonance.

Solution

(a) $$f_r = \frac{1}{2\pi \sqrt{LC}} = \frac{1}{2\pi \sqrt{(40 \times 10^{-6} \text{ H})(140 \times 10^{-12} \text{ F})}} = 2.13 \text{ MHz}$$

(b) $$Z_r = \frac{L}{RC} = \frac{40 \times 10^{-6} \text{ H}}{(6 \ \Omega)(140 \times 10^{-12} \text{ F})} = 47.62 \text{ k}\Omega$$

(c) $$I = \frac{E}{Z} = \frac{12 \text{ V}}{47.62 \times 10^3 \ \Omega} = 0.25 \text{ mA}$$

The frequency-response curves for a parallel *RLC* circuit are shown in Figure 17-9. In Figure 17-9(a) the impedance is at its maximum value, Z_r, at resonance. At frequencies below resonance, the parallel circuit is inductive. At frequencies above resonance, the parallel circuit is capacitive. In Figure 17-9(b), the current is at its minimum value, I_r, at resonance. At very high frequencies, the capacitor acts as a short circuit and the current becomes very large.

At resonance, the current in an *RLC* circuit is determined entirely by the resistance of the circuit. The capacitive and inductive currents can be of any magnitude and their net value will be zero because of phase displacement. Therefore, in a parallel-resonant circuit the capacitive and inductive currents can be considerably larger than the total circuit current and still not violate Kirchhoff's current law. In a parallel-resonant circuit, the quality factor is referred to as the **current magnification factor.** The total circuit current is magnified so that it is *Q*-times as large in either reactive component.

(a)

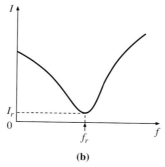

(b)

FIGURE 17-9   Parallel-resonant circuit frequency-response curves. (a) Impedance. (b) Current.

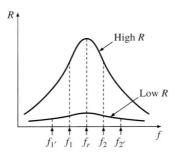

FIGURE 17–10   *Effect of resistance on bandwidth and selectivity.*

The resistance of a parallel-resonant circuit will have a pronounced effect on the bandwidth and selectivity, as shown in Figure 17–10. The high-resistance response curve shows that the bandwidth is narrower and the selectivity is greater than for a low-resistance parallel-resonant circuit. The points f_1 and f_2 represent the cutoff frequencies of the high resistance circuit, while f_1' and f_2' indicate the cutoff frequencies of the low resistance response curve. The bandwidth of a parallel-resonant circuit is measured in the same manner as series-resonant circuits. Like its series counterpart, the parallel-resonant circuit draws peak power at resonance and has half-power frequencies above and below the resonant frequency.

Because resonant frequency depends entirely on the reactive components in the circuit, by varying R, the Q of the circuit can be adjusted for different applications. For example, in audio equipment that is sensitive to radio-frequency interference, the circuit Q can be reduced by varying the resistance of the parallel-resonant circuit. This effect is known as **damping,** and variable resistors in parallel-resonant circuits are often referred to as **damping resistors.**

If a resistor is placed across a parallel-resonant circuit, the circuit voltage will be prevented from rising to the point it would otherwise. Essentially, the parallel resistor, R_p, acts as a load on the oscillation of electrons and any imbalance in the voltage across the tank capacitor will start to redistribute itself through R_p. In electronic circuits where wide ranges of signals are encountered, it is often necessary to increase the frequency response characteristics of a resonant circuit by adding a damping resistor in parallel. Increasing the value of the parallel resistance will increase the Q and decrease the bandwidth. In equation form,

$$Q = \frac{R_p}{X_L} \qquad (17.10)$$

where Q = figure of merit
R_p = resistance in parallel with inductor and capacitor, in ohms
X_L = inductive reactance, in ohms

_____ **EXAMPLE 17–8** _____

For the circuit shown in Figure 17–11, determine the

(a) Resonant frequency.
(b) Circuit Q.
(c) Bandwidth.

$R = 33\ \text{k}\Omega$ $L = 100\ \mu\text{H}$ $C = 200\ \text{pF}$

FIGURE 17–11   *Circuit for Example 17–8.*

Solution

(a)
$$f_r = \frac{1}{2\pi \sqrt{LC}} = \frac{1}{2\pi \sqrt{(100 \ \mu H) (200 \ pF)}} = 1.125 \ \text{MHz}$$

(b)
$$X_L \text{ (at resonance)} = 2\pi f_r L = 2\pi (1.125 \ \text{Mhz}) (100 \ \mu H) = 707 \ \Omega$$

$$Q = \frac{R_p}{X_L} = \frac{33 \ k\Omega}{707 \ \Omega} = 46.7$$

(c)
$$BW = \frac{f_r}{Q} = \frac{1.125 \ \text{MHz}}{46.7} = 24.09 \ \text{kHz}$$

Section Review

1. When is the impedance of a parallel-resonant circuit at its highest value?
2. What are the oscillations of charge and energy that occur at the natural LC resonant frequency known as?
3. What is another name for the current magnification factor of a parallel-resonant circuit?

17–6
TUNING CIRCUITS

Tuning is defined as the adjustment of the frequency of a circuit or system to secure optimum performance. In electronic circuits, tuning is generally used to achieve maximum current, voltage, and power. A tuning circuit performs three basic functions:

1. It allows a desired frequency to be selected.
2. It rejects all undesired signals.
3. It will increase the voltage of the circuit.

Series-resonant circuits in portable radio receivers are tuned by means of a variable capacitor. Automobile radios typically use variable inductors for tuning purposes. In a television receiver, the inductance is varied as a *coarse* adjustment, and the circuit is *fine tuned* by varying the capacitance of the resonant circuit. To tune in a specific station using a portable radio receiver, it is necessary to vary the capacitance until its reactance at that frequency becomes equal in magnitude to the inductive reactance of the coil in series with it. The desired signal will then send current at its own frequency through the circuit and it will be considerably larger than currents at any other frequency.

Other signals from broadcast stations that operate at different frequencies on adjacent channels will also appear as voltages applied to the same series-resonant circuit. If the circuit has a relatively large Q, that is, if the circuit is *selective,* the current versus frequency response curve will be quite steep on both sides of the peak value. Figure 17–12 shows three different values of Q. From this response curve it can be seen that the higher the Q, the higher the current peak at resonance, and the steeper the sides of the curve. In a radio circuit, the high Q value results in unwanted currents from other stations being too small to cause interference with reception of the resonant-frequency signal.

Figure 17–13 shows a simplified schematic of the first radio frequency (RF) stage of a radio receiver. Radio signals are broadcast from a transmitter via electromagnetic waves that multiply as they travel through the atmosphere. The action of the electromagnetic waves cutting across the receiving antenna induce small voltages. When a voltage is

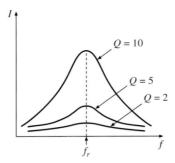

FIGURE 17–12 *Effect of Q on tuned circuit.*

FIGURE 17–13 *Tuned series-resonant circuit.*

induced into the antenna, it causes a current to flow in the primary winding of the transformer. As a result, a voltage is also induced in the secondary winding of the transformer. By adjusting the variable, or *tuning* capacitor, the circuit may be made to resonate at any one of a wide variety of radio frequencies.

When a radio signal voltage is induced in the coil of a series circuit tuned to the frequency of the signal voltage, the voltage across the variable capacitor is "picked off" and applied to the input of a voltage amplifier. The resonant frequency causes a resonance rise in voltage to appear across the tuning capacitor. This capacitor voltage, which is Q times the induced voltage of the inductor, has the same frequency. Consequently, a very small signal of a few microvolts can be converted to a much larger signal before being amplified by the radio's electronic circuitry.

_____ **EXAMPLE 17–9** _____

In the circuit of Figure 17–13, $L_2 = 100\ \mu H$. Determine the range of adjustment of the tuning capacitor that will allow the circuit to resonate at frequencies ranging from 535 kHz to 1605 kHz.

Solution At $f_r = 535$ kHz,

$$C = \frac{1}{4\pi^2(535 \times 10^3\ \text{Hz})^2(100 \times 10^{-6}\ \text{H})} = 884.98\ \text{pF}$$

At $f_r = 1605$ kHz,

$$C = \frac{1}{4\pi^2(1605 \times 10^3\ \text{Hz})^2(100 \times 10^{-6}\ \text{H})} = 98.33\ \text{pF}$$

Section Review

1. What is tuning?
2. A circuit with a large Q is said to be selective (True/False).
3. What is a tuning capacitor used for in radio receivers?

The logarithm of the ratio of two voltages, currents, or power levels is typically measured in a unit known as the **bel.** Named after Alexander Graham Bell, this unit of measure was originally used to provide comparisons between the power-loss properties of various telephone lines. In most electronic circuits, the bel represents too large a unit for practical applications. For this reason, the **decibel, dB,** which is one-tenth of a bel, is used.

$$dB = 10 \log_{10}\left(\frac{P_{out}}{P_{in}}\right) \qquad \text{(17.11)}$$

The decibel is a common unit of measure when discussing cutoff frequencies in resonant circuits. For example, the upper and lower cutoff frequencies of a bandwidth are also known as the *−3 dB frequencies*. The cutoff frequencies are also known as half-power frequencies because the output power at these points is reduced by 50% of the input power. Equation 17.11 can be used to prove the relationship between one-half power at resonance and −3 dB.

$$dB = 10 \log_{10}\left(\frac{P_{out}}{P_{in}}\right)$$

$$= 10 \log_{10}\left(\frac{0.5}{1}\right)$$

$$= -3 \text{ dB}$$

When the output power and the input power are equal, the power ratio is 1 and the decibel level is 0 dB. In situations where the input power is greater than the output power, the power ratio is less than 1 and the decibel level is negative. The reduction in power level of a signal is referred to as **attenuation.** When the power ratio is greater than 1, the increase in power level of the signal is called **gain.**

_____ EXAMPLE 17-10 _____

Solution

$$dB = 10 \log_{10}\left(\frac{P_{out}}{P_{in}}\right)$$

$$= 10 \log_{10}\left(\frac{20 \text{ W}}{0.1 \text{ W}}\right)$$

$$= 23.01 \text{ dB}$$

The input to a power amplifier is 0.1 W and the output is 20 W. What is the gain of the amplifier in dB?

Because decibels are logarithmic, the gains expressed in dB can be added instead of multiplied. The following example illustrates this procedure.

EXAMPLE 17–11

A directional antenna has a gain of 4 dB and the lead-in coaxial cable has a loss of 1.5 dB. The cable is fed to a tuner-pream-plifier with a gain of 54 dB, and then to an amplifier with a 10 dB gain. The system is con-nected to a speaker with an 8 dB loss. Determine the total system gain.

Solution

$$
\begin{aligned}
\text{Antenna gain} &= \quad +4.0 \text{ dB} \\
\text{Coaxial cable} &= \quad -1.5 \text{ dB} \\
\text{Tuner-preamp} &= +54.0 \text{ dB} \\
\text{Amplifier} &= +10.0 \text{ dB} \\
\text{Speaker} &= \quad -8.0 \text{ dB} \\
\hline
\text{Total gain} &= \quad 58.5 \text{ dB}
\end{aligned}
$$

While decibels are used primarily in expressing power gain and loss, you will also see them used on many occasions to express the voltage or current gain or loss in a circuit. The gain of an amplifier is often expressed as a voltage ratio rather than a power ratio. The formulas for expressing the ratio of two voltages or two currents in decibels are

$$ \text{dB} = 20 \log_{10} \left(\frac{V_{\text{out}}}{V_{\text{in}}} \right) \tag{17.12} $$

$$ \text{dB} = 20 \log_{10} \left(\frac{I_{\text{out}}}{I_{\text{in}}} \right) \tag{17.13} $$

The most important thing to remember about using current and voltage values is that the dB figure is meaningless unless the input and output impedances are equal.

EXAMPLE 17–12

At a particular frequency the input voltage to the four-terminal network shown in Figure 17–14 is 5 V and the output voltage is 18 V. If the input and output impedances are equal, determine the volt-age ratio in decibels.

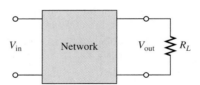

FIGURE 17–14 *Four-terminal network.*

Solution

$$
\begin{aligned}
\text{dB} &= 20 \log_{10} \left(\frac{V_{\text{out}}}{V_{\text{in}}} \right) \\
&= 20 \log_{10} \left(\frac{18 \text{ V}}{5 \text{ V}} \right) \\
&= 11.13 \text{ dB}
\end{aligned}
$$

Loudness is a term often associated with resonant audio circuits. Loudness is some-times called *volume* and represents the level of sound intensity perceived by the human ear.

TABLE 17-1 *Typical sound pressure levels.*

Sound Source	Sound Pressure Level
Jet engine at 10 m	160 dB
Jet taking off at 500 m	120 dB
Amplified rock music	110 dB
City traffic at 16 m	70 dB
Normal speech at 1 m	60 dB
Library	40 dB
Recording studio	20 dB
Breathing	10 dB

Loudness is measured in decibels, where the threshold of hearing is assigned the value 0 dB. The 0 dB level represents the reference level for audio signals. The **sound pressure level,** SPL, describes the relationship between the level of sound and the reference level. Table 17–1 shows examples of SPLs of common sounds. At 0 dB, the sound level is inaudible. SPLs greater than 120 dB can cause permanent hearing damage if sustained for long periods of time.

Section Review

1. When the power ratio of a circuit is greater than 1, the decibel level is negative (True/False).
2. What is loudness measured in?
3. What is sound pressure level (SPL)?

_____ 17-8 ___

PRACTICAL APPLICATION SOLUTION

In the chapter opener, you were assigned two tasks: one was to solve for the capacitance required to produce a resonant frequency of 4060 MHz, and the other task was to determine the input power to the satellite receiver. The following steps illustrate the method of solution for this practical application of resonance in AC circuits.

STEP 1   Determine the capacitance required to produce resonant frequency.

$$C = \frac{1}{4\pi^2 f_r^2 L} = \frac{1}{4\pi^2 (4060 \text{ MHz})^2 (0.1 \text{ μH})} = 0.015 \text{ pF}$$

STEP 2   Determine the dB loss between transmitter and receiver.

$$dB = 10 \log_{10}\left(\frac{P_{out}}{P_{in}}\right) = 10 \log_{10}\left(\frac{22 \text{ W}}{0.00015 \text{ μW}}\right) = 111.66 \text{ dB}$$

_____ Summary ___

1. Resonance is caused by the flow of energy stored in magnetic fields and energy stored in electric fields.
2. At resonant frequency, the inductive and capacitive reactances cancel each other out.
3. The Q factor is a measure of the quality of a resonance circuit.

4. The Q factor is also known as selectivity and represents the ability of a tuned circuit to select a group of frequencies while rejecting other frequencies.
5. The two frequencies that are 0.707 of the maximum current are called *half-power frequencies.*
6. The impedance of a parallel-resonant circuit will be at its maximum when X_L and X_C are equal.
7. The adjustment of the frequency of a circuit to secure optimum performance is called *tuning.*
8. When the output power and the input power are equal, the power ratio is 1 and the decibel level is 0 dB.
9. The reduction in power level of a signal is called the *attenuation.*
10. An increase in power level is called *gain.*

___ ## Answers to Section Reviews _____

Section 17–1

1. False
2. A circuit that contains both inductance and capacitance
3. The degree that the circuit accepts desired signal frequencies and rejects undesired signal frequencies

Section 17–2

1. True
2. At resonant frequency
3. False

Section 17–3

1. The ratio between the reactive power of the capacitance or inductance at resonance and the real power of a resonant circuit
2. Selectivity
3. Skin effect can change the circuit Q by contributing resistance in radio-frequency circuits.

Section 17–4

1. The total number of cycles below and above the resonant frequency for which the current is equal to or greater than 70.7% of its resonant value
2. False
3. $BW = f_r/Q$

Section 17–5

1. When the inductive reactance and capacitive reactance are equal
2. Flywheel effect
3. Q factor

Section 17–6

1. The adjustment of the frequency of a circuit or system to secure optimum performance
2. True
3. Tuning capacitors allow a circuit to resonate at any one of a wide variety of radio frequencies.

Section 17–7

1. False
2. Decibels
3. SPL describes the relationship between the level of sound and the reference level.

Review Questions

Multiple Choice Questions

17–1 When an object loses energy as a result of vibration, it is
(a) Because of resonance **(b)** Caused by selectivity
(c) Said to be damped **(d)** Called gain

17–2 For a given value of inductance and capacitance, resonance
(a) Occurs at more than one frequency **(b)** Is determined by selectivity
(c) Is also known as critical frequency **(d)** Occurs at only one frequency

17–3 A series-resonant circuit has _____ and _____ at reso-nance.
(a) High impedance, high current **(b)** Low impedance, high current
(c) High impedance, low current **(d)** Low impedance, low current

17–4 The ratio between the reactive power of the capacitance or inductance at resonance and the real power of a resonant circuit is called
(a) Q factor **(b)** Bandwidth
(c) Resonant frequency **(d)** Tank circuit

17–5 The Q factor of a circuit is also known as the
(a) Power factor **(b)** Bandwidth
(c) Selectivity **(d)** Resonant frequency

17–6 The total number of cycles below and above the resonant frequency for which the current is equal to or greater than 70.7% of its resonant value is called
(a) Selectivity **(b)** Bandwidth
(c) Bandstop **(d)** Critical frequency

17–7 Skin effect is the _____ encountered at very high frequencies.
(a) Resistance **(b)** Inductance
(c) Capacitance **(d)** Resonant frequency

17–8 Parallel-resonant circuits are used
(a) Whenever maximum current is required for a specific frequency or band of frequencies
(b) To provide low impedance and high current at resonance
(c) Whenever the signal strength of any one frequency or band of frequencies is to be reduced to a minimum
(d) To reduce the impedance of a circuit

17–9 The oscillations of charge and energy that occur at the natural LC resonant fre-quency is known as
(a) Damping **(b)** Tank resonance
(c) Tuning **(d)** Flywheel effect

17–10 In a parallel-resonant circuit, the Q factor is also called
(a) Current magnification factor **(b)** Flywheel effect
(c) Voltage magnification factor **(d)** Bandwidth

17–11   If a resonant circuit is selective, it will have a
(a) Small Q value   **(b)** Steep response curve
(c) Gradual response curve   **(d)** Large bandwidth

17–12   The reduction in power level of a signal is referred to as
(a) Gain   **(b)** Loudness
(c) Decibel   **(d)** Attenuation

17–13   The sound pressure level (SPL) describes the relationship between
(a) Level of sound and volume   **(b)** Threshold of sound and reference level
(c) Loudness and volume   **(d)** Level of sound and reference level

Practice Problems

17–14   A 24 V series *RLC* AC circuit has the following values: $R = 8\ \Omega$, $L = 5$ H, $C = 0.068\ \mu$F. Determine the frequency at which the circuit resonates.

17–15   A coil with an inductance of 5 mH and a resistance of 1.2 Ω is connected in series with a capacitor and a 12 V, 1 kHz supply. Determine the
(a) Value of capacitance that will cause the circuit to be in resonance
(b) Current at resonant frequency

17–16   Determine the Q factor of an AC circuit with a 20 mH inductor having an internal resistance of 5 Ω connected in series with a 22 μF capacitor.

17–17   An AC series *RLC* circuit consists of the following components: $R = 80\ \Omega$, $L = 0.2$ mH, $C = 100$ pF. Calculate the bandwidth at resonant frequency.

17–18   For the circuit shown in Figure 17–15, determine the following values:
(a) Resonant frequency
(b) Total impedance
(c) Line current

FIGURE 17–15

17–19   For the circuit shown in Figure 17–16, determine
(a) Resonant frequency
(b) Circuit Q
(c) Bandwidth

FIGURE 17–16

17–20   The input to a power amplifier is 1.5 W and the output is 35 W. Determine the gain of the amplifier in dB.

17–21 Determine the power gain in dB of an amplifier in which the input power is 20 mW and the output power is 100 W.

17–22 The input power to a communications cable is 1 W and the output is 200 mW. Determine the cable loss in decibels.

17–23 How much output power will be produced by an amplifier capable of 30 dB gain if fed an input of 20 mW?

17–24 A power amplifier increases a signal level by 12 dB. If the output signal level is 40 W, calculate the input level.

Essay Questions

17–25 Define resonance.

17–26 Explain what happens to the inductive and capacitive reactances of a series *RLC* circuit at resonance.

17–27 What is the main difference between a series-resonant circuit and a parallel-resonant circuit?

17–28 Describe the effect of *Q* on a resonant circuit.

17–29 Define bandwidth.

17–30 What are the three resonant conditions for a parallel *RLC* circuit?

17–31 Explain the flywheel effect of a resonant circuit.

17–32 List the three basic functions performed by a tuning circuit.

17–33 Describe the difference between gain and attenuation.

18

Coupling and Filter Circuits

Learning Objectives

Upon completion of this chapter you will be able to

- Define the terms *filter* and *coupler*.
- Explain the two basic types of coupling.
- Name two disadvantages of capacitive coupling.
- Describe the principles of transformer coupling.
- Define insertion loss.
- List the four types of filters.
- Explain the difference between passive filters and active filters.
- Understand how low-pass filters can be used to smooth the output of a pulsating DC signal.
- Draw a basic Bode plot.
- List four characteristics of an ideal op-amp.

Practical Application

As an electronic technician for a large medical center, you are responsible for maintaining a wide variety of electronic equipment. One of your responsibilities is calibrating video receivers. The medical center has hundreds of these receivers that require periodic maintenance. The video receiver consists of a video amplifier, audio amplifier, FM detector, band-pass and band-stop filters. The band-pass and band-stop filters are parallel resonant-types with inductors and capacitors.

Your task is to calibrate the band-stop filter by calculating the resonant frequency of the filter and comparing this value with the measured value. The band-stop filter consists of a 2 µH inductor and 0.001 µF capacitor. If the calculated and measured values are not approximately equal, you must replace the 0.001 µF capacitor.

INTRODUCTION

In Chapter 13, the term coupling was defined as a means of transferring energy from one circuit to another. A **filter,** as the name implies, is a device that removes, or filters, unwanted signals. The terms coupling and filtering are often used interchangeably because they perform similar functions. That is, coupling circuits are often used to filter out frequencies, and filter circuits are used to couple circuits together. The devices used to couple circuits together include capacitors, inductors, resistors, and transformers. Two circuits are said to be coupled when they have a common impedance that allows the transfer of energy from one circuit to another. The common impedance is called a *coupler* and represents a device used to achieve optimum transfer of power between electronic circuits.

Section Review

1. The terms coupling and filtering are often used interchangeably because they perform similar functions (True/False).
2. What is a coupler?
3. Filters remove unwanted frequencies (True/False).

_____ 18–2_____

DIRECT COUPLING

There are two types of coupling—direct and indirect. Circuits that are direct coupled have the current of the input circuit flowing through the common impedance where it creates a voltage drop. This voltage drop represents the output voltage of the circuit. Direct coupling is characterized by a wide-band frequency response because there is, essentially, no circuit tuning taking place. Figure 18–1 shows three typical direct-coupled circuits. In the inductive-coupled circuit of Figure 18–1(b), the proportion of energy transferred increases as the frequency increases. In the capacitive-coupled circuit of Figure 18–1(c), the proportion of energy transferred decreases as the frequency increases. The capacitor behaves in the same manner as the bypass capacitor discussed in Chapter 13.

Direct-coupled circuits can be used to transfer all frequencies from one circuit to another. When it is necessary to amplify DC signals, or very-low-frequency signals, direct-

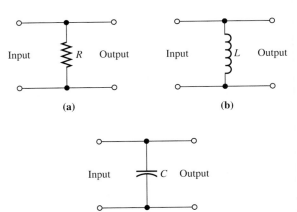

FIGURE 18–1 *Three examples of direct-coupled circuits. (a) Resistor coupled. (b) Inductor coupled. (c) Capacitor coupled.*

FIGURE 18–2 *Capacitive-coupled circuit.*

Input R Output

coupled amplifiers are used. Direct coupling transmits the alternating-current and direct-current components of a signal. In situations where only the alternating current is required to be passed, it is necessary to use indirect coupling.

Section Review

1. What is direct coupling?
2. Direct coupling is characterized by narrow-band frequency response (True/False).
3. Indirect coupling does not pass DC signals (True/False).

18–3
CAPACITIVE COUPLING

In addition to providing a means of direct-coupling circuits, capacitors can also be used to indirectly couple networks. As mentioned in Chapter 13, when a capacitor is connected in series between a source and load, the DC voltage is blocked and the AC signal passes through the capacitor. Figure 18–2 shows an example of indirect coupling using a capacitor and a resistor. If the reactance of the coupling capacitor is small compared with R, practically all of the AC input voltage appears across R. As a general rule of thumb, X_C should be less than 10% of R in order for capacitive coupling to occur.

The main disadvantage of capacitive coupling is the difficulty in making a good impedance match between circuits, or stages. Consequently, power transfer can become quite inefficient in capacitive-coupled networks. Capacitive coupling also transfers unwanted harmonic signals in addition to the fundamental or desired frequency. Tuned circuits are often used in conjunction with capacitive coupling to reduce these inefficiencies.

Section Review

1. What is the main disadvantage of capacitive coupling?
2. Capacitive coupling provides highly efficient power transfer (True/False).
3. State the general rule of thumb for capacitive coupling.

18–4
TRANSFORMER COUPLING

Circuits that are coupled using transformers are considered to be indirectly coupled because there is no physical connection between the input and output. In this type of circuit, energy is transferred as a result of the alternating current of the input circuit flowing through the primary winding and establishing an alternating magnetic field. The flux of this field links the secondary winding and induces a voltage that supplies the energy for the output circuit. Figure 18–3(a) illustrates a transformer-coupled circuit. The waveshape of Figure 18–3(b) shows the effect of combining DC and AC voltage sources. The output

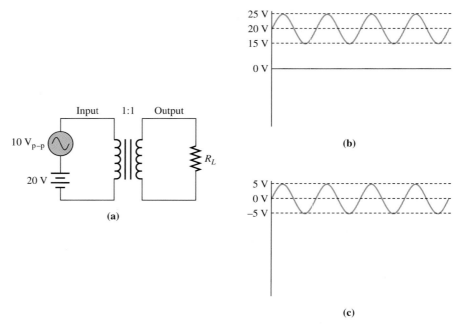

FIGURE 18–3 *(a) Transformer-coupled circuit. (b) Input waveshape. (c) Output voltage.*

wave of Figure 18–3(b) shows that transformer coupling will also block DC voltages and pass AC signals.

Transformer coupling is popular in radio-frequency circuits because it provides good impedance matching and gain and allows for a tuned bypass. In high-frequency applications, transformers, or a combination of transformers and capacitors, are often used to provide frequency-selective coupling between stages. In many cases, the transformers are tuned to the right frequency range by means of a fixed capacitor in parallel with the primary winding and an adjustable core that will vary the inductance of the circuit. In transmitter power amplifiers the output is almost always coupled to the antenna by means of a tuned transformer. This helps to minimize unwanted harmonic radiation.

Section Review

1. Circuits that are coupled using transformers are considered to be indirectly coupled because there is no physical connection between the input and output (True/False).
2. Why is transformer coupling popular in radio-frequency circuits?
3. Harmonic radiation is reduced by using a tuned transformer (True/False).

— 18–5

FILTER CIRCUITS

Filters are used to block or pass a specific range of frequencies and are either passive or active. **Passive filters** use resistors, capacitors, and/or inductors and do not contain an amplifying device. Passive filters never have a power output equal to or greater than the input because the filter always has **insertion loss.** This loss represents the difference between the power received at the load before insertion (installation) of a filter and the power after insertion. Insertion loss is stated as the log of a ratio of power output to power input and is the result of power loss because of resistance in the circuit. An **active filter** uses some type of amplifying device, such as a transistor and/or operational amplifier in combination with resistors and capacitors to obtain the desired filtering effect. Active filters have very low insertion loss.

Filters can be classified into four main types depending on which frequency components of the input signal are passed on to the output signal. The four types of filters are low pass, high pass, band pass, and band stop.

Section Review

1. Filters are classified as passive or active (True/False).
2. What is a passive filter?
3. Active filters have high insertion loss (True/False).

LOW-PASS FILTERS

A **low-pass filter** is a circuit that has a constant output voltage up to a cutoff, or critical, frequency, f_c. The frequencies above f_c are effectively shorted to ground and are described as being in the attenuation band, or **stopband.** The frequencies below f_c are said to be in the **passband.** Low-pass filters are designed using combinations of resistors, inductors, and capacitors. Figure 18–4 shows a basic low-pass filter using an RC network. At low frequencies, X_c is greater than R, and most of the input signal will appear at the output. The cutoff frequency occurs when $X_c = R$. At this frequency, $V_{out} = 0.707\ V_{in}$.

Figure 18–5 shows a low-pass filter using an inductor and resistor. In this circuit, the inductor is shorted at low frequencies, and most of the output voltage of the circuit is dropped across R. At high frequencies the input acts as an open, and the majority of the input signal is dropped across L. Therefore, the RL circuit passes low frequencies and blocks high frequencies.

The response curve for a low-pass filter is shown in Figure 18–6. When the output voltage is 0.707 of the input, the power output is 50% of the input power. This condition exists at f_c, and, for this reason, the cutoff frequency is also referred to as the **half-power point.** If $P_{out} = 1/2 P_{in}$, then a 3 dB loss of signal has occurred.

The cutoff frequency of a low-pass filter is found by the following equations:

$$f_c = \frac{1}{2\pi RC} \quad \text{(Hz)} \tag{18.1}$$

or

$$f_c = \frac{R}{2\pi L} \quad \text{(Hz)} \tag{18.2}$$

FIGURE 18–4 RC low-pass filter.

FIGURE 18–5 RL low-pass filter.

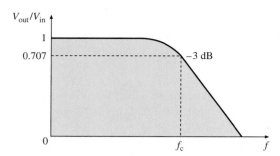

FIGURE 18–6 *Response curve for low-pass filter.*

_____ **EXAMPLE 18–1** _____

Determine the cutoff frequency of the circuit of Figure 18–7.

FIGURE 18–7 *Circuit for Example 18–1.*

Solution

$$f_c = \frac{1}{2\pi RC}$$

$$= \frac{1}{2\pi(2200 \ \Omega)(0.01 \times 10^{-6} \ \text{F})}$$

$$= 7.23 \ \text{kHz}$$

_____ **EXAMPLE 18–2** _____

For the circuit of Figure 18–8 calculate the half-power frequency.

FIGURE 18–8 *Circuit for Example 18–2.*

Solution

$$f_c = \frac{R}{2\pi L}$$

$$= \frac{33 \ \text{k}\Omega}{2\pi(100 \times 10^{-3} \ \text{H})}$$

$$= 52.52 \ \text{kHz}$$

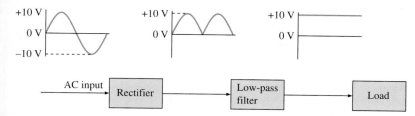

FIGURE 18–9 *A block diagram of a low-pass filter and rectifier.*

Low-pass filters have a wide variety of applications in electronic circuits. One common application is in smoothing the output of a pulsating DC signal. Essentially, the action of a capacitor in an *RC* circuit is to resist a change in voltage across the capacitor. If the direct current in a circuit has a ripple, the action of the *RC* circuit is to suppress or absorb the variations and to deliver a filtered DC current and voltage to the load. Figure 18–9 shows a block diagram of a rectifier, low-pass filter, and load. The rectifier converts the AC signal into pulsating DC, and the filter smooths out the ripple voltage. Rectifiers and filters are used in conjunction with each other in almost all power supply circuits. A more complete discussion of rectifiers occurs in Chapter 19.

Section Review

1. What is a low-pass filter?
2. The frequencies below the cutoff frequency of a low-pass filter are said to be in the passband (True/False).
3. What is the cutoff frequency for a low-pass filter consisting of a 0.68 µF capacitor and a 1 kΩ resistor?

HIGH-PASS FILTERS

A **high-pass filter** allows frequencies above the cutoff frequency to appear at the output, but blocks or shorts to ground all those frequencies below the −3 dB point on the curve. By interchanging the two circuit elements of the low-pass *RC* filter, we obtain the equivalent high-pass filter shown in Figure 18–10. Once again the capacitive reactance is large at low frequencies and small at high frequencies. The capacitor in Figure 18–10 acts as a coupling capacitor. The cutoff frequency for an *RC* high-pass filter is determined using the same equation that is used for the *RC* low-pass filter. That is,

$$f_c = \frac{1}{2\pi RC}$$

FIGURE 18–10 *RC high-pass filter.*

FIGURE 18–11 RL high-pass filter.

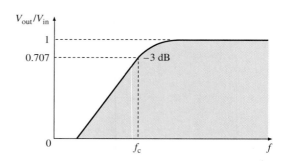

FIGURE 18–12 High-pass filter response curve.

An example of an *RL* high-pass filter is shown in Figure 18–11. In this circuit, there is a high value of inductive reactance at high frequencies. The result is that the high frequencies produce a large output voltage across the inductor. At low frequencies, X_L will be small and the low frequencies will be effectively shorted to ground. It should be noted that *RL* filters are not often used in active circuits because of their size, weight, and losses.

The equation for determining f_c is the same for both the high-pass and low-pass *RL* circuits. That is,

$$f_c = \frac{R}{2\pi L}$$

Figure 18–12 shows the response curve for a high-pass filter circuit. The cutoff frequency on the high-pass response curve is called the *low-cutoff frequency*.

--- **EXAMPLE 18–3** ---

Design an *RL* high-pass filter that will have a cut-off frequency of 1500 Hz.

Solution   Assume $R = 1 \text{ k}\Omega$

$$f_c = \frac{R}{2\pi L}$$

$$L = \frac{R}{2\pi f_c} = \frac{1000 \ \Omega}{2\pi(1500 \ \text{Hz})} = 106.1 \ \text{mH}$$

Section Review

1. What is a high-pass filter?
2. State the equation for finding the critical frequency of a high-pass *RC* filter.
3. In an *RL* high-pass filter, the inductive reactance will be small at high frequencies (True/False).

--- **18–8** ---

BAND-PASS FILTERS

A **band-pass filter** is a circuit that allows a certain range or band of frequencies to pass through it relatively unattenuated. Band-pass filters are designed to have a very sharp, defined frequency response. A band-pass filter is equivalent to combining a low-pass filter and a high-pass filter. The resulting response curve would appear as shown in Fig-

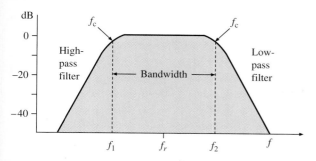

FIGURE 18–13   Response curve of a band-pass filter.

FIGURE 18–14   Series-resonant band-pass filter.

ure 18–13. The cutoff frequency of the high-pass section becomes the lower frequency limit in the passband, f_1. The upper frequency in the passband, f_2, is the result of the cutoff frequency in the low-pass section. The passband, or bandwidth, is the difference between f_2 and f_1. At resonant frequency, f_r, $X_L = X_C$, and the output voltage is equal to the input signal.

An example of a series-resonant, band-pass filter is shown in Figure 18–14. At resonance, the impedance of the circuit is at its minimum. The impedance of the circuit increases as the frequencies increase and decrease from the resonant frequency. As a result, only the frequencies close to resonance are passed. All others are blocked because of high circuit impedance. The resonant frequency for the circuit of Figure 18–14 is found using Equation 17.1. That is,

$$f_r = \frac{1}{2\pi \sqrt{LC}}$$

_____ EXAMPLE 18–4 _____

FIGURE 18–15   Circuit for Example 18–4.

For the circuit of Figure 18–15, calculate the resonant frequency and bandwidth.

Solution

$$f_r = \frac{1}{2\pi \sqrt{LC}} = \frac{1}{2\pi \sqrt{(100 \text{ mH})(0.1 \text{ μF})}} = 1591.55 \text{ Hz}$$

Using Equation 17.4 for series-resonant circuits,

$$Q = \left(\frac{1}{R}\right)\sqrt{L/C} = \left(\frac{1}{100 \text{ Ω}}\right)\sqrt{100 \text{ mH}/0.1 \text{ μF}} = 10$$

$$BW = \frac{f_r}{Q} = \frac{1591.55 \text{ Hz}}{10} = 159.16 \text{ Hz}$$

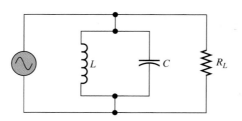

FIGURE 18-16 *Parallel-resonant band-pass filter.*

Figure 18–16 shows a parallel-resonant circuit used as a band-pass filter. As mentioned in Chapter 17, the impedance of a parallel-resonant circuit reaches a peak value when the resonant frequency is applied. At resonance, the high impedance of the parallel · *LC* circuit causes the current to be shunted through the load resistor. When the impedance of the parallel *LC* circuit is low, the current is passed through the reactive components. The magnitude of impedance decreases sharply with frequencies either below or above resonance. The band of frequencies between the half-power points will be developed across the load resistor. All other frequencies will find the *LC* circuit impedance to be quite low. Consequently, the undesirable frequencies are effectively eliminated while the desired band is retained with very little loss in signal amplitude.

_____ **EXAMPLE 18–5** _____

A parallel-resonant band-pass filter has a lower cutoff frequency of 6.3 kHz Hz and an upper cutoff frequency of 8 kHz. Determine the bandwidth and resonant frequency.

Solution

$$BW = f_{c1} - f_{c2} = 8 \text{ kHz} - 6.3 \text{ kHz} = 1.7 \text{ kHz}$$
$$f_r = \sqrt{f_{c1} f_{c2}} = \sqrt{(8 \text{ kHz})(6.3 \text{ kHz})} = 7.1 \text{ kHz}$$

Depending on the design and purpose of the band-pass filter, the bandwidth may be very wide or very narrow. Some of the common uses of band-pass filters are as follows:

1. Audio—to equalize sound levels
2. Communications—to select a narrow band of radio frequencies
3. Audio—to produce speaker crossover networks

Crossover networks are used in loudspeakers because individual speakers are more efficient in some frequency ranges than others. In other words, a large-diameter speaker produces low-frequency signals, or tones, more efficiently than high-frequency signals. Crossover networks prevent any signals outside a certain frequency range from being applied to the speaker. If a loudspeaker system has only one crossover frequency, it is called a *two-way system* because it divides the signal into two bands. A *three-way system* has two crossover frequencies and divides the signal into three bands.

Figure 18–17 shows an example of a three-way crossover network. In this system, the low-pass filter passes signals below 630 Hz to the 12 in. bass speaker, or woofer. The band-pass filter, consisting of a high-pass and low-pass filter, passes signals between 630 Hz and 8 kHz to the 5 in. midrange speaker. The high-frequency signals are filtered by the high-pass filter, and frequencies above 8 kHz are passed on to the tweeter, which is a small diaphragm-type speaker designed for high-frequency applications.

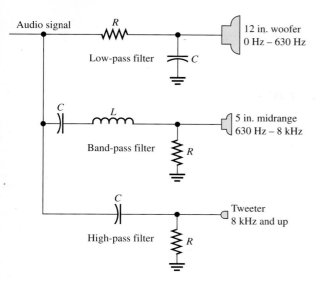

FIGURE 18–17 *Passive three-way crossover.*

The disadvantage of the circuit shown in Figure 18–17 is that there is significant power loss caused by the passive filters. In other words, the insertion loss is quite high in these circuits. For this reason, active filters are often used to isolate frequencies in audio electronic circuits.

Section Review

1. What is a band-pass filter?
2. Band-pass filters are designed to have a very defined resonant frequency (True/False).
3. How many crossover frequencies does a three-way audio system have?

18–9

BAND-STOP FILTERS

A **band-stop filter** rejects signals at frequencies within a specified band and passes signals at all other frequencies. A band-stop filter is also referred to as a band-reject, notch, wave-trap, band-elimination, or band-suppression filter. Like the band-pass filter, the band-stop filter may also be formed by combining a low-pass and high-pass filter. The band-stop filter is designed so that the cutoff frequency of the low-pass filter is below that of the high-pass filter, as shown in Figure 18–18. In this response curve, the band-pass filter rejects frequencies between 100 Hz and 10 kHz.

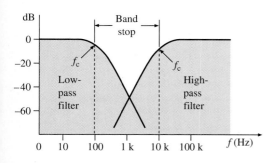

FIGURE 18–18 *Response curve of a band-stop filter.*

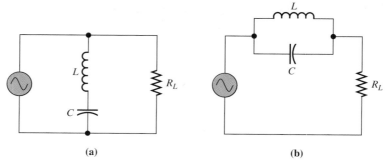

FIGURE 18–19 *Resonant band-reject filters. (a) Series resonant. (b) Parallel resonant.*

Band-reject filters are popular in a variety of applications, particularly audio electronics. These filters can be used in groups to provide equalization of the entire audio spectrum. For example, a typical one-third octave equalizer will contain 28 band-rejection filters. Each filter section is designed to remove a specific band of frequencies and provides up to 15 dB attenuation at its resonant frequency. The center frequencies typically range from 31.5 Hz to 16,000 Hz.

Figure 18–19(a) shows an example of a series-resonant band-reject filter. When a series-resonant *LC* circuit is connected in parallel with a load resistor, the *LC* network provides a low-impedance shunt path. At resonance, the *LC* circuit acts as a short circuit and the signal does not pass through R_L. At frequencies above or below resonance, the *LC* circuit offers high resistance and current flows through the load. Figure 18–19(b) shows a parallel-resonant band-reject filter. In this circuit, when the parallel *LC* network is at resonance, the impedance is very high and virtually no current flows to the load. The impedance decreases above and below resonance, so all other signals are passed through the filter.

EXAMPLE 18–6

For the circuit shown in Figure 18–20, calculate the resonant frequency.

FIGURE 18–20 *Circuit for Example 18–6.*

Solution

$$f_r = \frac{1}{2\pi \sqrt{LC}} = \frac{1}{2\pi \sqrt{(200 \text{ mH})(0.01 \text{ μF})}} = 3.56 \text{ kHz}$$

FIGURE 18–21 *Tank circuit used as a band-stop, or band reject, filter.*

A parallel-resonant circuit can also be made to function as a rejection filter by connecting a resistor in series with a tank circuit, as shown in Figure 18–21. In this circuit, the band of frequencies between the half-power points is blocked by the high impedance of the tank circuit. In this band, frequencies cause little or no change of current through the output resistor. Consequently, these frequencies are effectively eliminated from the output.

Section Review

1. What is a band-stop filter?
2. Explain the difference between a band-stop filter and a band-reject filter.
3. What is the resonant frequency of a band-stop filter consisting of a 50 mH inductor and a 0.047 μF capacitor?

18–10

BODE PLOTS

Figure 18–22 represents a straight-line approximation of the magnitude and phase angle response to frequency for a low-pass filter. These approximations of actual curves are called **Bode plots,** named after Hendrik W. Bode. The sharp corners, which occur at cutoff frequency, are called **breakpoints.** The breakpoints are the points at which the gain drops to 3 dB below the mid-frequency gain. The mid-frequency gain of the plot is the center section that has a fairly constant gain.

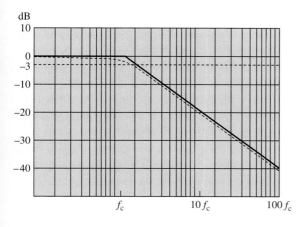

FIGURE 18–22 **Bode plot for low-pass filter.**

To properly draw a Bode plot, you must use *semilog graph paper*. The **semilogarithmic scale** has linear graduations on the vertical axis and logarithmic graduations on the horizontal axis. The linear axis is labeled in units of decibels and the log axis is labeled in units of frequency, radians per second, or angular velocity. The Bode diagram is plotted by drawing the mid-frequency gain from breakpoint to breakpoint. From each breakpoint straight lines are then drawn with slopes of −20 dB per decade. A **decade** represents a tenfold change in frequency. Therefore, the filter response of the Bode plot of Figure 18–22 is said to slope, or roll off, at 20 dB per decade.

A slope of −20 dB/decade is also equal to a change of −6 dB/octave. An **octave** is a doubling of frequency. For example, if the frequency changes from 1000 to 2000 Hz, it has changed one octave. The *attenuation rate* of a filter signifies how decisive the filter action is. In Bode plots, the attenuation rate is given in terms of dBs per octave, and is also known as *roll off.*

Section Review

1. What is a Bode plot?
2. What are breakpoints?
3. A decade represents a tenfold change in frequency (True/False).
4. What is the attenuation rate of a filter?

— 18–11

ACTIVE FILTERS

Active filters are rapidly replacing *RC* and *RL* passive filters in almost every application. Although passive *LC* filters are capable of producing sharp attenuation curves, the inductors are bulky and expensive. An increasingly viable alternative is the **active filter,** which has the following characteristics:

1. High input resistance and low output resistance
2. No attenuation in the passband
3. Possibility of filtering without using inductors

Active filters generally use **operational amplifiers (op-amps)** as the active device. An op-amp consists of many direct-coupled transistors built into one integrated circuit (IC). The introduction of the op-amp has completely eliminated the need for inductors in filter circuits, resulting in a large reduction in space and cost. The design of active filters to fit several specifications has been reduced to a matter of looking up circuit component values in tables. At low frequencies active filters are far superior to passive filters. However, active filters require one or more power supplies to function properly. Because they are active, these types of components are usually less reliable than passive circuits. A complete discussion of op-amps and active filters is presented in Chapter 22.

Section Review

1. Active filters have low-input resistance and high-output resistance (True/False).
2. At low frequencies, active filters are far superior to passive filters (True/False).

PRACTICAL APPLICATION SOLUTION

In the chapter opener, one of your responsibilities as an electronic technician in a large medical center was to test the band-stop filters in video monitors requiring servicing. To successfully complete this practical application of coupling and filter circuits, the following steps are followed.

STEP 1   Measure and record the resonant frequency using an oscilloscope or frequency counter.

STEP 2   Calculate the resonant frequency.

$$f_r = \frac{1}{2\pi \sqrt{LC}} = \frac{1}{2\pi \sqrt{(2 \ \mu H)(0.001 \ \mu F)}} = 3.56 \ \text{MHz}$$

STEP 3   Compare the value determined in Step 2 with the measured value in Step 1. If the two values are not approximately equal, replace the capacitor and retest the circuit.

_____ *Summary* ___

1. Direct coupling transmits the AC and DC components of a signal.
2. Capacitive coupling blocks DC signals but allows AC signals to pass.
3. Capacitive coupling transfers unwanted harmonic signals in addition to the fundamental frequency.
4. Transformer coupling provides good impedance matching and gain and allows for a tuned bypass.
5. Filters are used to block or pass a specific range of frequencies.
6. Low-pass filters have a constant output voltage up to the cutoff frequency.
7. A high-pass filter has a constant output voltage above the cutoff frequency.
8. Band-pass filters allow a certain range of frequencies to pass, while band-stop filters attenuate a specific band of frequencies.
9. Bode plots represent straight-line approximations of the frequency response of a filter circuit.
10. Active filters have high-input resistance, low-output resistance, and no attenuation in the passband.

_____ *Answers to Section Reviews* ___

Section 18–1

1. True
2. The common impedance between two circuits and basically a device used to achieve optimum transfer of power between electronic circuits
3. True

Section 18–2

1. When the current of the input circuit flows through a common impedance where it creates a voltage drop

2. False

3. True

Section 18–3

1. The difficulty created in making a good impedance match between circuits

2. False

3. The capacitive reactance should be less than 10% of the resistance for capacitive coupling to occur.

Section 18–4

1. True

2. Because it provides good impedance matching and gain and allows for a tuned bypass

3. True

Section 18–5

1. True

2. Filters that do not use an amplifying device

3. False

Section 18–6

1. A circuit that has a constant output voltage up to a cutoff, or critical, frequency

2. True

3. $f_c = 1/[2\pi(1\ k\Omega)(0.68\ \mu F)] = 234$ Hz

Section 18–7

1. A circuit that allows frequencies above the cutoff frequency to appear at the output, but blocks all other frequencies

2. $f_c = 1/2\pi RC$

3. False

Section 18–8

1. A circuit that allows a certain range of frequencies to pass through it unattenuated

2. True

3. Two

Section 18–9

1. A filter that rejects signals at frequencies within a specified band and passes signals at all other frequencies

2. No difference

3. $f_r = 1/[2\pi\ \sqrt{(50\ mH)(0.047\ \mu F)}] = 3.28$ kHz

Section 18–10

1. A straight-line approximation of a frequency response curve

2. The sharp corners that occur at cutoff frequency

3. True

4. The attenuation rate of a filter signifies how decisive the filter action is.

Section 18–11

1. False

2. True

Multiple Choice Questions

18–1   A filter is a device that
(a) Is used to achieve minimum transfer of power between electronic circuits
(b) Removes unwanted signals
(c) Has no physical connection between the input and output
(d) Has no insertion loss

18–2   There are two types of coupling:
(a) High pass and low pass   **(b)** Band pass and band stop
(c) Direct and indirect   **(d)** Capacitive and transformer

18–3   The main disadvantage of capacitive coupling is
(a) Frequency response   **(b)** Inefficient power transfer
(c) Both AC and DC signals are passed   **(d)** High frequencies are blocked

18–4   Filters are classified as either
(a) High pass or low pass   **(b)** Band pass or band stop
(c) _RC_ or _RL_   **(d)** Active or passive

18–5   The difference between the power received at the load before installation of a filter and the power after installation is called
(a) Insertion loss   **(b)** Stop band   **(c)** Attenuation   **(d)** Gain

18–6   A low-pass filter is a circuit that has
(a) A constant voltage above the cutoff frequency
(b) Two critical frequencies
(c) A constant voltage up to the cutoff frequency
(d) A short circuit at resonance

18–7   When the output voltage of a low-pass filter is 0.707 of input, the power output is
(a) 25% of the input power   **(b)** 50% of the input power
(c) Equal to the input power   **(d)** Double the input power

18–8   A high-pass filter
(a) Blocks signals above the −3 dB point
(b) Allows frequencies below f_c to pass
(c) Has a low-value of capacitive reactance at low frequencies
(d) Allows frequencies above f_c to appear at the output

18–9   At resonance, the impedance of a series-resonant band-pass filter is
(a) At its minimum   **(b)** Equal to the load impedance
(c) At its maximum   **(d)** The same as a parallel-resonant band-pass filter

18–10   A band-stop filter
(a) Passes signals at frequencies within a specified band and rejects signals at all other frequencies
(b) Has only one cutoff frequency
(c) Has no insertion loss
(d) Rejects signals at frequencies within a specified band and passes signals at all other frequencies

18–11   An octave is
(a) A tripling of frequency   **(b)** A tenfold change in frequency
(c) A doubling of frequency   **(d)** The same as a decade

18–12 In Bode plots, the attenuation rate of a filter is measured in
(a) dB/decade   **(b)** dB/octave   **(c)** dB/volt   **(d)** dB/hertz

18–13 Active filters do not use
(a) Inductors   **(b)** Capacitors   **(c)** Resistors   **(d)** Op-amps

18–14 Which of the following properties is associated with an active filter?
(a) Low-input resistance and high-output resistance
(b) High attenuation in the passband
(c) Filtering impossible without inductors
(d) High-input resistance and low-output resistance

18–15 A decade represents a
(a) Tenfold change in frequency   **(b)** 100% increase in frequency
(c) 1/10 change in frequency   **(d)** 1/100 change in frequency

Practice Problems

18–16 Determine the cutoff frequency of the circuit of Figure 18–23.

$R = 1\ k\Omega$

FIGURE 18–23

$C = 0.1\ \mu F$

18–17 An RL high-pass filter consists of a resistor $R = 470\ \Omega$ and an inductor $L = 50$ mH. Calculate the cutoff frequency.

18–18 For the circuit of Figure 18–24, calculate the half-power frequency.

$L = 50$ mH

FIGURE 18–24

$R = 100\ k\Omega$

18–19 Design an RL high-pass filter that will have a cutoff frequency of 8 kHz.

18–20 A band-pass filter is to be constructed using a high-pass RC filter and a low-pass RC filter. The low-pass filter has a resistor $R = 2\ k\Omega$ and capacitor $C = 10$ nF. The high-pass filter has $R = 3\ k\Omega$ and $C = 40$ nF. Determine the bandwidth of the circuit.

18–21 For the circuit of Figure 18–25, calculate the resonant frequency and bandwidth.

$C = 0.68\ \mu F$ $L = 25$ mH

FIGURE 18–25

$R_L = 20\ \Omega$

18–22   A parallel-resonant band-pass filter has a lower cutoff frequency of 7.5 kHz and an upper cutoff frequency of 10 kHz. Determine the bandwidth and resonant frequency.

18–23   For the circuit shown in Figure 18–26, calculate the resonant frequency.

FIGURE 18–26

Essay Questions

18–24   Define the terms *filter* and *coupler*.

18–25   Describe the two types of coupling used in electronic circuits.

18–26   What is capacitive coupling?

18–27   Why is transformer coupling more efficient than capacitive coupling?

18–28   List the four main types of filters.

18–29   Define insertion loss.

18–30   Name one common application for a low-pass filter.

18–31   What is the difference between a band-pass filter and a band-stop filter?

18–32   List three characteristics associated with an active filter.

19

Semiconductor Fundamentals

Learning Objectives

Upon completion of this chapter you will be able to

- Explain the atomic structure of semiconductors.
- Differentiate between P-type and N-type semiconductors.
- Define the term *PN junction.*
- Describe how a PN junction is forward biased and reverse biased.
- Name the two leads of a semiconductor diode.
- Explain the purpose of diode ratings.
- Troubleshoot diodes and rectifier circuits.
- Understand the basic operation of half-wave and full-wave rectifiers.
- Differentiate between percent voltage regulation and percent ripple.
- Describe the operating characteristics of zener diodes.
- Name two types of optoelectronic devices and explain their operation.

As a service technician for a large manufacturer of consumer electronic products, you are required to test the average output voltage of a power supply consisting of a silicon full-wave bridge rectifier with a 560 μF capacitor connected in parallel with a 100 Ω load. Your task is to calculate the average output voltage and compare this value with the value measured with a DC voltmeter. The full-wave bridge rectifier is connected to a step-down transformer that converts the 120 V supply to a 12 V_{rms}, 60 Hz signal that is fed to the rectifier. Include the voltage drop across the diodes in your calculation.

19-1
INTRODUCTION

In this chapter on semiconductor fundamentals, you will learn how and why semiconductors are used in electronics. A semiconductor material is a material having a conductivity lower than that of a conductor, but higher than that of an insulator. The electrons in the outer shell of the atoms of a semiconductor can be removed when some form of energy is applied to the material. The energy may be in the form of light, heat, or an electric field. Components that are made of semiconductor materials are often called *semiconductor devices,* or *solid-state components* because they are made from solid materials such as germanium or silicon. The three most commonly used semiconductor devices are diodes, transistors, and integrated circuits.

Section Review

1. Semiconductors have a conductivity lower than that of an insulator, but higher than that of a conductor (True/False).
2. What is a solid-state device?
3. List the three most commonly used semiconductor devices.

19-2
ATOMIC STRUCTURE OF SEMICONDUCTORS

The semiconductors used in electronic devices are used in their crystalline form. In other words, the atoms are joined together in a regular pattern that is repeated over and over. This repeated pattern is called a *crystal lattice*. The crystal lattice is the basic building block of solids. It consists of a three-dimensional arrangement of atoms, which in its simplest form takes the shape of a cube. This simple cubic structure has an atom located at each corner. Figure 19-1 shows a three-dimensional model of a crystal lattice.

The fundamental materials used for semiconductors are elements from Group IV of the periodic table. Group IV elements are characterized by having eight permissible energy states in the outer subshell and four electrons in the outer subshell. To become a stable grouping of atoms that form molecules, these elements must either lose four electrons, gain four electrons, or share four electrons. The result is that atoms of **silicon** or **germanium**

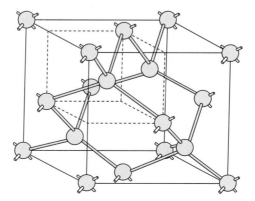

FIGURE 19-1 *Three-dimensional model of a crystal lattice.*

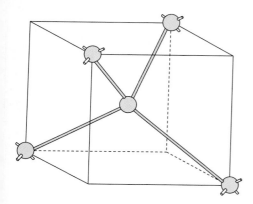

FIGURE 19-2 *Covalent bonding in a silicon or germanium crystal.*

neither release nor acquire electrons. Instead, they enter into a sharing of electrons called *covalent bonding.* Figure 19-2 shows an example of covalent bonding in a silicon or germanium crystal. The basic structure is governed by the fact that each atom will form four covalent bonds with four of its nearest neighbors.

In Chapter 1, a copper conductor was shown to have an abundance of conduction band electrons. These electrons are loosely bound to parent atoms and move freely under the influence of any external electric field. In contrast an insulator, such as ceramic, has virtually all its electrons tightly bound to parent atoms. We also stated that a conductor will release free electrons with very little additional energy, but insulators require large amounts of added energy to release free electrons. Free electrons are the basis of conductivity. Because the normal position of the electron is in its orbit, it is said to be in its chemically ready, or valence state. When a material is heated, electrons are released from their valence state, or band of energy, and are said to move into their conduction state, or conduction band of energy. Each thermally excited electron produces two charge carriers, one positive and one negative—the electron itself and the hole from which it came. These opposite charges can recombine within the crystal to form a covalent bond.

The net electrical charge of a pure silicon or germanium crystal is zero. Although the valence electrons of one atom pair up with those of its neighbors to form covalent bonds, the positive charges of its nucleus still neutralize the negative charges of its electrons. The number of electron-hole pairs developed within a semiconductor material will increase with temperature. As the temperature of the semiconductor material is increased, the valence electrons become agitated and some of them will break away from the covalent bonds. As a result, a small percentage of electrons will be free to drift from one atom to another in a random direction. Some of these free-moving electrons are absorbed by holes, and the electron simply jumps from one shell to another shell that contains a hole. This process can continue so that the hole drifts throughout the crystal lattice structure and, in doing so, conducts electricity. Because a hole is always created when an electron leaves a shell, the hole appears to move in the opposite direction of the electron.

Although most semiconductor devices are made of silicon crystals, pure silicon cannot be used because it would act as an insulator. When a semiconductor is in its purest form, it is referred to as an **intrinsic semiconductor.** The high-purity silicon is then processed by adding certain chemical elements, or impurities. The process of adding impurities is called **doping.** These impurities are added in precise amounts to alter the electrical conductivity of the material. For example, doping germanium with either the element indium or antimony will reduce the resistance of the material. The conductivity is then determined by the volume density of the added element and is referred to as **extrinsic conductivity.**

Section Review

1. A crystal lattice is the basic building block of solids (True/False).
2. What is covalent bonding?
3. What is the net electrical charge of a pure silicon or germanium crystal?
4. What is an intrinsic semiconductor?
5. What is doping?

— 19–3

P-TYPE AND N-TYPE SEMICONDUCTORS

Conduction resulting from impurities is the basic principle of current flow in semiconductor devices. Semiconductor properties can be obtained by either adding impurities with one more valence electron or one less valence electron. For example, germanium has four valence electron. If an impurity, such as antimony, with five valence electrons is added to germanium, four of the five electrons will bond. The fifth electron will be held in place by the relatively weak electric charge of the nucleus. Consequently, these extra electrons are essentially free to roam through the crystal lattice structure and conduct electricity. A semiconductor having an excess of electrons is called *N-type* because the conduction is by electrons or negative carriers.

When an impurity has five valence electrons, it is called a **pentavalent impurity.** Arsenic, antimony, bismuth, and phosphorous are all pentavalent impurities. These materials are also known as *donor* impurities because these materials give up, or donate, one electron.

It is also possible to create a semiconductor by using impurities having one less valence electron. Indium is commonly used with germanium because indium has one less valence electron. As an indium atom enters the crystal lattice structure, there will be a shortage of one electron, and a hole is created in the covalent pair bonds. This type of impurity tends to compensate for its deficiency of one valence electron by acquiring an electron from its neighbor. Because a hole represents the absence of an electron, it acts as a positive charge and the semiconductor is called *P-type*.

As previously mentioned, whenever an atom loses an electron, a nearby bound electron may receive enough energy from some external source to jump into the position left by the lost electron. Impurities such as indium, gallium, and boron have only three valence electrons and are called **trivalent impurities.** These materials are also referred to as *acceptor impurities* because they accept valence electrons. As acceptor impurity atoms are added to silicon or germanium, the number of positive charge carriers (holes) is increased. It is interesting to note that hole flow exists only in the valence band of the semiconductor, and electron flow exists only in the conduction band.

Both holes and electrons are current *carriers*. Electrons are negative carriers and holes are positive carriers. The current carrier that is present in the largest quantity represents the **majority carrier** and the smaller carrier is known as the **minority carrier.** A typical N-type material has a large number of free electrons and a small number of holes. The electrons in the N-type material are considered majority carriers because the major portion of any current is electron flow, and the holes are called minority carriers. In a P-type material, the holes are dominant because of the effect of the impurity. Consequently, the holes are the majority carriers, and the relatively small number of free electrons, created by normal thermal energy, are the minority carriers.

Conduction in an N-type semiconductor is similar to that of a copper conductor. When a voltage is applied to the semiconductor material, conduction will exist through the

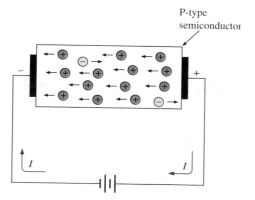

FIGURE 19–3 *Current flow in an N-type semiconductor material.*

FIGURE 19–4 *Current flow in a P-type semiconductor material.*

crystal and will occur as a result of carriers moving through the conduction band. However, fundamental differences exist between an N-type semiconductor and a copper conductor. One of these differences is that semiconductors have a negative temperature coefficient. When a material has a negative temperature coefficient, an increase in temperature causes a decrease in resistance. Current flow through an N-type material is shown in Figure 19–3. In this circuit, electrons are the majority carriers, although some minority carriers, or hole current, also exist. As a result, the total current in N-type material is the sum of the electron current in the conduction band and the hole current in the valence band.

Current flow through a P-type material is illustrated in Figure 19–4. The majority carriers in this type of material are holes, or positive charge carriers that shift toward the negative terminal of the supply voltage. In hole flow, the movement of electrons into holes near the positive terminal of the voltage source results in a continual movement of holes toward the negative terminal of the source. In Figure 19–4, the flow of electrons in the external circuit consists of electrons moving out of the negative terminal and into the positive terminal.

Section Review

1. What is a pentavalent impurity?
2. Why are pentavalent impurities also called donor impurities?
3. Holes and electrons are both current carriers (True/False).
4. What are the majority carriers in a P-type semiconductor material?

19–4
THE PN JUNCTION

Rectification and amplification using semiconductor devices is made possible as a result of the barrier existing between two types of impure silicon. A piece of N-type material, or P-type material, has no new properties when placed in an electrical circuit. It generally behaves in the same manner as an intrinsic material. For example, when a voltage is applied to the material, a drift current is produced, and the drift current may heat the material causing the material resistivity to decrease. The amount of drift current is greater in the extrinsic material because the resistivity is lower.

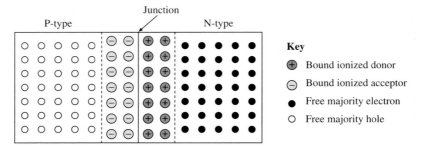

FIGURE 19–5 *Depletion region barrier produced by formation of a PN junction.*

When a single piece of semiconductor material is formed with both types of extrinsic materials, a **PN junction** is created. The junction formation requires that the material be started as one type and that the impurity be changed to the other type, forming a molecular transition. At the boundary where a PN junction is formed, there is a difference of energy levels and, consequently, a movement of charges takes place. The N-type material has a higher energy level and the P-type material has a lower energy level. As a result, the loosely bound valence electrons in the N-type material move across the boundary and combine with the holes provided by the P-type material. This movement creates a potential difference across the junction. Remember that both P- and N-type materials are electrically neutral when kept apart, but when a junction is formed, the P-side of the junction acquires electrons and, therefore, a negative charge, and the N-side acquires a positive charge because of the electrons that left to fill the holes in the P-type material. This interchange of holes and electrons cancels out the current carriers in the vicinity of the junction, forming a **depletion region.**

Figure 19–5 shows an example of a PN junction with a depletion region. The only charges in the depletion region are those associated with the impurity atoms, or ions, that have lost their neutralizing carriers. These ions are fixed within the crystal structure and, therefore, are unable to contribute to current flow. Throughout the remainder of the crystal, each bound charge, donor or acceptor, is compensated for by a free majority carrier, producing a neutral charge everywhere but in the depletion region.

The ions located in the depletion region of Figure 19–5 are said to be *uncompensated.* In other words, they are no longer neutralized by an equal number of carrier charges of opposite polarity. Consequently, there is a net negative charge in the depletion region on the P-side of the junction and a net positive charge in the depletion region on the N-side of the junction. The difference of potential, or charge, at the junction is called the **barrier potential.** At ambient temperature, the barrier potential for silicon is approximately 0.7 V, and for germanium it is 0.3 V. The barrier potential behaves inversely with temperature; as the temperature increases, the barrier voltage decreases.

Section Review

1. How is a PN junction created?
2. The ions located in the depletion region of a PN junction are said to be compensated (True/False).
3. What is barrier potential?

BIASING THE PN JUNCTION

Any voltage that is connected across a PN junction is called a **bias voltage.** When a bias voltage is connected across a PN junction, as shown in Figure 19–6, the polarity of the source forces the majority carriers toward the junction. This action reduces the extent of the depletion region, and permits majority carriers to cross the junction. When electrons cross the PN junction, they recombine with holes in the P-type material. For each recombination that occurs, an electron leaves the bias voltage source and enters the N-type material. At the same instant, an electron is attracted away from the P-type material and returns to the voltage source. This bias condition is characterized by high conduction and is referred to as **forward bias** of the PN junction. The resulting current is called *forward current*. Because the resistance of a forward-biased PN junction is quite low, a current limiting resistor, R, is used in the circuit of Figure 19–6.

Figure 19–7 illustrates the characteristic curve for a forward-biased silicon PN junction. The forward current flowing through the junction is shown on the vertical axis and the voltage on the horizontal axis. Until the voltage reaches 0.7 V, the curve is shown to be quite linear. That is, the resistance and current increase proportionately. However, once the barrier or **breakover voltage** is exceeded, the current increases dramatically and the voltage drop remains fairly constant. Although this appears to violate Ohm's law, it does not because the resistance of the PN junction is *decreasing* as the current increases. This is one of the unique qualities of semiconductor materials. The total resistance of a PN junction is actually the sum of two resistances: the *bulk resistance* and the *junction resistance*. The bulk resistance represents the resistance of the basic semiconductor material. The junction resistance is the resistance of the barrier region and will vary inversely as the current through the junction varies.

When the positive terminal of the voltage source of Figure 19–8 is connected to the N-side of a PN junction and the negative terminal to the P-side, the PN junction is said to be **reverse biased.** The barrier region contains no free carriers and it has a much higher

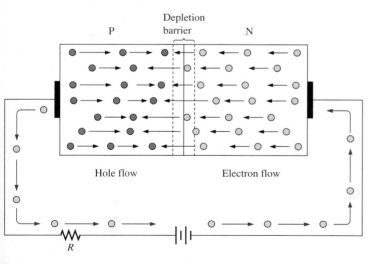

FIGURE 19–6 *Forward-biased PN junction.*

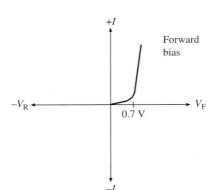

FIGURE 19–7 *Characteristic curve for forward-biased PN junction.*

resistance than the remainder of the crystal on either side of the barrier. The external voltage source in Figure 19–8 attracts electrons and holes away from the junction, producing a wider ion layer at the junction. This causes the depletion layer to widen, which charge carriers must overcome in traversing the PN junction. Because the majority carriers are concentrated further away from the junction, majority current will not flow. The small amount of current that does flow is the result of minority carriers and is often called **leakage current.**

When reverse bias is increased to values beyond the tolerances of the semiconductor material, minority-carrier electrons acquire sufficient energy in crossing the boundary region to release valence-band electrons out of the semiconductor atoms. These electrons add to the normal electron flow and rapidly increase the current flow through the PN junction. This phenomenon is called **avalanche breakdown.** Figure 19–9 shows the characteristic curve for a reverse-biased PN junction. As the applied voltage is increased, the acceleration of majority carriers is also increased. At a certain critical value of voltage, known as the **breakdown voltage,** V_{BR}, the acceleration is so great that the force of impact between carriers and atoms is large enough to break some of the covalent bonds, resulting in more electrons and holes. This process is cumulative and results in a sharp increase in reverse current flow.

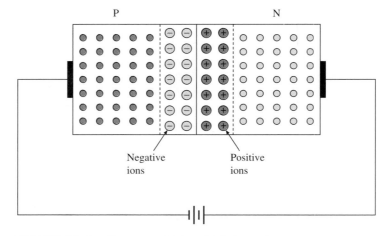

FIGURE 19–8 *Reverse-biased PN junction.*

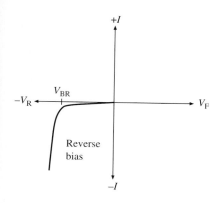

FIGURE 19–9   *Characteristic curve for reverse-biased PN junction.*

Section Review

1. What is a bias voltage?
2. The resistance of a forward-biased PN junction is quite low (True/False).
3. What is the bulk resistance of a PN junction?
4. Define leakage current.

19–6

SEMICONDUCTOR DIODES

A semiconductor **diode** is a device comprised of a single PN junction with typically glass or metal packaging. Diodes that are designed for large values of current and high-power dissipation are mounted on metal studs for connection to heat sinks. Figure 19–10 shows some typical silicon and germanium diodes. The earliest types of diodes used in electronic circuits were manufactured using electron tubes. Semiconductor diodes are much more

FIGURE 19–10   *Silicon and germanium diodes.*

FIGURE 19–11 *(a) Diode schematic symbol. (b) Equivalent PN junction. (c) Band identifying cathode. (d) Diode symbol marked on diode casing.*

efficient than tube types, have higher current-handling capabilities, and are considerably less expensive. Consequently, tube-type diodes are now virtually obsolete and semiconductor diodes are used almost exclusively in electronics applications.

Figure 19–11(a) shows the schematic symbol for a semiconductor diode. The two leads of the device are labeled **anode** and **cathode.** The anode is composed of P-type material and the cathode is constructed of N-type material. The equivalent PN junction is illustrated in Figure 19–11(b). The cathode marking on the diode is identified by a band, as shown in Figure 19–11(c). Diodes with large current-handling capabilities are often marked with the schematic symbol for a diode, as indicated in Figure 19–11(d).

Diode ratings represent the limiting values assigned by a manufacturer to various parameters of a diode. These parameters include temperature, voltage, current, and power. If any of these assigned values are exceeded, permanent damage may result to the diode. Table 19–1 shows some typical ratings for diodes. The seven diodes in this table are all prefixed by "1N," such as 1N4001 or 1N4004. Most diodes are identified by 1N, regardless of the manufacturer. The reverse voltage capability of a diode is often called the *peak inverse voltage* (PIV) or the *peak reverse voltage* (PRV). This voltage rating represents the maximum allowable instantaneous reverse voltage that may be applied across the diode under specific operating conditions. The average DC current, I_0, is shown as 1A for the

TABLE 19–1 *Diode ratings for typical silicon diodes.*

	Characteristic	1N4001	1N4002	1N4003	1N4004	1N4005	1N4006	1N4007	Unit
PIV	Peak reverse voltage from −65° C to +175° C	50	100	200	400	600	800	100	V
V_R	Maximum reverse voltage from 25° C to 75° C	50	100	200	400	600	800	100	V
I_0	Average DC forward current at 25° C				1				A
I_F	Peak forward current at (or below) 75° C				10				A
T_A	Ambient temperature range				−65 to +175				°C
V_F	Forward voltage drop				0.8				V
I_R	Leakage current				30				μA

diodes in Table 19–1. The peak forward current, I_F, represents the maximum forward current that the diode can sustain at or below 75° C. The forward voltage, V_F, indicates the average forward voltage drop across the diode. The diodes in Table 19–1 are made of silicon and have an average forward voltage drop of 0.8 V. I_R represents reverse, or leakage, current of the diode. As mentioned in Section 19–5, a reverse-biased PN junction will exhibit a small amount of leakage current. The diodes in Table 19–1 have typical leakage currents of 30 μA.

Most diode characteristics are affected by temperature. The forward voltage drop and reverse leakage current are particularly susceptible to temperature variations. In a typical diode, a 10° C increase in temperature will cause the reverse current to nearly double. The forward voltage drop and temperature of the diode are inversely proportional; as the temperature increases, the voltage drop decreases. Typical diode ratings are specified at ambient temperature, or 25° C.

When a diode is forward biased, as shown in Figure 19–12(a), the forward breakover voltage has been exceeded and current flows through the diode. Ideally, a forward-biased diode will have a voltage drop of 0.7 V if it is made of silicon and 0.3 V if made of germanium. A forward-biased diode behaves in the same manner as the closed switch in Figure 19–12(b). When the switch is closed, maximum current is supplied to the load. For a diode to be forward biased, the anode must be more positive than the cathode by a value of at least the breakover voltage. This does not mean that the anode must have a positive voltage and the cathode a negative voltage. As long as the anode has a positive potential equal to or greater than the breakover voltage, with respect to the cathode, the diode will conduct current. The circuit of Figure 19–12(a) is represented by the following Kirchhoff voltage equation:

$$E_T = V_F + V_L$$

The equivalent circuit of Figure 19–12(b) assumes that the voltage across the diode is negligible, and the Kirchhoff equation is

$$E_T \approx V_L$$

Figure 19–13(a) shows a reverse-biased diode. In this circuit, the anode is more positive than the cathode, and current flow is effectively blocked. The diode will prevent the flow of current to the load as long as the PIV rating of the diode is not exceeded. For example, if the PIV of the diode is 50 V and a voltage source of 100 V is connected so that the diode is reverse biased, the diode will probably be permanently damaged. The open switch of Figure 19–13(b) is the equivalent of a reverse-biased diode. Although all diodes have a small value of leakage current, the equivalent circuit of Figure 19–13(b) indicates an infinite value of resistance, and no current flows in the circuit.

(a)   (b)

FIGURE 19–12 *(a) Forward-biased diode. (b) Equivalent circuit.*

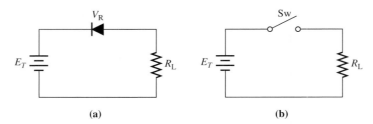

(a)   (b)

FIGURE 19–13 *(a) Reverse-biased diode.*
(b) Equivalent circuit.

───── **EXAMPLE 19–1** ─────

For the circuit of Figure
19–14, calculate the value
of current flowing through
the load resistor.

FIGURE 19–14 *Circuit for Example 19–1.*

Solution   Because the diode is made of silicon, a forward voltage drop of 0.7 V is assumed. The voltage dropped across the load is found using Kirchhoff's voltage law.

$$V_{RL} = E_T - V_F = 5 \text{ V} - 0.7 \text{ V} = 4.3 \text{ V}$$

The current flowing through the load is found by Ohm's law.

$$I_L = \frac{V_{RL}}{R_L} = \frac{4.3 \text{V}}{10 \text{ }\Omega} = 0.43 \text{ A}$$

───── **EXAMPLE 19–2** ─────

The circuit of Figure
19–15 shows a diode with
a leakage current of 50 µA.
Determine the voltage
drop across the load.

FIGURE 19–15 *Circuit for Example 19–2.*

Solution   The voltage dropped across R_L when the diode is reverse-biased is solved using Ohm's law.

$$V_{RL} = I_R R_L = (50 \text{ µA}) (22 \text{ k}\Omega) = 1.1 \text{ V}$$

Section Review

1. What is a semiconductor diode?
2. A cathode is constructed using N-type semiconductor material (True/False).
3. What are diode ratings?

19–7

TROUBLESHOOTING DIODES

When troubleshooting diodes, in most cases we are testing for continuity. A properly functioning diode will conduct current in one direction only and block current in the other direction. As with virtually any electronic device, a diode will generally fail as either an open circuit or a short circuit. Some of the most common causes of diode failure are excess current, temperature, and reverse voltage.

All semiconductor devices are subject to the possibility of *thermal runaway*. As the temperature of a diode increases, its internal resistance decreases, allowing more current to flow through the device. An increase in current, results in more heat being dissipated and an even further decrease in resistance. This cycle can continue until the semiconductor material begins to burn and the device fails. As long as semiconductor devices are operated within their rated values, the possibility of device failure is greatly reduced.

The most accurate method of testing a diode is with the device removed from the circuit. Although diodes can be tested "in-circuit" with biasing voltages applied, inaccurate voltage readings may result. When a diode is tested in-circuit using an ohmmeter, it may be difficult to detect an open circuit if any other devices are connected in parallel with the diode.

Many digital multimeters (DMMs) have a specific diode setting on the meter to allow for out-of-circuit tests of diodes. These meters generally have a diode symbol on the faceplate. When the selector switch is on the diode setting, the red lead is connected to the anode and the black lead to the cathode. An audible "beep" will indicate a forward-biased condition. If the leads of the meter are reversed, there should be a reading of infinity on the meter's seven-segment display. Infinite resistance is often shown as "OL" on the meter. If the meter beeps when the leads are reversed, the diode is shorted out. If the diode reads infinity, or OL, in both directions it implies an open circuit.

Figure 19–16 shows an DMM with a semiconductor setting to test forward- and reverse-biased conditions. In Figure 19–16(a) the meter is connected with the positive lead connected to the anode and the negative lead connected to the cathode. An audible beep is produced by the meter and a small voltage drop (under 1 V) is displayed. When testing for reverse bias, the leads of the meter are reversed, as shown in Figure 19–16(b).

When using a VOM on the ohmmeter setting, a properly functioning diode will typically exhibit a total resistance of several hundred ohms when forward biased and over 2 MΩ when reverse biased. It should be noted that the true resistance of a semiconductor device is difficult to determine with a typical ohmmeter. Some ohmmeters have "range" settings that are used to increase the accuracy of the resistance reading. The disadvantage of these meters when testing semiconductors is that the resistance reading can change dramatically for each setting. Because the ohmmeter is being used to check continuity in the device, a setting that produces a few hundred ohms when the diode is forward biased, and at least 2 MΩ when reverse biased will be sufficient.

When a diode is short circuited, as shown in Figure 19–17, it will produce a reading of zero ohms. When the device offers no opposition to current flow, it has been damaged beyond repair and will require replacement.

FIGURE 19–16 *DMM test of a diode. (a) Forward-biased test. (b) Reverse-biased test.*

An open-circuited diode will produce an infinite value of resistance in both directions, as shown in Figure 19–18. In Figure 19–18(a) the meter is reading infinity when the positive lead is connected to the anode and the negative connected to the cathode. An infinite value of resistance is also shown in Figure 19–18(b) when the leads are reversed.

When replacing a diode, it is important to find a replacement diode with the same operating characteristics. Because diodes have a wide variety of current and voltage ratings, the replacement diode must have equivalent characteristics. Care also must be taken to ensure that the replacement diode is installed in the circuit with the anode and cathode properly positioned. Diodes are designed to conduct current in one direction only. If improperly inserted the new device may be damaged when power is applied to the circuit.

FIGURE 19–17 *Diode with a short circuit. (a) Forward-biased test. (b) Reverse-biased test.*

FIGURE 19-18 *Diode with an open circuit. (a) Forward-biased test. (b) Reverse-biased test.*

Section Review

1. As the temperature of a diode increases, its internal resistance increases (True/False).
2. What is the most accurate method of testing a diode?
3. A shorted diode will produce an ohmmeter reading of zero ohms (True/False).

19-8
HALF-WAVE RECTIFIERS

An ideal diode allows current flow in one direction only. This characteristic enables diodes to be used as rectifiers. As defined in Chapter 2, a rectifier is a device that converts alternating current to direct current. An ideal rectifier diode should have zero forward resistance, infinite reverse resistance, and infinite breakdown voltage. A **half-wave rectifier** is a device that allows current to flow in one direction during the positive half-cycle of input voltage and blocks reverse flow during the negative half-cycle.

Figure 19–19(a) shows a typical half-wave rectifier circuit. Because many loads controlled by semiconductor devices operate at relatively low-power levels, a transformer is used to step down the voltage. The AC input waveform is shown in Figure 19–19(b). As the input voltage begins to rise, the breakover voltage of the diode is reached, and the diode begins to conduct, as shown in Figure 19–19(c). The diode is now acting as a closed switch, and the load voltage is approximately equal to the input voltage. As the input voltage passes the 90° point, the voltage falls toward zero. When the input voltage falls below the breakover voltage, the diode becomes reverse biased and current flow is interrupted. If we ignore the leakage current in the circuit, the diode basically is acting as an open switch. Consequently, the applied voltage is now equal to the voltage dropped across the diode, and the voltage drop across the load is zero. The load voltage waveform of Figure 19–19(d) shows the half-wave output from the rectifier.

When a DC voltmeter is connected across the load resistor in the circuit of Figure 19–19(a), the voltmeter will read 0.3185 of the peak voltage. In Chapter 11, it was shown that if the sine values for all angles up to 180° are added, the average equals 0.637 of the

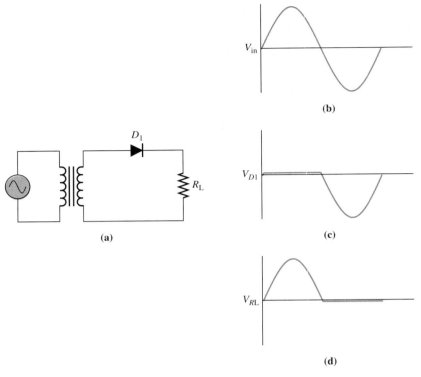

(b)

(c)

(d)

FIGURE 19–19 *(a) Half-wave rectifier. (b) Input signal. (c) Voltage drop across the diode. (d) Voltage drop across the load.*

peak. Because the half-wave rectifier is conducting only for 180° of a possible 360°, the average voltage is one-half, or 0.3185 of V_p. The inverse of 0.3185 is π. Therefore,

$$V_{DC} = \frac{V_p}{\pi} \tag{19.1}$$

The AC input signal for the circuit of Figure 19–19(a) is a sine wave that repeats every 360°. The waveshape appearing across the load also repeats every 360°. The diode conducts for 180° and blocks for 180°, repeating each cycle in synchronism with the input signal. Because frequency represents the number of cycles in a given unit of time, the input frequency and output frequency are considered equal for a half-wave rectifier.

EXAMPLE 19–3

For the circuit of Figure 19–20, determine the DC load voltage. Ignore the voltage drop across the diode and assume a negligible value of leakage current.

FIGURE 19–20 *Circuit for Example 19–3.*

Solution The input voltage on the primary winding of the transformer is given as 120 V_{rms}. The peak voltage is found by dividing the rms by 0.707.

$$V_{p(pri)} = \frac{V_{rms}}{0.707} = \frac{120 \text{ V}}{0.707} = 169.7 \text{ V}$$

Because the circuit of Figure 19–20 uses a 10:1 step-down transformer, the secondary peak voltage is found by dividing the primary voltage by 10.

$$V_{p(sec)} = \frac{V_{p(pri)}}{10} = \frac{169.7 \text{ V}}{10} = 16.97 \text{ V}$$

The DC voltage is now found using Equation 19.1.

$$V_{DC} = \frac{V_{p(sec)}}{\pi} = \frac{16.97 \text{ V}}{\pi} = 5.4 \text{ V}$$

EXAMPLE 19–4

Silicon diode

$E_T = 5 \text{ V}_p$ R_L

FIGURE 19–21 *Circuit for Example 19–4.*

The half-wave rectifier circuit shown in Figure 19–21 is connected to an AC supply voltage with a peak value of 5 V. Determine the average output voltage. Include the forward voltage drop of the diode.

Solution According to Kirchhoff's voltage law, the peak output voltage is equal to the peak input voltage minus the voltage drop across the diode.

$$V_p = E_T - V_F = 5 \text{ V} - 0.7 \text{ V} = 4.3 \text{ V}$$

The average voltage is found using Equation 19.1

$$V_{DC} = \frac{V_p}{\pi} = \frac{4.3 \text{ V}}{\pi} = 1.37 \text{ V}$$

Section Review

1. What is a half-wave rectifier?
2. State the equation for finding the average output voltage of a half-wave rectifier.
3. If the input frequency for a half-wave rectifier is 60 Hz, what is the output frequency?

19–9

FULL-WAVE RECTIFIERS

A **full-wave rectifier** is a device that converts alternating current to direct current for 360° per cycle. Full-wave rectifiers are much more common than half-wave rectifiers because of their ability to provide higher values of DC voltage more efficiently. Full-wave rectifiers

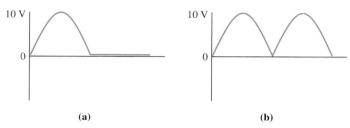

FIGURE 19–22 *Rectifier waveshapes. (a) Half wave. (b) Full wave.*

are also more popular because they produce a smoother output waveform. Figure 19–22 shows a comparison between half-wave and full-wave rectifier output signals. Figure 19–22(a) illustrates a half-wave output with a peak voltage of 10 V. The output waveform of Figure 19–22(b) is from a full-wave rectifier that also has a peak voltage of 10 V.

The voltage waveform of Figure 19–22(b) is occurring twice as often as the wave shown in Figure 19–22(a). As a result, the average voltage is twice as high in a full-wave rectifier as it is in a half-wave rectifier.

$$V_{DC} = \frac{2V_p}{\pi} \tag{19.2}$$

───── **EXAMPLE 19–5** ─────

Determine the average voltages for the two waveforms shown in Figure 19–22.

Solution The waveshape of Figure 19–22(a) is a half-wave rectifier with a peak voltage of 10 V. Equation 19.1 is used to solve for V_{DC}.

$$V_{DC} = \frac{V_p}{\pi} = \frac{10 \text{ V}}{\pi} = 3.18 \text{ V}$$

The average voltage of the waveshape shown in Figure 19–22(b) is found using Equation 19.2.

$$V_{DC} = \frac{2V_p}{\pi} = \frac{2(10 \text{ V})}{\pi} = 6.37 \text{ V}$$

The two most common types of full-wave rectifiers are the **center-tap** and **bridge rectifier.** The center-tap full-wave rectifier uses a center-tap transformer and two diodes to produce a waveform every 180°. Figure 19–23(a) shows a typical center-tap rectifier circuit. The center tap of the secondary winding of the transformer is grounded. This permits only one-half of the secondary peak voltage to be delivered to either diode. If the transformer has a 1:1 transformation ratio, the maximum amplitude to either diode is only one-half of the peak voltage of the primary winding. Because transformers are manufactured with a wide variety of transformation ratios, this is not a serious disadvantage. If a higher-peak voltage is required at the secondary, a transformer is selected with a higher number of turns on the secondary.

The waveform of Figure 19–23(b) shows the AC input signal to the transformer. As the input signal begins to rise, point A in Figure 19–23(a) goes positive, and diode D_1 conducts. When D_1 becomes forward biased, current flows from the ground through R_L and develops a positive alternation between point C and ground. While the voltage at point A is

FIGURE 19–23   (a) Full-wave rectifier using a center-tap transformer.
(b) Input waveform. (c) Output waveform.

positive, the voltage at point B is negative, which reverse biases diode D_2. When the AC input crosses the zero axis and begins to go negative, D_2 is now forward biased and begins to conduct. Point A is now negative and diode D_1 is reverse biased. Current flows through the load resistor in the same direction and through the forward-biased diode, D_2. The resulting waveshape is shown in Figure 19–23(c). Each diode is conducting for one alternation and blocking for one alternation.

The waveform shown in Figure 19–23(c) occurs twice as often as the AC input signal of Figure 19–23(b). Consequently, the output frequency of a full-wave rectifier is twice as high as the input frequency.

$$\text{output Hz} = 2(\text{input Hz}) \qquad (19.3)$$

The peak inverse voltage (PIV) ratings of the diodes used in a full-wave center-tap rectifier must be at least double the peak voltage of the load. Therefore,

$$\text{PIV} = 2V_{p(\text{out})} \qquad (19.4)$$

_____ EXAMPLE 19–6 _____

Solution

(a) $V_{DC} = 0.636V_p = 0.636\,(100\ \text{V}) = 63.6\ \text{V}$

(b) $\text{PIV} = 2V_p = 2(100\ \text{V}) = 200\ \text{V}$

(c) $\text{Hz}_{\text{out}} = 2\text{Hz}_{\text{in}} = 2(60\ \text{Hz}) = 120\ \text{Hz}$

The peak rectified output voltage of a full-wave center-tap rectifier is 100 V. The supply frequency is 60 Hz. Calculate the

(a) DC voltage.
(b) PIV rating of diodes.
(c) Output frequency.

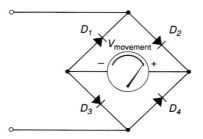

FIGURE 19–24 *Full-wave bridge rectifier.*

The full-wave bridge rectifier is the most common type of rectifier used in electronic circuits. This rectifier is small, lightweight, and highly efficient. Figure 19–24 shows some typical full-wave bridge rectifiers. The bridge rectifier consists of four semiconductor diodes interconnected so that an AC input signal is converted into a full-wave DC output.

Figure 19–25(a) shows a typical full-wave bridge rectifier connected to an AC supply and a load resistor. During the first 180° of the input cycle, point A in the circuit is pos-

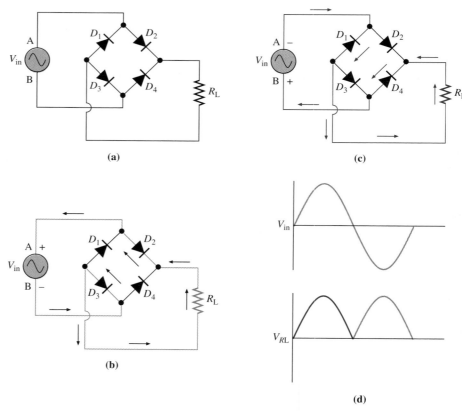

FIGURE 19–25 *(a) Full-wave bridge rectifier. (b) Current flow during first 180°. (c) Current flow during second 180°. (d) Relationship between input and output waves.*

itive and point B is negative. The positive voltage forward biases diode D_2 and reverse biases D_1, while the negative voltage reverse biases D_4 and forward biases D_3. The current flow as a result of the two forward-biased diodes is shown in Figure 19–25(b). When the AC input signal crosses the zero axis and begins to go negative, the polarity of the signal generator reverses; point A is now negative and point B is positive. Diodes D_1 and D_4 are now forward biased and D_2 and D_3 are reverse biased. The flow of current during the second 180° is shown in Figure 19–25(c). The waveforms shown in Figure 19–25(d) illustrate the phase relationship between the input and output voltages. During the first 180° of the input cycle, the current through the load resistor flows from the bottom of the resistor to the top, as shown in Figure 19–25(b). During the second 180°, current flow through the resistor is in exactly the same direction, as indicated in Figure 19–25(c). Because the current flow through the load is in the same direction for the entire 360° cycle, the polarity of the voltage at the output remains constant.

EXAMPLE 19–7

FIGURE 19–26 *Circuit for Example 19–7.*

Determine the average output voltage for the circuit shown in Figure 19–26. Include the voltage drop across the silicon diodes in the calculation.

Solution   During each half cycle, two silicon diodes are in series with the load. Therefore, the peak output voltage is equal to the applied voltage minus 1.4 V.

$$V_p = V_{in} - V_{D1} - V_{D2}$$
$$= 10 \text{ V} - 0.7 \text{ V} - 0.7 \text{ V} = 8.6 \text{ V}$$

The average output voltage is now determined using Equation 19.2.

$$V_{DC} = \frac{2V_p}{\pi} = \frac{2(8.6 \text{ V})}{\pi} = 5.47 \text{ V}$$

Section Review

1. What is a full-wave rectifier?
2. Why are full-wave rectifiers more popular than half-wave rectifiers?
3. State the equation for finding the average output voltage of a full-wave rectifier.
4. What is a center-tap full-wave rectifier?
5. If the input frequency to a full-wave rectifier is 1 kHz, what is the output ripple frequency?

RECTIFIER FILTERS

Rectifier filters are used to reduce the AC component of a pulsating DC signal. Because most electronic circuits require DC voltage levels that are as smooth as possible, a capacitor is often connected in parallel with the load to reduce the pulsating DC produced by a rectifier. Figure 19–27(a) shows a half-wave rectifier with a rectifier filter. When the AC input signal begins to rise, the diode is forward biased. The capacitor begins to charge until it reaches a peak value at 90°, as shown in Figure 19–27(c). As the input voltage passes the 90° point, the diode becomes reverse biased, because the voltage across the capacitor equals the input voltage. The capacitor now begins to discharge through the load resistor, R_L, as shown in Figure 19–27(b). The diode remains reverse biased from the 90° point to the 360° point. During this time, the voltage supplied to the load is from the discharging capacitor. The rate that the capacitor discharges is determined by the size of the capacitor and load resistor. If the capacitor is very large, it will take longer to discharge through the load, and the discharge curve will be gradual. If the capacitor is very small, it will discharge quickly through the load.

Figure 19–28(a) shows a full-wave bridge rectifier connected to a load resistor and rectifier filter. In this circuit, the capacitor charges up to the peak value of the input signal. As the output from the full-wave rectifier drops toward zero, the capacitor discharges through R_L until the voltage from the rectifier rises to a value greater than the capacitor voltage. At this point, two diodes will again be forward biased and the capacitor is recharged. The output wave for the filtered output is shown in Figure 19–28(b).

Ideally, the average output voltage produced by a rectifier filter circuit should equal the peak voltage. Although this is not feasible in a practical circuit, it is possible for the average voltage to be very close to the peak input signal. The equation for calculating average voltage in a rectifier filter is

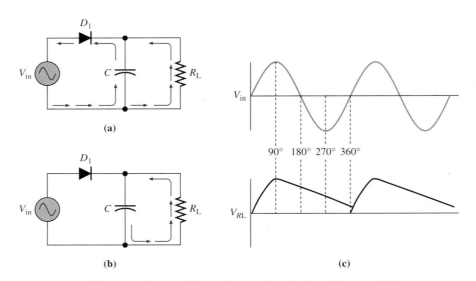

FIGURE 19–27 *(a) Current flow through a forward-biased diode in a rectifier filter. (b) Current flow when the diode is reverse-biased. (c) Input and output waveforms.*

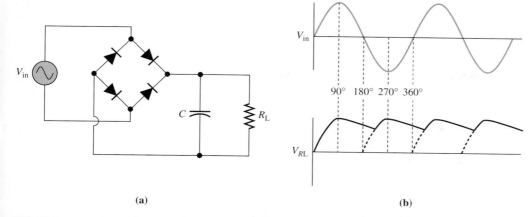

(a)   (b)

FIGURE 19–28 *(a) Full-wave bridge rectifier with filter. (b) Input and output waveforms.*

$$V_{DC} = \frac{V_p}{1 + 1/(2fR_LC)} \qquad (19.5)$$

where f is the output ripple frequency.

EXAMPLE 19–8

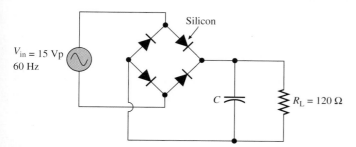

FIGURE 19–29 *Circuit for Example 19–8.*

For the circuit shown in Figure 19–29, calculate the average output voltage when

(a) A 100 µF filter capacitor is used.
(b) A 1000 µF capacitor is used.

Include the voltage drop across the diodes.

Solution

$$V_p = V_{in} - 2V_F = 15\ V - 2(0.7\ V) = 13.6\ V$$

(a) The output frequency, f, is double the input frequency; therefore, $f = 120$ Hz.

$$V_{DC} = \frac{V_p}{1 + 1/(2fR_LC)}$$

$$= \frac{13.6\ V}{1 + 1/[2(120\ Hz)(120\ \Omega)(100\ \mu F)]}$$

$$= 10.09\ V$$

(b) $$V_{DC} = \frac{V_p}{1 + 1/(2fR_LC)}$$

$$= \frac{13.6\ V}{1 + 1/[2(120\ Hz)(120\ \Omega)(1000\ \mu F)]}$$

$$= 13.14\ V$$

FIGURE 19–30 *Waveform for 120 Ω resistor and 100 μF capacitor.*

From Example 19–8, it is apparent that as the size of the filter capacitor increases, the average voltage also increases. Rectifier filters generally use very large capacitors to increase the DC output voltage as close to peak as possible. The difference between the average output voltage and the peak output voltage represents the peak voltage of the AC component. If an AC voltmeter were connected across the load resistor of Figure 19–29, it would read an rms value based on this peak voltage. Figure 19–30 shows the output waveform for a rectifier filter with a 120 Ω load resistor and a 100 μF capacitor [Example 19–8(a)].

In Figure 19–30, the ripple component present in the output wave is shown as "riding" on the 10.09 V_{DC} component. The peak value of the AC ripple is the difference between 13.6 V and 10.09 V.

$$V_{p(of\ ripple)} = V_p - V_{DC} = 13.6\ V - 10.09\ V = 3.51\ V$$

The rms component of the ripple voltage is approximated by multiplying the peak of the ripple voltage by 0.707. This represents the voltage that an AC voltmeter would read if connected across the output of the rectifier filter. For Example 19–8, the rms component in the rectifier filter with a 100 μF capacitor would be

$$
\begin{aligned}
V_{rms(of\ ripple)} &= V_{p(of\ ripple)} \times 0.707 \\
&= (3.51\ V)(0.707) \\
&= 2.48\ V
\end{aligned}
$$

The **percent ripple** of a rectifier filter represents the ratio between the rms of the AC component and the average DC output voltage. An ideal rectifier filter will have a percent ripple of zero percent. As the amount of ripple voltage increases, the percent ripple also increases. In equation form,

$$\%\ ripple = \left[\frac{V_{rms(of\ ripple)}}{V_{DC}} \right] 100\% \qquad \textbf{(19.6)}$$

Section Review

1. Rectifier filters are used to reduce the DC component of a pulsating AC signal (True/False).
2. Ideally, what should the average voltage of a rectifier filter be equal to?
3. Filter capacitors are always connected in series with the load (True/False).
4. What is the percent ripple of a rectifier filter?

___ 19–11 _____

TROUBLESHOOTING RECTIFIER CIRCUITS

Faults in rectifier circuits generally occur as a result of open or short circuits in diodes or filter capacitors. Figure 19–31(a) shows a half-wave rectifier with a filter capacitor. Figure 19–31(b) shows the effect of an open-circuited diode. The diode acts as an open switch and no current flows through the load. The output voltage for a short-circuited diode is shown in Figure 19–31(c). In this condition, no rectification occurs and the input voltage is approximately equal to the output voltage. The circuit current will increase when the diode is shorted. If the rectifier is fused properly, the fuse should open in the event of a shorted diode. The output waveform for a half-wave rectifier with an open capacitor is shown in Figure 19–31(d). When the capacitor is open, no filtering takes place and the load voltage is a result of the half-wave rectifier, and the average output voltage is substantially reduced. The waveform of Figure 19–31(e) shows the effect of a shorted capacitor on the load voltage. The short circuit causes the circuit current to be shunted around the load, and the load voltage is zero. A leaky capacitor behaves as a partial short and will cause the average output voltage to be reduced. The DC output voltage may fall below the rated value, but is not necessarily zero volts. The AC ripple voltage may increase as a result of a leaky capacitor, because of the reduction in the filtering ability of the capacitor.

If the fuse shown in Figure 19–31(a) is rated properly, it will blow when a short circuit occurs in any of the components. For example, if the load resistor becomes shorted, a path with 0Ω resistance is placed across the transformer's secondary winding when the diode is conducting. This causes excess current to flow and the fuse should open.

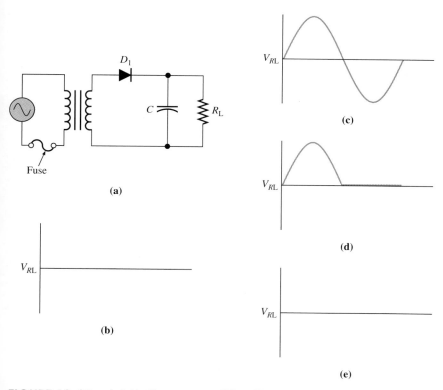

FIGURE 19–31 (a) Half-wave rectifier filter circuit. (b) Load voltage waveform for open-circuited diode. (c) Waveform for a short-circuited diode. (d) Waveform for an open capacitor. (e) Waveform for a shorted capacitor.

TABLE 19–2 Troubleshooting chart for half-wave rectifier.

Problem	DC Output Voltage	AC Ripple Voltage	Fuse
Diode shorted	Zero (because of excess current)	Zero	Blown
Diode open	Zero	Zero	Good
Capacitor shorted	Zero (because of excess current)	Zero	Blown
Capacitor open	Below rated value	Above rated value	Good
Capacitor leaky	Below rated value	Above rated value	Good

Table 19–2 shows some typical half-wave rectifier problems and symptoms. If the diode is shorted, excess current will flow in the circuit because of a lack of rectification, and the fuse should blow. If the diode is open, no current flows through the load. If the capacitor is open, the DC voltage is reduced and the AC component is increased. A shorted capacitor should cause the fuse to open. If the capacitor is leaky, the ripple voltage should increase and the average voltage should decrease.

Figure 19–32(a) shows a full-wave, center-tap rectifier filter circuit. If one of the diodes should become open, the circuit would behave as a filtered half-wave rectifier, with a waveshape similar to that of Figure 19–32(b). The average output voltage would be reduced, and the AC ripple voltage would increase. If either of the diodes becomes shorted, a large value of current will flow that should cause the fuse to open. Figure 19–32(c) shows the result of an open-circuited filter capacitor. When filtering is removed, the percent ripple of the rectifier increases. If the capacitor were shorted, the fuse should also open because of excess current flow.

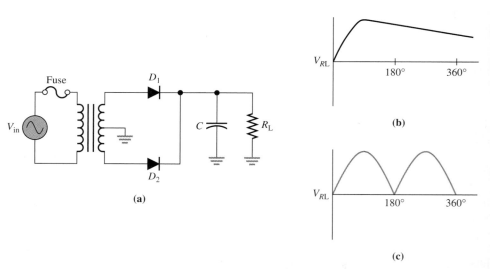

FIGURE 19–32 *(a) Full-wave center-tap rectifier with filter. (b) Output waveshape if one diode is open. (c) Output waveshape if capacitor is open.*

TABLE 19–3 *Troubleshooting chart for full-wave center-tap and full-wave bridge rectifiers.*

Problem	DC Output Voltage	AC Ripple Voltage	Fuse
Diode shorted	Zero (because of excess current)	Zero	Blown
Diode open	Below rated value	Above rated value	Good
Capacitor shorted	Zero (because of excess current)	Zero	Blown
Capacitor open	Below rated value	Above rated value	Good
Capacitor leaky	Below rated value	Above rated value	Good

Full-wave bridge rectifiers respond to open- and short-circuited diodes and capacitors in a manner very similar to full-wave center-tap rectifiers. If any one of the diodes opens, the circuit behaves as a half-wave rectifier. Table 19–3 shows five typical problems in full-wave rectifier circuits. If the rectifiers are properly fused, excess current flow will open fuses or trip circuit breakers. However, if short-circuit current occurs for prolonged periods of time, other components in series with the short may also become damaged. The fuse is a good place to start when troubleshooting. The fuse provides a good indication of whether short-circuit current was flowing at the time of the fault. A visual inspection of the rectifier circuit will also provide clues to component failure. If a diode has failed, it may appear burned or melted. If the fuse is not blown and the circuit produces a DC voltage below its rated value, it is generally caused by an open-circuited component.

Section Review

1. A leaky capacitor behaves like a partial short (True/False).
2. A leaky filter capacitor will cause the ripple voltage to decrease (True/False).

19–12
ZENER DIODES

The reverse breakdown that occurs in a reverse-biased PN junction is either the result of the avalanche effect or the zener effect. In 1934, physicist Clarence Zener noted that under very intense electric fields, electrons are pulled out of the valence band and traverse the forbidden energy gap. Although Zener's theory was originally applied to dielectric materials, the same effect occurs when the applied voltage is large enough to cause the breakup of covalent bonds in the depletion region of a semiconductor. The result is an excess of mobile charge carriers and the current increases substantially with a small increase in voltage beyond the breakdown voltage. The **zener diode** is a PN junction that is based on the zener effect. Zener diodes are heavily doped and are designed to operate in the breakdown region at V_z, as shown in Figure 19–33.

The diodes discussed previously in this chapter are based on the avalanche effect and have relatively high breakdown voltages. Typical zener diodes are characterized as having low breakdown voltages, which range from two or three volts to several hundred volts, and

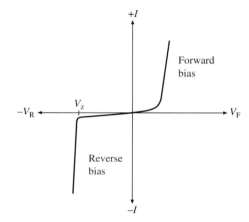

FIGURE 19–33 *Zener diode characteristic curve.*

their design can be based on both the avalanche effect and zener effect. Diodes with breakdown voltages below 5 V are generally based on the zener effect, and those with breakdown voltages above 5 V predominantly are based on the avalanche effect. In Figure 19–33, V_z represents the breakdown voltage for the zener diode. When the diode is operated in the zener region of the curve, the voltage across the zener remains fairly constant as the reverse current, I_z, increases. The internal resistance of the zener decreases to compensate for increased current flow. If the current flow increases above its rated value, the semiconductor material eventually will be destroyed. Zener diodes are also rated in terms of their power-handling capabilities. The power dissipation rating is given for a specific operating temperature, typically 25° C. If the temperature of the zener increases above the rated value, the power that can be safely dissipated by the zener decreases. The maximum current flow through a diode is determined by dividing the maximum power, P_z, by the breakdown voltage,

$$I_z = \frac{P_z}{V_z}$$

(19.7)

_____ EXAMPLE 19–9 _____

Determine the maximum current for a 5 W zener diode with a breakdown voltage of 6.2 V

Solution

$$I_z = \frac{P_z}{V_z} = \frac{5 \text{ W}}{6.2 \text{ V}} = 0.81 \text{ A}$$

Although zener diodes are operated almost exclusively in the reverse-bias region, the zener is occasionally used in the forward-bias condition. As the graph of Figure 19–33 indicates, when the zener is forward biased, it behaves like any other PN junction. That is, it will act as a closed switch and conduct a large value of current with minimal resistance. Figure 19–34(a) shows the schematic symbol for a zener diode, and Figure 19–34(b) represents an equivalent circuit. The circuit of Figure 19–34(b) is for a reverse-biased zener diode and consists of a relatively small value of internal resistance in series with an ideal battery.

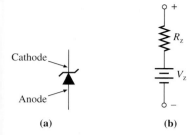

FIGURE 19–34   *(a) Zener diode symbol. (b) Equivalent circuit.*

Zener diodes are used primarily as **voltage regulators.** In a voltage regulator, the voltage across the load is kept at a constant value over a wide range of varying load conditions. Ideally, the output of a regulator should be a constant voltage. If the load resistance changes, or the unregulated input voltage changes, the regulator should maintain a constant output value. The **percent voltage regulation** is a ratio of how much the output voltage changes over a range of load resistance values. This ratio is expressed by the following equation:

$$\% \text{ voltage regulation} = \left(\frac{E_{nl} - E_{fl}}{E_{fl}} \right) 100\% \qquad (19.8)$$

where E_{nl} = no-load voltage of the regulator
        E_{fl} = full-load voltage of the regulator

<hr>

EXAMPLE 19–10

Solution

$$\% \text{ voltage regulation} = \left(\frac{E_{nl} - E_{fl}}{E_{fl}} \right) 100\%$$

$$= \left(\frac{24 \text{ V} - 22 \text{ V}}{22 \text{ V}} \right) 100\%$$

$$= 9.09\%$$

A power supply produces 24 V when the load current is zero. If the output voltage drops to 22 V when full-load current flows, what is the percent voltage regulation of the circuit?

<hr>

Ideally, a voltage regulator should provide zero percent voltage regulation. In other words, the output voltage should be constant under all load conditions. By using a voltage regulator, the percent voltage regulation can be held to a very low value.

Figure 19–35 shows a basic voltage regulator. The input voltage is connected so that the zener diode is reverse biased. The series resistor, R_s, allows enough current to flow through the zener diode so that it operates within its zener breakdown region. The voltage

FIGURE 19–35   *Simple regulator circuit.*

FIGURE 19–36 *Current flow in a voltage regulator.*

across the resistor will be equal to the difference between the diode's zener voltage and the unregulated DC input voltage.

$$V_{Rs} = V_{in} - V_z$$

According to Kirchhoff's voltage law, the voltage across the zener is the output voltage for the circuit. When a load is connected to the output of the regulator circuit, the regulator must supply current as well as voltage to the load. The input current is the sum of I_z and I_L.

$$I_{in} = I_z + I_L$$

Whenever I_L increases, I_z decreases by approximately the same amount to hold $I_z + I_L$ almost constant and, therefore, maintain a constant output voltage. Figure 19–36 shows the relationship between the three currents in a simple regulator circuit.

Table 19–4 shows the effect of an increase in the unregulated input voltage on the voltages and currents in the regulator circuit of Figure 19–35. If V_{in} increases, the voltage drop across the series resistor also increases. This causes more current to flow into the regulator and I_s increases. The excess current flows through the zener, maintaining a constant current through the load. The zener holds the load voltage at a constant value, so the difference between the input voltage and the output is dropped across R_s.

Section Review

1. What is the zener effect?
2. Zener diodes are lightly doped and designed to operate in the breakdown region (True/False).

TABLE 19–4 *Effect of an increase in unregulated input voltage on the regulator circuit components.*

	Increase	Decrease	Remain Constant
V_s	X		
V_z			X
V_{RL}			X
I_s	X		
I_z	X		
I_L			X

3. What is a voltage regulator?
4. What is the percent regulation of a voltage regulator?
5. One hundred percent voltage regulation is ideal (True/False).

19–13
VARACTOR DIODES

A **varactor diode** is a PN junction semiconductor diode designed for low losses at high frequencies. Varactors are popular in a variety of electronic applications and are often used in radio circuits for FM modulation. Varactor diodes are also known as *variable-voltage capacitors* because the capacitance of the PN junction varies with the applied voltage. Unlike the zener diode, a varactor is generally operated with a reverse-bias voltage that is less than its breakdown voltage rating. As the reverse voltage increases, the depletion region within the device widens and, essentially, acts as a wide dielectric between the P- and N-type materials. Figure 19–37 shows the relationship between the diode's junction capacitance and reverse voltage.

The schematic symbol for a varactor diode is shown in Figure 19–38(a), and the equivalent circuit is shown in Figure 19–38(b). C_v represents the variable capacitance and R_s is the equivalent series resistance. Varactor diodes, instead of mechanically variable capacitors, often are used as a means of providing solid-state tuning. Because the capacitance of the varactor varies with voltage, the frequency of a resonant circuit can be controlled by varying the diode's reverse-bias voltage.

Section Review

1. A varactor diode is designed for low losses at low frequencies (True/False).
2. Why are varactor diodes also known as variable-voltage capacitors?

FIGURE 19–37 *Typical voltage versus capacitance characteristics of a varactor diode.*

FIGURE 19–38 *Varactor diode. (a) Schematic symbol. (b) Equivalent circuit.*

(a)　　　　(b)

LEDs AND PHOTODIODES

A **light-emitting diode, LED,** is a PN junction device that radiates energy primarily as light instead of as heat. In a typical forward-biased diode, as free electrons cross the junction and fall from a higher energy level to a lower level, they radiate energy in the form of heat. In an LED, this energy is radiated in the form of light. Instead of using silicon, LEDs are constructed using elements like gallium, phosphorus, and arsenic. By varying the quantities of these elements, it is possible to produce light of different wavelengths with colors that include red, green, yellow, blue, and infrared (invisible). For example, when an LED is manufactured using gallium arsenide with an energy gap of 1.37 eV, it will produce a red color light. If the LED is made with gallium phosphide with an energy gap of 2.25 eV, it will radiate green light.

Figure 19–39 shows a cutaway view of a typical LED. This type of LED is a small, round device approximately 1/4 in. wide. The cathode of the LED is identified by a notch, or flat area, in the side of the device. The LED chip is extremely small and radiates a low-energy level of light. The light from the LED chip is magnified and directed by the plastic lens.

The forward breakover voltage for an LED is typically 1.6 V, with a forward current of 20 mA. The LED is considerably more efficient than incandescent lighting for low-power applications. Figure 19–40 is a graph representing the relationship between radiated light and forward current. From this curve, it is apparent that the light intensity and current flow are directly proportional.

Plastic lens

LED chip

Flat side

Cathode   Anode

FIGURE 19–39 *Cutaway view of a LED.*

FIGURE 19–40 *Radiated light versus forward current characteristic curve.*

Radiated Light (mW) vs Forward Current (mA)

FIGURE 19–41 *Schematic symbol for LED.*

Anode ━━━━▶│◀ Cathode

FIGURE 19–42 *Seven-segment display in IC package.*

Figure 19–41 shows the schematic symbol for an LED. The arrows are shown as pointing away from the diode, indicating that energy is being radiated by the device when forward biased. If the LED is reverse biased, it will behave as any other reverse-biased PN junction. That is, it will oppose current flow and act as an open switch. If the breakdown voltage is exceeded for any length of time, the device will be destroyed eventually.

LEDs are often grouped together to form seven-segment displays. Figure 19–42 shows a typical seven-segment display. Each of the seven segments is controlled by an LED. Instead of a plastic lens, these LEDs use a light tube to focus the beam of light. The seven LEDs are mounted in an integrated circuit (IC) package for durability and convenience. Because LEDs operate at low voltages and low currents, they are ideally suited for use in IC packages.

The seven-segment LED display is found in many electronics applications, including test equipment, digital clocks and radios, automobile dashboards, and so forth. Figure 19–43(a) shows the front of a seven-segment display. By energizing various combinations of the seven LEDs, it is possible to reproduce decimal numbers. The letters a through g identify each of the seven segments. Figure 19–43(b) shows the equivalent circuit consisting of resistors and LEDs. In Figure 19–43(b), the anodes of all seven LEDs are connected to a common, positive voltage source. This configuration is known as *common-anode*. Seven-segment displays are also available in *common-cathode* packages. When one of the seven LEDs is forward biased, it will illuminate one of the light tubes. In Figure 19–43(b), forward biasing of the diodes is accomplished by grounding points a through g.

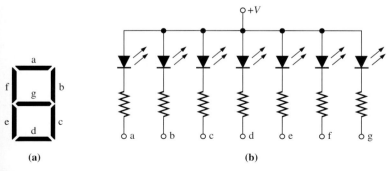

(a)   **(b)**

FIGURE 19–43 *(a) Seven-segment display. (b) Schematic diagram.*

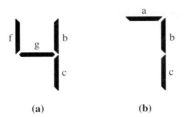

FIGURE 19–44 *Seven-segment display producing the numbers (a) 4 and (b) 7.*

(a)   (b)

By energizing different combinations of the seven light tubes, it is possible to reproduce the numbers 0 through 9. For example, if the number 4 were required, diodes f, g, b, and c would be forward biased, as shown in Figure 14–44(a). The number 7 can be reproduced, as shown in Figure 19–44(b), by energizing points a, b, and c

The LED is classified as an **optoelectronic device** because of its ability to give off light. A device that depends on light to operate is also referred to as optoelectronic. One of the more popular optoelectronic devices that require light is the **photodiode.** The photodiode is constructed so that light can be directed to the junction of the P and N regions. When light falls on this region, the resistance of the depletion region is lowered and current through the diode increases.

If no external bias voltage is applied across the photodiode, the diode will produce its own open-circuit potential, or photovoltage, when the PN junction is illuminated. This type of photovoltaic cell is the basic element of the solar cell and is used in a many applications. When the photodiode is connected to an external forward-biased voltage, it acts as a light-variable resistor. A typical photodiode package is shown in Figure 19–45. This package consists of a PN junction mounted on an insulated substrate and sealed inside a metal case. A glass window is mounted on top of the case to allow light to enter and strike the PN junction. The two leads extending from the case are labeled anode and cathode. The cathode is typically identified by a tab protruding from the side of the case.

Photodiodes are used mainly in control circuit applications because of their low current-handling capabilities. Reverse-current flow in a photodiode is generally measured in microamperes. These diodes find many applications in counter circuits, alarm systems, and in a variety of light-controlled circuits. Figure 19–46(a) shows the schematic symbol for a photodiode. The photodiode is typically operated in its reverse-bias condition, as shown in Figure 19–46(b). When there is no light, the current through the load, R_L, is negligible. As the light intensity increases, the internal resistance of the photodiode decreases, and the current through the load increases.

Window

FIGURE 19–45 *Typical photodiode package.*

Cathode   Anode

FIGURE 19–46 *Photodiode.*
(a) Schematic symbol. (b)
Reverse-biased operation.

(a)   (b)

Section Review

1. What is an LED?
2. What is the forward voltage drop of a typical LED?
3. Why is the LED classified as an optoelectronic device?
4. Photodiodes are designed to emit energy as light rather than heat (True/False).

19–15
PRACTICAL APPLICATION SOLUTION

In the chapter opener, you were required to test the average output voltage of a filtered power supply. The power supply was described as a full-wave bridge rectifier constructed of silicon diodes with a 560 µF capacitor connected in parallel with a 100 Ω load. The rectifier supply voltage is 120 V_{rms} at 60 Hz, stepped down to 12 V. Your task was to calculate the average output voltage and compare this value with the measured value taken with a DC voltmeter. The following steps illustrate the solution for this practical application of semiconductor fundamentals.

STEP 1   Use a voltmeter to measure and record the average output voltage of the rectifier filter.

STEP 2   Calculate the peak input voltage for the filter.

$$V_p = \frac{V_{rms}}{0.707 - 2V_F} = \frac{12 \text{ V}}{0.707 - 1.4 \text{ V}} = 15.57 \text{ V}$$

STEP 3   Determine the output ripple frequency of the power supply. The output frequency for a full-wave bridge rectifier is double the input frequency. Therefore,

$$Hz_{out} = 2Hz_{in} = 2 \ (60 \text{ Hz}) = 120 \text{ Hz}$$

STEP 4   Calculate the average output voltage.

$$V_{DC} = \frac{V_p}{1 + 1/(2fR_L C)} = \frac{15.57 \text{ V}}{1 + 1/[2(120 \text{ Hz})(100 \text{ Ω})(560 \text{ µF})]} = 14.49 \text{ V}$$

STEP 5   Compare the calculated value with the measured average voltage. These two values should be approximately equal. If they are not, systematically check the values of each component in the power supply and replace any devices that do not meet the manufacturer's tolerances.

Summary

1. A semiconductor having an excess of electrons is called *N-type*.
2. When a hole is created by adding an impurity, it is called a *P-type* material.
3. PN junctions are created by forming a semiconductor with P- and N-type materials.
4. PN junctions can be either forward biased or reverse biased.
5. The two leads of the diode are labeled anode and cathode.
6. A properly functioning diode will conduct current in one direction only and block current in the other direction.
7. A full-wave rectifier converts AC to DC for 360° per cycle.
8. Rectifier filters are used to reduce the AC component of a pulsating DC signal.
9. Zener diodes are used primarily as voltage regulators.
10. LEDs radiate energy as light instead of as heat.

Answers to Section Reviews

Section 19–1

1. False
2. A component that is made of semiconductor material
3. Diodes, transistors, and integrated circuits.

Section 19–2

1. True
2. A unique sharing of electrons that occurs between atoms of silicon or germanium
3. Zero
4. A semiconductor in its purest form
5. The process of adding impurities

Section 19–3

1. An impurity with five valence electrons
2. Because these materials give up, or donate, one electron per atom
3. True
4. Holes

Section 19–4

1. When a single piece of semiconductor material is formed with both types of extrinsic materials
2. False
3. The difference of potential, or charge, at the junction

Section 19–5

1. Any voltage that is connected across a PN junction.
2. True
3. The resistance of the basic semiconductor material
4. The small amount of current that flows through a reverse-biased PN junction

Section 19–6

1. A device comprised of a single PN junction, which typically is packaged in glass or metal.

2. True

3. The limiting values assigned by a manufacturer to various parameters of a diode

Section 19–7

1. False

2. With the device removed from the circuit

3. True

Section 19–8

1. A device that allows current to flow in one direction during one half-cycle of the input while blocking reverse flow during the other half-cycle

2. $V_{DC} = V_p/\pi$

3. 60 Hz

Section 19–9

1. A device that converts alternating current to direct current for 360° per cycle

2. Because full-wave rectifiers provide higher values of DC voltage more efficiently and produce a smoother output wave

3. $V_{DC} = 2V_p/\pi$

4. A device that uses a center-tap transformer and two diodes to produce a waveform every 180°

5. 2 kHz

Section 19–10

1. False

2. The peak voltage

3. False

4. The ratio between the rms of the AC component and the average DC output voltage

Section 19–11

1. True

2. False

Section 19–12

1. When the applied voltage is large enough to cause the breakup of covalent bonds in the depletion region of a semiconductor

2. False

3. A device that maintains a constant load voltage under a wide range of varying load conditions

4. A ratio of how much the output voltage changes over a range of load resistance values

5. False

Section 19–13

1. False

2. Because the capacitance of the PN junction varies with the applied voltage

Section 19–14

1. A PN junction device that radiates energy as light instead of as heat

2. 1.6 V

3. Because of its ability to give off light

4. False

Multiple Choice Questions

19–1 A three-dimensional arrangement of atoms is also known as
(a) Covalent bonding   **(b)** Doping   **(c)** A crystal lattice   **(d)** A PN junction

19–2 Group IV elements from the periodic table are characterized by having
(a) Four permissible energy states and four electrons in the outer subshell
(b) Eight permissible energy states and four electrons in the outer subshell
(c) Eight permissible energy states and eight electrons in the outer subshell
(d) Four permissible energy states and eight electrons in the outer subshell

19–3 The net electrical charge of a pure silicon or germanium crystal is
(a) Zero   **(b)** Negative   **(c)** Positive   **(d)** Impossible to determine

19–4 When silicon is in its purest form, it is called
(a) Extrinsic silicon   **(b)** Covalent bonding
(c) Intrinsic silicon   **(d)** Doping

19–5 The conductivity of an impure semiconductor is referred to as
(a) Hole flow   **(b)** Intrinsic conductivity
(c) Majority carriers   **(d)** Extrinsic conductivity

19–6 When an impurity has five valence electrons, it is called
(a) P-type   **(b)** N-type   **(c)** Pentavalent   **(d)** Trivalent

19–7 The electrons in N-type semiconductor material are
(a) Majority carriers   **(b)** Minority carriers
(c) Positive charge carriers   **(d)** Present only in the valence band

19–8 When a PN junction is created, the interchange of holes and electrons cancels out the current carriers in the vicinity of the junction, forming a(n)
(a) Forward-biased junction   **(b)** Reverse-biased junction
(c) Depletion region   **(d)** Avalanche breakdown

19–9 The difference of voltage across the PN junction is called the
(a) Depletion region   **(b)** Barrier potential
(c) Forward current   **(d)** Bulk resistance

19–10 Any external voltage that is connected across a PN junction is called a
(a) Barrier voltage   **(b)** Forward voltage
(c) Reverse voltage   **(d)** Bias voltage

19–11 When a PN junction is forward biased, the
(a) Junction resistance will vary inversely as the current through the PN junction varies
(b) Junction resistance is directly proportional to the junction current
(c) Bulk resistance will vary inversely as the current through the PN junction varies
(d) Bulk resistance is directly proportional to the junction current

19–12 The current that flows through a PN junction when reverse biased is called
(a) Majority carriers   **(b)** Breakover current
(c) Leakage current   **(d)** PIV

19–13 The two leads of a diode are labeled
(a) P and N   **(b)** Positive and negative
(c) Base and emitter   **(d)** Anode and cathode

19–14 Most diode characteristics are affected by
(a) Temperature   **(b)** Voltage   **(c)** Current   **(d)** Power

19–15   A reverse-biased diode
(a) Has a high value of current flow   **(b)** Acts like a closed switch
(c) Has a small voltage drop   **(d)** Acts like an open switch

19–16   The most popular instrument for testing diodes is the
(a) Voltmeter   **(b)** Ohmmeter   **(c)** Ammeter   **(d)** Wattmeter

19–17   A full-wave rectifier is a device that converts
(a) Direct current to alternating current for 180° per cycle
(b) Direct current to alternating current for 360° per cycle
(c) Alternating current to direct current for 180° per cycle
(d) Alternating current to direct current for 360° per cycle

19–18   The center-tap full-wave rectifier uses
(a) Two diodes and a bridge   **(b)** Four diodes
(c) Two diodes and a transformer   **(d)** One diode

19–19   The output frequency of a full-wave rectifier is
(a) Equal to the input frequency   **(b)** Double the input frequency
(c) One-half the input frequency   **(d)** Triple the input frequency

19–20   Rectifier filters are used to
(a) Increase ripple frequency   **(b)** Reduce the DC component
(c) Reduce the AC component   **(d)** Increase the rms voltage

19–21   Ideally, the average output voltage produced by a rectifier should
(a) Equal the peak voltage   **(b)** Exceed the peak voltage
(c) Equal the rms voltage   **(d)** Have a high ripple frequency

19–22   Zener diodes are used primarily as
(a) Rectifiers   **(b)** Filters   **(c)** Optoelectronic devies   **(d)** Regulators

19–23   Varactor diodes are also known as
(a) Variable-voltage inductors   **(b)** Optoelectronic devices
(c) Filters   **(d)** Variable-voltage capacitors

19–24   A light-emitting diode is a PN junction device that
(a) Absorbs light when reverse biased   **(b)** Radiates light when reverse biased
(c) Absorbs light when forward biased   **(d)** Radiates light when forward biased

Practice Problems

19–25   Determine the current flowing in the circuit of Figure 19–47.

Silicon diode   **FIGURE 19–47**

$E_T = 10$ V $I = ?$ $R_L = 1$ kΩ

19–26   The circuit of Figure 19–48 shows a diode with a leakage current of 20 μA. Determine the voltage drop across the load.

FIGURE 19-48

$I_R = 20\ \mu\text{A}$

$R_L = 15\ \text{k}\Omega$

19-27 For the circuit of Figure 19–49, determine the DC load voltage. Include the voltage drop across the silicon diode.

8:1

D_1

FIGURE 19-49

$V_{in} = 120\ V_{rms}$

R_L

19-28 Determine the output frequency of the circuit shown in Figure 19–49 if the input frequency is 60 Hz.

19-29 Determine the average voltages for the two waveforms shown in Figure 19–50.

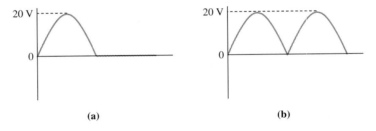

20 V

0

(a)

20 V

0

(b)

FIGURE 19-50

19-30 The peak rectified output voltage of a full-wave center-tap rectifier is 85 V. The input frequency is 60 Hz. Calculate the

(a) DC voltage **(b)** PIV rating of the diodes **(c)** Output frequency

19-31 Determine the average output voltage for the circuit shown in Figure 19–51. Include the voltage drop across the diodes in this calculation.

Silicon diodes **FIGURE 19-51**

12 V$_p$

R_L

19–32 For the circuit shown in Figure 19–52, calculate the average output voltage. Include the voltage drop across the diodes.

FIGURE 19–52

19–33 For the waveshape of Figure 19–53, calculate the percent ripple.

FIGURE 19–53

19–34 Determine the maximum current for a 1 W zener diode with a breakdown voltage of 5.1 V.

19–35 A power supply produces 12 V when the load current is zero. If the output voltage drops to 10 V when full-load current flows, what is the percent voltage regulation of the circuit?

Essay Questions

19–36 What is a crystal lattice?

19–37 Explain the difference between intrinsic conductivity and extrinsic conductivity.

19–38 List three types of pentavalent impurities.

19–39 How is a PN junction made?

19–40 Define the terms *depletion region* and *barrier potential*.

19–41 Describe how PN junctions are biased.

19–42 Name the two types of resistance in a PN junction.

19–43 What is avalanche breakdown?

19–44 How are the two leads of a diode identified?

19–45 Describe the effect of thermal runaway on a semiconductor device.

19–46 What is the difference between a half-wave rectifier and a full-wave rectifier?

19–47 Name the two most common types of full-wave rectifiers.

19–48 Explain the basic function of a rectifier filter.

19–49 Describe the two main types of faults that occur in rectifier circuits.

19–50 What is a varactor?

20

Transistors and Thyristors

Learning Objectives

Upon completion of this chapter you will be able to

- Describe the basic operation of a transistor.
- Explain how transistors are biased.
- List two types of transistors.
- Explain the relationship between current, voltage, and power in a transistor.
- Discuss the purpose of voltage divider biasing.
- Test bipolar transistors.
- Differentiate between FETs and BJTs.
- Define transconductance.
- Test FETs and thyristors.
- List three types of thyristors.
- Explain how SCRs and triacs are used for phase angle control.
- Describe the basic principles of a relaxation oscillator.

Practical Application

As an electronic technician for a defense contractor, you are presently refurbishing the microprocessor-controlled gun turrets on a destroyer-class ship. Each gun turret axis is rotated by a large DC motor with a solid-state variable-speed drive. The variable-speed drives use SCRs to provide large amounts of current to the DC motors. One of the gun turrets fails to respond to signals from the control console. After using a voltmeter to systematically check the control circuit and power circuit voltages and comparing these measured values with the values in the technical literature supplied by the manufacturer, you have determined that the variable-speed drive is receiving the correct voltage levels and that the problem appears to be in one of the SCRs.

Your task is to systematically test the SCRs in the drive system and determine whether the failure is caused by one of the SCRs.

20-1
INTRODUCTION

The **transistor** is a semiconductor device that basically acts as a valve to control the flow of current. The name transistor is derived from the words *transfer resistor*. The transistor has replaced vacuum tube technology to become the standard device used in amplifying and switching circuitry. Transistors are capable of amplifying electrical signals by transferring the signal from a low-resistance to a high-resistance circuit. There are two basic types of transistors—bipolar junction and field effect. Bipolar junction transistors control the current by varying the number of charge carriers in the semiconductor material, and field effect transistors vary the shape of the conducting volume.

In this chapter, we will examine the theory and construction of transistors and thyristors. A **thyristor** is a semiconductor device that is used primarily in switching circuits. The two most common types of thyristors are **triacs** and **silicon-controlled rectifiers, SCRs.**

Section Review

1. Transistors replaced vacuum tubes to become the standard device used in amplifying and switching circuitry (True/False).
2. What are the two basic types of transistors?
3. What are the two most popular types of thyristors?

20-2
BIPOLAR JUNCTION TRANSISTORS (BJTs)

The **bipolar junction transistor, BJT,** is a semiconductor device consisting of two junctions in a continuous crystal. BJTs are made of either silicon or germanium, although silicon transistors are the most common. Figure 20–1 shows some typical BJTs. The devices vary in size according to their current-handling capabilities. BJTs that handle large amounts of power are called *power transistors* and generally are mounted on heat sinks to dissipate heat.

There are two basic types of BJTs: NPN and PNP. Figure 20–2(a) illustrates a basic NPN transistor, and Figure 20–2(b) shows the schematic symbol. In Figure 20–2(a), a layer of P-type semiconductor material is sandwiched between two layers of N-type material. The middle, or P region, is called the *base* of the transistor and the outer or N regions are called the *emitter* and *collector* of the transistor. The depletion region between the emitter and base is called the *emitter-base junction,* and the junction between the collector and base is referred to as the *collector-base junction.* The schematic symbol of Figure 20–2(b) uses an arrow to identify the emitter. The direction of the arrowhead corresponds to the direction of hole flow relative to the base. When the emitter is made of N-type material, holes flow from the P-type base toward the emitter. This is indicated by the arrowhead pointing away from the base in the NPN transistor symbol.

Figure 20–3(a) shows a basic PNP transistor. The N-type material is now wedged between two layers of P-type material. In this device, the base is made of N-type material and the collector and emitter are made of P-type material. The schematic symbol of Figure 20–3(b) again uses an arrowhead to identify the emitter. Because the base is now made of N-type material, the arrowhead is shown pointing toward the base. With a P-type emitter, holes flow from the emitter toward the base.

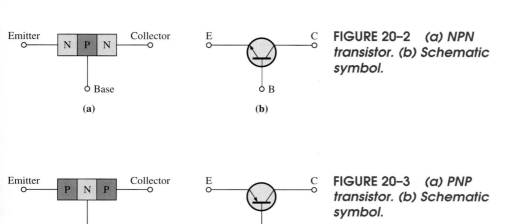

FIGURE 20–1 *Typical bipolar junction transistors (a) Common packaging configurations. (b) Construction views and examples of pin arrangements.*

FIGURE 20–2 *(a) NPN transistor. (b) Schematic symbol.*

FIGURE 20–3 *(a) PNP transistor. (b) Schematic symbol.*

FIGURE 20–4 *A properly biased NPN transistor.*

For a transistor to operate as an amplifier or a switch, it must be properly biased. In the BJT, this is accomplished by forward biasing the emitter-base junction, and reverse biasing the collector-base junction. NPN and PNP transistors are both biased the same way, although the biasing voltages have opposite polarities for the two devices. Figure 20–4 shows a properly biased NPN transistor. In this circuit, a battery is connected across the emitter-base junction so that the voltage at the emitter is negative with respect to the base. Another battery is connected across the collector-base junction to ensure that the voltage at the collector is more positive than the voltage at the base.

When the emitter-base junction of an NPN transistor is forward biased, the barrier potential of the junction is low and the depletion region is quite narrow. This allows electrons to flow easily from the emitter to the base. At the same time, holes from the base flow into the emitter. Because the base region is narrow and has a low number of holes, most of the electrons reach the depletion region of the collector-base junction. Once across the collector-base junction, the electrons are attracted to the positive potential of the reverse-bias battery. Although most of the electrons injected into the base region from the emitter pass through the base to the collector, a small fraction of the injected carriers are drawn towards the forward-biased voltage connected to the base terminal. Consequently, the flow of current from the emitter branches into two currents: one that flows to the collector and one that flows to the base. However, because the base is lightly doped, the majority of the current flows directly from the emitter to collector. If a steady source of electrons is connected to the emitter, a continuous flow of current will occur between the collector and emitter of a properly biased BJT.

Figure 20–5 shows a properly biased PNP transistor. Once again the emitter-base junction is forward biased and the collector-base junction is reverse biased. However, because the emitter is now made of P-type material and the base is N-type, the biasing voltage is connected so that the emitter is more positive than the base. The collector-base junction of the PNP transistor is reverse biased by connecting a negative voltage to the collector with respect to the base.

When a transistor is properly biased, current will flow through each of the three leads. Figure 20–6(a) shows the relationship between the three transistor currents. The current flowing into the emitter is called the *emitter current,* I_E, the current flowing from the base is called the *base current,* I_B, and the current flowing from the collector is the *collec-*

FIGURE 20–5 *A properly biased PNP transistor.*

(a)

(b)

FIGURE 20–6 *(a) Current flow in an NPN transistor. (b) Current flow in a PNP transistor.*

tor current, I_C. The flow of current in a transistor always obeys Kirchhoff's current law. That is, the sum of the currents flowing into the device always must equal the sum of the currents leaving. Therefore,

$$I_E = I_C + I_B \qquad (20.1)$$

EXAMPLE 20–1

$I_E = 7.25$ mA

$I_B = 500$ μA

FIGURE 20–7 *Circuit for Example 20–1.*

Determine the collector current for the transistor circuit shown in Figure 20–7.

Solution

$$I_E = I_C + I_B$$

Therefore, $\qquad I_C = I_E - I_B = 7.25 \text{ mA} - 500 \text{ μA} = 6.75 \text{ mA}$

As mentioned earlier, more electrons leave the emitter than arrive at the collector. Consequently, the emitter current is always greater than the collector current. However, in many practical applications, the base current is very small. In fact, this current is often so small that it can be ignored when calculating circuit values such as voltage, current, and power. When the value of base current is ignored, the collector and emitter currents are said to be approximately equal to each other.

$$I_C \approx I_E \qquad (20.2)$$

The relationship between the amount of collector current that flows in a transistor compared to the emitter current flow is known as the DC alpha, α_{DC}.

$$\alpha_{DC} = \frac{I_C}{I_E} \qquad (20.3)$$

Because I_C and I_E are approximately equal to each other, the alpha in a typical BJT is nearly 1, usually between 0.95 and 0.99. However, α_{DC} will never be equal to or greater than 1. The α_{DC} of a transistor is a parameter set by the manufacturer and cannot be changed. It is simply a design ratio that specifies how much collector current will flow for a given value of emitter current. The α_{DC} of a transistor is also known as the *forward current transfer ratio* because it represents the amount of current actually transferred from the emitter to collector. Another parameter for the BJT specifies the ratio between base current and collector current and is called the DC beta, β_{DC}.

$$\beta_{DC} = \frac{I_C}{I_B} \qquad (20.4)$$

The DC beta of a transistor is also stated as the h_{FE} on the transistor data sheets. Because I_B is very small compared to I_C, the β_{DC} of a BJT is relatively high. Typical beta values range from 20 to 200. If the β_{DC} of a transistor is known, it is possible to calculate the α_{DC} by using the following equation:

$$\alpha = \frac{\beta}{\beta + 1} \tag{20.5}$$

EXAMPLE 20–2

When the base current of a BJT is 10 µA, a collector current of 2 mA flows. Determine the

(a) β_{DC}
(b) α_{DC}

Solution

(a)

$$\beta = \frac{I_C}{I_B} = \frac{2 \text{ mA}}{10 \text{ µA}} = 200$$

(b)

$$\alpha = \frac{\beta}{\beta + 1} = \frac{200}{200 + 1} = 0.995$$

Because α and β are ratios, these numbers are not expressed in units, although they are an indication of the transistor's amplifying capabilities. For example, if the base current were considered the input current and the collector were the output current, the transistor in Example 20–2 would be able to provide a current gain of 200. In other words, the output current would be 200 times greater than the input current.

Figure 20–8 shows a transistor circuit with biasing voltages V_{BB} and V_{CC}. The three voltages, V_E, V_C, and V_B represent the voltages at the emitter, collector, and base with respect to ground. The emitter voltage is calculated as follows:

$$V_E = I_E R_E \tag{20.6}$$

The base voltage, V_B, is found using Kirchhoff's voltage law. The voltage at the base (with respect to ground) is equal to the voltage drop across the forward-biased, emitter-base junction plus the voltage drop across R_E.

$$V_B = V_E + V_{BE} \tag{20.7}$$

The voltage at the collector (with respect to ground) is equal to the difference between the applied voltage, V_{CC}, and the voltage drop across resistor R_C.

$$V_C = V_{CC} - I_C R_C \tag{20.8}$$

The voltage drop between the collector and emitter, V_{CE}, represents the difference between the collector voltage, V_C, and the emitter voltage, V_E.

$$V_{CE} = V_C - V_E \tag{20.9}$$

FIGURE 20–8 *Transistor circuit with biasing voltages.*

EXAMPLE 20–3

FIGURE 20–9   *Circuit for Example 20–3.*

The transistor shown in Figure 20–9 has a β_{DC} of 100. Find the values of

(a) I_B
(b) I_C
(c) I_E

Assume that the transistor is made of silicon.

Solution

(a) The base current is solved using Ohm's law. Because the transistor is silicon, a forward voltage drop of 0.7 V is assumed for V_{BE}. The emitter is connected to ground, so the voltage at the base is equal to V_{BE}. The voltage dropped across R_B is the difference between V_{BB} and V_B.

$$I_B = \frac{V_{BB} - V_{BE}}{R_B} = \frac{2.5\ V - 0.7\ V}{39\ k\Omega} = 46.15\ \mu A$$

(b)
$$\beta_{DC} = \frac{I_C}{I_B}$$

Therefore,
$$I_C = \beta_{DC}I_B = (100)(46.15\ \mu A) = 4.62\ mA$$

(c)
$$I_E = I_C + I_B = 4.62\ mA + 46.15\ \mu A = 4.66\ mA$$

Section Review

1. What are the two basic types of BJTs?
2. What are the two depletion regions of the BJT called?
3. How is a BJT properly biased?
4. State the Kirchhoff current equation for the three currents in a BJT.

20–3
VOLTAGE DIVIDER BIASING

The biasing voltages that are used in transistor circuits are always steady DC values with low percent ripple. Although biasing voltages in transistor circuits are often represented by sources, such as batteries, it is also common to see a voltage "drop" used to bias a transistor. When biasing resistors are connected across the PN junctions, the voltage at one point can be made more or less positive than the other. One popular method of biasing a transistor is called *voltage divider biasing*. Figure 20–10 shows a typical voltage divider bias circuit. Resistor R_1 is connected across the collector-base junction and R_2 is connected across the emitter-base junction. The DC power supply is labeled V_{CC}. The voltage drop across R_2 provides the necessary forward bias for the emitter-base junction. The inclusion of R_E is required for proper circuit operation because this resistor essentially determines the value of emitter current, I_E. Consequently, this voltage divider configuration provides a stable value of I_E that is virtually independent of any fluctuations in the value of β_{DC}.

FIGURE 20-10   *Voltage divider bias circuit.*

Voltage divider bias circuits are classified as self-biasing because only one voltage source is required to forward bias the emitter-base junction and reverse bias the collector-base junction. A common approximation when determining the total resistance of two parallel resistances is that if two resistors are in parallel and one is at least ten times the other, the total resistance is considered to be approximately equal to the smallest value. A rule of thumb for voltage divider biased circuits is that if resistor R_2 is ten times smaller than the input resistance, R_{in}, of the transistor, R_{in} can be ignored. The voltage at the base of the transistor, V_B, is equal to the voltage dropped across resistor R_2 in Figure 20-10. Using the voltage divider rule,

$$V_B \approx V_{R2} = \left(\frac{R_2}{R_1 + R_2} \right) V_{CC} \qquad (20.10)$$

EXAMPLE 20-4

For the circuit of Figure 20-11, calculate the following values:

(a) V_B
(b) V_E
(c) I_E
(d) V_C
(e) V_{CE}

Assume that I_C and I_E are approximately equal and that the transistor is made of silicon.

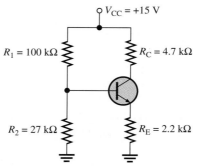

FIGURE 20-11   *Circuit for Example 20-4.*

Solution

(a) $\quad V_B \approx V_{R2} = \left(\dfrac{R_2}{R_1 + R_2} \right) V_{CC} = \left(\dfrac{27 \text{ k}\Omega}{100 \text{ k}\Omega + 27 \text{ k}\Omega} \right) 15 \text{ V} = 3.19 \text{ V}$

(b) $\quad V_E = V_B - V_{BE} = 3.19 \text{ V} - 0.7 \text{ V} = 2.49 \text{ V}$

(c) $\quad I_E = \dfrac{V_E}{R_E} = \dfrac{2.49 \text{ V}}{2.2 \text{ k}\Omega} = 1.13 \text{ mA}$

(d)
$$I_E \approx I_C$$
$$V_C = V_{CC} - I_C R_C = 15 \text{ V} - (1.13 \text{ mA})(4.7 \text{ k}\Omega) = 9.68 \text{ V}$$
(e)
$$V_{CE} = V_C - V_E = 9.68 \text{ V} - 2.49 \text{ V} = 7.19 \text{ V}$$

Section Review

1. Voltage drops are often used instead of batteries to bias a transistor (True/False).
2. What is voltage divider biasing?

20–4

THE BJT AS AN AMPLIFIER

When the transistor is used as an amplifier, it is essentially acting as a valve to control the amount of current flowing at the output. The flow of current through the transistor is varied by increasing or decreasing the amount of current flow through the emitter-base junction.

Figure 20–12 shows a BJT with the emitter connected to a variable DC input voltage and the collector connected to a load resistor. As the input signal increases, the voltage at the base becomes more positive. This causes an increase in the flow of electrons from the emitter to the base, resulting in an increase in the flow of holes from the base to the emitter. Because the base is lightly doped, less holes are available to move from the base to emitter than there are electrons to move from the emitter to the base. As a result, the majority of this current flow is from the emitter to the base. The base region is very narrow and, because of the mutual repulsion between electrons, practically all the electrons entering the base region from the emitter drift across the collector-base junction and flow through the collector region to the load.

In Figure 20–12, if V_{in} increases, it aids V_{EE}, causing the input current I_E to increase, and the output current I_C to also increase. This will cause the voltage drop across the load resistor to increase.

If the input voltage decreases, or becomes less positive, the bias voltage V_{BE} is reduced and current flow to the emitter and collector decreases. Consequently, the output voltage across the load resistor R_L will also decrease. The input impedance for the circuit of Figure 20–12 is fairly low because the emitter-base junction is forward biased. The output impedance for the circuit is very high because of the reverse bias of the collector-base junction. The biasing voltage V_{EE} is used to provide a constant forward-biased voltage for the emitter-base junction. The biasing voltage V_{CC} performs two functions: (1) It provides reverse-biased voltage for the collector-base junction and (2) it supplies the output current

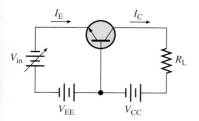

FIGURE 20–12 *Transistor amplifier circuit.*

and voltage for the load resistor. Because I_E is approximately equal to I_C, the transistor circuit shown in Figure 20–12 is not amplifying the current. However, the output voltage can be greater than the input because the input resistance is low and the output resistance is high.

Transistors are capable of amplifying three quantities: voltage, current, and power. The gain or amplification factor represents a ratio of the output signal to the input and is represented by the symbol $\mathcal{A}$. The current gain, $\mathcal{A}_i$, of an amplifier is a comparison between the output current and the input current.

$$\mathcal{A}_i = \frac{I_{out}}{I_{in}} \tag{20.11}$$

Figure 20–13 shows a voltage divider bias circuit with a load resistor, R_L, connected to the collector. The collector current, I_C, represents the load current, and I_B is the input. The base current acts as a controlling valve to limit the amount of output current flowing at the collector. As the base current increases, the collector current also increases. Because the collector current is much higher than the base current, the transistor circuit of Figure 20–13 provides excellent current gain. The current gain for this circuit is the same as the β_{DC} for the transistor. That is,

$$\mathcal{A}_i = \frac{I_{out}}{I_{in}} = \frac{I_C}{I_B} = \beta_{DC}$$

The voltage gain, $\mathcal{A}_v$, in an amplifier is an indication of the output voltage developed for a given value of input voltage.

$$\mathcal{A}_v = \frac{V_{out}}{V_{in}} \tag{20.12}$$

The voltage gain for the circuit shown in Figure 20–13 is found by dividing the collector resistance by the emitter resistance. Because the base voltage and emitter voltage are approximately equal, we can assume that R_E is the input resistance, and the voltage gain is a ratio of the output resistance to the input resistance. The input resistance is actually the sum of the internal resistance of the transistor at the emitter plus the external resistor R_E. Because the internal resistance is very low compared to R_E, it can be ignored in most calculations.

$$\mathcal{A}_v = \frac{R_C}{R_E} \tag{20.13}$$

FIGURE 20–13 *Current flow in voltage divider bias circuit.*

EXAMPLE 20–5

FIGURE 20–14 Circuit for Example 20–5.

For the circuit shown in Figure 20–14, determine the current gain and voltage gain.

Solution Because the load resistor is shown connected to the collector in Figure 20–14, the collector current represents the output current and the base current is the input.

$$I_C = I_E - I_B = 10 \text{ mA} - 200 \text{ μA} = 9.8 \text{ mA}$$

$$A_i = \frac{I_{out}}{I_{in}} = \frac{I_C}{I_B} = \frac{9.8 \text{ mA}}{200 \text{ μA}} = 49$$

$$A_v = \frac{R_C}{R_E} = \frac{4.7 \text{ kΩ}}{2.2 \text{ kΩ}} = 2.14$$

In electronics circuits, power is measured in watts and is found by multiplying the current times the voltage. The power gain, A_p, is a ratio of the output power produced by the transistor to the input power required to control the device.

$$A_p = \frac{P_{out}}{P_{in}} \tag{20.14}$$

The power gain can also be determined by finding the product of the current gain and voltage gain.

$$A_p = A_v A_i \tag{20.15}$$

EXAMPLE 20–6

Solution In Example 20–5, the voltage gain was found to equal 2.14 and the current gain was 49. The power gain for the circuit is determined using Equation 20.15.

Determine the power gain for the amplifier shown in Figure 20–14.

$$A_p = A_v A_i = (2.14)(49) = 105$$

When the emitter current is zero, the transistor is said to be *cutoff*. The collector current then consists of only the small current I_{CBO}. This value of current is also known as the *collector leakage current*. When the emitter current is at its absolute maximum, the transistor is referred to as being *saturated*. A BJT that is used as an amplifier will operate between these two extremes. A **load line** is a graphical representation of the operating range of a transistor between saturation and cutoff. Load lines are often used to analyze large-signal amplifiers, although the process will also provide information relevant to small-signal circuits.

Figure 20–15(a) shows a BJT with biasing voltages and a load resistor. In this circuit, the maximum potential difference between collector and emitter can never exceed the applied voltage, 10 V. This represents one extreme of the load line, as shown in Figure 20–15(b). The characteristic curves of Figure 20–15(b) are an indication of the amount of collector current flowing for a given value of base current and collector-emitter voltage. Each curve represents a different value of base current. When the base current is zero, the collector current is essentially zero. This is because the collector acts as a reverse-biased diode with only a small value of leakage current flowing. As the base current is increased in 0.1 mA steps, the collector current increases in 5 mA increments. Therefore, $\beta = 50$ for this device.

The other extreme of the load line shown in Figure 20–15(b) occurs when the transistor is fully on, or saturated. In this condition, the emitter-base junction is forward biased, and the base region is saturated with majority carriers from the emitter. When saturated, the collector current is limited only by V_{CC} and any series resistance, such as R_L. In Figure 20–15(a), the maximum collector current is found by Ohm's law.

$$I_{C(max)} = \frac{V_{CC}}{R_L} = \frac{10 \text{ V}}{500 \text{ }\Omega} = 20 \text{ mA}$$

When a load line is constructed, it is possible to predict all values of amplifier output voltage and current for any given input signal. Load current is the same as collector current in Figure 20–15(a), because all current leaving the collector does so via the load. The **quiescent,** or *Q point*, is specified as the point on the load line that indicates the steady-state condition of V_{CE} when no input signal is applied to the circuit. The Q point is typically located near the center of the load line to minimize signal distortion. In other words, the amplifier would be biased so that the collector voltage is one-half of the supply voltage,

(a)   (b)

FIGURE 20–15 *(a) Amplifier circuit. (b) Collector characteristic curves.*

V_{CC}. In Figure 20–15(b), the Q point is at 5 V. If a straight line is drawn up from the 5 V mark on the horizontal axis, it will intersect the load line at the point marked Q.

A manufacturer's specification sheet will often provide the characteristic curves for the transistor. In actual practice, these characteristic curves are average curves and give only typical values. Because transistor characteristics vary widely, the only way to get accurate curves for a particular transistor is to use an oscilloscope curve tracer.

Section Review

1. The flow of current through a BJT is varied by increasing or decreasing the amount of current flow across the emitter-base junction (True/False).
2. What three quantities are transistors capable of amplifying?
3. State the DC beta of a transistor in equation form.
4. The voltage gain of an amplifier is an indication of the output voltage developed for a given value of input voltage (True/False).
5. What is a load line?

20–5

THE TRANSISTOR AS A SWITCH

When a transistor is used as a switch, it is either driven into saturation or cutoff by interrupting the flow of current to the base. Figure 20–16 shows a switch connected in series with the input voltage and the base. When the switch is open, as in Figure 20–16(a), the base current is zero, and the voltage drop between the collector and emitter is approximately equal to V_{CC}, the applied voltage.

Figure 20–16(b) shows the input switch closed and current flowing from the collector through the load. The BJT is said to be saturated when the voltage drop across the transistor's collector and emitter terminals is approximately equal to zero volts. This means that the entire bias voltage, V_{CC}, appears across the load R_L.

$$I_C R_L = V_{CC}$$

The maximum collector, or saturation, current is found by Ohm's law.

$$I_C = \frac{V_{CC}}{R_L}$$

To determine the minimum amount of base current required to produce saturation in a BJT, the beta Equation (20.4) is used.

(a)   (b)

FIGURE 20–16   Transistor as a switch. (a) Cutoff. (b) Saturated.

$$\beta = \frac{I_C}{I_B}$$

Therefore,

$$I_B = \frac{I_C}{\beta}$$

The input voltage required to produce saturation is found using Kirchhoff's voltage law. In Figure 20–16, the input voltage must be large enough to overcome the forward-biased voltage drop of the emitter-base junction and the voltage drop across the base resistor, R_B.

$$V_{in} = I_B R_B + V_{BE} \tag{20.16}$$

EXAMPLE 20–7

Calculate the minimum base current and input voltage required to fully turn on the transistor shown in Figure 20–17.

FIGURE 20–17   *Circuit for Example 20–7.*

Solution

$$I_{C(max)} = \frac{V_{CC}}{R_L} = \frac{18\ \text{V}}{1\ \text{k}\Omega} = 18\ \text{mA}$$

$$I_B = \frac{I_C}{\beta} = \frac{18\ \text{mA}}{80} = 0.225\ \text{mA}$$

$$V_{in} = I_B R_B + V_{BE} = (0.225\ \text{mA})(10\ \text{k}\Omega) + 0.7\ \text{V} = 2.95\ \text{V}$$

Section Review

1. Switching transistors operate in either cutoff or saturation (True/False).
2. A BJT is saturated when V_{CE} equals the applied voltage (True/False).

20–6

TESTING BIPOLAR TRANSISTORS

BJTs are extremely rugged, durable, solid-state devices that are capable of operating for years without failure. However, like any semiconductor device, they are susceptible to excess voltage, current, and power. When a transistor is overheated, the high temperatures can cause permanent damage, resulting in a short- or open-circuit condition inside the device. Although shorts and opens are the two most common types of transistor failure, damaged transistors are also capable of operating with different characteristics than those stated in the manufacturer's specification sheet. For example, if the semiconductor material inside the transistor case is melted slightly by excess heat, the transistor may still con-

In-circuit
test leads

Out-of-circuit
test socket

Out-of-circuit
test socket

In-circuit test
sockets

FIGURE 20–18 *Transistor testers.*

duct but with different values of base and collector current. Consequently, the beta of the
transistor is altered, and the BJT no longer operates at its rated values.

Transistors can be tested with a variety of meters. Figure 20–18 shows transistor
testers that can be used to determine operating characteristics including continuity, leakage
current, and beta.

Because the majority of faults that occur in a transistor are either a short- or open-circuit condition, the ohmmeter is popular as a test instrument. Figure 20–19(a) shows the
proper forward-biased DMM readings for an NPN transistor. When the selector switch is
on the semiconductor test position, it will beep and indicate a voltage of under 1 V for a
proper forward-biased reading. The positive lead of the meter is connected to the base of
the transistor and the negative lead is connected to either the emitter or collector. If using

(a)

(b)

FIGURE 20–19 *Testing NPN transistor with an ohmmeter. (a) Forward-
biased test. (b) Reverse-biased test.*

an ohmmeter, the meter readings should correspond to those of a forward-biased diode and be approximately 300 Ω to 700 Ω. If the meter reading is zero ohms, a short circuit exists.

Figure 20–19(b) shows the DMM readings when the leads are reversed to test for reverse-biased conditions. The letters "OL" are displayed to indicate a reverse-biased condition. When the positive lead of an ohmmeter is connected to the N-type material of the emitter or collector and the negative lead is connected to the P-type base, the meter reading should be in the range of several megohms. Ideally, the ohmmeter should read infinity. If the meter reads zero ohms, a short is present. If the reading is low, such as hundreds or even thousands of ohms, the transistor may be leaky. That is, it will pass substantial current when cutoff.

Figure 20–20(a) shows the DMM readings for testing the forward-bias characteristics of the two PN junctions in the PNP transistor. Because the base is now made of N-type material, the negative lead of the meter is connected to the base and the positive lead is connected to the emitter or collector. Once again, when using an ohmmeter, the meter readings should never be zero ohms. Instead, the reading should be in the range of several hundred ohms. An open circuit would be indicated by infinity.

In Figure 20–20(b), the correct meter readings are shown for testing a PNP transistor's reverse-bias characteristics. The positive lead of the meter is connected to the base and the negative lead is connected to either the emitter or collector. For a DMM on the semiconductor test position, a reverse-biased condition will be indicated by the letters "OL." When using an ohmmeter, this reading should be in the order of several megohms of resistance.

In some situations, it is necessary to perform in-circuit tests of transistors. When a transistor is still connected in an energized circuit, it is possible to test for current, voltage, and power using ammeters, voltmeters, and oscilloscopes. Always, great care should be taken when performing tests on energized equipment. Never touch any component unless you have verified that it is at ground potential. Although components like heat sinks and transistor cases may seem to be grounded, they may in fact be carrying current. Power transistors often use the case of the device as the collector, and the transistor is connected to a heat sink via the case. Consequently, the heat sink may have a voltage equal to the supply voltage for the circuit.

(a)   (b)

FIGURE 20–20 *Testing PNP transistor with an ohmmeter. (a) Forward-biased test. (b) Reverse-biased test.*

FIGURE 20–21 *In-circuit test of BJT using voltmeter. (a) Voltage reading taken between emitter and collector. (b) Forward bias removed by jumper wire.*

Figure 20–21 shows a simple in-circuit test to determine if a transistor is functioning. In Figure 20–21(a) a voltmeter is connected between the emitter and collector of an energized circuit. The voltmeter reading is recorded. A jumper wire is then connected between the base and emitter, as shown in Figure 20–21(b). The purpose of the jumper wire is to create a short circuit at the base and shunt the base current around the transistor through the jumper. By removing the forward bias, the transistor should turn off. The voltmeter reading should indicate a value approximately equal to the supply voltage.

Section Review

1. BJTs are fragile, highly sensitive devices (True/False).
2. BJTs cannot be tested using an ohmmeter (True/False).

_____ **20–7** ___

FIELD EFFECT TRANSISTORS (FETs)

The **field effect transistor, FET** is a three-terminal, voltage-controlled semiconductor device. The three leads of the FET are called drain, source, and gate. The gate is typically used as the input. The majority of current flow in the FET occurs between the source and drain. The path connecting these two leads is called the **channel.** The main difference between a BJT and a FET is that the BJT is a current-operated device and the FET is voltage operated. Because the BJT is based on the principle of current flowing into or out of the input terminal, the input resistance to the device is relatively low. Ideally, the input resistance to an amplifier should be infinity. The input resistance of the FET is very high, which makes it more suitable for certain amplifier applications.

The FET is considered unipolar because it has one PN junction instead of the two that are found in bipolar transistors. In addition to having a higher input resistance than BJTs, the FET also provides better isolation between input and output circuits and has less internal noise than BJTs. *Noise* is a term associated with random signal fluctuations caused by the movement of electrons in the semiconductor material. Another advantage of the

FET is that unlike the BJT, the FET is not subject to thermal runaway when the temperature of the device increases.

As any semiconductor device, the FET must be properly biased in order to operate. The BJT is turned on by increasing the input current, while the FET is turned off by increasing the input biasing voltage. As the external bias voltage of the FET increases, the internal resistance increases, and the transistor is eventually cutoff. Field effect transistors are grouped into two categories: junction field-effect transistors (JFETs) and metal oxide semiconductor field-effect transistors (MOSFETs). The MOSFET is also known as an insulated-gate field effect transistor (IGFET) because of its construction. FETs are available in two types: N-channel and P-channel.

Section Review

1. The FET is a three-terminal current-operated device (True/False).
2. What are the three leads of the FET called?
3. What are the two types of FETs?

— 20–8

JUNCTION FIELD EFFECT TRANSISTORS (JFETs)

Figure 20–22 shows the basic structure of an N-channel **JFET.** The channel is made of N-type material and each of the P-type regions is called a gate. Although most JFETs have the two gates connected internally to produce a single external gate lead, special types of JFETs are designed with two external gates. These devices are called *dual-gate JFETs,* and are occasionally used in communications equipment. Because the single-gate JFET is by far the most common JFET used in industry, we will limit our discussion of FETs to this type.

For proper biasing, the source of an N-channel JFET is connected to a negative voltage or to ground and the drain is connected to a positive potential. Figure 20–23(a) shows a properly biased N-channel JFET. In this transistor the majority carriers consist of electrons flowing through the channel from the source to the drain. The gate is connected to a negative potential, which forms a strong electrostatic field around the center of the channel. As the gate voltage is made more negative, the depletion layer widens and the flow of

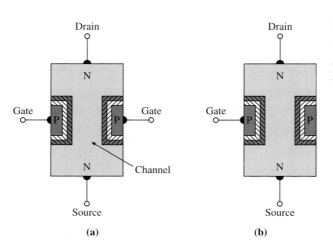

FIGURE 20–22   *An N-channel JFET. (a) Dual-gate JFET. (b) Single-gate JFET.*

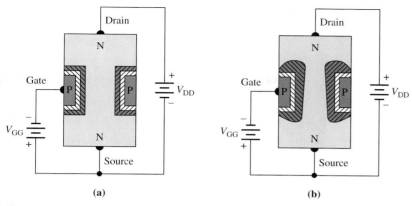

FIGURE 20–23 *(a) A properly biased N-channel JFET. (b) The depletion layer widens as the input voltage becomes more negative.*

current through the channel is restricted. Figure 20–23(b) illustrates how the flow of current is restricted in the channel by an increase in the depletion region. Because the two gates are at the same potential, both depletion layers widen by an equal amount. As the gap between the two depletion layers becomes narrower, the device begins to act as a valve, limiting the amount of current flow through the FET.

Figure 20–24(a) shows the field effect structure of a P-channel JFET. In this device, the majority current carriers are holes. The gate is made of N-type material and is reverse biased by applying a positive gate voltage. Figure 20–24(b) illustrates a properly biased P-channel transistor.

Figure 20–25 shows the schematic symbols for the JFET. The N-channel JFET is represented by the symbol shown in Figure 20–25(a), and the P-channel is illustrated in Figure 20–25(b). In both diagrams, the arrow is shown on the gate lead. In Figure 20–25(a), the arrow points towards the N-type channel; In Figure 20–25(b), the arrow points towards the N-type gate and away from the P-type channel.

Figure 20–26(a) shows a self-biased N-channel JFET with a single voltage source. In this circuit, both the gate and source are connected to ground. Consequently, the gate-to-

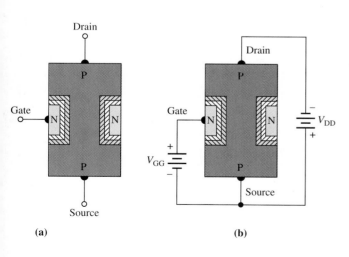

(a)

(b)

FIGURE 20–24 *(a) A P-channel JFET. (b) A properly biased P-channel JFET.*

FIGURE 20–25 *JFET schematic symbols. (a) N-channel. (b) P-channel.*

Drain

Gate

Source

(a)

Drain

Gate

Source

(b)

source voltage, V_{GS}, would be zero. As the supply voltage, V_{DD}, increases, the current flow increases and the depletion layers widen. The velocity of the electrons in the channel also increases until the channel is saturated with electrons rushing to the positive potential of the supply voltage. When the channel is saturated, any further increase in supply voltage will have no effect on the drain current, I_D. The depletion region will expand to a point where the channel becomes constricted and only high-velocity electrons are capable of pushing through.

The graph of Figure 20–26(b) is a characteristic curve for the drain current flowing in the JFET. The point where the transistor becomes saturated is called the **pinchoff point** and represents the point where the rate of current change approaches zero. We should note that the drain current at the pinchoff point is not zero; it is a constant DC current labeled I_{DSS}. When the drain-to-source voltage, V_{DS}, increases beyond the pinchoff point, I_{DSS} remains constant until the breakdown voltage, V_{BR}, is reached. The amount of voltage required to produce pinchoff is called the **pinchoff voltage, V_p.** FETs that are biased to operate at drain current levels below I_{DSS} are said to operate in the **depletion mode.** All JFETs are depletion mode devices. However other FETs, such as some MOSFETs, are biased to operate at drain currents higher than I_{DSS} and are said to operate in the **enhancement mode.**

Figure 20–27(a) shows a FET circuit with a variable negative voltage, V_{GS}, connected between the gate and source. When V_{GS} is zero, the drain current will be at its maximum value. As V_{GS} goes negative, the drain current will decrease, because the increase in reverse bias on the gate pinches off the channel even further. Figure 20–27(b) shows the family of characteristic curves for drain current at various values of V_{GS}. In Figure 20–27(b) a gate-to-source voltage of -4 V will produce cutoff and the drain current will be at its minimum value.

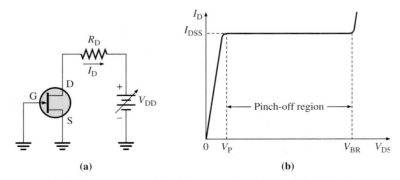

(a)

(b)

FIGURE 20–26 *(a) A JFET biased with $V_{GS} = 0$ V. (b) Drain characteristic curve for $V_{GS} = 0$ V.*

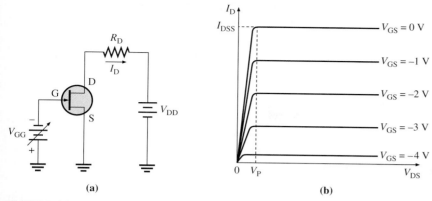

(a)

(b)

FIGURE 20–27 *(a) An N-channel FET circuit. (b) Characteristic curves for various values of V$_{GS}$.*

The relative ability of a gate to control the drain current of a FET is called **transconductance,** g_m. In Figure 20–27, a change in V_{GS} produced a change in I_D when the drain-to-source voltage was held constant. The amplifying ability of a FET is measured in terms of a change in input voltage producing a change in output current. Ideally, a small change in input voltage should produce a large change in output current. The conductance of a circuit was defined as a ratio of current divided by voltage and was measured in siemens, S. The transconductance, or mutual conductance, of a FET is a ratio between the change in drain current ΔI_D, produced by a change in gate-to-source voltage ΔV_{GS}.

$$g_m = \frac{\Delta I_D}{\Delta V_{GS}} \tag{20.17}$$

Figure 20–28 shows a transconductance curve for a typical JFET. The curve plots I_D versus V_{GS} and is obtained by holding V_{DS} constant while V_{GS} is allowed to vary. This type of curve, called a *transfer curve,* is very useful in determining the Q point of the FET.

Section Review

1. Explain the pinchoff point of a FET.
2. The drain current is zero at the pinchoff point of a FET (True/False).
3. What is transconductance?

FIGURE 20–28 *Transfer characteristic curve for a JFET.*

METAL OXIDE SEMICONDUCTOR FETs (MOSFETs)

MOSFETs are FETs in which the gate is electrically insulated from the source-drain channel. The input to a MOSFET acts as a capacitor. MOSFETs use silicon dioxide as a dielectric. These devices have an extremely high input impedance developed by using a "capacitor" instead of a reverse-biased PN junction. The main disadvantage of MOSFET devices is that static electricity can easily destroy the thin insulating material. As a result, MOSFETs are shipped with a shorting ring wrapped around the terminals of the transistor. Great care always should be taken when handling any MOS-type device.

There are two basic types of MOSFETs—N-channel and P-channel. MOSFETs are also classified by the conducting state when a bias voltage of zero volts is applied. When the device is conducting with zero bias, it is said to be a *depletion-mode device.* If it is not conducting with zero bias, it is referred to as an *enhancement-mode MOSFET.* The name *enhancement MOSFET* is derived from the fact that a voltage is needed on the gate to enhance, or increase, current flow in the channel. Enhancement-mode MOSFETs can be either P-channel or N-channel. If the device is an enhancement-mode MOSFET, no current will flow between the drain and source when V_{GS} is zero.

The depletion MOSFET derives its name from the fact that a voltage on the gate is used to deplete, or reduce, channel current flow. N-channel MOSFETs have a positive layer of fixed charge located near the surface in the silicon dioxide. Figure 20–29(a) shows the basic structure of an N-channel, depletion-mode MOSFET. This device consists of two N-type regions diffused onto a P-type substrate. The two heavily doped N-type regions are called the *source* and *drain.* A moderately doped N-type channel runs between the source and the drain. The channel is insulated from the gate by a layer of oxide (SiO_2). The gate is constructed of a thin layer of aluminum located in the center of the channel. If a positive voltage is applied to the drain and a negative voltage applied to the gate, the channel would appear as shown in Figure 20–29(b). The negative gate potential will attract holes from the P-type substrate and repel or neutralize the electrons in the moderately doped N-type channel. In Figure 20–29(b), the channel is shown as being depleted of carriers and, as a result, the drain current will decrease.

Because the gate of the MOSFET is electrically isolated from the channel, it is also possible to operate the device with a positive gate-to-source voltage. As V_{GS} increases, the number of free electrons flowing through the channel increases. Consequently, the current

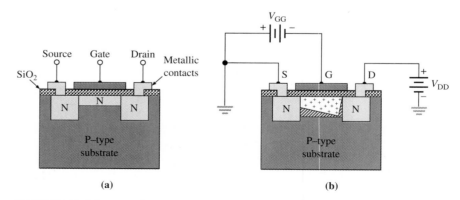

(a)   **(b)**

FIGURE 20–29 *(a) An N-channel depletion-mode MOSFET. (b) Channel depleted of carriers.*

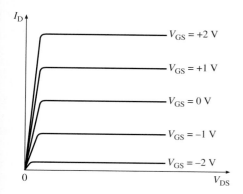

FIGURE 20–30 *Drain curves for depletion-enhancement MOSFET.*

flow increases as V_{GS} is made more positive. Figure 20–30 shows the drain curves for a typical N-channel depletion-mode MOSFET. The curves are formed by adjusting V_{GS} to various positive and negative values and observing the relationship between V_{DS} and I_D for each value of V_{GS}. From these curves it is apparent that the MOSFET operates in the depletion mode when a negative V_{GS} is applied and in the enhancement mode when a positive V_{GS} is applied. As a result, the depletion MOSFET is also known as *depletion-enhancement (DE) MOSFET.*

A P-channel depletion-mode MOSFET consists of two heavily doped P-regions, which have been diffused into a lightly doped N-type substrate. The gate is once again separated from the channel by a thin layer of silicon dioxide. The basic operation of the P-channel depletion-mode MOSFET is essentially the same as the N-channel device. The main difference is that the majority carriers in the P-type channel are holes instead of electrons. Also, the drain lead of the P-channel device must be made negative with respect to its source lead. As a result, the drain current, I_D is opposite in direction to that of the N-channel depletion-mode MOSFET. The schematic symbols for depletion-mode MOSFETs are shown in Figure 20–31. Like the JFET, the arrow is used to indicate whether the device is an N-channel or P-channel. MOSFETs are also available as four-terminal devices with the substrate brought out as a separate lead. In Figure 20–31, the substrate is shown connected to the source. This connection can be made internally or externally, depending on whether the device has three leads or four.

Figure 20–32(a) shows the construction of an N-channel enhancement-type MOSFET. The main difference between the enhancement-mode MOSFET and the depletion-mode MOSFET is the enhancement-mode device has no channel. For this transistor to conduct, the gate must be biased with the same polarity as the drain. Figure 20–32(b) illustrates the proper biasing for an N-channel enhancement-mode MOSFET. As V_{GS} is increased, the conductivity of the channel is enhanced and electrons are concentrated in the substrate just below the layer of insulation under the gate. Eventually, a channel is formed between the source and drain as the minority electron carriers are drawn from the substrate to the positive gate voltage. This principle is very similar to that of a typical capacitor. In the N-channel enhancement-mode MOSFET, the gate area represents the top plate of a capacitor, the P-

(a)

(b)

FIGURE 20–31 *Schematic symbols for depletion-enhancement MOSFET. (a) N-channel. (b) P-channel.*

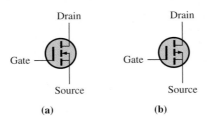

FIGURE 20–32 *(a) Construction of N-channel enhancement-mode MOSFET. (b) Biasing voltages required to create the channel.*

FIGURE 20–33 *Schematic symbols for enhancement-mode MOSFET. (a) N-channel. (b) P-channel.*

type substrate forms the bottom plate, and the silicon dioxide is the dielectric. The amount of gate voltage required to produce a channel is called the **threshold voltage.** If the bias voltage drops below the threshold needed to enhance the channel, current flow stops. A negative bias voltage will also produce cutoff in an N-channel enhancement-mode MOSFET.

Enhancement-mode MOSFETs can also be designed using P-type material for the drain and source and an N-type substrate. In this device, a negative gate-to-source voltage will attract holes toward the gate to form a P-channel, enhancement-mode MOSFET. For this device to attract channel carriers, the drain voltage must be negative with respect to the source. Figure 20–33 shows the schematic symbols for the two types of enhancement-mode MOSFETs. The schematic symbol of Figure 20–33(a) represents an N-channel device, and Figure 20–33(b) represents a P-channel enhancement-mode MOSFET.

Section Review

1. What is a MOSFET?
2. How are MOSFETs classified?
3. What is an enhancement-mode MOSFET?
4. Enhancement-mode MOSFETs can be either N-channel or P-channel (True/False).
5. What is threshold voltage?

___ 20–10 _____

TESTING FIELD EFFECT TRANSISTORS

As mentioned in Section 20–6, the ohmmeter is well suited for testing all types of transistors. When testing a FET with an ohmmeter, the device should be removed from the circuit. The resistance of the channel between drain and source should be approximately 10

FIGURE 20–34 *Testing an N-channel JFET with an ohmmeter. (a) Forward-biased test. (b) Reverse-biased test.*

kΩ for a JFET and over 2 MΩ for a MOSFET. A reading of zero ohms would indicate that the device is shorted, and a reading of infinity could mean that an open circuit exists. If the device is a MOSFET, many digital ohmmeters will indicate infinity (OL) because of the inherently high resistance between the drain and source. Figure 20–34 shows the connections for testing a JFET with a DMM. The N-channel JFET has a gate made of P-type material and should indicate a forward-bias condition when the positive lead of the ohmmeter is connected to the gate and the negative lead is connected to either the source or the drain, as indicated in Figure 20–34(a). When the leads are reversed, as shown in Figure 20–34(b), the DMM should indicate infinity (OL).

An ideal MOSFET should have infinite resistance between the gate and any other terminal. The resistance between the source and drain of an enhancement-mode MOSFET should be several megohms because there is no actual channel between the two terminals. A reading of several hundred ohms is typical for the drain-to-source resistance of an depletion-mode MOSFET. Figure 20–35 illustrates the ohmmeter tests for an N-channel, depletion-mode MOSFET. Because the gate is isolated from the channel, the gate-to-source resistance and gate-to-drain resistance should be several megohms regardless of how the leads are connected. Figure 20–35(a) shows the positive lead of the meter connected to the gate and the negative lead connected to the source and drain. In Figure 20–35(b), the leads are reversed and the meter readings do not change.

For continuity checks on FETs, only ohmmeters with a low-ohm setting should be used. Otherwise, the ohmmeter's battery voltage could switch on the transistor and produce a false reading. Also, devices such as MOSFETs are extremely sensitive to voltage and could be destroyed by ohmmeter tests. FETs are often tested during in-circuit operation using a voltmeter or oscilloscope. An oscilloscope allows the input and output waveforms to be monitored and compared against those in the service literature. MOSFETs should be tested in-circuit whenever possible because of their high susceptibility to electrostatic damage during insertion and removal. When MOSFETs are installed or removed, the leads should be shorted together, and great care should be taken when handling these devices. As previously mentioned, all MOS-type devices are sensitive to static electricity. Environments with low humidity levels that have synthetic floor coverings, such as carpeting, can produce particularly high levels of electrostatic discharge (ESD). In some circum-

FIGURE 20–35 *Testing an N-channel depletion-mode MOSFET with an ohmmeter. (a) Forward-biased test. (b) Reverse-biased test.*

stances, a person can be charged up to 20 kV in these environments. For this reason, it is good practice to discharge the static electricity in your body before handling any MOS devices. Although a ground strap should be used when testing MOSFETs, static electricity can be discharged to a safe level by simply wetting your fingers and touching a grounded metal object.

Section Review

1. An ideal MOSFET should have infinite resistance between the gate and any other terminal (True/False).
2. For continuity checks on FETs, what type of ohmmeter should be used?

— 20–11
THYRISTORS

The term **thyristor** is applied to a family of semiconductor devices that are used extensively in high-power electronic switching circuits. A thyristor is basically a semiconductor device that exhibits the regenerative switching characteristic of a four-layer or PNPN arrangement. Although BJTs and FETs are capable of performing switching functions, these devices are limited in terms of their current-handling capabilities. In industrial electronic circuits, load currents can often be in the hundreds, or even thousands, of amperes. Thyristors provide a highly efficient means of switching such large values of current and have excellent heat dissipation qualities. The two most popular types of thyristors used in industry are the **silicon controlled rectifier, SCR** and **triac.** Figure 20–36 shows some typical thyristors.

SCRs

The three leads of the SCR are called anode, cathode, and gate. Although the device is basically a switchable diode, the theory of its operation has certain similarities with BJTs. The SCR is constructed using four layers of semiconductor material arranged in a PNPN

FIGURE 20–36 Typical triacs.

configuration as illustrated in Figure 20–37(a). Similar to a typical diode, the anode of the SCR is composed of P-type material, and the cathode is composed of N-type material as shown in Figure 20–37(b). The gate of the SCR is made of P-type substrate. Figure 20–37(c) shows an equivalent circuit consisting of an NPN transistor and a PNP transistor. The SCR is a latching-type device that requires a small value of positive gate current for a brief instant to switch on. Once the SCR has been switched on, it will remain on as long as a minimum value of current, called the **holding current,** flows between the anode and cathode.

In Figure 20–37(c), when a positive current is applied to the base of transistor Q_2 it causes Q_2 to turn on. The collector of Q_2 drives the base of Q_1 negative, forward biasing the emitter-base junction of Q_1. Current now flows at the base of Q_1 and Q_1 conducts. If the gate is now disconnected, current will continue to flow at the base of Q_2 through the collector of Q_1. Therefore, the two transistors hold each other in the conducting state, allowing a continuous flow of current between the anode and cathode.

Figure 20–38 shows the schematic symbol for an SCR. This symbol is the same as that of a diode except an extra lead for the gate has been added. The cathode is generally represented by the letter "K" on schematic diagrams and even on the device itself. The SCR is a current-operated device and requires a certain value of positive gate current to switch it into a conducting state. The amount of gate current required to switch on an SCR is called the **gate trigger current,** I_{GT}, and is required for only a brief instant (10 to 50 μs). To switch off an SCR, the holding current must be reduced below its minimum value. This switching-off process is called *commutation.*

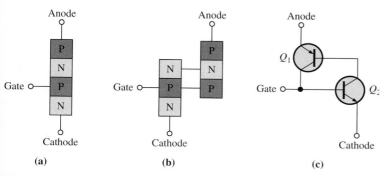

(a)   (b)   (c)

FIGURE 20–37   SCR equivalent circuits. (a) PNPN structure. (b) NPN and PNP interconnections. (c) Transistor equivalent circuit.

Anode (A)

Gate (G)

Cathode (K)

FIGURE 20–38 *Schematic symbol for SCR.*

Figure 20–39 shows the characteristic curve for an SCR. If current is to flow through the device, it must be properly biased. That is, the anode must be more positive than the cathode. If the SCR is reverse biased, current will not flow even when the gate is triggered. The characteristic curve shows that, when the forward voltage, V_F, is increased, the SCR conducts only a small amount of leakage current initially. As V_F continues to increase, eventually the breakover voltage is reached, and the device latches into a conducting state. The forward breakover voltage represents the point where the SCR switches from a non-conducting state to a conducting state with no triggering gate signal. The SCR is almost always triggered at a value below the breakover voltage. Unlike the diode, the SCR has a breakover point that can be changed by increasing the gate current. The reverse breakdown voltage is equivalent to the zener or avalanche region of a typical diode. This voltage is usually not exceeded. If the device is operated in the zener region for prolonged periods, the SCR may be damaged permanently.

SCRs are primarily used to vary the amount of current supplied to a load. Figure 20–40(a) shows an SCR circuit that rectifies and controls the amount of current supplied to the load resistor R_L. Resistor R_1 is a variable resistor that is used to control the point in the sine wave where I_{GT} is reached. Diode D_1 is used to ensure that only positive current is applied to the gate. Figure 20–40(b) illustrates the AC input wave for the circuit. The rectified signal of Figure 20–40(c) represents the gate current for the SCR. As E_{in} goes positive, the SCR is forward biased, but will not conduct until the trigger current, I_{GT}, is exceeded (point X). When the SCR turns on, maximum current flows through the circuit and the load voltage is approximately equal to the input voltage, as shown in Figure 20–40(d). The SCR acts as a closed switch when triggered, and the anode-to-cathode voltage, V_{AK}, is approximately equal to zero, as indicated in Figure 20–40(e). The SCR remains on until E_{in} decreases to the point where the load current is below the SCR holding current. When E_{in} decreases below the holding current, the SCR turns off and remains off because its anode-cathode is reverse biased. Because the SCR is acting as an open switch when reverse biased, the load voltage is approximately zero during this period.

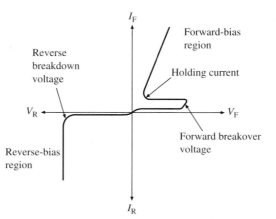

FIGURE 20–39 *Characteristic curve for SCR.*

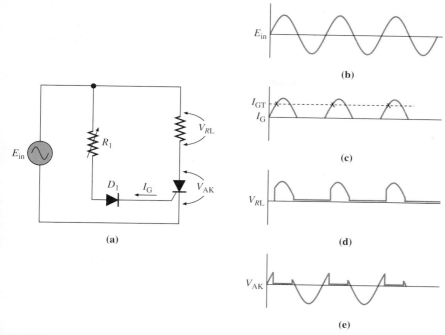

(b)

(c)

(d)

(e)

FIGURE 20–40 *SCR phase angle control circuit and waveforms.*

FIGURE 20–41 *SCR configuration for phase angle control of a complete sine wave.*

The load voltage waveform shown in Figure 20–40(d) can be controlled by varying resistor R_1. If R_1 is increased, the gate current will reach its trigger value, I_{GT}, at a greater value of E_{in} and the SCR will trigger at a later point in the E_{in} positive half-cycle. Therefore, the conduction time per cycle, or conduction angle, will decrease and the amount of current supplied to the load will decrease. This type of control, called *phase angle control,* is popular in a variety of control systems. The circuit shown in Figure 20–40(a) provides limited phase angle control. In fact, the load waveform can only be controlled between zero and 90° in this circuit. To achieve phase angle control for a full 360° of the input cycle, it is necessary to use two SCRs connected anode-to-cathode, as shown in Figure 20–41. By using two SCRs, it is possible to control the positive input voltage with one SCR and the negative input voltage with the other.

Triacs

The triac is a semiconductor device that functions in essentially the same manner as the circuit shown in Figure 20–41. The schematic symbol for a triac is shown in Figure 20–42. The symbol consists of two parallel diodes connected in opposite directions with a single gate lead attached. The main difference between the triac and the SCR is that the triac will allow current flow regardless of the biasing voltage and regardless of the polarity of the trigger pulses at the gate. Because there is no longer a specific anode or cathode, these

MT 2

Gate

MT 1

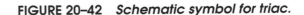

FIGURE 20–42 *Schematic symbol for triac.*

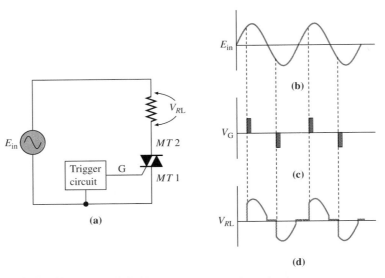

FIGURE 20–43 *(a) Triac circuit. (b) Waveforms*

leads are referred to as main terminals *MT*1 and *MT*2. Triacs are turned on by applying either a positive or negative voltage at the gate of the device. Once the triac has been latched into conduction, the gate has no further control over the device. The triac is commutated by lowering the holding current below its rated value.

Figure 20–43 shows a typical triac circuit and its waveforms. The trigger circuit shown in Figure 20–43(a) provides positive and negative input pulses to the gate of the triac. The AC input waveform of Figure 20–43(b) shows the point where the triac latches into conduction. When the triac is latched on, current flows through the load and the triac remains latched until the holding current falls below minimum. At this point, load voltage is zero and the device remains off until another trigger pulse is applied, as shown in Figure 20–43(c). The waveform of Figure 20–43(d) shows the output waveform for the load voltage.

Diacs

The **diac** is a two-terminal semiconductor device that is also classified as a thyristor. The diac is basically a triac that has no gate lead. Figure 20–44 shows the schematic symbol for a diac. The diac consists of a three-layer "sandwich" of alternate P-type and N-type semiconductors, similar to a typical BJT. In operation, when the voltage applied to the diac exceeds the breakover voltage, the diac latches on and remains in the conducting state until the anode voltage is removed. The diac is a bilateral device; that is, it will function with voltage of either positive or negative polarity applied to it. The diac is used in a variety of applications including triggering circuits for SCRs and triacs.

MT 2

FIGURE 20–44 *Schematic symbol for diac.*

MT 1

Section Review

1. What is a silicon-controlled rectifier (SCR)?
2. What are the three leads of an SCR called?
3. How is an SCR switched off?

_____ **20–12** ___

TESTING THYRISTORS

Testing four-layer thyristors such as SCRs and triacs is generally more difficult than testing a diode or transistor. Four-layer semiconductor devices have high values of internal resistance and the breakover voltages for thyristors are often higher than those provided by a typical ohmmeter. Figure 20–45 shows a standard method of testing an SCR. When functioning properly, the SCR should behave as a diode between the gate and cathode, and the anode and cathode ideally should have infinite resistance between its terminals. In Figure 20–45(a) a resistance reading is taken between the anode and cathode with the gate open, or unconnected. A high value of resistance is observed. The test circuit of Figure 20–45(b) shows a jumper wire connected between the gate and anode, and the ohmmeter indicates a low-resistance value. This indicates that the SCR has latched from the current supplied from the ohmmeter. If the jumper is now removed, the ohmmeter reading should still be low from the SCR remaining latched in the conducting state. If either ohmmeter lead is removed and reconnected, the ohmmeter reading should rise to a very high reading.

The procedure for testing triacs is essentially the same as the method used for testing SCRs. An ohmmeter reading between *MT*1 and *MT*2 is obtained with the gate open and the

(a)   (b)

FIGURE 20–45 *Ohmmeter connections for checking an SCR.*
(a) High-resistance reading. (b) Low-resistance reading.

(a)

(b)   (c)

FIGURE 20–46 *Triac test connections. (a) Nonconducting state. (b) Conducting state. (c) Latched state.*

triac in a nonconducting state, as shown in Figure 20–46(a). The ohmmeter should read a very high resistance. If the gate is shorted to either *MT*1 or *MT*2, the ohmmeter reading should immediately drop to a much lower value, as shown in Figure 20–46(b). If the jumper wire is removed, the resistance reading should remain low. The triac is then commutated, or switched off, by removing the jumper wire and momentarily disconnecting the ohmmeter. The jumper is now connected between the gate and the other main terminal, as shown in Figure 20–46(c). The ohmmeter reading should be low, and it should remain low when the jumper is removed.

Section Review

1. When functioning properly, an SCR should behave as a diode between the gate and cathode (True/False).
2. Ideally, the resistance between the anode and cathode terminals of an SCR should be zero (True/False).

The **unijunction transistor, UJT** is a three-terminal, single-junction, field-effect device. Despite its name, the basic function of a UJT more closely resembles a thyristor than a transistor. UJTs are never used as amplifiers but they are popular in switching circuits. The UJT is used primarily in electronic circuits such as oscillators, timing circuits, and as triac and SCR trigger circuits. Figure 20–47(a) shows the basic construction of a UJT. The three leads of the device are identified as base 1, base 2, and emitter. The emitter is made of P-type material, and the substrate between base 1 and base 2 is N-type material. The N-type block or bar between base 1 and base 2 is lightly doped and, therefore, has few majority carriers to support current flow. The equivalent circuit for a UJT is shown in Figure 20–47(b), and the schematic symbol is shown in Figure 20–47(c).

Internally, the UJT acts as a voltage divider consisting of the two resistors R_{B1} and R_{B2}, and the single PN junction is shown as a diode. Resistor R_{B1} is shown as a variable resistor, as its ohmic value is dependent on the amount of emitter current flowing. The internal resistance ratio of R_{B1} and R_{B2} is called the **intrinsic standoff ratio,** and is represented by the Greek letter η (eta).

$$\text{intrinsic standoff ratio} = \eta = \frac{R_{B1}}{R_{B1} + R_{B2}} \qquad (20.18)$$

When the voltage applied between the emitter and B_1 reaches a certain value, the UJT will turn on. The amount of emitter voltage required to switch the UJT on is called the **standoff voltage, V_p.** The standoff voltage is found by multiplying the applied voltage, V_{BB}, by the intrinsic standoff ratio. The forward-voltage drop of the PN junction is also included in the calculation for accuracy.

$$V_p = \eta V_{BB} + V_F \qquad (20.19)$$

When V_p is reached, the UJT latches in the conducting state and current flows between E and B_1. The UJT is a latching device that will remain in the conducting state as long as the emitter current remains above a minimum value called the **holding current.** The resistance R_{B1} varies as a function of the emitter current. If the emitter current increases, R_{B1} decreases. This effect is because of the negative resistance characteristic of semiconductors and is shown in the characteristic curve of Figure 20–48. When the emitter voltage reaches V_p, the PN junction becomes forward biased and I_E begins to increase. Holes are injected into the N-type bar from the P-type emitter. As a result, the conductivity between

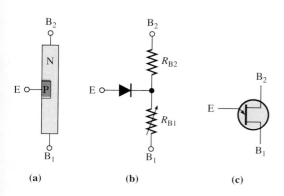

FIGURE 20–47 *(a) UJT construction. (b) Equivalent circuit. (c) Schematic symbol.*

(a)   (b)   (c)

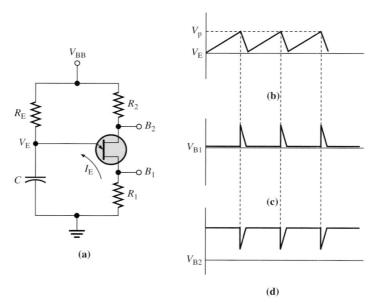

FIGURE 20–48 *Characteristic curve for UJT.*

the emitter and B_1 increases and R_{B1} decreases. When the UJT latches, R_{B1} rapidly decreases and less emitter voltage is required to produce more emitter current. Consequently, V_E decreases as I_E increases. When R_{B1} has decreased to its absolute minimum, the UJT is saturated. On the characteristic curve, the point of saturation is identified as the *valley point*. After reaching the valley point, any further increase in emitter current produces a fairly linear curve. That is, a small increase in V_E causes a small increase in I_E.

Figure 20–49(a) shows a typical application for a UJT in an oscillator circuit. This particular circuit is called a **relaxation oscillator** and is capable of providing voltage pulses at fixed frequencies for triac and SCR triggering circuits. The resistor and capacitor connected to the emitter form an *RC* timing circuit. The size of these two components determines the rate of charge across the capacitor. As the capacitor is initially charged, the voltage at the emitter, V_E, begins to rise, as shown in Figure 20–49(b). When the voltage at the emitter reaches the standoff voltage, the UJT is triggered into a conducting state and a voltage drop instantly appears across resistor R_1. This voltage appears as a positive-going spike, as illustrated in Figure 20–49(c). As the internal resistance R_{B1} decreases, the cur-

FIGURE 20–49 *UJT relaxation oscillator and waveforms.*

rent flowing through R_2 increases and the voltage drop V_{R2} increases. Therefore, the voltage between B_2 and ground is momentarily reduced, as shown in Figure 20–49(d). The capacitor is rapidly discharged when the UJT latches, and the holding current quickly falls below its minimum value to maintain conduction. The UJT automatically switches off and the charging process begins again. The intrinsic standoff ratio will affect the frequency of the voltage spikes produced. The period, T, of the oscillations is found using the following equation:

$$T = R_E C \ln\left(\frac{1}{1-\eta}\right) \qquad\qquad (20.20)$$

EXAMPLE 20–8

FIGURE 20–50 Circuit for Example 20–8.

The circuit shown in Figure 20–50 uses a variable resistor to change the frequency of pulses delivered at V_{out}. The variable resistor is initially set at 5 kΩ, and is then adjusted to 10 kΩ. Determine the frequency of the voltage spikes produced for the

(a) 5 kΩ setting.
(b) 10 kΩ setting.

Solution

(a)
$$T = R_E C \ln\left(\frac{1}{1-\eta}\right)$$
$$= (5 \text{ k}\Omega)(0.2 \text{ }\mu\text{F}) \ln\left(\frac{1}{1-0.54}\right) = 0.78 \text{ ms}$$
$$f = \frac{1}{T} = \frac{1}{0.78 \text{ ms}} = 1282.05 \text{ Hz}$$

(b)
$$T = R_E C \ln\left(\frac{1}{1-\eta}\right)$$
$$= (10 \text{ k}\Omega)(0.2 \text{ }\mu\text{F}) \ln\left(\frac{1}{1-0.54}\right) = 1.55 \text{ ms}$$
$$f = \frac{1}{T} = \frac{1}{1.55\text{ms}} = 645.16 \text{ Hz}$$

Section Review

1. UJTs are often used as amplifiers (True/False).
2. What is the intrinsic standoff ratio?
3. What is the standoff voltage?

TABLE 20–1 *Test results.*

	SCR 1	SCR 2	SCR 3	SCR 4
Anode-cathode Ω (gate open)	High	High	High	High
Anode-cathode Ω (jumper connected across anode and cathode)	Low	Low	High	Low
Anode-cathode Ω (jumper removed)	Low	Low	High	Low
Anode-cathode Ω (DMM removed and reconnected)	High	High	High	High

___ 20–14 ___
PRACTICAL APPLICATION SOLUTION

In the chapter opener, your task was described as troubleshooting the electronic drive system for a large DC motor aboard a military ship. After using a voltmeter to check the control circuit and power circuit voltages for the DC drive and comparing these values with the manufacturer's specifications, you have determined that the problem is caused by one of the SCRs in the drive system. The following steps illustrate the test procedure used for this practical application of transistors and thyristors.

STEP 1   Remove each SCR from the circuit and test with an ohmmeter or semiconductor tester. Table 20–1 shows the test results for four SCRs tested out-of-circuit with a digital multimeter with a PN junction tester. This tester indicates *low* for a forward-biased resistance, or shorted PN junction, and *high* for a reverse-biased resistance, or open PN junction.

STEP 2   Evaluate the data obtained from testing each SCR. From Table 20–1, it is apparent that SCR3 is defective because it does not latch when its gate is triggered.

STEP 3   Replace the defective SCR and retest the circuit.

___ Summary ___

1. Bipolar junction transistors (BJTs) are semiconductor devices consisting of two junctions in a continuous crystal.
2. There are two types of BJTs—NPN and PNP.
3. BJTs are biased by forward biasing the emitter-base junction and reverse biasing the collector-base junction.
4. The biasing voltages used in transistor circuits are steady DC values with low percent ripple.
5. The flow of current through the BJT is varied by increasing or decreasing the amount of current flow through the emitter-base junction.
6. The FET is a voltage-controlled semiconductor device.
7. The ability of a FET's gate to control the drain current is called *transconductance*.
8. MOSFETs have a gate that is electrically insulated from the source-drain channel.
9. Two of the most popular types of thyristors are SCRs and triacs.
10. UJTs are three-terminal, single junction, field-effect devices.

Section 20–1

1. True
2. Bipolar junction transistors (BJTs) and field effect transistors (FETs)
3. Triacs and silicon-controlled rectifiers

Section 20–2

1. NPN and PNP
2. Emitter-base junction and collector-base junction
3. By forward-biasing the emitter-base junction and reverse-biasing the collector-base junction
4. $I_E = I_C + I_B$

Section 20–3

1. True
2. When two resistors are connected to bias a BJT so that one resistor is across the collector-base junction and the other resistor is across the emitter-base junction

Section 20–4

1. True
2. Current, voltage, and power
3. $\beta = I_C/I_B$
4. True
5. A graphical representation of the operating range of a transistor between saturation and cutoff.

Section 20–5

1. True
2. False

Section 20–6

1. False
2. False

Section 20–7

1. False
2. Drain, source, and gate
3. N-channel and P-channel

Section 20–8

1. The point where the transistor becomes saturated
2. False
3. The relative ability of a gate to control the drain current of a FET

Section 20–9

1. A semiconductor device in which the gate is electrically insulated from the source-drain channel
2. By the conducting state when a bias voltage of zero volts is applied
3. A device where a voltage is needed on the gate to enhance, or increase, current flow in the channel
4. True
5. The amount of gate voltage required to produce a channel

Section 20-10

1. True
2. An ohmmeter with a low-ohm setting

Section 20-11

1. A semiconductor device that exhibits the regenerative-switching characteristic of a four-layer or PNPN arrangement
2. Anode, cathode, and gate
3. By reducing the holding current below its rated value

Section 20-12

1. True
2. False

Section 20-13

1. False
2. The internal resistance ratio of a UJT
3. The amount of voltage required to trigger a UJT into a conducting state

Review Questions

Multiple Choice Questions

20–1   Transistors have replaced _____ in most electronic circuit applications.
(a) Thyristors   **(b)** Vacuum tubes   **(c)** Transformers   **(d)** Resistors

20–2   The two basic types of transistors are
(a) NPN and PNP   **(b)** JFETs and MOSFETs
(c) BJTs and FETs   **(d)** BJTs and thyristors

20–3   The arrow on the schematic symbol for a BJT identifies the
(a) Emitter   **(b)** Collector   **(c)** Base   **(d)** Gate

20–4   For a BJT to operate as an amplifier or a switch, the
(a) Emitter-base junction is reverse biased and the collector-base junction is reverse biased
(b) Emitter-base junction is forward biased and the collector-base junction is forward biased
(c) Emitter-base junction is reverse biased and collector-base junction is forward biased
(d) Emitter-base junction is forward biased and collector-base junction is reverse biased

20–5   Which equation is correct for the relationship between I_E, I_C, and I_B in a BJT?
(a) $I_B = I_C + I_E$ **(b)** $I_E = I_C - I_B$
(c) $I_C = I_E + I_B$ **(d)** $I_E = I_C + I_B$

20–6   The relationship between the amount of collector current that flows in a transistor compared to the emitter current is known as
(a) Beta   **(b)** Gamma   **(c)** Alpha   **(d)** Bias

20–7   The forward current transfer ratio is commonly referred to as
(a) Beta   **(b)** Hole flow   **(c)** Alpha   **(d)** Forward bias

20–8   Self-biasing transistor circuits use
(a) One voltage source   **(b)** Two voltage sources
(c) AC bias voltages   **(d)** NPN transistors only

20–9   When the emitter current is zero, a BJT is said to be
(a) Conducting   **(b)** Cutoff   **(c)** Saturated   **(d)** Damaged

20–10 The point on a load line that indicates the steady-state condition of V_{CE} when no input signal is applied is called
(a) Beta   **(b)** Saturation   **(c)** Q point   **(d)** Cutoff

20–11 The three leads of a FET are called
(a) Anode, cathode, gate   **(b)** Emitter, collector, gate
(c) Emitter, collector, base   **(d)** Drain, source, gate

20–12 The main difference between a BJT and a FET is
(a) FETs are current operated   **(b)** BJTs are voltage operated
(c) FETs cannot be used as switches   **(d)** FETs are voltage operated

20–13 The two most common types of FETs are
(a) NPN and PNP   **(b)** N-channel and P-channel
(c) Enhancement and depletion   **(d)** Single gate and dual gate

20–14 For proper biasing, the source of a(n)
(a) N-channel JFET is connected to negative voltage
(b) N-channel JFET is connected to positive voltage
(c) P-channel JFET is connected to negative voltage
(d) N-channel JFET is connected to the same polarity voltage source as the drain

20–15 The relative ability of a gate to control the drain current of a FET is called
(a) Beta   **(b)** Alpha   **(c)** Transfer curve   **(d)** Transconductance

20–16 The input to a MOSFET acts as a(n)
(a) Capacitor   **(b)** Inductor   **(c)** Resistor   **(d)** Short-circuit

20–17 The two most popular types of thyristors used in industry are
(a) MOSFETs and JFETs   **(b)** NPN and PNP
(c) Depletion and enhancement   **(d)** SCRs and triacs

20–18 To switch off an SCR, it must be
(a) Saturated   **(b)** Triggered   **(c)** Commutated   **(d)** Biased

20–19 The main difference between a triac and an SCR is
(a) Triacs only respond to positive gate voltages
(b) SCRs control current in both directions
(c) Triacs allow current flow regardless of the biasing voltage and polarity of gate current
(d) SCRs are used as amplifiers

20–20 A diac is a(n)
(a) Three-terminal transistor   **(b)** Two-terminal thyristor
(c) Unilateral device   **(d)** Device used to trigger transistors

20–21 Unijunction transistors are
(a) Used primarily in amplifier circuits   **(b)** Similar to BJTs
(c) Similar to FETs   **(d)** Used primarily in switching circuits

Practice Problems

20–22 Determine the emitter current for the circuit shown in Figure 20–51.

$I_C = 1.35$ mA   **FIGURE 20–51**

$I_B = 20$ μA

20–23 When the base current of a BJT is 50 μA, a collector current of 10 mA flows. Determine the

(a) β_{DC} **(b)** α_{DC}

20–24 The transistor shown in Figure 20–52 has a β_{DC} of 80. Find the values of I_B, I_C, and I_E. Assume the transistor is made of silicon.

FIGURE 20–52

20–25 For the circuit of Figure 20–53, calculate the following values:

(a) V_B **(b)** V_E **(c)** I_E **(d)** V_C **(e)** V_{CE}

Assume that I_C and I_E are approximately equal.

FIGURE 20–53

20–26 For the circuit shown in Figure 20–54, determine the current gain.

FIGURE 20–54

20–27 Determine the power gain of the amplifier shown in Figure 20–54.

20–28 Calculate the minimum base current and input voltage required to fully turn on the transistor shown in Figure 20–55.

FIGURE 20-55

$R_L = 2.2\ k\Omega$

$R_B = 15\ k\Omega$

$\beta = 100$

$V_{CC} = 15\ V$

20–29 For the circuit shown in Figure 20–56, determine the standoff voltage and the approximate frequency of the voltage spikes. Assume the UJT is made of silicon with a V_F of 0.7 V.

$V_{BB} = 20\ V$

FIGURE 20-56

$R_E = 20\ k\Omega$

R_2

$\eta = 0.57$

$C = 0.5\ \mu F$

R_1

Essay Questions

20–30 Describe the basic operation of a BJT.

20–31 How is a BJT biased?

20–32 What is the forward current transfer ratio of a BJT?

20–33 Explain the difference between α and β.

20–34 Why is voltage divider biasing used?

20–35 Describe the basic principle of amplification.

20–36 What is the Q point of a transistor?

20–37 List two types of FETs.

20–38 Explain the main difference between a FET and a BJT.

20–39 How is an N-channel FET biased?

20–40 Define the terms *depletion mode* and *enhancement mode*.

20–41 What is a thyristor?

20–42 How are thyristors commutated?

20–43 Describe the fundamental difference between a triac and an SCR.

20–44 Explain the basic operation of a relaxation oscillator.

21

Amplifier Circuits

Learning Objectives

Upon completion of this chapter you will be able to

- List three main characteristics of linear amplifiers.
- Describe the effect of AC signals on an amplifier.
- Name three configurations for BJT amplifiers.
- Explain why coupling capacitors and bypass capacitors are used in amplifier circuits.
- List three configurations for FET amplifiers.
- Discuss the advantages and disadvantages of direct coupling, capacitor coupling, and transformer coupling.
- Differentiate between class A, B, and C amplifiers.
- Define crossover distortion.
- Troubleshoot amplifier circuits.

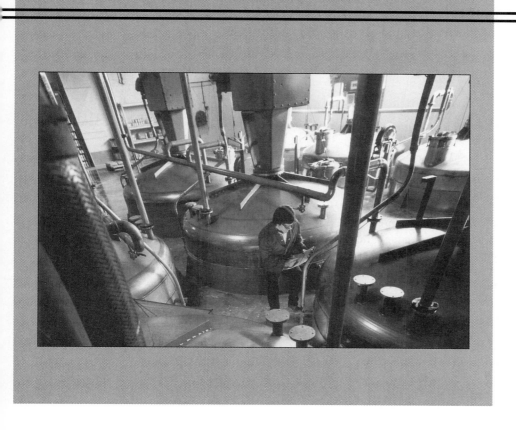

As a technician for a company specializing in instrumentation control for industrial applications, you are asked to design a circuit combining a darlington amplifier and a transistor switch with two electrodes to provide resistance measurement of a liquid to indicate level. Resistance measurements are a popular method of industrial electronic control. The electrical resistance of a substance can depend on many physical parameters, such as temperature, chemical composition, pressure, and so on.

The circuit that you are required to design uses an amplifier and a switch to monitor the level of liquid in a tank and to activate an alarm if the level rises above a certain predetermined value. A potentiometer is to be included in the design to allow for adjustment of the reference value for the liquid level. Two transistors are to be connected in a darlington configuration to provide a high-current gain. The third transistor is a high-current switching transistor that is connected to the output of the darlington. One of the electrodes is common to the collector of all three BJTs, and the other electrode is connected to the input of the darlington. A relay coil is to be connected to the output of the circuit to activate an alarm when the liquid level is too high. Included in the design is a method of dissipating the counter-emf stored in the relay coil. The electrodes are connected so that, if any conductive liquid is present between the electrodes, a current will flow between them. This current will divide at the input of the darlington amplifier with a portion flowing into the darlington and the rest flowing through the potentiometer.

21-1

INTRODUCTION

An **amplifier** is a device capable of increasing the magnitude of a signal. An ideal amplifier provides infinite gain, zero output resistance, and infinite input resistance. Although most transistors are capable of performing amplification, the term *amplifier* usually refers to the complete circuit in which the transistor operates. For example, the biasing resistors connected to a transistor would also be considered part of an amplifier circuit. A single amplifier circuit is often called a **stage.** Electronic circuits that combine more than one amplifier are called **multistage amplifiers.** One advantage of using multistage amplifiers is that it is possible to combine the characteristics of one amplifier with those of another to produce a circuit with a combination of qualities. If an amplifier with a high input impedance is connected in series with an amplifier with a low output impedance, the result is a circuit with two excellent qualities. Another advantage is that the overall gain of a circuit can be increased by combining two amplifiers. When the output of one amplifier is connected to the input of another amplifier, it is called **cascading.**

Amplifiers are generally classified in terms of their characteristics and inherent limitations. The most typical classification for an amplifier is frequency. There are three general divisions within this category: audio-frequency, radio-frequency, and video-frequency amplifiers. The frequency range of audio amplifiers is from approximately 15 Hz to 20,000 Hz. Frequencies above 20 kHz are considered radio frequencies. Video amplifiers range from 5 Hz to over 10 MHz. Amplifiers are also classified in terms of their voltage and power-handling capabilities. In this chapter, we will examine standard configurations for transistors used in linear amplifier circuits. The main characteristics of linear amplifiers are as follows:

1. The waveshape of the output signal is identical to the input signal.
2. The transistor is usually biased near the center of its operating limits.
3. The transistor is never driven to either cutoff or saturation.

Amplifiers also are often classified in terms of signal strength. A signal can be any varying voltage. Because many signals used in amplifier circuits involve sine waves, we will first consider how an amplifier responds to AC input signals.

Section Review

1. A single amplifier is often called a stage (True/False).
2. What are two advantages of using multistage amplifiers?
3. What is cascading?

EFFECT OF AC SIGNALS ON AN AMPLIFIER

AC amplifier circuits contain both AC and DC voltages. The DC voltages are always present in the circuit to provide biasing for the transistor, and the AC voltage represents the input signal to be amplified. For a transistor to respond properly to AC input signals, it must be biased correctly. Because an AC input voltage alternates between positive and negative polarities, a single transistor must have its operating point shifted so that the transistor does not shut off when the input signal crosses the zero axis. Figure 21–1(a) shows a self-biased NPN transistor. A coupling capacitor is used between the AC input source and the DC biasing voltages. The purpose of this capacitor is to block any DC voltage that might be present at V_{in} from being applied to the base of the transistor as bias. The presence of this voltage could saturate the transistor and adversely effect the operation of the amplifier. In this circuit, resistor R_1 is adjusted until the voltage dropped across resistor R_E is 5 V. This voltage level represents a DC **reference voltage** for the base of 5.7 V, as shown by the dotted line in Figure 21–1(b). When an AC input signal is applied, the AC will ride on this reference point. When the AC input goes positive, it will add to the positive DC reference voltage. When the AC input goes negative, it will subtract from the positive biasing voltage and lower the current flowing at the base.

Figure 21–1(c) shows the voltage waveform at the emitter, with respect to ground. The DC biasing voltage is again represented by a dotted line. As the input wave goes positive, the current flow at the base increases. This results in an increase in emitter current. According to Ohm's law the voltage drop across R_E will increase if I_E increases, and the resulting wave is shown as increasing and decreasing in proportion with the input current.

The waveform of Figure 21–1(d) represents the voltage at the collector (with respect to ground). As the base current increases, the transistor begins to conduct "harder," and the

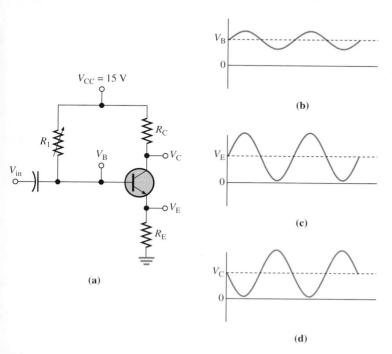

(a)

(b)

(c)

(d)

FIGURE 21–1 *AC amplifier and output waveforms.*

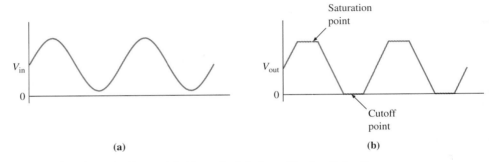

FIGURE 21-2 *(a) Sinusoidal input. (b) Amplitude distortion.*

currents at the collector and emitter increase. Because V_C equals $V_{CC} - I_C R_C$, the voltage at the collector decreases as the collector current increases. Conversely, if the base current decreases, the transistor begins to shut off and I_C decreases. When I_C decreases, the voltage drop across R_C decreases and the voltage at the collector increases. The waveforms in Figure 21–1 show that the collector voltage is 180° out of phase with both the emitter and base voltages. When signals are 180° out of phase with each other, they are said to be *inverted*.

When a transistor is not properly biased, the device may be driven into saturation or cutoff. This results in distortion at the output. In practical amplifiers, the output wave is never exactly the same as the input wave. As a result, the output wave may appear distorted if viewed with an oscilloscope. There are three main types of amplifier distortion: frequency distortion, phase distortion, and amplitude distortion. Frequency and phase distortion are caused by input signals that are not sinusoidal and contain various frequencies, or harmonics. Amplitude distortion occurs when equal changes in amplitude of the input signal do not produce proportional changes in the output signal. Amplitude distortion is primarily caused by improper biasing and excessive input signals. The result is a **clipping** of the waveshape, as shown in Figure 21–2(b). Although the input signal of Figure 21–2(a) is sinusoidal, the output is distorted because the output current reaches saturation on the positive cycle and cutoff on the negative cycle. When the operating point is selected properly, the input signal will be amplified and distortion will be kept to a minimum.

Section Review

1. AC amplifiers contain both AC and DC voltages (True/False).
2. Explain the phase relationship between the emitter and collector voltages for a BJT amplifier.
3. Name the three main types of amplifier distortion.
4. What are the main causes of clipping?

— 21-3 —

COMMON-BASE (CB) AMPLIFIERS

Bipolar junction transistor amplifiers are classified as either common-base, common-emitter, or common-collector amplifiers. The term *common* identifies which terminal is connected to both the input and the output of the transistor. Figure 21–3(a) shows a typical common-base NPN transistor circuit with biasing voltages and a load resistor. In this circuit, the emitter current I_E is shown as the input current and the collector current I_C is

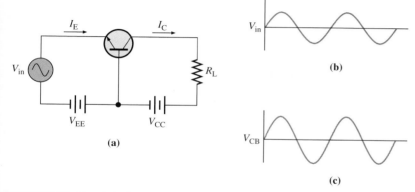

FIGURE 21–3 *(a) Common-base configuration. (b) Input signal. (c) Output signal.*

shown as the output, or load, current. Therefore, the current gain of a **common-base amplifier** is the same as the alpha for the transistor used in the circuit.

$$\mathcal{A}_i = \frac{I_{\text{out}}}{I_{\text{in}}} = \frac{I_C}{I_E} = \alpha$$

Because I_C is always less than I_E, the current gain for a common-base (CB) amplifier is less than 1. However, a CB amplifier is capable of providing voltage gain. In Figure 21–3(a), the input resistance of the circuit is a forward-biased PN junction between the emitter and base. The output resistance is very large because of the reverse-biased PN junction between the collector and base terminals. If the input and output currents are approximately equal but the input resistance is lower than the output resistance, then the output voltage must be greater than the input voltage. The same relationship is true for power between the input and output. Therefore, although the output current is somewhat lower than the input current for the CB configuration, both voltage and power gains can be achieved.

The waveforms in Figures 21–3(b) and (c) represent the input voltage produced by the AC signal and the output voltage developed across the collector-base junction. When the input voltage is positive going, it opposes the forward bias produced by the base-emitter battery, and lowers the value of input current. As I_E decreases, I_C decreases, and the top of the load resistor becomes less negative (or more positive) with respect to the bottom of the load resistor. When the input voltage is negative going, the transistor conducts "harder" and more current flows at the collector causing the top of the load resistor to become more negative. Therefore, the input signal and output signal for the CB circuit are said to be in phase with each other.

Voltage amplification in a CB circuit occurs as a result of the difference between input resistance and output resistance. When no emitter resistor is connected, the voltage gain is a result of the internal emitter resistance r_e and the load resistor connected at the collector. The internal emitter resistance, also known as the **dynamic resistance,** is not a fixed value and is dependent on current flow in the circuit. As a result, it is necessary to approximate r_e in order to determine the voltage gain of a CB circuit. r_e is solved by Ohm's law, assuming a voltage drop of 25 mV.

$$r_e = \frac{25 \text{ mV}}{I_E} \qquad\qquad (21.1)$$

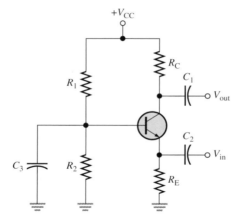

FIGURE 21-4 *Common-base amplifier with voltage divider biasing.*

Because voltage gain is essentially a ratio of input and output resistances, the voltage gain for a CB circuit is approximately

$$\mathcal{A}_v = \frac{R_C}{r_e} \tag{21.2}$$

Figure 21-4 shows a CB amplifier with voltage divider biasing. The transistor is driven from the emitter, and the output is taken at the collector. The two coupling capacitors, C_1 and C_2, isolate the DC biasing voltages from the AC input and output signals. Capacitor C_3 is a bypass capacitor that has a low reactance at the operating frequency of the amplifier. In Chapter 13, it was shown how a bypass capacitor was capable of removing the AC component of a signal while allowing the DC voltage to remain. Consequently, the AC voltages across R_1 and R_2 are effectively bypassed to ground, while the DC biasing voltages are unaffected. When an emitter resistor is connected in a CB circuit, it is generally much higher in value than the internal emitter resistance. Because R_E is basically in parallel with r_e, biasing resistor R_E can be ignored when calculating voltage gain of a CB amplifier.

_____ **EXAMPLE 21-1** _____

Calculate the voltage gain of the CB amplifier shown in Figure 21-5.

FIGURE 21-5 *Circuit for Example 21-1.*

Solution The first step is to determine the base voltage V_B. This voltage is found by applying the voltage divider rule.

$$V_B = \left(\frac{R_2}{R_1 + R_2}\right) V_{CC} = \left(\frac{30\ k\Omega}{100\ k\Omega + 30\ k\Omega}\right) 10\ V = 2.31\ V$$

Assuming a silicon transistor, the emitter voltage is found by applying Kirchhoff's voltage law and subtracting V_{BE} from V_B.

$$V_E = V_B - V_{BE} = 2.31\ V - 0.7\ V = 1.61\ V$$

The emitter current is determined by applying Ohm's law.

$$I_E = \frac{V_E}{R_E} = \frac{1.61\ V}{2.7\ k\Omega} = 0.596\ mA$$

Equation 21.1 is now used to find the internal emitter resistance.

$$r_e = \frac{25\ mV}{0.596\ mA} = 41.95\ \Omega$$

The voltage gain of the circuit can now be determined using Equation 21.2.

$$\mathcal{A}_v = \frac{R_C}{r_e} = \frac{5.6\ k\Omega}{41.95\ \Omega} = 133.5$$

Table 21–1 is a summary of important CB circuit characteristics. These values are for a typical BJT connected in a CB configuration.

Section Review

1. State the DC current gain for a common-base amplifier in equation form.
2. The current gain for a common-base amplifier is greater than 1 (True/False).
3. What is dynamic resistance?
4. State the voltage gain for a common-base amplifier in equation form.

TABLE 21–1 *Common-base circuit characteristics.*

Characteristic	Typical Value
High voltage gain	Up to 1000
Low current gain	0.99
Medium power gain	Up to 1000
Low input resistance	10 Ω to 150 Ω
High output resistance	Up to 1 MΩ

COMMON-EMITTER (CE) AMPLIFIERS

The **common-emitter amplifier** is the most popular configuration for BJT amplifier circuits. The CE amplifier is capable of providing high values of both voltage and current gain, which makes it well suited for a variety of applications. In the CE connection shown in Figure 21–6(a), the signal is introduced into the base-emitter portion of the circuit and is extracted from the collector.

When the input signal is applied to the base, as shown in Figure 21–6(b), a positive-going input signal will increase the base-emitter bias of the BJT, and the collector current increases. The increased electron flow through R_L then causes the output voltage to become negative, as illustrated in Figure 21–6(c). During the second half-cycle of the waveform, the process is reversed, and a phase inversion is seen between the input and output waveforms. The reason that the common emitter produces an inverted output compared to the common base is that in the CE circuit, the input signal is applied from the base to emitter, whereas in the CB circuit, the signal is applied from emitter to base.

The CE configuration has more moderate input and output impedances than the CB circuit. The voltage gain is less than that of the CB circuit because of the higher input impedance, but this is offset by the high values of current gain obtainable with the CE configuration. Because the base current now represents the input and the output current is taken at the collector, the current gain of the CE amplifier is based on the collector-to-base current ratio beta (β_{AC}). Therefore, a small change in base current controls a large change in collector current.

$$\mathcal{A}_i = \beta_{AC} = \frac{\Delta I_C}{\Delta I_B} \tag{21.3}$$

The DC beta and AC beta of a transistor are not usually identical, although they are close enough to be considered approximately equal. β_{DC} is defined in terms of total collector current, while β_{AC} represents small changes around a certain operating point. Manufacturers generally provide separate data for the two quantities. However, for our purposes we will consider β_{DC} and β_{AC} to be equal. The above equation is applicable only when the collector-emitter voltage, V_{CE}, is held constant and the change in collector current is noted for each change in base current.

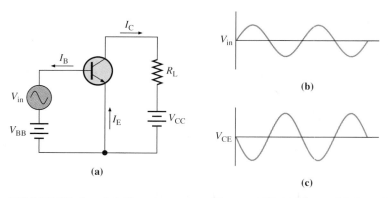

FIGURE 21–6 *(a) Common-emitter configuration. (b) Input signal. (c) Output signal.*

The following example illustrates how the current gain is calculated using values of current obtained from CE output curves.

_____ EXAMPLE 21–2 _____

TABLE 21–2 Data for Example 21–2.

I_B	I_C	V_{CE}
20 μA	1 mA	5 V
40 μA	2 mA	5 V
60 μA	3 mA	5 V
80 μA	4 mA	5 V

Determine the B_{AC} and B_{DC} for a CE amplifier using data from Table 21–2.

Solution Equation 21.3 states that B is obtained by taking the ratio of the collector current change to the base current change while holding V_{CE} constant. The gain is found using the first two values of I_B and I_C from Table 21–2.

$$\beta_{AC} = \frac{\Delta I_C}{\Delta_B} = \frac{2 \text{ mA} - 1 \text{ mA}}{40 \text{ μA} - 20 \text{ μA}} = 50$$

$$\beta_{DC} = \frac{I_C}{I_B} = \frac{1 \text{ mA}}{20 \text{ μA}} = 50$$

The fact that a 180° phase displacement occurs between the input and output of a CE circuit, does not restrict its use in most applications. The waveshape is not distorted by this inversion, and the peak and rms voltages are not affected. In many electronic circuits, it makes no difference whether the output signal is inverted or not. The main disadvantage of the CE amplifier is that it has relatively low input impedance. As mentioned previously, an ideal amplifier should have infinite input impedance.

Figure 21–7(a) shows an NPN common-emitter circuit with voltage divider biasing, two coupling capacitors, and a bypass capacitor C_3. The waveshapes shown in Figures 21–7(b) and (c) represent the input and output signals. The purpose of the bypass capacitor is to increase the AC voltage gain of the circuit. In Figure 21–7(a), the DC voltage drop across R_E is necessary for biasing, but any AC signal that is dropped across R_E represents a direct loss in AC output voltage. The voltage at the emitter is shown in Figure 21–7(d) as a DC biasing voltage of fixed magnitude.

If capacitor C_3 were removed from the circuit, the AC voltage drop across the emitter would increase and, according to Kirchhoff's voltage law, the AC output signal would decrease.

The resistance between ground and V_{in} in a CE circuit consists of the emitter resistor R_E and the internal resistance r_e. If a bypass capacitor is not used, resistor R_E is included when determining the resistance to AC signals.

$$\mathcal{A}_v = \frac{R_C}{r_e + R_E} \tag{21.4}$$

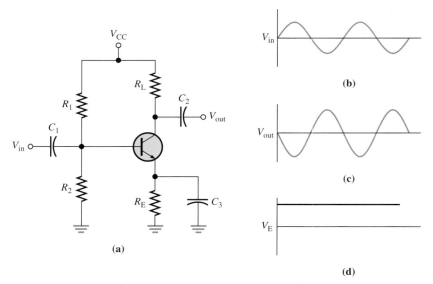

(a)

(b)

(c)

(d)

FIGURE 21–7 *(a) Common-emitter circuit with coupling capacitors and bypass capacitor. (b) Input signal. (c) Output signal. (d) DC biasing voltage at emitter.*

EXAMPLE 21–3

For the circuit shown in Figure 21–8, calculate the voltage gain of the circuit

(a) With the bypass capacitor connected across R_E.
(b) With the bypass capacitor removed.

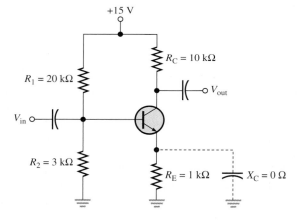

FIGURE 21–8 *Circuit for Example 21–3.*

Solution

(a)
$$V_B = \left(\frac{R_2}{R_1 + R_2}\right) V_{CC} = \left(\frac{3\ k\Omega}{20\ k\Omega + 3\ k\Omega}\right) 15\ V = 1.96\ V$$

$$V_E = V_B - V_{BE} = 1.96\ V - 0.7\ V = 1.26\ V$$

$$I_E = \frac{V_E}{R_E} = \frac{1.26\ V}{1\ k\Omega} = 1.26\ mA$$

Equation 21.1 is now used to find the internal emitter resistance.

$$r_e = \frac{25\ mV}{1.26\ mA} = 19.84\ \Omega$$

With the bypass capacitor connected, the voltage gain can be determined using Equation 21.2.

$$\mathcal{A}_v = \frac{R_C}{r_e} = \frac{10 \text{ k}\Omega}{19.84 \text{ }\Omega} = 504$$

(b) With the bypass capacitor disconnected, the emitter resistor R_E is added to the internal resistance to determine the voltage gain.

$$\mathcal{A}_v = \frac{R_C}{R_E + r_e} = \frac{10 \text{ k}\Omega}{1 \text{ k}\Omega + 19.84 \text{ }\Omega} = 9.8$$

From Example 21–3, it is apparent that the addition of a bypass capacitor makes a tremendous difference to the voltage gain of the CE circuit. When the bypass capacitor was removed, the voltage gain decreased from 504 to 9.8.

Table 21–3 shows some typical values for various CE circuit characteristics. Although the current, voltage, and power gains are all excellent, the input and output resistances are opposite that of an ideal amplifier. As mentioned previously, an ideal amplifier provides infinite gain, zero output resistance, and infinite input resistance.

Section Review

1. Which is the most popular BJT amplifier configuration?
2. The common-emitter configuration has more moderate input and output impedances than the common-base circuit (True/False).
3. State the DC current gain for a common-emitter amplifier in equation form.

21–5

COMMON-COLLECTOR (CC) AMPLIFIERS

The third type of connection, shown in Figure 21–9, is the **common-collector amplifier.** In this circuit, the signal is introduced into the base-collector portion of the transistor and extracted at the emitter. The CC amplifier is also known as an **emitter follower,** because the emitter always follows, or copies, the base. If the base voltage increases, the emitter voltage will also increase. The input resistance of a CC amplifier is very high because the input signal is now applied across a reverse-biased collector-base junction. Because the CC amplifier has a high input resistance and a relatively low output resistance, it is capable of amplifying current but does not produce an increase in voltage.

TABLE 21–3 *Common-emitter circuit characteristics.*

Characteristic	Typical Value
High voltage gain	Up to 500
High current gain	50 to 200
High power gain	Up to 10,000
Low input resistance	1 kΩ
High output resistance	100 kΩ

FIGURE 21-9   *Common-collector amplifier.*

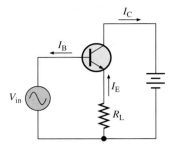

The input signal to the circuit of Figure 21–9 is applied to the base, and the output signal is developed across R_L. As in other circuits, the input signal either aids or opposes the input bias voltage. However, the CC amplifier differs from CE and CB amplifiers in that the collector current has no real significance. The collector is basically used in this circuit to allow emitter current to flow. Because the emitter current is now the output and the base current is the input, the current gain for a CC amplifier is found as a ratio of these two values.

$$\mathcal{A}_i = \frac{I_{out}}{I_{in}} = \frac{I_E}{I_B}$$

Because $I_E = I_C + I_B$, the current gain for a CC amplifier is also determined using the following equation:

$$\mathcal{A}_i = \frac{I_C + I_B}{I_B} = \frac{I_C}{I_B} + \frac{I_B}{I_B} = \frac{I_C}{I_B} + 1 = \beta + 1 \tag{21.5}$$

Figure 21–10(a) shows a CC amplifier with voltage divider bias. The only differences between this circuit and the CE amplifier with voltage divider bias are that the CC amplifier has no collector resistor, and the output is taken from the emitter instead of the collector. The input and output waveforms are shown in Figures 21–10(b) and (c), respectively. When the input signal is positive, the forward bias is increased and the emitter current increases, producing a positive-going output voltage. On the negative input cycle, the

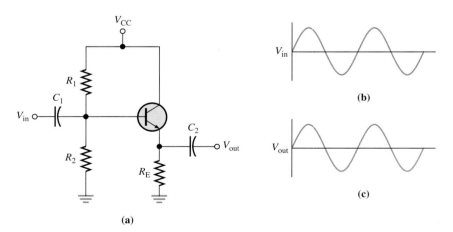

(a)

(b)

(c)

FIGURE 21-10   *(a) Common-collector amplifier with coupling capacitors. (b) Input signal. (c) Output signal.*

emitter output is negative and the forward bias is reduced. Consequently, the emitter and collector currents are reduced, producing a decreased voltage drop across the emitter output resistor.

It is often important to know the input resistance for a CC amplifier because its main application is impedance matching where a high-resistance source is used to drive a low-resistance load. This function is essentially the same as that of an impedance-matching transformer. A circuit that performs impedance matching or isolation functions is often referred to as a **buffer.**

The AC input resistance of the CC amplifier is $(\beta + 1)$ times greater than the sum of R_E and r_e. If R_E is at least ten times greater than r_e, then r_e can be ignored because of its negligible effect on the circuit. Because r_e is generally less than 100 Ω, when R_E is greater than 1 kΩ, the effect of r_e is ignored. Therefore, the approximate input resistance for a CC amplifier is calculated as follows:

$$R_{in} = (\beta + 1)\,(R_E) \tag{21.6}$$

Any biasing resistors connected to the base of the CC amplifier must be considered as parallel resistors to R_{in}. The following example illustrates how the total input resistance is determined in a voltage divider biased CC circuit.

_____ **EXAMPLE 21–4** _____

For the circuit shown in Figure 21–11, find the total input impedance.

FIGURE 21–11 *Circuit for Example 21–4.*

Solution

$$R_{in} = (\beta + 1)(R_E) = (101)(2\ k\Omega) = 202\ k\Omega$$

$$R_{in(T)} = R_{in}\|R_1\|R_2 = 202\ k\Omega\|100\ k\Omega\|100\ k\Omega$$

$$= \frac{1}{1/202\ k\Omega + 1/100\ k\Omega + 1/100\ k\Omega} = 40.08\ k\Omega$$

The fundamental characteristics for a CC amplifier are summarized in Table 21–4. The CC configuration provides low power gain but has excellent characteristics in terms of input and output resistances compared to the CB and CE amplifiers.

TABLE 21–4 *Common-collector circuit characteristics.*

Characteristic	Typical Value
Low voltage gain	0.99
High current gain	50 to 200
Low power gain	50 to 200
High input resistance	Up to 1 MΩ
Low output resistance	1 kΩ

Section Review

1. In which amplifier configuration is the signal introduced into the base-collector portion of the transistor and extracted at the emitter?
2. Why is the common-collector amplifier also known as an emitter follower?
3. State the DC current gain for a common-collector amplifier in equation form.
4. Circuits that perform impedance matching are often referred to as buffers (True/False).

— 21–6

FET AMPLIFIERS

Like the BJT, the FET has three terminals and may be connected in three configurations: common-source, common-drain, and common-gate. These configurations apply to both JFETs and MOSFETs. The **common-source amplifier** is the most popular FET circuit and is similar in operation to the BJT common-emitter configuration. The common-source connection, shown in Figure 21–12(a), provides a high input impedance, medium to high output impedance, and voltage gain greater than 1. The input signal is applied between the gate and source, and the output signal is taken between the drain and source. The gate-to-

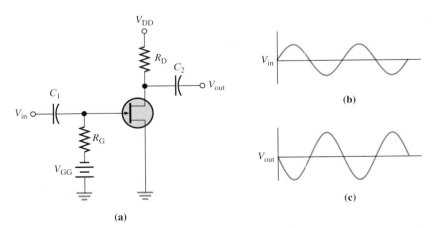

FIGURE 21–12 *(a) Common-source amplifier. (b) Input signal. (c) Output signal.*

source bias voltage is provided by the voltage source V_{GG}. Resistor R_G has a high value of resistance, so the input resistance of the circuit is high. Because R_G is essentially in parallel with the gate-to-source resistance of the FET, a low value of R_G would lower the input resistance of the amplifier.

Figures 21–12(b) and (c) show the input and output waves for a common-source configuration. As the AC input signal goes positive, it reduces the reverse-bias voltage across the gate-channel junction, which in turn reduces the size of the depletion region. As a result, the channel current, I_D, increases and the voltage drop across the drain resistor increases. As the voltage across R_D increases, the voltage at the drain (with respect to ground) decreases, which creates a 180° phase shift between the input and output signals. A negative-going input signal increases the reverse bias of the PN junction, causing the depletion region to increase and the channel resistance to increase. This action reduces the drain current and the voltage at the drain goes positive.

The voltage gain of a common-source FET is dependent on the transconductance of the FET and the size of drain resistor connected in the circuit. In Chapter 20, transconductance was defined as an indication of how much gate voltage signal is transferred to a drain current signal. Because of the small values of current involved, transconductance is generally measured in microsiemens (μS). The voltage gain of a common-source amplifier is found using the following equation:

$$\mathscr{A}_v = g_m R_D \tag{21.7}$$

where g_m is transconductance in microsiemens.

When a DE MOSFET is used in a common-source configuration, a negative-biasing voltage is no longer required at the gate. Because this type of MOSFET is capable of operating in both the depletion mode and the enhancement mode, it will operate with a gate-to-source voltage of 0 V_{DC}. Figure 21–13 shows a DE MOSFET in a common-source configuration. In this particular circuit, no coupling capacitor is required at the input because no DC biasing voltage is present. Therefore, the AC input and the amplifier are said to be *direct coupled*. Although coupling capacitors are quite popular in amplifier circuits, they have certain disadvantages in low-frequency applications. Unlike capacitor coupling, circuits that use direct coupling are capable of responding to signals below 10 Hz. When it is necessary to amplify DC signals, direct-coupled amplifiers must be used. Because pure DC signals have no frequency, coupling capacitors cannot pass DC input signals.

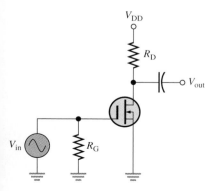

FIGURE 21–13 *DE MOSFET common-source amplifier.*

EXAMPLE 21–5

For the circuit shown in Figure 21–14, determine the AC output voltage if the AC input signal is 10 mV and the FET has a transconductance of 5000 μS.

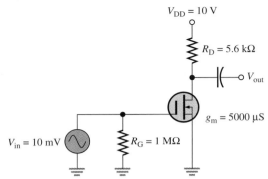

FIGURE 21–14 *Circuit for Example 21–5.*

Solution

$$\mathscr{A}_v = \frac{V_{out}}{V_{in}} = g_m R_D$$

Therefore, $V_{out} = V_{in} g_m R_D = (10 \text{ mV})(5000 \text{ μS})(5.6 \text{ kΩ}) = 280 \text{ mV}$

The **common-drain amplifier** is similar to the common-collector BJT configuration. A typical common-drain circuit is shown in Figure 21–15(a). This circuit is also known as a **source follower** because the output signal, which appears at the source, follows the input signal in both amplitude and phase. Like the emitter follower, the source follower has a voltage gain of less than 1. However, the high input impedance makes this configuration well suited for many applications.

The amplifier shown in Figure 21–15(a) is self-biased because it uses only one biasing voltage. Resistor R_S is used to produce the self-bias to establish the Q point and to develop the AC output signal. The input signal for the common-drain amplifier is applied to the gate, and the output signal appears across the source resistor. When the input to the N-channel JFET goes positive, as shown in Figure 21–15(b), it opposes the reverse bias and increases the flow of channel current. This produces a positive-going output signal, as illustrated in Figure 21–15(c).

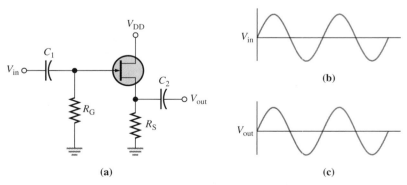

FIGURE 21–15 *(a) Common-drain amplifier. (b) Input signal. (c) Output signal.*

The input voltage to the common-drain is larger than the output voltage because V_{in} = $V_{GS} + I_D R_S$. The voltage gain for the common-drain is found using the following:

$$\mathcal{A}_v = \frac{g_m R_S}{1 + g_m R_S} \qquad (21.8)$$

EXAMPLE 21-6

$g_m = 5000 \ \mu S$

$V_{in} = 20 \ mV$

C_1

$R_G = 4.7 \ M\Omega$

C_2

V_{out}

$R_S = 1 \ k\Omega$

FIGURE 21-16 *Circuit for Example 21-6.*

The source follower shown in Figure 21-16 uses a FET with a g_m of 5000 μS. Determine the output voltage if the input signal is 20 mV.

Solution

$$\mathcal{A}_v = \frac{g_m R_S}{1 + g_m R_S} = \frac{(5000 \ \mu S)(1 \ k\Omega)}{1 + (5000 \ \mu S)(1 \ k\Omega)} = 0.83$$

$$V_{out} = \mathcal{A}_v V_{in} = (0.83)(20 \ mV) = 16.67 \ mV$$

The **common-gate amplifier** is used to transform a low input impedance to a high output impedance. A typical common-gate amplifier is shown in Figure 21-17(a). In this circuit, the input signal is applied to the source, and the output is taken from the drain. In this circuit connection, the input resistance is low, the output signal is not inverted, and the output resistance is the same as in the common-source amplifier. When the AC input signal of Figure 21-17(b) is applied, the positive-going signal aids the reverse bias to decrease the drain current flow. Because $V_D = V_{DD} - I_D R_D$, as I_D decreases, the output voltage developed at the drain (with respect to ground) increases, as shown in Figure 21-17(c). Consequently, the output voltage and the input voltage are in phase with each other.

(a)

$+V_{DD}$

R_D

C_1

V_{out}

C_2

V_{in}

R_S

(b)

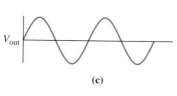

(c)

FIGURE 21-17
(a) Common-gate amplifier. (b) Input signal. (c) Output signal.

V_{in}

V_{out}

The input impedance for a common-gate amplifier is very low, often around 200 Ω, which is a major disadvantage for this type of amplifier. Because the drain current is the output current and the gate-to-source voltage is the input signal, the input impedance is found by Ohm's law: $R_{in} = V_{GS}/I_D$, which is the inverse of transconductance. Therefore,

$$R_{in} = \frac{1}{g_m} \tag{21.9}$$

Although the source resistor in Figure 21–17 is shown in parallel with the gate and source of the transistor, it can be ignored when approximating the input impedance of the circuit. Because R_{in} is generally quite low, as long as R_S is above 10 kΩ, the source resistor can be ignored when determining input impedance in most practical applications. The voltage gain of a common-gate FET amplifier is found in the same manner as the voltage gain for the common-source amplifier. That is,

$$\mathcal{A}_v = g_m R_D$$

EXAMPLE 21–7

For the common-gate amplifier shown in Figure 21–18, find the input impedance and the voltage gain.

$+V_{DD}$

$R_D = 20\ k\Omega$

V_{out}

$g_m = 7500\ \mu S$

V_{in}

$R_S = 10\ k\Omega$

FIGURE 21–18 *Circuit for Example 21–7.*

Solution

$$R_{in} = \frac{1}{g_m} = \frac{1}{7500\ \mu S} = 133.33\ \Omega$$

$$\mathcal{A}_v = g_m R_D = (7500\ \mu S)(20\ k\Omega) = 150$$

Section Review

1. What is the most popular FET amplifier configuration?
2. State the equation for finding the voltage gain of a common-source amplifier.
3. A common-gate amplifier is similar to the common-collector BJT configuration (True/False).
4. Which FET configuration is also known as a source follower?
5. Which FET amplifier is used to transform a low input impedance to a high output impedance?

In many situations, the total gain of an amplifier circuit is greater than that produced by a single transistor. Under these conditions, the output of one amplifier may be connected to the input of another amplifier to increase the overall gain of the circuit. These configurations are called **multistage amplifiers** or *cascaded amplifiers*. The term *cascade* is applied when several amplifiers or signal paths are connected in such a way that the signal is forced to flow through all of them to reach its destination. A multistage amplifier is identified according to the method used to couple one amplifier stage to the next. As mentioned in Chapter 18, there are three basic methods of coupling circuits together: direct coupling, capacitor coupling, and transformer coupling.

When amplifier stages are coupled together, their combined gain is essentially the gains of the individual stages multiplied together. This is true for the overall voltage, current, and power gains of a multistage amplifier. An example of a direct-coupled multistage amplifier is shown in Figure 21–19. This particular circuit is known as a **darlington amplifier** and consists of two CC amplifiers cascaded together. The configuration of Figure 21–19 has a high input impedance, low output impedance, and high current gain. Because of the high input impedance, darlington amplifiers require low input signals to operate. This makes the circuit well suited for the audio output section in radio receivers. Darlington amplifiers are constructed using two separate transistors or by combining two transistors in a single package. The total current gain is the product of the individual gains of the two CC amplifier stages. Therefore, the current gain for a darlington amplifier is

$$\mathscr{A}_i = (\mathscr{A}_{i1})(\mathscr{A}_{i2}) = (\beta_1 + 1)(\beta_2 + 1) \tag{21.10}$$

EXAMPLE 21-8

Solution

Determine the current gain of the darlington amplifier shown in Figure 21–19 if both transistors have a beta of 80.

$$\mathscr{A}_i = (\beta_1 + 1)(\beta_2 + 1) = (81)(81) = 6561$$

FIGURE 21-19 *Darlington amplifier.*

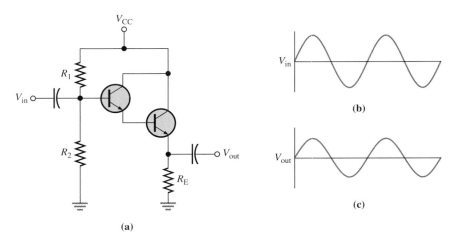

FIGURE 21–20 *(a) Darlington amplifier with voltage divider biasing. (b) Input signal. (c) Output signal.*

As with any emitter-follower circuit, the output voltage of the darlington is less than the input voltage. According to Kirchhoff's voltage law, the input voltage to the circuit shown in Figure 21–19 is equal to the voltage drops across the two base-emitter junctions of Q_1 and Q_2 plus the voltage drop across the load resistor R_E. Figure 21–20(a) shows a darlington amplifier with voltage divider biasing. As the input signal in Figure 21–20(b) goes positive, the base of Q_1 goes positive and the emitter of Q_1 follows the base. Because the emitter of Q_1 is the base of Q_2 , when the base of Q_1 goes positive, the base of Q_2 does the same. Consequently, the output voltage shown in Figure 21–20(c) is in phase with the input voltage.

Figure 21–21(a) shows an example of a multistage amplifier with capacitor coupling. The output of stage 1 is taken at the collector of Q_1 and fed into the base of Q_2 via the coupling capacitor C_2. The coupling capacitor isolates the DC voltages, allowing the AC signal to pass through to the output of the amplifier. Figure 21–21(b) shows the AC input signal to the multistage amplifier. As the input signal goes positive, the output at the

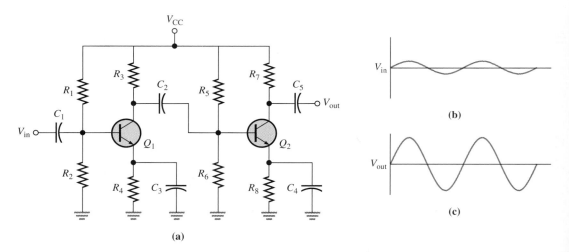

FIGURE 21–21 *(a) Multistage amplifier with capacitor coupling. (b) Input signal. (c) Output signal.*

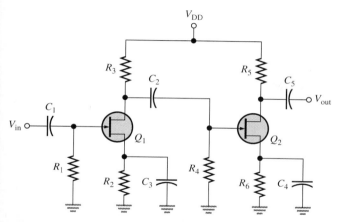

FIGURE 21-22 *Multistage FET amplifier.*

collector of Q_1 goes negative. Because the collector of Q_1 is essentially the base of Q_2, as the base of Q_1 goes positive, the base of Q_2 goes negative. Consequently, the output signal for the multistage amplifier is inverted twice. The double-inversion cancels out, and the output signal is shown in Figure 21–21(c) as being in phase with the input signal.

The voltage gain of the circuit shown in Figure 21–21 is a product of the individual voltage gains. However, the gain can no longer be calculated as R_C/r_e because of the **loading effect** on the output of Q_1. The input impedance of Q_2 is seen by the collector of Q_1 as a load resistor in parallel with the biasing resistors connected to the base of Q_2. This loading effect can drastically reduce the overall voltage gain of a circuit.

The multistage FET amplifier shown in Figure 21–22 has very high values of input impedance and is not adversely affected by loading. The total voltage gain of a two-stage FET is also the product of the individual voltage gains. That is,

$$\mathcal{A}_v = (g_{m1}R_{D1})(g_{m2}R_{D2}) \tag{21.11}$$

EXAMPLE 21-9

Determine the output voltage for the circuit shown in Figure 21–23.

+10 V

$R_2 = 2.2\ \text{k}\Omega$

C_2

$R_5 = 2.2\ \text{k}\Omega$

C_4

V_{out}

$V_{\text{in}} = 10\ \text{mV}$

C_1

$g_m = 6000\ \mu\text{S}$

$g_m = 6000\ \mu\text{S}$

$R_1 = 100\ \text{M}\Omega$

$R_3 = 330\ \Omega$ C_3

$R_4 = 100\ \text{M}\Omega$

$R_6 = 330\ \Omega$ C_5

FIGURE 21-23 *Circuit for Example 21-9.*

Solution   The first step is to determine the voltage gain for the circuit. Because the two stages of the amplifier are identical, the individual voltage gains are equal.

$$\mathcal{A}_{v1} = \mathcal{A}_{v2} = g_m R_D = (6000 \ \mu S)(2.2 \ k\Omega) = 13.2$$

$$\mathcal{A}_v = (\mathcal{A}_{v1})(\mathcal{A}_{v2}) = (13.2)(13.2) = 174.24$$

$$V_{out} = (\mathcal{A}_v)(V_{in}) = (174.24)(10 \ mV) = 1.74 \ V$$

Transformer coupling between amplifier stages is more efficient than capacitor coupling and has the advantage of providing impedance matching by varying the turns ratio of the transformer. The main disadvantages of transformer coupling are cost and size. However, in some applications, such as audio amplifiers and RF amplifiers, these disadvantages are outweighed by the benefits of DC isolation, impedance matching, and efficiency. Recall that when impedances are matched, maximum power transfer occurs. The transformer is the only passive component that can provide impedance matching and maximum power transfer.

Figure 21–24 shows a basic transformer-coupled multistage amplifier. The transistors are biased by resistors R_1, R_2, R_3, and R_4. Transformer T_1 is connected so that its primary winding acts as the output impedance for the first stage. The secondary winding of T_1 is connected in series with the base of the second stage. Because the collector of Q_1 and the base of Q_2 are coupled electromagnetically, the DC biasing voltages are isolated and the AC signal passes from the primary to the secondary of T_1.

Section Review

1. The total gain of a multistage amplifier is the sum of the individual gains (True/False)
2. What is the loading effect of a multistage amplifier?
3. Transformer coupling between amplifier stages is less efficient than capacitor coupling (True/False).

FIGURE 21–24   *Multistage amplifier with transformer coupling.*

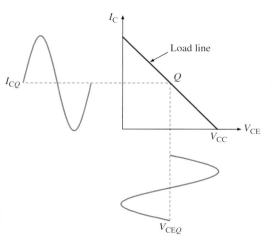

FIGURE 21–25   *Class A bias point.*

_____ **21–8** _____

CLASS A AMPLIFIERS

The amplifier circuits discussed in this chapter have been biased so that the output voltage can vary in response to both positive and negative input signals. An amplifier that is capable of responding to positive and negative signals without becoming saturated or cutoff is called a **class A amplifier.** The operating point in class A amplifiers is located approximately halfway between cutoff and saturation on the load line. This permits maximum obtainable output signal variation with minimum distortion. Figure 21–25 illustrates the typical Q point for a class A amplifier. When no input signal is applied, the collector current is I_{CQ} and the collector-emitter voltage, V_{CEQ}, is at its reference voltage. Because the total power for the circuit is equal to I_{CQ} times V_{CC}, the circuit is drawing full input power even when there is no output power.

The efficiency of an amplifier is an indication of how well an amplifier converts DC input power to AC output power. An amplifier with 100% efficiency would have no losses. This is impossible in a practical amplifier because of the power losses in resistors and transistors in the amplifier circuit. Class A amplifiers have poor efficiency, typically below 25%. The efficiency of an amplifier is a ratio of the power delivered to the load, P_L, divided by the input power to the circuit $V_{CC}I_{CQ}$.

$$\eta = \left(\frac{P_L}{V_{CC}I_{CQ}} \right) 100\% \qquad (21.12)$$

Section Review

1. What is a class A amplifier?
2. On a load line for a class A amplifier, where is the operating point?
3. What is the typical efficiency of a class A amplifier?

_____ **21–9** _____

CLASS B AMPLIFIERS

The main disadvantage of class A amplifiers is that they draw power even when no input signal is applied. This is caused by the bias point being in the center of the active region. If an amplifier has its Q point biased at cutoff, it is called a **class B amplifier** and said to be

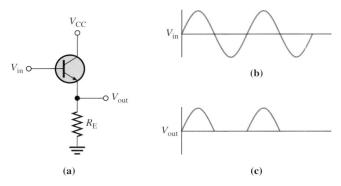

(a)   (b)   (c)

FIGURE 21–26   *(a) Class B amplifier. (b) Input signal. (c) Output signal.*

operating in the class B mode. When the bias point is in cutoff, no power is drawn except when an AC input signal is applied. Figure 21–26(a) shows a simple class B amplifier. When no signal is applied, the transistor is in the cutoff mode. As the AC input signal of Figure 21–26(b) goes positive, the transistor conducts. A negative-going input signal moves the operating point further into cutoff and no output occurs, as shown in Figure 21–26(c). Consequently, a class B amplifier only conducts for one-half of the input cycle.

Class B amplifiers produce large amounts of distortion because one-half of the input signal is cut off. In practical class B amplifiers, two transistors are used to drive a load. This configuration is called a **push-pull amplifier** and is illustrated in Figure 21–27(a). The push-pull circuit is also referred to as a *complementary amplifier,* because one transistor is an NPN and the other is a PNP. When the input signal of Figure 21–27(b) is positive, the NPN transistor conducts and the PNP is off. When the signal goes negative, the PNP transistor conducts and the NPN is off. The resulting waveshape is shown in Figure 21–27(c). Note that this wave is not a pure sine wave.

Although the push-pull class B amplifier is very efficient compared to class A amplifiers, there is one major drawback to using this configuration. This disadvantage is called **crossover distortion** and is shown in Figure 21–27(c). Recall that in order for a PN junction to conduct current, the forward breakover voltage must be exceeded. This voltage

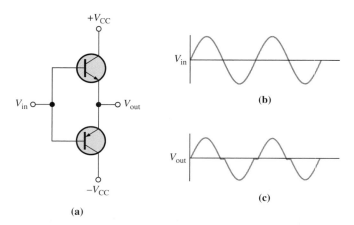

(a)   (b)   (c)

FIGURE 21–27   *(a) Push-pull amplifier. (b) Input signal. (c) Output signal with crossover distortion.*

FIGURE 21-28 *Class AB amplifier with reduced crossover distortion.*

is typically 0.7 V for silicon transistors. In the push-pull amplifier of Figure 21–27(a), the input signal must reach a certain minimum value before either transistor begins conducting. This means that both transistors are off between 0 V and 0.7 V on both the positive- and negative-going alternations. Crossover distortion can be minimized by operating the transistors at a slightly forward bias. The amount of forward bias needed to overcome crossover distortion is about 0.7 V.

Figure 21–28 shows a biasing method using two diodes to bias the push-pull amplifier and eliminate crossover distortion. Resistors R_1 and R_2 and diodes D_1 and D_2 form a voltage divider network that holds both transistors at the threshold of conduction when no input signal is applied. Coupling capacitors are required for this circuit because DC biasing voltage is present at the base of both transistors. Diodes D_1 and D_2 "pre-bias" the emitter-base junctions of the transistors, ensuring that the base of Q_1 is more positive than the base of Q_2. The circuit of Figure 21–28 is also called a **class AB amplifier,** because it is neither a class A nor class B configuration. A true class B amplifier is biased at cutoff, while the circuit shown in Figure 21–28 is biased partially on. Class B and AB amplifiers are characterized as having relatively high efficiencies of 60% to 65%. Most push-pull power amplifiers will operate as class AB instead of class B to reduce crossover distortion.

Section Review

1. What is the main disadvantage of class A amplifiers?
2. If an amplifier has its Q point biased at saturation, it is said to be operating in the class B mode (True/False).
3. What is a class AB amplifier?

21-10
CLASS C AMPLIFIERS

Class C amplifiers produce the highest efficiency and the highest distortion of the three classes of operation. A class C amplifier produces a highly distorted wave that conducts for less than 180°. Because the transistor is biased off during most of the input cycle, very low power is dissipated by the transistor. Class C amplifiers are used primarily in high-power, high-frequency applications, such as radio-frequency transmitters. Figure 21–29(a) shows

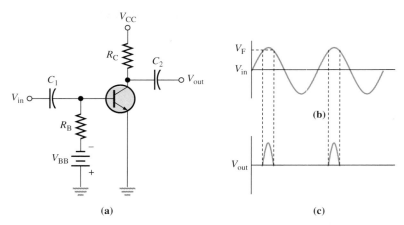

FIGURE 21–29 *(a) Class C amplifier. (b) Input signal.*
(c) Output waveshape.

a simple class C amplifier. For the transistor to begin conducting, it must overcome the negative bias supplied by V_{BB} and the forward voltage drop across the base-emitter junction (0.7 V).

$$V_F = V_{BB} + V_{BE} \qquad (21.13)$$

The voltage V_F is the amount of input voltage required to cause the transistor to begin conducting. The length of time the transistor is in the conducting state is determined by V_F. In Figure 21–29(b), the input voltage is shown to rise above V_F at approximately the 60° point. The transistor remains in the conducting state as long as V_F is exceeded. When the input voltage falls below V_F, the transistor shuts off. The output waveform of Figure 21–29(c) shows the conduction interval for the amplifier. The output is essentially a series of pulses, which can be varied by changing V_{BB}. For example, if V_{BB} was made more negative, the conduction interval would be shorter. The output pulses produced by the class C amplifier are ideal for oscillator circuits, including thyristor control.

Section Review

1. Which class of amplifiers produces the highest efficiency and the highest distortion?
2. The output wave of a class C amplifier conducts for less than 180° (True/False).
3. Class C amplifiers are often used in high-power high-frequency applications (True/False).

— 21-11

DIFFERENTIAL AMPLIFIERS

The **differential amplifier** consists of two direct-coupled transistors and is designed to amplify the difference between the two input terminals. Figure 21–30 shows a basic differential amplifier. This type of amplifier is very popular as the input stage for an operational amplifier (op amp). Two power supplies are required for the differential amplifier—one positive and one negative. Although two input terminals are available, the differential amplifier can be operated as a single-input device by grounding one of the input terminals. There are three possible output connection points for the differential amplifier: from either

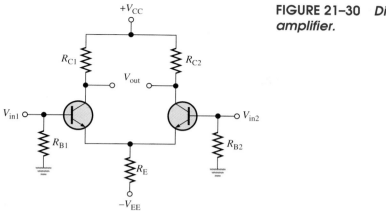

FIGURE 21–30 *Differential amplifier.*

collector to ground, and between the two collectors. The output signal between the two collectors is called the *differential signal.* Ideally, when the two input signals are the same, the differential signal would be 0 V. If the input signals are not identical, the output is proportional to the difference between the inputs.

$$V_{out} = \mathcal{A}_v(V_{in1} - V_{in2}) \tag{21.14}$$

The differential amplifier is essentially two common-emitter configurations. The voltage gain for the circuit of Figure 21–30 is approximated using the following equation:

$$\mathcal{A}_v = \frac{R_C}{r_e} \tag{21.15}$$

Although the voltage gain using the above equation is not an exact value, it is very close for most applications. The current flowing through resistor R_E is the sum of the emitter current for both transistors. The emitter voltage is approximately equal to ground (-0.7 V). The emitter current is solved as follows:

$$I_E = \frac{V_E - V_{EE}}{R_E}$$

EXAMPLE 21–10

For the circuit shown in Figure 21–31, calculate the output voltage.

$V_{CC} = +15$ V

$R_{C1} = 5.6$ kΩ $R_{C2} = 5.6$ kΩ

V_{out}

$V_{in1} = 20$ mV $V_{in2} = 10$ mV

$R_{B1} = 1$ kΩ $R_E = 4.7$ kΩ $R_{B2} = 1$ kΩ

$V_{EE} = -15$ V

FIGURE 21–31 *Circuit for Example 21–10.*

Solution First, solve for the total emitter current.

$$I_E = \frac{V_E - V_{EE}}{R_E} = \frac{-0.7 - (-15 \text{ V})}{4.7 \text{ k}\Omega} = 3.04 \text{ mA}$$

$$r_e = \frac{25 \text{ mV}}{I_E/2} = \frac{25 \text{ mV}}{1.52 \text{ mA}} = 16.45 \Omega$$

$$\mathcal{A}_v = \frac{R_C}{r_e} = \frac{5.6 \text{k}\Omega}{16.45 \Omega} = 340$$

$$V_{out} = \mathcal{A}_v(V_{in1} - V_{in2}) = (340)(20 \text{ mV} - 10 \text{ mV}) = 3.4 \text{ V}$$

There are two different types of gain associated with differential amplifiers: *differential gain* and *common mode gain*. The differential gain, $\mathcal{A}_{vd}$, is the voltage gain found using Equation 21.15. A signal that is common to both inputs is called *common mode*. A measure of a differential amplifier's ability to eliminate, or reject, these common signals is called the **common mode rejection ratio, CMRR.** Often, the common signal represents unwanted noise or hum. If this type of distortion is present at both inputs, it will be canceled. An ideal differential amplifier will completely cancel all common mode signals and, therefore, will have no effect on the output signals. As a result, the common mode gain, $\mathcal{A}_{vcm}$, will be zero ideally. The CMRR of a differential amplifier is the ratio of the differential voltage gain to the common mode gain.

$$\text{CMRR} = \frac{\mathcal{A}_{vd}}{\mathcal{A}_{vcm}} \tag{21.16}$$

Ideally, the CMRR of a differential amplifier is infinity and the only signal being amplified is the difference between the two inputs. Typical values for CMRR range from 1000 to 10,000. CMRR is commonly stated in decibels as

$$\text{CMRR (dB)} = 20 \log_{10}\left(\frac{\mathcal{A}_{vd}}{\mathcal{A}_{vcm}}\right) \tag{21.17}$$

$$= 20 \log_{10} \text{CMRR}$$

_____ **EXAMPLE 21–11** _____

Determine the CMRR and CMRR(dB) for a differential amplifier with a differential voltage gain of 800 and a common mode gain of 0.1

Solution

$$\text{CMRR} = \frac{\mathcal{A}_{vd}}{\mathcal{A}_{vcm}} = \frac{800}{0.1} = 8000$$

$$\text{CMRR(dB)} = 20 \log_{10} \text{CMRR} = 20 \log_{10}(8000) = 78 \text{ dB}$$

When only one output from a differential amplifier is used, it is called a **single-ended output amplifier.** A basic single-ended output differential amplifier is illustrated in Figure 21–32. Because only one-half of the available output is used by a single-ended configuration, the voltage gain is reduced by 50%.

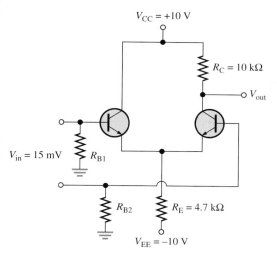

$V_{CC} = +10 \text{ V}$

$R_C = 10 \text{ k}\Omega$

V_{out}

$V_{in} = 15 \text{ mV}$ R_{B1}

R_{B2} $R_E = 4.7 \text{ k}\Omega$

$V_{EE} = -10 \text{ V}$

FIGURE 21–32 *Single-ended output differential amplifier.*

EXAMPLE 21–12

Determine the approximate output voltage of the differential amplifier shown in Figure 21–32.

Solution   The total emitter current is found first.

$$I_E = \frac{V_E - V_{EE}}{R_E} = \frac{10 \text{ V} - 0.7 \text{ V}}{4.7 \text{ k}\Omega} = 1.98 \text{ mA}$$

Next, solve for r_e.

$$r_e = \frac{25 \text{ mV}}{I_E/2} = \frac{25 \text{ mV}}{0.99 \text{ mA}} = 25.25 \text{ }\Omega$$

The voltage gain for a differential output is found using Equation 21.15. Because a single-ended output is one-half of a differential output, the voltage gain is

$$\mathcal{A}_v = \left(\frac{R_C}{2r_e}\right) = \left[\frac{10 \text{ k}\Omega}{2(25.25 \text{ }\Omega)}\right] = 198$$

The output voltage can now be determined.

$$V_{out} = \mathcal{A}_v V_{in} = (198)(15 \text{ mV}) = 2.97 \text{ V}$$

Figure 21–33(a) shows a differential amplifier with one input grounded and the signal voltage applied to the other input. This configuration is known as a **single-ended input amplifier.** The input signal is shown in Figure 21–33(b). The output voltage for transistor Q_1 is illustrated in Figure 21–33(c). As with any common-emitter amplifier, the voltage at the collector of Q_1 is 180° out of phase with the base. Because the base of Q_2 is grounded, it is essentially acting as a CB amplifier. Consequently, the voltage at the emitter of Q_2 will be amplified and will appear in phase with the collector of Q_2. Because the emitters of the two transistors are connected together, when the emitter of Q_1 goes negative, the collector of Q_2 will also go negative. The waveshape for the collector of Q_2 is shown in Figure 21–33(d). Because the two collectors are 180° out of phase with each other, the amplitudes of the two signals will add, producing a signal that is twice as large as either individual output.

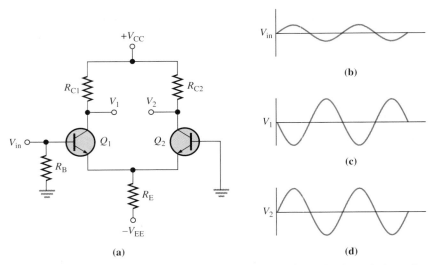

FIGURE 21–33 *(a) Single-ended input amplifier. (b) Input signal. (c) Signal at collector of Q_1. (d) Signal at collector of Q_2.*

Section Review

1. What is a differential amplifier?
2. What are the two types of gain associated with differential amplifiers?
3. A signal that is common to both inputs of a differential amplifier is called common mode (True/False).

___ 21-12 _____

TROUBLESHOOTING AMPLIFIERS

The troubleshooting techniques discussed in previous chapters form the basis for testing amplifier circuits. The main difference between testing an amplifier and testing an individual component is that amplifiers consist of a group of components that must remain interconnected for accurate readings. When no short circuit is present, amplifiers can be tested with the circuit energized. Often, the schematic diagram supplied by the manufacturer will provide test voltages for various points in the circuit. The test voltages provided in the circuit documentation may include both the proper DC-biasing voltages as well as the correct signal voltages. These voltage values are then checked using a voltmeter or oscilloscope. When an abnormal voltage reading or waveshape is detected, the problem can then be isolated to a specific area in the amplifier circuit. When no documentation is available, it is also possible to determine what the correct voltages should be in an amplifier and to compare the calculated values to those read by a voltmeter.

One of the main causes of amplifier failure is improper biasing. If a component such as a resistor or capacitor should fail, the transistor could be driven into saturation or cutoff. Saturation is particularly dangerous because it can cause excess values of current to flow in the amplifier circuit. When troubleshooting an amplifier circuit, the best place to start is the fuse. If the fuse is blown, we can assume a short circuit is present at some point in the amplifier. Simply replacing a blown fuse will not solve this type of problem. If the fuse is not blown, then an open circuit is probably contributing to the circuit failure. The following examples illustrate the use of voltmeters and oscilloscopes on energized circuits.

EXAMPLE 21–13

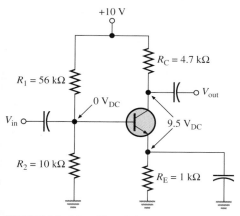

FIGURE 21–34 *Circuit for Example 21–13.*

The amplifier circuit shown in Figure 21–34 is producing below its rated power and the output signal is highly distorted. Determine what the problem is.

Solution   The DC-biasing voltages are tested first. The base voltage V_B indicates $0\ V_{DC}$. The base voltage is calculated as

$$V_B = \left(\frac{R_2}{R_1 + R_2}\right) V_{CC} = \left(\frac{10\ k\Omega}{56\ k\Omega + 10\ k\Omega}\right) 10\ V = 1.5\ V_{DC}$$

Because a DC-biasing voltage is not at the base, the biasing current flowing through R_1 has been interrupted. This could be caused by R_1 open circuiting, as shown in Figure 21–35.

FIGURE 21–35 *Resistor R₁ open circuited.*

Multistage amplifiers consist of several single-stage amplifiers. To troubleshoot a multistage amplifier, we first must determine which stage in the amplifier is malfunctioning. This is generally done by injecting a signal using a signal generator and tracing the signal with either an oscilloscope or a voltmeter.

EXAMPLE 21–14

The two-stage FET amplifier shown in Figure 21–36 has an AC output voltage of 0 V. The DC-biasing voltage at the drain of Q_2 is 8 V and the AC voltage at the drain is 5 V. What is wrong with this circuit?

FIGURE 21–36 *Circuit for Example 21–14.*

Solution Because an AC voltage is present at the drain of Q_2 and not at the load, it is apparent that the AC signal is not getting through the coupling capacitor C_5. Therefore, C_5 must be open circuited.

Section Review

1. What is one of the main causes of amplifier failure?
2. When troubleshooting an amplifier, the best place to start is the fuse (True/False).

— 21–13

PRACTICAL APPLICATION SOLUTION

In the chapter opener, you were asked to design a liquid-level controller using a transistor amplifier and switch. The circuit was to measure the resistance between two electrodes submerged in a tank of conductive liquid. If the liquid level rose above a predetermined value, an alarm, controlled by a relay coil, was to be activated. The following steps show how the solution for this practical application of amplifier circuits was found.

STEP 1   Draw a sketch of the system. Figure 21–37 shows the basic circuit design, as outlined in the chapter opener. If the level in the tank is low, the resistance between the electrodes is high, and the current flowing into Q_1 is low. The amplified current flowing from Q_2 into the base of Q_3 will be insufficient to turn Q_3 on. The relay coil connected in series with Q_3 will also be off because no current is flowing at the collector of Q_3. As the liquid level rises, the resistance between the electrodes decreases, and the output current from the darlington amplifier increases. Eventually, the current at the emitter of Q_2 will be sufficient to turn on Q_3. This transistor will then draw a heavy current to energize the relay. When the relay coil energizes, the associated contacts will change state, and an alarm condition is signaled.

STEP 2   Use a diode to dissipate energy stored in the control relay coil. Diode D_1 in Figure 21–37 is a device designed to discharge the energy that remains stored in an

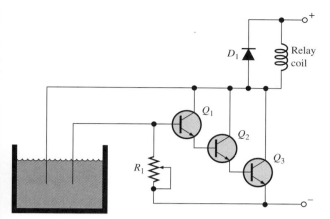

FIGURE 21-37 *Liquid-level controller using transistor amplifier and switch.*

inductor once the coil has been switched off. When transistor Q_3 switches off, the energy stored in the coil, because of inductance, will have a polarity that is opposite to the applied voltage. The reverse polarity across the coil will now forward bias D_1, allowing the energy stored in the coil to discharge. D_1 remains forward biased until the coil has discharged.

Summary

1. AC amplifiers contain both AC and DC voltages.
2. BJT amplifiers are classified as common-emitter, common-base, or common-collector amplifiers.
3. Amplifiers that perform impedance matching or isolation functions are often called *buffers.*
4. FET amplifiers can be connected in three configurations: common-source, common-drain, and common-gate amplifiers.
5. When amplifier stages are coupled together, the overall gain is a product of the individual gains.
6. Class A amplifiers have efficiencies below 25%.
7. If an amplifier has its Q point biased at cutoff, it is operating in the class B mode.
8. Class C amplifiers produce the highest efficiency and the highest distortion of the three classes of operation.
9. Differential amplifiers consist of two direct-coupled transistors and are designed to amplify the difference between the two input terminals.
10. One of the main causes of amplifier failure is improper biasing.

Answers to Section Reviews

Section 21-1

1. True
2. To combine the characteristics of one amplifier with those of another to produce a circuit with a combination of qualities, such as high input impedance and low output impedance, and to increase the overall gain of the circuit
3. When the output of one amplifier is connected to the input of another amplifier

Section 21–2

1. True
2. The emitter and collector waveforms are 180° out of phase with each other.
3. Frequency distortion, phase distortion, and amplitude distortion
4. Improper biasing and excessive input signals

Section 21–3

1. Current gain = I_C/I_E
2. False
3. The internal emitter resistance, which is dependent on the current flow in the circuit
4. Voltage gain = R_C/r_e

Section 21–4

1. Common-emitter
2. True
3. Current gain = I_C/I_B

Section 21–5

1. Common-collector amplifier
2. Because the emitter always follows, or copies, the base
3. Current gain = I_E/I_B
4. True

Section 21–6

1. Common-source
2. Voltage gain = $g_m R_D$
3. False
4. Common-drain amplifier
5. Common-gate amplifier

Section 21–7

1. False
2. When the input impedance of an amplifier is seen by the previous stage as a load resistor in parallel with the biasing resistors connected to the base of the amplifier
3. False

Section 21–8

1. An amplifier that is capable of responding to positive and negative signals without becoming saturated or cutoff
2. Approximately half-way between cutoff and saturation
3. Below 25%

Section 21–9

1. They draw power even when no input signal is applied.
2. False
3. An amplifier that is biased partially on

Section 21–10

1. Class C amplifiers
2. True
3. True

Section 21-11

1. An amplifier designed to amplify the difference between the two input terminals
2. Differential gain and common mode gain
3. True

Section 21-12

1. Improper biasing
2. True

Review Questions

Multiple Choice Questions

21-1 A single amplifier circuit is also known as a(n)
(a) Transistor   **(b)** Stage   **(c)** Inverter   **(d)** Cascading circuit

21-2 In a linear amplifier circuit,
(a) The waveshape of the output signal is not the same as the input
(b) The transistor is driven to either cutoff or saturation
(c) The transistor is usually biased near the center of its operating limits
(d) The output signal is not actually amplified

21-3 AC amplifier circuits consist of
(a) AC input signals and DC biasing voltages
(b) Only AC voltages
(c) Only DC voltages
(d) An AC power supply and a DC input signal

21-4 When the input signal and output signal are displaced by 180°, they are said to be
(a) Cascaded   **(b)** In phase   **(c)** Synchronized   **(d)** Inverted

21-5 When an amplifier is not properly biased, the output signal may be
(a) Inverted   **(b)** Cascaded   **(c)** Distorted   **(d)** Shifted

21-6 The current gain for a common-base amplifier is
(a) Less than one   **(b)** The same as a common-emitter amplifier
(c) Very high   **(d)** Generally between 1 and 10

21-7 In a common-base circuit, the input current is the
(a) Emitter current and the output is the collector
(b) Base current and the output is the collector
(c) Base current and the output is the emitter
(d) Collector current and the output is the emitter

21-8 The most popular configuration for a BJT amplifier is the
(a) Common base   **(b)** Common emitter
(c) Common collector   **(d)** Common source

21-9 The current gain of a common-emitter circuit is
(a) Less than one   **(b)** The same as a common-base circuit
(c) Very high   **(d)** Generally between 1 and 10

21-10 Bypass capacitors are used in amplifier circuits to
(a) Short circuit DC voltages   **(b)** Isolate AC voltages
(c) Increase voltage gain   **(d)** Reduce voltage gain

21-11 The emitter-follower amplifier is also known as the
(a) Common-base amplifier   **(b)** Common-emitter amplifier
(c) Common-collector amplifier   **(d)** Common-source amplifier

21–12 The voltage gain of a common-collector circuit is
(a) Less than one   (b) The same as a common-emitter circuit
(c) Very high   (d) Generally between 1 and 10

21–13 Circuits that perform impedance matching or isolating functions are often referred to as
(a) Impedance amplifiers   (b) Voltage amplifiers
(c) Inverters   (d) Buffers

21–14 The voltage gain of a common-source FET amplifier is dependent on the
(a) Beta and size of drain resistor
(b) Transconductance and size of drain resistor
(c) Beta and size of source resistor
(d) Transconductance and size of source resistor

21–15 The common-drain amplifier is similar to the
(a) Common base   (b) Common emitter
(c) Common collector   (d) Common source

21–16 An amplifier that delivers a large amount of current to a low impedance load is known as
(a) A voltage amplifier   (b) A current amplifier
(c) An impedance amplifier   (d) A low-output impedance amplifier

21–17 When amplifier stages are coupled together, the total gain is the
(a) Product of the individual gains
(b) Difference between the individual gains
(c) Sum of the individual gains
(d) Square root of the individual gains

21–18 When the operating point is located approximately halfway between cutoff and saturation, the amplifier is classified as a
(a) Class A amplifier   (b) Class B amplifier
(c) Class AB amplifier   (d) Class C amplifier

21–19 The main disadvantage of class B amplifiers is
(a) Efficiency   (b) Impedance matching
(c) Crossover distortion   (d) The output signal conducts for only 90°

21–20 The common-mode rejection ratio (CMRR) of a differential amplifier is a ratio of the
(a) Voltage gain to the current gain
(b) Differential gain to the voltage gain
(c) Two impedances
(d) Differential voltage gain to the common mode gain

Practice Problems

21–21 Calculate the voltage gain of the circuit shown in Figure 21–38.

21–22 For the circuit shown in Figure 21–39, calculate the voltage gain of the circuit with the
(a) Bypass capacitor connected
(b) Bypass capacitor disconnected

21–23 For the circuit shown in Figure 21–40, find the total input impedance.

FIGURE 21-38

FIGURE 21-39

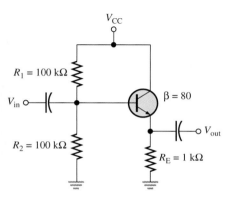

FIGURE 21-40

21–24 Determine the AC output voltage for the circuit of Figure 21–41 if the AC input signal is 20 mV and the FET has a transconductance of 3500 μS.

FIGURE 21–41

21–25 The circuit shown in Figure 21–42 consists of a FET with a g_m of 6000 μS. Determine the output voltage if the input signal is 10 mV.

FIGURE 21–42

21–26 Find the input impedance and voltage gain for the circuit shown in Figure 21–43.

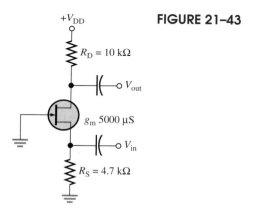

FIGURE 21–43

21–27 Determine the current gain of the amplifier shown in Figure 21–44 if both transistors have a beta of 100.

FIGURE 21-44

21-28 Calculate the output voltage for the circuit shown in Figure 21–45.

FIGURE 21-45

21-29 For the circuit shown in Figure 21–46, calculate the output voltage.

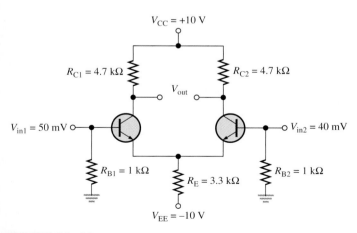

FIGURE 21-46

21–30   Determine the CMRR and CMRR(dB) for a differential amplifier with a differential voltage gain of 1000 and a common mode gain of 0.2.

Essay Questions

21–31   What is a multistage amplifier?

21–32   List the three main characteristics of a linear amplifier.

21–33   Why are DC voltages necessary in an AC amplifier?

21–34   Describe the three most common types of amplifier distortion.

21–35   Name the three configurations for BJT amplifiers.

21–36   Which configuration for a BJT amplifier is the most popular?

21–37   Why are coupling capacitors and bypass capacitors used in amplifier circuits?

21–38   What is a buffer?

21–39   List the three configurations for FET amplifiers.

21–40   Define the term *loading effect.*

21–41   Name three methods of coupling in amplifier circuits.

21–42   Describe the main difference between a class A amplifier and a class B amplifier.

21–43   What is crossover distortion and how can it be minimized?

21–44   Which class of amplifier has the highest efficiency?

21–45   How is the CMRR of a differential amplifier obtained?

21–46   What is one of the main causes of amplifier failure?

22

Operational Amplifiers (Op-Amps)

Learning Objectives

Upon completion of this chapter you will be able to

- List three characteristics of an ideal op-amp.
- Define slew rate.
- Describe the purpose of feedback in op-amp circuits.
- Determine the voltage gain of inverting and noninverting amplifiers.
- Explain the purpose of voltage followers.
- Name two applications of summing amplifiers.
- Describe the basic operation of a comparator.
- List two types of op-amp voltage regulators.
- Determine the resonant frequency of an oscillator.
- Name three types of multivibrators.

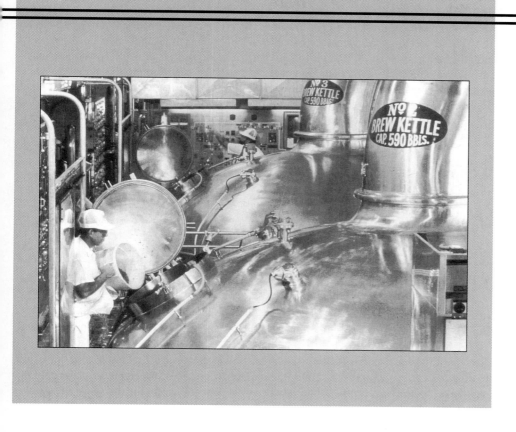

Practical Application

As an electronic technician in a large brewery, you are asked to design an electronic-weighing system for a grain hopper. The hopper is controlled by an electronic dump valve that controls the grain flow from the hopper to a vessel. The vessel's contents are weighed by a group of four load cells with differential amplifiers. A load cell is a device that converts force to a proportional electrical signal. The outputs of the load cells are to be fed into a summing amplifier. The output of the summing amplifier is then fed to the input of a comparator. The comparator produces a signal that closes the dump valve when the vessel's contents reach a desired weight.

Another design requirement is that the operator be able to adjust the weight using a potentiometer, or dial, on the control console. When the weight of the vessel reaches the value set by the operator, the electrically operated dump valve receives a signal from the output of the comparator to close. The valve remains closed as long as the contents of the vessel are below the reference value.

22-1

INTRODUCTION

The semiconductor devices discussed up to this point have generally been discrete in nature. That is, each device is packaged individually. When groups of electronic devices are combined into entire circuits on a single silicon chip, it is called an *integrated circuit* (IC). These solid-state devices are capable of containing thousands or even millions of miniaturized transistors. Although ICs may consist of combinations of resistors, capacitors, diodes, and transistors, they are generally classified according to the number of transistors they contain. ICs are normally operated at low power levels and are not capable of dissipating large amounts of heat. There are two basic types of integrated circuits: analog and digital. In this chapter we will examine the most popular type of analog-, or linear-, integrated circuit—the op-amp.

Section Review

1. What is an integrated circuit (IC)?
2. ICs are generally operated at high power levels (True/False).
3. What is the most popular type of linear-integrated circuit?

22-2

OP-AMP CHARACTERISTICS

An **operational amplifier, op-amp** is a high-gain, direct-coupled multistage amplifier with the capability of amplifying signals ranging from DC (0 Hz) to several megahertz. The op-amp is essentially three different amplifiers cascaded together. The input stage consists of a differential amplifier, the second stage is a multistage darlington amplifier, and the output stage is generally a common collector to provide low output impedance. Op-amps have five basic terminals: one for the positive supply, one for the negative supply, two for the input signals, and one for the output signal. The positive and negative power supplies connected to the op-amp allow the output voltage to swing positive or negative (with respect to ground). Figure 22–1 shows the schematic diagram for the components in a typical op-amp. The two inputs are shown connected to the base of transistors Q_1 and Q_2 and are labeled + and −. The + input is called the *noninverting input* and the − input is called the *inverting input*. Any voltage applied to the inverting input terminal appears at the output terminal with a reversed polarity. Any voltage applied to the noninverting input appears at the output with the same polarity. It should be noted that the + and − symbols at the op-amp input terminals have nothing to do with polarities of voltages at the inputs. The voltage at either input can be positive or negative.

An ideal op-amp has infinite input impedance, zero output impedance, and infinite voltage gain. When an amplifier has a high input impedance, it will draw a low value of current from the signal source. The low output impedance allows the amplifier to provide a constant output voltage, independent of the current drawn from the source. Figure 22–2 shows the schematic symbols for an op-amp. In Figure 22–2(a) the power supply connections are included. Often, the power supply connections are ignored, as in Figure 22–2(b). In our study of op-amps, we will omit the power supply connections and assume that the power supplies are connected. An ideal op-amp will produce an output voltage of 0 V when the voltage applied between the two input terminals is zero. To make this possible, the differential amplifier at the input must be perfectly matched. In a practical op-amp the

FIGURE 22–1 *Schematic diagram of typical op-amp.*

output is never exactly 0 V, although it is close enough for most applications. If the output voltage is required to be exactly zero when both inputs are zero, it is possible to apply a small DC bias voltage to correct, or offset, the discrepancy. This compensating voltage is called the **input offset voltage.**

The **slew rate, SR,** is another important characteristic of the op-amp. Basically, the slew rate is an indication of how fast the output voltage can change. The base unit for slew rate is volts per second, although in high-speed devices such as op-amps, it is generally specified in volts per microsecond, V/µs. A typical op-amp will have a slew rate of 0.5 V/µs. In other words, the output voltage can change 1/2 V in 1 µs. Slew rate is an impor-

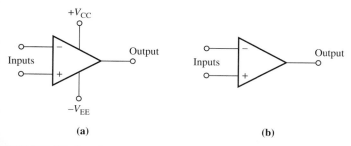

FIGURE 22–2 *Schematic symbols for op-amp.*
(a) Power supplies shown. (b) Power supply connection assumed.

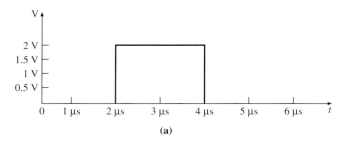

FIGURE 22–3 *Effect of slew-rate on op-amp output signal. (a) Input pulse. (b) Distorted output.*

(a)

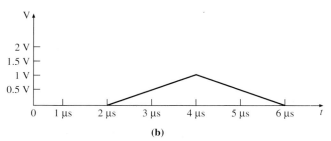

(b)

tant parameter because it provides a good indication of how accurately the op-amp can amplify and reproduce input signals. When the input signal is changing very quickly, it is essential that the output change in exact synchronism. Figure 22–3(a) shows a high-frequency, pulse-type input signal. The period of this signal is 4 μs. The output waveshape for an op-amp with a slew rate of 0.5 V/μs is shown in Figure 22–3(b). The output signal is highly distorted and only reaches 1/2 of the input voltage. During the 2 μs that the input signal is at 2 V, the output is rising at a rate of 0.5 V/μs. As a result, the output voltage has only reached 1 V when the input falls back to zero.

The output signal of an op-amp will be distorted if the rate of change of the input signal is greater than the slew rate. The maximum frequency that can be obtained without distortion with a peak voltage, V_p, is found using the following equation:

$$f_{max} = \frac{SR}{2\pi V_p} \tag{22.1}$$

where f_{max} = maximum output frequency before distortion occurs
V_p = peak input voltage, in volts
SR = slew rate, in volts per second

───── **EXAMPLE 22–1** ─────

Determine the maximum frequency that can be obtained from an op-amp if the slew rate is 0.5 V/μs and the peak input voltage is 5 V.

Solution   The given slew rate must first be converted into its base unit of V/s in order to use Equation 22.1.

$$SR = 0.5 \text{ V/μs} = 500 \text{ kV/s}$$

$$f_{max} = \frac{SR}{2\pi V_p} = \frac{500 \text{ kV/s}}{2\pi(5 \text{ V})} = 15.92 \text{ kHz}$$

Op-amps often use feedback to establish a specific voltage gain. **Feedback** is defined as taking a portion of the output signal and feeding it back to the input. When an op-amp has its output connected to one of its inputs, it is referred to as a *closed-loop configuration*.

An op-amp with no feedback is called an *open-loop circuit.* The open-loop voltage gain of an op-amp is typically 200,000. The two basic types of feedback in electronic circuits are negative feedback and positive feedback. When the output voltage is in phase with the input voltage, the feedback signal will add to the input. This is called *positive feedback.* When the output signal is out of phase with the input, the feedback signal will subtract from the input. This is known as *negative feedback.* Because op-amps have very high open-loop gain, they are generally designed to operate in circuits that develop negative feedback.

Section Review

1. What is an op-amp?
2. What is input offset voltage?
3. Slew rate is an indication of how fast the input voltage can change (True/False).
4. What is the typical open-loop gain of an op-amp?

22-3

INVERTING AMPLIFIERS

An **inverting amplifier** performs the same function as a common-emitter BJT amplifier or a common-source FET amplifier. When an op-amp is used as an inverting amplifier, the input signal is fed into the inverting input and the noninverting input is grounded. Figure 22–4 shows a typical inverting amplifier. Resistor R_f provides a closed loop between the output and the input, and the output voltage is 180° out of phase with the input. Because the voltage drop across R_f is the opposite polarity to the applied voltage, the circuit of Figure 22–4 is said to have negative feedback. The voltage between the + and − inputs is essentially equal to 0 V. Because the + terminal is grounded, the − input terminal is 0 V and is referred to as a **virtual ground.** It is called virtual ground because the terminal is at ground potential but is not physically connected to the system or chassis ground.

Resistor R_1 in Figure 22–4 is connected in series with the signal source to prevent short circuiting of the feedback voltage if the source resistance is very low. Almost all of the input current, I, flows through the feedback resistor, R_f, because a negligible amount is drawn by the inverting input (−). Because the voltage at the − input is zero, the output voltage is found by Ohm's law.

$$V_{out} = -IR_f$$

The input voltage is dropped across R_1, which is in parallel with the input resistance of the op-amp. Therefore,

$$V_{in} = IR_1$$

FIGURE 22–4   Inverting amplifier.

Because the voltage gain of any circuit is a ratio of the output voltage to input voltage, the voltage gain of an inverting amplifier is

$$\mathscr{A}_v = \frac{V_{out}}{V_{in}} = \frac{-IR_f}{IR_1} = \frac{-R_f}{R_1} \tag{22.2}$$

Equation 22.2 is applicable for both AC and DC voltage gains. However, when AC voltage gain is calculated, the − sign is ignored.

EXAMPLE 22-2

For the circuit shown in Figure 22–5, determine the output voltage if the input signal is

(a) 1.2 V$_{rms}$
(b) −0.5 VDC

FIGURE 22-5 *Circuit for Example 22-2.*

Solution The voltage gain is found using Equation 22.2.

$$\mathscr{A}_v = \frac{-R_f}{R_1} = \frac{-100 \text{ k}\Omega}{10 \text{ k}\Omega} = -10$$

(a) When calculating AC voltage gain, the − sign is ignored.

$$V_{out} = V_{in}(\mathscr{A}_v) = (1.2 \text{ V})(10) = 12 \text{ V}_{rms}$$

(b) $$V_{out} = V_{in}(\mathscr{A}_v) = (-0.5 \text{ V})(-10) = +5 \text{ V}_{DC}$$

Section Review

1. What is virtual ground?
2. When the voltage at the inverting input of an inverting amplifier starts to go positive, the output goes negative (True/False).
3. What is the voltage gain of an inverting amplifier if the input resistor is 3.3 kΩ and the feedback resistor is 82 kΩ?

22-4

NONINVERTING AMPLIFIERS

When the op-amp is connected so that there is no phase displacement between the input and output signals, it is called a **noninverting amplifier.** A typical noninverting amplifier is shown in Figure 22–6. In this circuit, the input signal is taken at the noninverting input (+). If a positive DC input voltage is applied to this circuit, the output voltage will also go positive. The increase in output voltage will cause an increase in the voltage dropped across feedback resistor R_f. Resistors R_f and R_1 basically form a voltage divider at the inverting input (−). If the output voltage increases, the voltage at the inverting input will

FIGURE 22-6 Noninverting amplifier.

also increase. Because the voltage being amplified is the difference between the voltages at the two input terminals, the differential voltage will decrease when the output voltage increases. Consequently, the circuit of Figure 22–6 uses negative feedback.

According to Kirchhoff's voltage law, the output voltage is the sum of voltage drops across resistors R_f and R_1. If we ignore the negligible current flowing at the inverting input, the current, I, flowing through R_f is approximately equal to the current in R_1.

$$V_{out} = IR_f + IR_1$$

The input voltage includes the voltage dropped between the input terminals and the voltage dropped across resistor R_1. Because the voltage between the + and − terminals is approximately zero, the input voltage is found by multiplying the input current, I, by resistor R_1.

$$V_{in} = IR_1$$

The voltage gain of any circuit is a ratio of the output voltage to the input. Therefore,

$$\mathcal{A}_v = \frac{V_{out}}{V_{in}} = \frac{IR_f + IR_1}{IR_1} = \frac{IR_f}{IR_1} + \frac{IR_1}{IR_1} = \frac{R_f}{R_1} + 1 \qquad (22.3)$$

EXAMPLE 22-3

FIGURE 22-7 Circuit for Example 22-3.

Determine the output voltage for the circuit shown in Figure 22–7.

Solution

$$\mathcal{A}_v = \frac{R_f}{R_1} + 1 = \frac{100 \text{ k}\Omega}{20 \text{ k}\Omega} + 1 = 6$$

$$V_{out} = \mathcal{A}_v V_{in} = (6)(+0.5 \text{ V}) = +3 \text{ V}$$

The input resistance of the inverting amplifier was shown to equal that of the input resistor R_1. However, the noninverting amplifier has its input signal connected directly to the op-amp. As a result, the input impedance for the noninverting amplifier is equal to the impedance between the + and − terminals of the op-amp. Because this value is typically

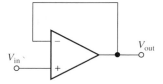

FIGURE 22–8 *Voltage follower.*

2 MΩ, a noninverting amplifier has a much higher input impedance than an inverting amplifier. Owing to their extremely high input impedance, noninverting amplifiers are often used as buffers in electronic circuits. These devices have no voltage gain and are used primarily for isolation between devices and circuits. Figure 22–8 shows a typical op-amp buffer, or **voltage follower.** In this circuit, no resistance is in the feedback loop, and the entire output signal is fed back to the inverting input. Because the voltage follower is a noninverting amplifier, Equation 22.3 can be applied to the circuit of Figure 22–8. According to this equation, if the feedback resistor is zero ohms, then the voltage gain is equal to 1.

$$\mathcal{A}_v = 1 \qquad\qquad (22.4)$$

Section Review

1. What is a noninverting amplifier?
2. State the symbol that represents the input for a noninverting amplifier.
3. Because of their extremely low input impedance, noninverting amplifiers are often used as buffers (True/False).
4. What is another name for a buffer?

— 22–5

SUMMING AMPLIFIERS

A **summing amplifier** is an op-amp circuit that is used to combine several input signals while maintaining a common ground. Summing amplifiers are popular in the broadcast and recording industries. By using two or more input devices, such as microphones or musical instruments, it is possible to combine, or mix, the input signals together to produce a single output. Figure 22–9 shows a summing amplifier that is used to mix the individual signals from three separate inputs, or channels, and produce an output signal that is based on the sum of these individual inputs. In the circuit of Figure 22–9, all the input currents flow through resistor R_f. This means that the current flowing through resistor R_1 does not affect the current flowing through R_2 or R_3. Because all the resistors are connected to the virtual

FIGURE 22–9 *Summing amplifier.*

ground, the input currents and input voltages are independent of each other. The voltage drop across each input resistor is determined using Equation 22.2. These voltage drops are then added together, as indicated in the following equation:

$$V_{out} = -\left(V_1 \frac{R_f}{R_1} + V_2 \frac{R_f}{R_2} + V_3 \frac{R_f}{R_3} \right)$$

(22.5)

EXAMPLE 22-4

$R_1 = 30 \text{ k}\Omega$
$V_1 = +3 \text{ V}$
$R_f = 100 \text{ k}\Omega$
$R_2 = 10 \text{ k}\Omega$
$V_2 = +2 \text{ V}$
$R_3 = 20 \text{ k}\Omega$
$V_3 = -4 \text{ V}$
V_{out}

FIGURE 22-10 *Circuit for Example 22-4.*

For the circuit of Figure 22-10, determine the output voltage for the following input condition: $V_1 = +3$ V, $V_2 = +2$ V, $V_3 = -4$ V.

Solution

$$
\begin{aligned}
V_{out} &= -\left(V_1 \frac{R_f}{R_1} + V_2 \frac{R_f}{R_2} + V_3 \frac{R_f}{R_3} \right) \\
&= -\left[(+3 \text{ V}) \frac{100 \text{ k}\Omega}{30 \text{ k}\Omega} + (+2 \text{ V}) \frac{100\text{k}\Omega}{10 \text{ k}\Omega} + (-4 \text{ V}) \frac{100 \text{ k}\Omega}{20 \text{ k}\Omega} \right] \\
&= -[(+10 \text{ V}) + (+20 \text{ V}) + (-20 \text{ V})] \\
&= -10 \text{ V}
\end{aligned}
$$

The summing amplifier is also popular in circuits that convert digital signals to analog signals. These devices, known as *digital-to-analog converters* (DACs), are used in many industrial electronic applications. Figure 22-11 shows a simplified version of a DAC using binary-weighted resistors. Four resistors are connected in parallel and each resistor has double the value of the previous resistor. Each input is driven by a simple high/low voltage level, which represents a binary 1 or 0 at the input terminal. Because the input resistors vary by a factor of 2, each value of input resistance produces a different value of

$R_1 = R_f$
A
$R_2 = 2R_f$
B
$R_f = 1 \text{ k}\Omega$
$R_3 = 4R_f$
C
$R_4 = 8R_f$
D
V_{out}

FIGURE 22-11 *Simplified digital-to-analog converter.*

output voltage when connected to a common source. Because there are four inputs in the circuit, there are 2^4, or 16, possible output voltages. In a summing amplifier, if more than one resistor is connected at the input, the voltage gain is determined by the feedback resistor, R_f, and the equivalent input resistance R_X. In equation form,

$$\mathcal{A}_v = -\frac{R_f}{R_X} \qquad (22.6)$$

Each combination of binary voltage levels for the circuit shown in Figure 22–11 produces a proportional change in the analog output. If we assume that terminal A is the most significant digit, or bit, for the DAC and terminal D is the least significant bit, the ABCD binary combinations would range from 0000 to 1111.

EXAMPLE 22–5

For the circuit of Figure 22–11, determine the output voltages when the ABCD binary input values are

(a) 1000
(b) 0100
(c) 1100
(d) 1111

Assume binary 1 = +5 V and binary 0 = 0 V.

Solution

(a)
$$\mathcal{A}_v = \frac{-R_f}{R_1} = \frac{-1\ k\Omega}{1\ k\Omega} = -1$$

$$V_{out} = \mathcal{A}_v V_{in} = (-1)(+5\ V) = -5\ V$$

(b)
$$\mathcal{A}_v = \frac{-R_f}{R_2} = \frac{-1\ k\Omega}{2\ k\Omega} = -0.5$$

$$V_{out} = \mathcal{A}_v V_{in} = (-0.5)(+5\ V) = -2.5\ V$$

(c)
$$R_X = \frac{R_1 R_2}{R_1 + R_2} = \frac{(1\ k\Omega)(2\ k\Omega)}{1\ k\Omega + 2\ k\Omega} = 0.67\ k\Omega$$

$$\mathcal{A}_v = \frac{-R_f}{R_X} = \frac{-1\ k\Omega}{0.67\ k\Omega} = -1.5$$

$$V_{out} = \mathcal{A}_v V_{in} = (-1.5)(+5\ V) = -7.5\ V$$

(d)
$$R_X = \frac{1}{1/R_1 + 1/R_2 + 1/R_3 + 1/R_4}$$

$$= \frac{1}{1/1\ k\Omega + 1/2\ k\Omega + 1/4\ k\Omega + 1/8\ k\Omega} = 0.533\ k\Omega$$

$$\mathcal{A}_v = \frac{-R_f}{R_X} = \frac{1\ k\Omega}{0.533\ k\Omega} = -1.88$$

$$V_{out} = \mathcal{A}_v V_{in} = (-1.88)(+5\ V) = -9.4\ V$$

Section Review

1. What is a summing amplifier?
2. Summing amplifiers are used in circuits that convert digital signals to analog signals (True/False).
3. State the equation for finding the voltage gain of a summing amplifier.

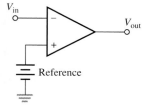

FIGURE 22-12 *Simple comparator circuit.*

COMPARATORS

A **comparator** is a circuit that is used to detect whether the voltage at one input is higher or lower than the voltage at another input. The comparator shown in Figure 22–12 is an open-loop application of the op-amp. A very small input signal drives the output into saturation because of the high open-loop gain of the op-amp. In Figure 22–12, a DC reference voltage is applied to the noninverting input. When the input voltage is equal to the reference voltage, the output is zero. If the input signal increases slightly, the output is driven into saturation. If the input signal is slightly less than the reference, the polarity of the output signal changes and the output is driven into saturation in the opposite direction. Although it is possible to use an ordinary op-amp in a comparator circuit, manufacturers produce a special op-amp for such applications. These special op-amps have a much higher slew rate, which makes them ideally suited for switching applications.

In some electronic systems, it is important to detect the instant a signal crosses the zero axis. Devices that are capable of detecting these zero-crossing points are called **zero-crossing detectors.** Figure 22–13(a) shows a basic zero-crossing detector. The waveform of Figure 22–13(b) represents the AC input voltage. The output voltage wave of Figure 22–13(c) shows how the op-amp is driven to saturation at the instant the input crosses the zero axis. From these waveforms it is apparent that the output voltage switches from one saturation level to the other when the input signal crosses 0 V.

When noise or other forms of signal disturbance occur on the input signal, false switching can occur on the output of a comparator. To eliminate this problem, the comparator can also be used in a closed-loop configuration known as a **Schmitt trigger.** The

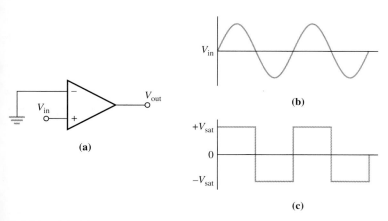

FIGURE 22-13 *(a) Zero-crossing detector. (b) Analog input signal. (c) Digital output signal.*

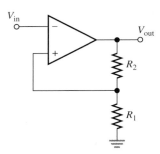

FIGURE 22–14 *Schmitt trigger circuit.*

circuit of Figure 22–14 is a Schmitt trigger that uses positive feedback. This means that the feedback voltage is aiding instead of opposing the input voltage. For example, if we assume that the inverting input voltage is slightly negative, a positive voltage will be produced at the output. The voltage divider feeds the positive voltage back to the noninverting input, which produces a larger positive signal at the output. As a result, the comparator is driven into saturation very quickly.

A typical application for a comparator is voltage regulation. In our discussion of zener diodes in Chapter 19, a voltage regulator was defined as a device capable of maintaining a constant output voltage over a wide range of varying load conditions. The two basic types of voltage regulators are shunt and series. Figure 22–15 shows a typical shunt regulator using an op-amp. Resistors R_2 and R_3 form a voltage divider that feeds a voltage proportional to the output back to the noninverting input of the op-amp. The voltage at the inverting input is held to a fixed value by the zener diode. If the output voltage increases, the voltage across R_3 increases and the output of the op-amp causes the BJT to conduct "harder," increasing the voltage drop across the series resistor R_s. As R_s increases, the voltage at the output decreases and the differential voltage at the op-amp is lowered.

Figure 22–16 shows an example of a series regulator that also uses a comparator to maintain a constant output voltage. This circuit has no series resistor and the difference between the input voltage and the output is dropped primarily across the BJT. If the output

FIGURE 22–15 *Shunt regulator.*

FIGURE 22–16 *Series regulator.*

voltage increases, the voltage dropped across R_3 increases and the voltage at the inverting input increases. This causes the output voltage of the op-amp to decrease and the base current for the NPN transistor is reduced. The BJT is then driven towards cutoff, increasing the collector-emitter voltage and decreasing the load voltage.

Section Review

1. A comparator is a closed-loop application of the op-amp (True/False).
2. What is a zero-crossing detector?
3. Schmitt triggers use negative feedback (True/False).
4. Name two types of op-amp regulators.

INTEGRATORS AND DIFFERENTIATORS

Integrators and **differentiators** are op-amp circuits that are frequently used in signal-processing applications. Integration is basically the process of finding the area under a curve. The mathematical operations of differentiation and integration are calculus functions and are beyond the scope of this book. However, it is possible to examine the basic operation of integrators and differentiators without any prior knowledge of calculus. The principle of integration in op-amp circuits is based on the relationship between resistance and capacitance. In Chapter 13, we noted that an RC network requires five time constants to charge or discharge a capacitor. The circuit of Figure 22–17(a) is an integrator that produces an output voltage that changes inversely with the input voltage. Figure 22–17(b) shows a square-wave input signal. When the input voltage instantly rises, the voltage across the capacitor cannot change instantaneously, so the output voltage gradually increases as the capacitor charges. The output waveform shown in Figure 22–17(c) provides a linear increase in output voltage that is inversely proportional to the RC time constant. When the polarity of the input signal changes, the capacitor begins to discharge and the output signal reverses.

The circuit illustrated in Figure 22–17 uses a capacitor as a feedback device. The disadvantage of this configuration is that there is no feedback at zero frequency. By connecting a feedback resistor in parallel with the capacitor, the gain of the integrator can be limited, as shown in Figure 22–18. At low frequencies, or DC, the capacitor acts as an open

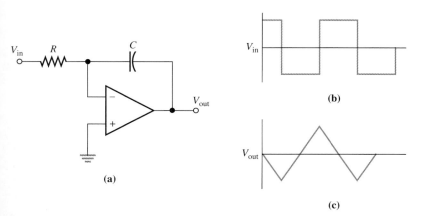

FIGURE 22–17 *(a) Integrator. (b) Input signal.*
(c) Output signal.

FIGURE 22-18 *Integrator with feedback resistor.*

circuit and the voltage gain is found by R_f/R_1. As the frequency increases, the capacitive reactance decreases, and the capacitor current increases.

All integrators have a critical frequency, f_c, where integration begins. If the input frequency is less than f_c, the circuit behaves as a simple inverting amplifier and no integration occurs. The following equation is used to calculate the critical frequency of an integrator circuit:

$$f_c = \frac{1}{2\pi R_f C} \qquad (22.7)$$

_____ **EXAMPLE 22-6** _____

Determine the lower frequency limit for the circuit of Figure 22-19.

FIGURE 22-19 *Circuit for Example 22-6.*

Solution

$$f_c = \frac{1}{2\pi R_f C} = \frac{1}{2\pi(100\ \text{k}\Omega)(0.02\ \mu\text{F})} = 79.6\ \text{Hz}$$

A differentiator is an electronic circuit that produces an output waveform whose value is equal to the rate of change of the input at any instant in time. Essentially, differentiation is the opposite of integration. If an integrator and a differentiator were cascaded together, the output signal would be the same as the input signal. Figure 22-20(a) shows a typical differentiator. If the triangular wave of Figure 22-20(b) is applied to the input of the differentiator, the output wave would appear as shown in Figure 22-20(c).

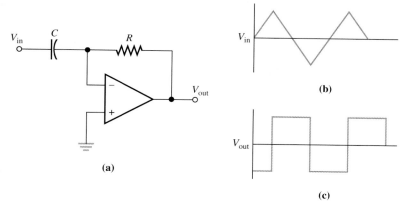

(a)

(b)

(c)

FIGURE 22–20   (a) Differentiator. (b) Input signal.
(c) Output signal.

Section Review

1. Integrators use a capacitor in the feedback loop (True/False).
2. State the equation for finding the critical frequency of an integrator.
3. What is a differentiator?

22-8

ACTIVE FILTERS

In Chapter 18, an active filter was described as a device having high input impedance, low output impedance, and very low signal loss. The op-amp is ideally suited for this application because of its operating characteristics. Figure 22–21(a) shows an example of a simple low-pass active filter. Because there is no resistance in the feedback loop, the voltage gain of this particular filter is unity. The op-amp provides the high input impedance and low output impedance required for most filtering applications. The critical, or cutoff, frequency for the low-pass active filter is found using a slight variation of Equation 22.7.

$$f_c = \frac{1}{2\pi RC} \qquad (22.8)$$

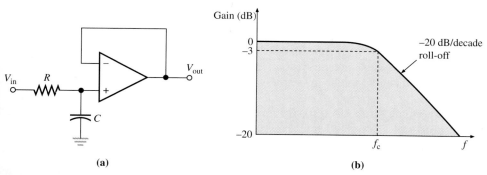

(a)

(b)

FIGURE 22–21   (a) Low-pass active filter. (b) Response curve.

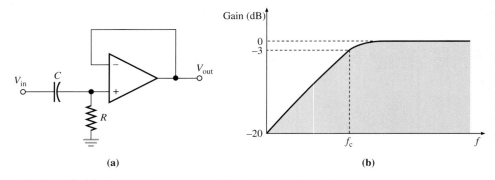

(a)

(b)

FIGURE 22–22 *(a) High-pass active filter. (b) Response curve.*

Figure 22–21(b) shows the response curve for a typical low-pass filter. The roll-off, or rate of attenuation, is indicated as –20 dB/decade. A filter with one RC network that produces a –20 dB/decade rate of attenuation is referred to as a **single-pole,** or first-order, **filter.**

A basic high-pass active filter is shown in Figure 22–22(a). In this circuit, the voltage gain approaches zero at very low frequencies and approaches unity at high frequencies. The response curve for a high-pass active filter, with gain indicated in decibels, is shown in Figure 22–22(b). This single-pole filter also produces a slope of –20 dB/decade. The cutoff frequency for this circuit is found using Equation 22.8.

_____ **EXAMPLE 22–7** _____

Determine the cutoff frequency for the circuit shown in Figure 22–23.

FIGURE 22–23 *Circuit for Example 22–7.*

Solution

$$f_c = \frac{1}{2\pi RC} = \frac{1}{2\pi(1 \text{ k}\Omega)(0.02 \text{ μF})} = 7.96 \text{ kHz}$$

Although a rate of attenuation of –20 dB/decade is quite acceptable for many applications, there are some situations where a steeper roll-off is required. A **two-pole filter** uses two RC networks and an op-amp to produce a roll-off rate of –40 dB/decade. Figure 22–24(a) shows an example of a two-pole, active low-pass filter. This filter is also known as a **Butterworth filter.** The response curve of Figure 22–24(b) shows an attenuation rate of –40 dB/decade. The cutoff frequency of a Butterworth filter is found using the following equation:

$$f_c = \frac{1}{2\pi \sqrt{R_1 R_2 C_1 C_2}} \tag{22.9}$$

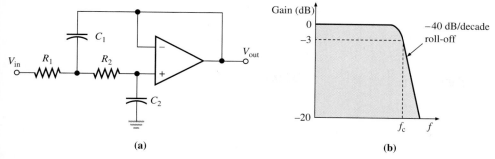

FIGURE 22–24   **(a) Butterworth low-pass filter. (b) Response curve.**

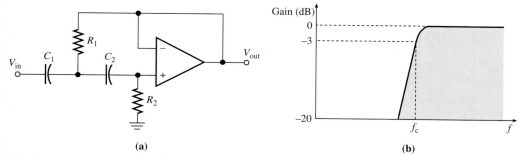

FIGURE 22–25   **(a) Butterworth high-pass filter. (b) Response curve.**

Figure 22–25(a) shows an example of a Butterworth high-pass filter. The response curve is illustrated in Figure 22–25(b). The two-pole, low-pass and high-pass filters have identical configurations; the resistors and capacitors are simply interchanged. Consequently, the cutoff frequency is also determined using Equation 22.9.

_____ **EXAMPLE 22–8** _____

Determine the cutoff frequency of the Butterworth low-pass filter shown in Figure 22–26.

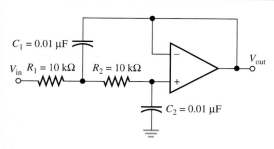

FIGURE 22–26   **Circuit for Example 22–8.**

Solution

$$f_c = \frac{1}{2\pi \sqrt{R_1 R_2 C_1 C_2}}$$

$$= \frac{1}{2\pi \sqrt{(10\ k\Omega)(10\ k\Omega)(0.01\ \mu F)(0.01\ \mu F)}}$$

$$= 1.59\ kHz$$

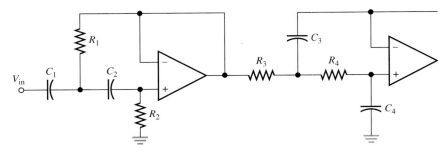

FIGURE 22–27   Butterworth band-pass filter.

An active band-pass filter can be constructed by combining Butterworth low-pass and high-pass filters, as shown in Figure 22–27. In this circuit, the bandwidth that passes through the filter represents the difference between the two values of cutoff frequencies, f_{c1} and f_{c2}. This configuration provides a roll-off of −40 dB/decade and unity voltage gain.

___ **EXAMPLE 22–9** ___

Calculate the bandwidth of the active filter shown in Figure 22–28.

FIGURE 22–28   Circuit for Example 22–9.

Solution

$$f_{c1} = \frac{1}{2\pi \sqrt{R_1 R_2 C_1 C_2}}$$

$$= \frac{1}{2\pi \sqrt{(22 \text{ k}\Omega)(22 \text{ k}\Omega)(0.02 \text{ μF})(0.02 \text{ μF})}} = 362 \text{ Hz}$$

$$f_{c2} = \frac{1}{2\pi \sqrt{R_1 R_2 C_1 C_2}}$$

$$= \frac{1}{2\pi \sqrt{(15 \text{ k}\Omega)(15 \text{ k}\Omega)(0.01 \text{ μF})(0.01 \text{ μF})}} = 1061 \text{ Hz}$$

$$BW = f_{c2} - f_{c1}$$
$$= 1061 \text{ Hz} - 362 \text{ Hz} = 699 \text{ Hz}$$

Section Review

1. If there is no resistance in the feedback loop, the voltage gain of an op-amp is zero (True/False).
2. What is a single-pole filter?
3. State the equation for finding the cutoff frequency of a Butterworth filter.

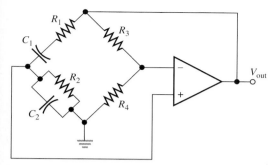

FIGURE 22-29 *Wien-bridge oscillator.*

OSCILLATOR CIRCUITS

An oscillator was defined in Chapter 12 as a device or circuit that produces sinusoidal or other repetitive waveforms. Op-amps are ideally suited for oscillator circuits. Oscillators are characterized as having some form of positive feedback. The effect of the positive feedback is to provide enough signal strength to sustain continuous oscillation. A typical sinusoidal oscillator is the **Wien-bridge oscillator** shown in Figure 22–29. This type of oscillator is standard for audio-frequency generators that produce frequencies in the range of 5 Hz to 1 MHz. The Wien-bridge oscillator utilizes both positive and negative feedback to produce a sinusoidal waveform. The bridge network shown in Figure 22–29 is basically a band-pass filter with a series RC and parallel RC configuration. At low frequencies, capacitor C_1 acts as an open circuit and there is no output. At very high frequencies, capacitor C_2 bypasses the output to ground and once again there is no output. At the resonant frequency, the output voltage is at its maximum.

The output of the op-amp in Figure 22–29 is fed back to the bridge input. Positive feedback is developed across R_1 and C_1 and applied to the noninverting input. Consequently, the positive feedback is in phase with the output signal. Negative feedback is developed across the voltage divider formed by resistors R_3 and R_4 and is applied to the inverting input. To maintain oscillations, the positive feedback must be greater than the negative feedback. If $R_1 = R_2$ and $C_1 = C_2$, the resonant frequency for a Wien-bridge oscillator is calculated using Equation 22.8.

$$f_r = \frac{1}{2\pi RC}$$

EXAMPLE 22-10

Determine the resonant frequency of the Wien-bridge oscillator shown in Figure 22–30.

$R_1 = 51\ k\Omega$
$C_1 = 0.001\ \mu F$
$R_3 = 200\ k\Omega$
$R_2 = 51\ k\Omega$
$R_4 = 100\ k\Omega$
$C_2 = 0.001\ \mu F$
V_{out}

FIGURE 22-30 *Circuit for Example 22-10.*

Solution

$$f_r = \frac{1}{2\pi RC} = \frac{1}{2\pi(51 \text{ k}\Omega)(0.001 \text{ }\mu\text{F})} = 3.12 \text{ kHz}$$

Oscillators that produce nonsinusoidal waveforms are called **relaxation oscillators,** or **multivibrators.** There are three basic types of multivibrators: astable, monostable, and bistable. Astable multivibrators produce a square wave output with a frequency determined by the values of the components connected to the op-amp. Monostable multivibrators, or one-shots, produce a constant output until the trigger pulse at the input causes the output to change state for a predetermined length of time. After this time period has elapsed, the output of the monostable multivibrator returns to its original voltage level. The bistable multivibrator, or flip-flop, is a digital device and will be discussed in Chapter 23.

An example of an astable multivibrator is shown in Figure 22–31(a). This circuit is basically a Schmitt trigger using an op-amp and an *RC* timing circuit formed by R_f and *C*. The voltage across the capacitor is shown in Figure 22–31(b) and represents the voltage at the inverting input. The voltage at the noninverting input is determined by the voltage divider network formed by R_1 and R_2. As the capacitor begins to charge, eventually it reaches a value where the inverting and noninverting voltages are equal. This causes the output voltage to reverse polarity and the capacitor begins to discharge. When the voltage at the inverting input falls below that of the noninverting input, the polarity of the output reverses and the charging cycle begins again. The rate that the capacitor charges and discharges is determined by the *RC* timing constant. The voltage gain of the circuit is selected so that the output is driven to saturation during both the positive and negative alternations, as shown in Figure 22–31(c).

Figure 22–32(a) shows an example of a monostable multivibrator. Monostable, or one-shot, circuits require an input trigger pulse to initiate the output pulse. Monostable multivibrators are often used in timing circuits that require a constant frequency with a variable pulse width. When a trigger signal, shown in Figure 22–32(b), is applied at the

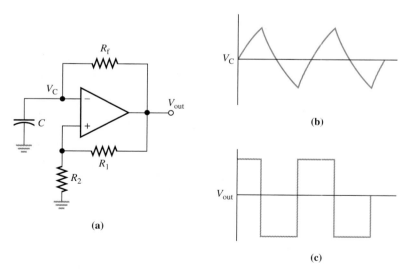

FIGURE 22–31 *(a) Astable multivibrator. (b) Voltage at inverting input. (c) Output voltage.*

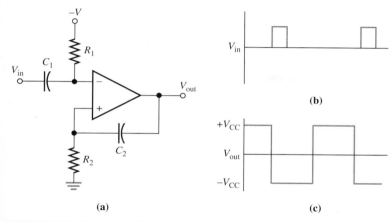

FIGURE 22–32 *(a) Monostable multivibrator. (b) Trigger pulse. (c) Output waveform.*

inverting input, the output voltage will change state, as illustrated in Figure 22–32(c). The trigger signals are applied through the high-pass filter consisting of R_1 and C_1. If a positive trigger signal is applied, C_1 cannot change instantaneously and the full voltage of the trigger pulse appears at the − terminal. The − terminal is now more positive than the + terminal, and the output voltage is driven to $-V_{CC}$. Capacitor C_2 charges toward $-V_{CC}$ through R_2, and eventually it reaches a point where the voltage drop across R_2 is equal to the voltage at the − input. At this point, the output voltage switches polarity and remains at $+V_{CC}$ until the next trigger pulse is applied.

Section Review

1. Oscillators are characterized as having some form of positive feedback (True/False).
2. What is a Wien-bridge oscillator?
3. What is a multivibrator?

_____ **22–10** __

PRACTICAL APPLICATION SOLUTION

In the chapter opener, you were assigned the task of designing a weight-detection system for a grain hopper in a brewery. The hopper feeds grain into a vessel via an electrically controlled dump valve. The weight of the vessel is to be measured by four load cells connected to the input of a summing amplifier. The output of the summing amplifier feeds a comparator, which compares this signal with a reference signal provided by the operating console. The output of the comparator is fed to an output amplifier, to increase the signal, and then to a dump valve. The following steps illustrate the solution for this practical application of operational amplifiers.

STEP 1   Draw a diagram of the circuit (Figure 22–33). The outputs of the load cells are connected to the inputs of a summing amplifier. The output of the summing amplifier is fed into a comparator. The output of the comparator is amplified by an output amplifier and fed into an electrically controlled dump valve.

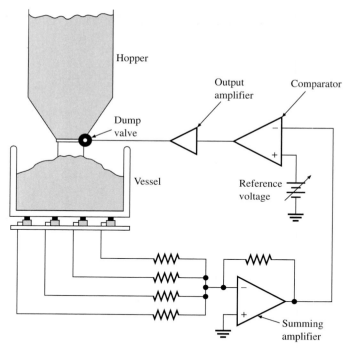

FIGURE 22–33 *Electronic weighing system.*

STEP 2   Include a reference voltage. The reference voltage shown in Figure 22–33 represents the proportional voltage from the dial on the operating console. This voltage is calibrated so that each weight setting on the console produces a specific reference voltage. This value is then compared with the output of the summing amplifier to produce an output signal for the dump valve.

Summary

1. An op-amp is a high-gain, direct-coupled multistage amplifier with the capability of amplifying signals from DC (0 Hz) to several MHz.
2. Feedback is defined as taking a portion of the output signal and feeding it back to the input.
3. An inverting amplifier performs the same function as a common-emitter BJT.
4. Noninverting amplifiers have no phase displacement between the input and output signals.
5. Summing amplifiers combine several input signals while maintaining a common ground.
6. Comparators are used to detect whether the voltage at one input is higher or lower than the voltage at another input.
7. A zero-crossing detector is a type of open-loop comparator that detects the instant a signal crosses the zero axis.
8. A differentiator is an op-amp circuit that produces an output waveform whose value is equal to the rate of change of the input at any instant in time.
9. A filter with one *RC* network that produces a −20 dB/decade rate of attenuation is referred to as a single-pole filter.
10. Oscillators are characterized as having some form of positive feedback.

Section 22–1

1. A device that combines entire electronic circuits on a single silicon chip.
2. False
3. The op-amp

Section 22–2

1. A high-gain, direct-coupled multistage amplifier
2. A small compensating voltage that is applied to reference the output voltage to zero
3. False
4. 200,000

Section 22–3

1. When the terminal of an op-amp is at ground potential but is not physically connected to the system or chassis ground
2. True
3. Voltage gain $= -82 \text{ k}\Omega/3.3 \text{ k}\Omega = -25$

Section 22–4

1. An op-amp circuit that has no phase displacement between the input and output signals
2. +
3. False
4. Voltage follower

Section 22–5

1. An op-amp circuit that is used to combine several input signals together while maintaining a common ground
2. True
3. Voltage gain $= -R_f/R_X$

Section 22–6

1. False
2. A device that is capable of detecting the zero-crossing points of a waveform
3. False
4. Shunt regulators and series regulators

Section 22–7

1. True
2. $f_c = 1/(2\pi R_f C)$
3. An electronic circuit that produces an output waveform whose value is equal to the rate of change of the input at any instant in time

Section 22–8

1. False
2. A filter with one RC network that produces a -20 dB/decade rate of attenuation
3. $f_c = 1/(2\pi \sqrt{R_1 R_2 C_1 C_2})$

Section 22–9

1. True
2. A sinusoidal oscillator that uses both positive and negative feedback
3. An oscillator that produces a nonsinusoidal waveform

Multiple Choice Questions

22–1 An op-amp is a
(a) High-gain, capacitor-coupled multistage amplifier
(b) Digital circuit
(c) High-gain, direct-coupled multistage amplifier
(d) Low-gain, direct-coupled multistage amplifier

22–2 An ideal op-amp has
(a) Zero input impedance, zero output impedance, and zero voltage gain
(b) Infinite input impedance, infinite output impedance, and infinite voltage gain
(c) Zero input impedance, infinite voltage gain, zero output impedance
(d) Infinite input impedance, zero output impedance, and infinite voltage gain

22–3 The first stage of an op-amp is a(n)
(a) Darlington amplifier   (b) Differential amplifier
(c) Emitter follower   (d) Summing amplifier

22–4 The slew rate of an op-amp is an indication of
(a) The voltage gain
(b) Power
(c) How fast the output voltage can change
(d) How the temperature changes with current

22–5 An op-amp circuit with no feedback is called
(a) Open loop   (b) Closed loop
(c) Positive feedback   (d) Negative feedback

22–6 Which of the following circuits use an op-amp in an open-loop configuration?
(a) Summing amplifiers   (b) Active filters
(c) Voltage followers   (d) Comparators

22–7 An inverting amplifier performs the same basic function as a
(a) Common-base amplifier   (b) Common-collector amplifier
(c) Common-emitter amplifier   (d) Common-drain amplifier

22–8 When the op-amp is connected so that there is no phase displacement between the input and output signals, it is called
(a) An inverting amplifier   (b) A noninverting amplifier
(c) Negative feedback   (d) Degenerative feedback

22–9 The circuit shown in Figure 22–34 has
(a) Unity gain, low input impedance, and high output impedance
(b) High gain, high input impedance, and low output impedance
(c) Unity gain, high input impedance, and low output impedance
(d) High gain, low input impedance, and high output impedance

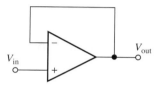

FIGURE 22–34

22–10 An op-amp circuit that is used to combine several input signals while maintaining a common ground is called a(n)
(a) Voltage follower   (b) Summing amplifier   (c) Active filter   (d) Oscillator

22–11 A circuit that is used to detect whether the voltage at one input is higher or lower than the voltage at another input is called a(n)
(a) Comparator
(b) Summing amplifier
(c) Voltage follower
(d) Active filter

22–12 An op-amp circuit that produces an output waveform with a value proportional to the rate of change of the input at any instant in time is called a(n)
(a) Comparator
(b) Integrator
(c) Voltage follower
(d) Differentiator

22–13 A filter with one RC network that produces a −20 dB/decade rate of attenuation is called a
(a) Butterworth filter
(b) Single-pole filter
(c) Second-order filter
(d) Two-pole filter

Practice Problems

22–14 Determine the maximum frequency that can be obtained from an op-amp if the slew rate is 0.2 V/μs and the peak input voltage is 10 V.

22–15 For the circuit shown in Figure 22–35, determine the output voltage if the input voltage is
(a) 1.5 V_{rms}
(b) −0.8 V_{DC}

FIGURE 22–35

22–16 Calculate the output voltage of the circuit shown in Figure 22–36.

FIGURE 22–36

22–17 For the circuit shown in Figure 22–37, determine the output voltage for the following input condition: $V_1 = +4$ V, $V_2 = -8$ V, $V_3 = +6$ V.

FIGURE 22–37

22–18 Determine the lower frequency limit for the circuit of Figure 22–38.

FIGURE 22–38

$R_f = 100 \text{ k}\Omega$

V_{in} $R_1 = 12 \text{ k}\Omega$ $C = 0.01 \text{ µF}$

V_{out}

22–19 Calculate the cutoff frequency for the circuit shown in Figure 22–39.

FIGURE 22–39

V_{in} $C = 0.022 \text{ µF}$

V_{out}

$R = 1.5 \text{ k}\Omega$

22–20 Determine the cutoff frequency of the Butterworth filter shown in Figure 22–40.

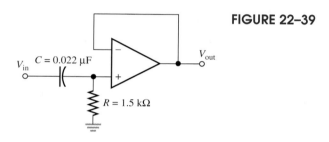

FIGURE 22–40

$R_1 = 15 \text{ k}\Omega$

V_{in} $C_1 = 0.039 \text{ µF}$ $C_2 = 0.039 \text{ µF}$

V_{out}

$R_2 = 20 \text{ k}\Omega$

22–21 Calculate the bandwidth of the active filter shown in Figure 22–41.

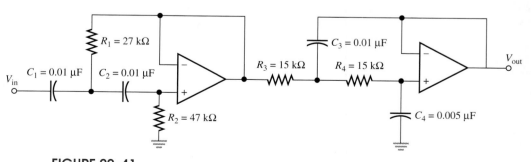

$R_1 = 27 \text{ k}\Omega$

$C_3 = 0.01 \text{ µF}$

$R_3 = 15 \text{ k}\Omega$ $R_4 = 15 \text{ k}\Omega$

V_{out}

V_{in} $C_1 = 0.01 \text{ µF}$ $C_2 = 0.01 \text{ µF}$

$R_2 = 47 \text{ k}\Omega$

$C_4 = 0.005 \text{ µF}$

FIGURE 22–41

22–22 For the oscillator circuit shown in Figure 22–42, determine the resonant frequency.

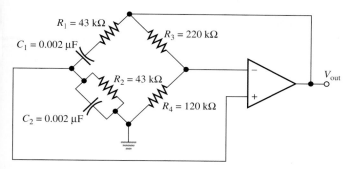

FIGURE 22–42

Essay Questions

22–23 Name the three main amplifiers in a typical op-amp.

22–24 How are the two inputs to the op-amp referred to and how are they identified on a schematic diagram?

22–25 Define the term *input offset voltage.*

22–26 Describe slew rate and explain how it affects the output of an op-amp.

22–27 What is feedback?

22–28 Why are voltage followers used in electronic circuits?

22–29 Name two applications for summing amplifiers.

22–30 What is a comparator?

22–31 Describe the basic difference between integrators and differentiators.

22–32 Why are active filters preferred over passive filters?

22–33 List three types of multivibrators.

23

Digital Electronics

Learning Objectives

Upon completion of this chapter, you will be able to

- Understand the binary number system.
- Convert binary numbers to decimal and decimal numbers to binary.
- Explain the hexadecimal number system
- Convert hexadecimal numbers to binary and binary numbers to hexadecimal.
- Differentiate between natural binary and binary coded decimal (BCD).
- Understand the ASCII code.
- Apply truth tables to troubleshooting digital circuits.
- List five logic gates.
- Describe the basic operation of an inverter.
- Explain the purpose of Boolean algebra.
- Understand logic gate combinations.
- Name eight Boolean theorems.
- List three types of flip-flops.
- Apply basic troubleshooting techniques to digital circuits.
- List four types of digital test instruments.

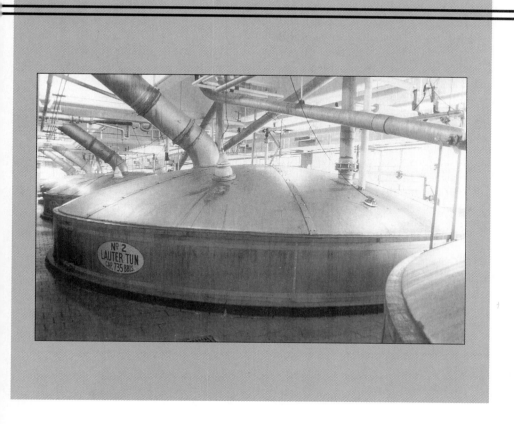

Practical Application

After successfully completing the task described in Chapter 22, your supervisor has instructed you to design and install a digital control circuit for a new mixing vat in the brewery. The mixing vat is to have three detectors: pressure, level, and temperature. The following three combinations must activate an alarm:

1. High level with high temperature
2. High level with high pressure
3. Low level with high temperature and high pressure

Your task is to design a combinational logic circuit that will activate an alarm when any of these three conditions occur.

_ 23-1 _____

INTRODUCTION

In this chapter, we will see how logic gates and flip-flops can be combined to form digital circuits. All digital equipment, regardless of its complexity, is constructed from combinations of logic gates. A logic gate is a decision-making element. The gate produces an output that is high or low, depending on its input conditions. In a logic gate circuit, input and output signals can only assume one of two possible logic levels, or states. Typically, a high signal is considered to represent +5 V, and a low signal implies 0 V. The two voltage levels in digital circuits are well suited for use in the binary number system. The binary number system has only two digits: 1 and 0. This chapter begins with a discussion of binary and hexadecimal numbers.

Truth tables will also be discussed in this chapter, as both a troubleshooting tool and a method of understanding the operation of logic gates. A **truth table** is basically a graphical illustration of the relationship between input and output logic levels of a gate circuit. The concept of Boolean algebra will also be introduced in this chapter. The five main logic gates that we will discuss in this chapter are the AND, OR, NOT, NAND, and NOR gates. We will also examine the use of flip-flops in digital circuits. A **flip-flop** is a memory device constructed of logic gates that is capable of storing binary data.

Section Review

1. A logic gate is a decision-making element (True/False).
2. What is a truth table?
3. What is a flip-flop?

_ 23-2 _____

NUMBER SYSTEMS

A number system is, essentially, a code consisting of symbols that are assigned for each individual quantity. Once a code is memorized, it is possible to count using this code. The decimal number system uses ten basic symbols, or digits—0 through 9. Each digit represents a specific quantity. When the digits are grouped together, larger quantities can be expressed.

The **radix,** or **base,** of a number system is the total number of individual symbols in that system. The largest-valued symbol always has a magnitude of one less than the radix. For example, the decimal number system has a radix of 10, so the largest single digit is $10 - 1$, or 9. All number systems use position weighting to represent the significance of an individual digit in a group of numbers. As the digits move to the left of a decimal point, the value of the digit increases by its base power. If a digit moves to the right of the decimal place, it decreases by its base power.

The decimal number system uses combinations of ten individual digits to represent any number imaginable. Because there are ten digits in this system, the decimal number system is referred to as a base 10 system. As the digits move to the left of the decimal place, they increase by a power of 10. As the digits move to the right, they decrease by a power of 10. For example, in the decimal number 5053.72, the digit to the immediate left of the decimal point has a value of 3×10^0, or 3; the next digit to the left of the decimal point has a value of 5×10^1, or 50; and the digit to the immediate right of the decimal place has a value of 7×10^{-1}, or 0.7.

The total value of the decimal number 5053.72 is determined as follows:

$$2 \times 10^{-2} = \quad 0.02$$
$$7 \times 10^{-1} = \quad 0.7$$
$$3 \times 10^{0} = \quad 3$$
$$5 \times 10^{1} = \quad 50$$
$$0 \times 10^{2} = \quad 0$$
$$5 \times 10^{3} = \underline{\quad 5000}$$
$$5,053.72$$

From the above example, it is apparent that each position in a decimal number has a value that is 10 times the value of the next position to the right. Therefore, each position value is a multiple of 10 and can be expressed as 10 raised to some exponent.

EXAMPLE 23–1

Express the decimal number 123 in exponential form.

Solution

$$123 = 1 \times 100 + 2 \times 10 + 3 \times 1$$
$$= 1 \times 10^{2} + 2 \times 10^{1} + 3 \times 10^{0}$$

Section Review

1. What is the base of a number system?
2. The largest-valued symbol in a number system always has a magnitude of one more than the radix (True/False).

23–3

BINARY NUMBER SYSTEM

As mentioned earlier, the **binary number system** is based on two digits: 1 and 0. Any number imaginable can be expressed in binary form as a combination of 1s and 0s. To process numbers in electronic circuitry, such as in a computer, it is necessary to represent numerical quantities by electrical signals. If digital equipment processed decimal numbers, each of the ten digits would have to be assigned a specific voltage level, and combinations of these ten different voltages would represent a decimal number. This would mean that a six-digit decimal number would require voltage levels on six different conductors. Having ten different voltage levels would considerably complicate the circuitry required to process data.

The binary number system is ideally suited for processing data because only two voltage levels are required to represent all the different digits in the binary system. Typically 5 V is used to represent binary 1, and 0 V is used to represent binary 0. A single binary digit is often referred to as a **bit,** and a combination of eight bits is called a **byte,** or word. The term *byte* is often associated with memory capacity of digital circuits. In computer systems, this memory capacity is extremely large and is often measured in terms of megabytes, Mb. A byte would be represented inside a digital circuit as a combination of voltage levels. For example, the binary word 10110111 would appear as the following voltage levels:

$$5\ V, \quad 0\ V, \quad 5\ V, \quad 5\ V, \quad 0\ V, \quad 5\ V, \quad 5\ V, \quad 5\ V$$

Because only two symbols, or digits, are used in the binary number system, the base, or radix, is 2. Therefore, the position weighting that is assigned to a binary symbol will double each position a digit is moved to the left of the decimal point, and it will decrease by 1/2 each position a digit is moved to the right of the decimal point. The digit of a binary number that has the lowest weight is called the **least significant bit,** or LSB, and the digit with the highest value is called the **most significant bit,** or MSB.

_____ EXAMPLE 23-2 _____

Express the binary number 10111 in exponential form.

Solution

$$10111 = 1 \times 2^0 + 1 \times 2^1 + 1 \times 2^2 + 0 \times 2^3 + 1 \times 2^4$$

Section Review

1. State the two digits used in the binary number system.
2. What is a _bit?_
3. How many bits in a byte?

23-4

BINARY-TO-DECIMAL AND DECIMAL-TO-BINARY CONVERSION

Although digital equipment processes data in binary form, the data must be converted to decimal values to be utilized in visual applications. For example, a calculator may perform math instructions in binary form, but to be relevant to the user, the data must be displayed in decimal form. The conversion of binary digits to decimal is referred to as **decoding.** In a digital machine, such as a calculator, there is an integrated circuit chip called a decoder that performs this function. When a number is manually converted from binary to decimal, the position weight of each binary digit is added together. The result is the equivalent decimal number. The following example illustrates the conversion of a binary number to decimal form. Convert

11001_2 to decimal

$$1 \times 2^0 = 1$$
$$0 \times 2^1 = 0$$
$$0 \times 2^2 = 0$$
$$1 \times 2^3 = 8$$
$$1 \times 2^4 = \underline{16}$$
$$25$$

The sum of $1 + 8 + 16 = 25$. Therefore, the decimal equivalent of binary number 11001 is 25. In the above example, the number 11001 is designated by the subscript number 2 to indentify the base, or radix, of the number system.

EXAMPLE 23–3

Solution

Convert binary number 101101 to decimal form.

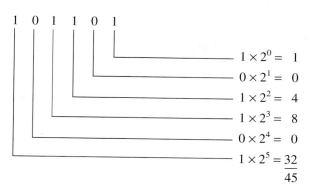

$$1 \times 2^0 = 1$$
$$0 \times 2^1 = 0$$
$$1 \times 2^2 = 4$$
$$1 \times 2^3 = 8$$
$$0 \times 2^4 = 0$$
$$1 \times 2^5 = 32$$
$$45$$

Decimal number 45 is the equivalent of binary number 101101.

Decimal values entered into a digital machine must be converted, or **encoded,** into binary form. This is accomplished by using an encoder chip. To convert a decimal number into binary form, divide the decimal number by 2 and record the remainders as shown below. Convert

$$25_{10} \text{ to binary}$$

$$\frac{25}{2} = 12 + \text{remainder of } 1$$

$$\frac{12}{2} = 6 + \text{remainder of } 0$$

$$\frac{6}{2} = 3 + \text{remainder of } 0$$

$$\frac{3}{2} = 1 + \text{remainder of } 1$$

$$\frac{1}{2} = 0 + \text{remainder of } 1$$

$$\begin{array}{cccccc} & 1 & 1 & 0 & 0 & 1 \\ & \text{MSB} & & & \text{LSB} \end{array}$$

The equivalent binary number of decimal number 25 is 11001. The number determined in decimal-to-binary conversion can be proven by converting the binary number back to decimal form.

Section Review

1. The conversion of binary digits to decimal is called encoding (True/False).
2. What is a decoder?
3. To convert a decimal number into binary form, divide the decimal number by two and record the remainders (True/False).

23-5

HEXADECIMAL NUMBER SYSTEM

The hexadecimal system uses base 16. This means it has 16 possible digit symbols. The **hexadecimal number system** uses the ten digits in the decimal system, 0 through 9, as well as the first six letters of the alphabet: A, B, C, D, E, and F. It is important to remember that the six alphabetic letters used are equivalent to decimal numbers 10, 11, 12, 13, 14, and 15. Table 23–1 shows a comparison between decimal, binary, and hexadecimal digits.

The use of letters to represent numbers may seem strange initially, although any number system is actually a set of symbols. The hex code is simple to interpret in terms of binary digits and is easy to read. This makes the hexadecimal system well suited for representing binary numbers and digital codes.

Any hexadecimal number can be converted to its decimal equivalent in a manner similar to converting a binary number to decimal. The hexadecimal system uses base 16, which means the position weighting moves to the left of the decimal place by the power of 16.

$$16^4 \quad 16^3 \quad 16^2 \quad 16^1 \quad 16^0 \ . \ 16^{-1} \quad 16^{-2} \quad 16^{-3} \quad 16^{-4}$$

To convert from a decimal number to a hexadecimal number, divide the decimal number by 16 and keep track of the remainders. The remainders will form the hexadecimal equivalent number. For example, to convert the decimal number 283 to hex,

$$\frac{283}{16} = 17 \text{ with a remainder of } 11 \quad \text{or} \quad B$$

$$\frac{17}{16} = 1 \text{ with a remainder of } 1$$

$$\frac{1}{16} = 0 \text{ with a remainder of } 1$$

TABLE 23–1 *Comparison of three number systems.*

Decimal	Binary	Hexadecimal
0	0000	0
1	0001	1
2	0010	2
3	0011	3
4	0100	4
5	0101	5
6	0110	6
7	0111	7
8	1000	8
9	1001	9
10	1010	A
11	1011	B
12	1100	C
13	1101	D
14	1110	E
15	1111	F

_____ EXAMPLE 23–4 _____

Solution

Convert the following decimal numbers to their hexadecimal equivalents.

(a) 2377
(b) 62
(c) 35599

(a)

$$\frac{2377}{16} = 148 \text{ with a remainder of } 9$$

$$\frac{148}{16} = 9 \text{ with a remainder of } 4$$

$$\frac{9}{16} = 0 \text{ with a remainder of } 9$$

$$2377_{10} = 949_{16}$$

(b)

$$\frac{62}{16} = 3 \text{ with a remainder of } 14 \quad \text{or} \quad E$$

$$\frac{3}{16} = 0 \text{ with a remainder of } 3$$

$$62_{10} = 3E_{16}$$

(c)

$$\frac{35599}{16} = 2224 \text{ with a remainder of } 15 \quad \text{or} \quad F$$

$$\frac{2224}{16} = 139 \text{ with a remainder of } 0$$

$$\frac{139}{16} = 8 \text{ with a remainder of } 11 \quad \text{or} \quad B$$

$$\frac{8}{16} = 0 \text{ with a remainder of } 8$$

$$35599_{10} = 8B0F_{16}$$

The hexadecimal equivalent of the decimal number 283_{10} is $11B_{16}$.

Hexidecimal numbers are converted to binary form by determining the four-bit equivalent binary digit for each hexadecimal digit. This method is also known as the 8-4-2-1 system, which will be discussed in the next section. The following example illustrates hex to binary conversion. Convert

$$3A5_{16} \text{ to binary}$$

8421	8421	8421
3	A	5
↓	↓	↓
0011	1010	0101

$$3A5_{16} = 0011\ 1010\ 0101 \quad \text{or} \quad 11\ 1010\ 0101$$

EXAMPLE 23–5

Convert the following hexadecimal numbers to their binary equivalents.

(a) E277
(b) 92F6
(c) FFA3B

Solution

(a)

E	2	7	7
↓	↓	↓	↓
1110	0010	0111	0111

(b)

9	2	F	6
↓	↓	↓	↓
1001	0010	1111	0110

(c)

F	F	A	3	B
↓	↓	↓	↓	↓
1111	1111	1010	0011	1011

Binary numbers can be converted to equivalent hexadecimal numbers by reversing the process discussed in the previous section. A binary word is divided into groups of four bits, and the equivalent hexadecimal number is determined. Zeros may be added to the most-significant-digit side of the binary number to form groups of four.

EXAMPLE 23–6

Convert the following binary numbers into hexadecimal numbers:

(a) 1110111010101011101
(b) 1010111010100110101
(c) 1100110010101100111

Solution

(a)

0001	1101	1101	0101	1101
↓	↓	↓	↓	↓
1	D	D	5	D

$1110111010101011101_2 = 1DD5D_{16}$

(b)

0101	0111	0101	0011	0101
↓	↓	↓	↓	↓
5	7	5	3	5

$1010111010100110101_2 = 57535_{16}$

(c)

0110	0110	0101	0110	0111
↓	↓	↓	↓	↓
6	6	5	6	7

$1100110010101100111_2 = 66567_{16}$

Section Review

1. How many digits in the hexadecimal number system?
2. Without referring to Table 23–1, what is the hexadecimal equivalent of the decimal number 12?
3. The hexadecimal equivalent of the decimal number 15 is F (True/False).

BINARY CODED DECIMAL (BCD)

The **binary coded decimal, BCD,** is a modification of the decimal number system where the decimal digits are independently coded as four-bit binary numbers. There are many types of BCD codes, with the 8–4–2–1 code being the most popular. In this discussion of BCD codes, we will consider the 8–4–2–1 system to be the only BCD system under consideration. The designation 8–4–2–1 indicates the binary weights of the four bits (2^3, 2^2, 2^1, 2^0). The BCD system was developed to make the conversion from decimal numbers to binary easier. Using four bits, it is possible to count from 0 to 15. However, in the BCD system, the six numbers over 9 are not valid. For example, 1011 in binary is not a legitimate BCD number because 1011 does not convert to a *single* decimal digit. Any number over 9 must be represented by two BCD numbers, or eight bits. Table 23–2 illustrates the relationship among the BCD system, the binary system, and the decimal number system.

To express any decimal number in BCD, simply replace each decimal digit by the appropriate four-bit code. To illustrate the BCD code, consider the number 375. Each digit is changed to its binary equivalent as follows:

3	7	5	(decimal)
↓	↓	↓	
0011	0111	0101	

In the above example, 0011 0111 0101 is the BCD equivalent of the decimal number 375. Because the 0s on the most significant digit side can be ignored, 11 0111 0101 is the correct expression for the decimal number 375.

TABLE 23–2 *BCD code with binary and decimal equivalents.*

Decimal	BCD	Binary
0	0000	0000
1	0001	0001
2	0010	0010
3	0011	0011
4	0100	0100
5	0101	0101
6	0110	0110
7	0111	0111
8	1000	1000
9	1001	1001
10	0001 0000	1010
11	0001 0001	1011
12	0001 0010	1100
13	0001 0011	1101
14	0001 0100	1110
15	0001 0101	1111

EXAMPLE 23–7

Convert the following decimal numbers into BCD:

(a) 27
(b) 458

Solution

(a)

2	7
↓	↓
0010	0111

(b)

4	5	8
↓	↓	↓
0100	0101	1000

It is also quite simple to determine a decimal number from a BCD number. Start at the least significant bit and break the code into groups of four bits. The decimal digit represented by each four-bit group is then written.

EXAMPLE 23–8

Convert the following BCD numbers into decimal numbers:

(a) 1001 0110 0101
(b) 1000 0001

Solution

(a)

1001	0110	0101
↓	↓	↓
9	6	5

(b)

1000	0001
↓	↓
8	1

Section Review

1. The BCD code is also known as the 4–2–1 code (True/False).
2. What is the BCD equivalent of the decimal number 85?
3. What is the decimal equivalent of the BCD number 0011 1001?

23–7

ASCII CODE

ASCII stands for American Standard Code for Information Interchange. It is an alphanumeric code that includes letters as well as numbers. The characters accessed by the ASCII code include the 10 numeric digits, the 26 lowercase and 26 uppercase letters of the alphabet, and about 25 special characters, including those found on a standard typewriter (@, #, $, %, *, etc.).

The ASCII code is a seven-bit code in which the decimal digits are represented by the 8–4–2–1 binary code preceded by 011. Uppercase letters are preceded by 100 or 101.

TABLE 23-3 *Partial listing of ASCII code.*

Character	7-Bit ASCII	Character	7-Bit ASCII
A	100 0001	S	101 0011
B	100 0010	T	101 0100
C	100 0011	U	101 0101
D	100 0100	V	101 0110
E	100 0101	W	101 0111
F	100 0110	X	101 1000
G	100 0111	Y	101 1001
H	100 1000	Z	101 1010
I	100 1001	0	011 0000
J	100 1010	1	011 0001
K	100 1011	2	011 0010
L	100 1100	3	011 0011
M	100 1101	4	011 0100
N	100 1110	5	011 0101
O	100 1111	6	011 0110
P	101 0000	7	011 0111
Q	101 0001	8	011 1000
R	101 0010	9	011 1001
		blank	010 0000

Lowercase letters are preceded by 110 or 111. Character symbols are preceded by 010, 011, 101, and 111. Table 23–3 is a partial listing of the ASCII code.

Section Review

1. What is ASCII an acronym for?
2. How many bits in an ASCII code?
3. The ASCII code is an alphanumeric code (True/False).

Electric circuits are composed of combinations of series- and parallel-connected elements. The same is true for digital circuits. Figure 23–1 shows two switches connected in series. For lamp Z to be energized, both switches must be closed. It can be stated, therefore, that for output Z to be energized, switch A *and* switch B both must be closed.

FIGURE 23-1 *Electrical series circuit.*

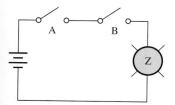

Switch A	Switch B	Output Z
Open	Open	OFF
Open	Closed	OFF
Closed	Open	OFF
Closed	Closed	ON

FIGURE 23-2 *Truth table for circuit of Figure 23-1.*

Switch A	Switch B	Output Z
0	0	0
0	1	0
1	0	0
1	1	1

FIGURE 23-3 *Truth table using binary numbers to represent the position of switches.*

A truth table shows how a circuit's output responds to various combinations of input signals. The truth table for the circuit of Figure 23–1 would be as shown in Figure 23–2.

If we assume binary 1 is a closed switch and binary 0 is an open switch, the truth table for Figure 23–1 would be as shown in Figure 23–3.

Boolean algebra is often used to express the mathematical function of a logic gate. Boolean algebra was developed by George Boole in 1847. However, it wasn't until 1938 that an engineer by the name of Claude Shannon showed how Boolean algebra could be used to analyze the two-state switching of relay logic circuits. The · symbol is used to express the AND operation in Boolean terms. The Boolean equation for a two-input AND gate would be as follows:

$$A \cdot B = Z \tag{23.1}$$

The · symbol is optional for the AND function. If two letters are together with no symbol, the AND operation is implied. For example, the above equation could also be expressed as

$$AB = Z$$

The equivalent logic gate for the circuit of Figure 23–1 would be the two-input AND gate shown in Figure 23–4. For any **AND gate,** regardless of the number of inputs, the output is high only when all the inputs are high. The total number of possible combinations of binary inputs is determined by the following equation:

$$N = 2^n \tag{23.2}$$

where N = the total possible combinations
             n = the number of input variables

_____ **EXAMPLE 23-9** _____

Determine the total possible combinations of input states for a four-input AND gate.

Solution

$$N = 2^n = 2^4 = 16$$

FIGURE 23-4 *Two-input AND gate.*

_____ EXAMPLE 23-10 _____

Solution The truth table for a three-input AND gate is shown in Figure 23–5.

A	B	C	Z
0	0	0	0
0	0	1	0
0	1	0	0
0	1	1	0
1	0	0	0
1	0	1	0
1	1	0	0
1	1	1	1

FIGURE 23–5 *Truth table for three-input AND gate.*

Section Review

1. What is the Boolean symbol for an AND function?
2. AND gates are the digital equivalent of an electrical parallel circuit (True/False).
3. State the equation for determining the number of possible combinations of binary inputs for a logic gate.
4. How many possible combinations of binary inputs exist for a three-input AND gate?

_____ **23-9** ____

OR GATES

Figure 23–6 shows two switches connected in parallel. For output Z to be energized, either or both switches must be closed. Therefore, switch A *or* switch B or both switches A and B must be closed to energized output Z.

The equivalent logic gate for the circuit shown in Figure 23–6 is the OR gate. The **OR gate** is essentially a parallel circuit that produces an output voltage when a voltage appears at any input. The schematic symbol for a two-input OR gate is shown in Figure 23–7. The truth table for a two-input OR gate is shown in Figure 23–8.

The + symbol is used to express the OR function in Boolean algebra. The Boolean equation for a two-input OR gate would be as follows:

$$A + B = Z \qquad (23.3)$$

FIGURE 23–6 *Parallel circuit.*

FIGURE 23–7 *Two-input OR gate.*

A	B	Z
0	0	0
0	1	1
1	0	1
1	1	1

FIGURE 23–8 *Truth table for two-input OR gate.*

EXAMPLE 23–11

Determine the output waveform for a two-input OR gate with the pulse waveforms shown in Figure 23–9 applied at the inputs.

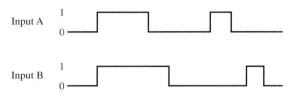

FIGURE 23–9 *Input waveforms for two-input OR gate for Example 23–11.*

Solution   Figure 23–10 shows the output waveform for the input pulses in Figure 23–9. When both inputs are zero, the output is zero. Whenever one of the inputs is high, the output is high. Because both inputs are initially zero in Figure 23–9, the output Z in Figure 23–10 is also zero. When either input changes to 1, the output goes high, and remains high until both inputs return to zero.

FIGURE 23–10   *Solution for Example 23–11.*

Section Review

1. OR gates are the digital equivalent of a series-parallel circuit (True/False).
2. What is the Boolean symbol to represent the OR logic function?
3. If both inputs to a two-input OR gate are high, what is the output?

23–10

INVERTERS

An **inverter** is a digital circuit that takes a digital input and produces an output that is the opposite state of the input. For example, if the input to an inverter were binary 0, the output would be binary 1. Conversely, a binary 1 input results in a binary 0 output.

The electrical circuit shown in Figure 23–11, is the equivalent of a digital inverter. If switch A is *not* closed, the output will be ON. Closing switch A turns OFF the output.

Figure 23–12 shows the two basic symbols that are used to represent an inverter. The inversion is indicated by the small circle on either the input or output side of the triangle. The triangle symbol itself is used to represent a buffer amplifier. In Figure 23–12(a), the inversion occurs at the input of the buffer. The inversion symbol in Figure 23–12(b) is shown at the output of the buffer. Both types of inverters have identical characteristics. In addition to inverting the signal, an inverter also provides the excellent isolation characteristics associated with a buffer, that is, high input impedance and low output impedance.

FIGURE 23-11
Equivalent circuit for an inverter.

(a)   (b)

FIGURE 23-12
Schematic symbols for an inverter.

A	Z
0	1
1	0

FIGURE 23-13
Truth table for an inverter.

The Boolean equation for an inverter with A designated as the input and Z as the output is as follows:

$$A = \overline{Z} \qquad\qquad (23.4)$$

or

$$Z = \overline{A}$$

The line above the output implies an inversion. The truth table for the inverter is shown in Figure 23-13.

EXAMPLE 23-12

Input A

FIGURE 23-14   *Input waveform for an inverter.*

Figure 23-14 shows an example of an input waveform for an inverter. Determine the output waveform.

Solution   The output waveform is the inverse of the input waveform, as shown in Figure 23-15.

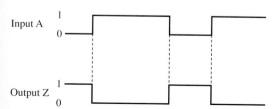

Input A

Output Z

FIGURE 23-15   *Solution for Example 23-12.*

Section Review

1. If the input to an inverter is low, the output is high (True/False).
2. Inverters provide excellent isolation characteristics (True/False).

23-11
NAND GATES

The **NAND gate** is simply an AND gate with an inverter connected to the output, as shown in Figure 23-16(a). The schematic symbol for a NAND gate is shown in Figure 23-16(b). The circle on the output of the AND gate shown in Figure 23-16(b) denotes the inversion operation.

FIGURE 23–16   (a) AND gate with inverter.
(b) NAND gate.

A	B	Z
0	0	1
0	1	1
1	0	1
1	1	0

FIGURE 23–17   Truth table for two-input NAND gate.

The output of the NAND gate is exactly the opposite of the AND gate for the same combinations of inputs. Therefore, the output of the NAND will only be low when *all* inputs are high. When *any* of the inputs are low, the output will be high. The truth table for a two-input NAND gate is shown in Figure 23–17.

The Boolean equation for a two-input NAND gate with inputs labeled A and B and the output labeled Z, would be as follows:

$$A \cdot B = \overline{Z} \qquad (23.5)$$

or

$$\overline{A \cdot B} = Z$$

Both of the above equations would adequately represent a two-input NAND gate. Inverting either the entire input or entire output would produce a similar mathematical result.

_____ EXAMPLE 23–13 _____

Determine the output waveform for a two-input NAND gate for the input waveforms shown in Figure 23–18.

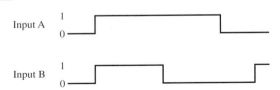

FIGURE 23–18   Input waveforms for two-input NAND gate.

Solution   The output waveform is shown in Figure 23–19.

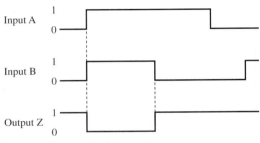

FIGURE 23–19   Solution for Example 23–13.

FIGURE 23–20   (a) OR gate with inverter.
(b) NOR gate.

A	B	Z
0	0	1
0	1	0
1	0	0
1	1	0

FIGURE 23–21   *Truth table for two-input NOR gate.*

Section Review

1. NAND gates are constructed of AND gates and inverters (True/False).
2. When is the output of a NAND gate low?
3. When any of the inputs to a NAND gate are low, the output is high (True/False).

23-12
NOR GATES

The **NOR gate** is an OR gate with an inverter connected to the output, as shown in Figure 23–20(a). The schematic symbol for the NOR gate is shown in Figure 23–20(b). The output of the NOR gate is exactly the opposite of the OR gate for the same combinations of inputs. The output of the NOR gate is low when *any* of the inputs are high. The output is high only when *all* the inputs are low. The truth table for a two-input NOR gate is shown in Figure 23–21.

The Boolean equation for a two-input NOR gate with inputs A and B and the output labeled Z would be as follows:

$$A + B = \overline{Z}$$ (23.6)

or

$$\overline{A + B} = Z$$

EXAMPLE 23–14

Determine the output waveform for a two-input NOR gate for the input waveforms shown in Figure 23–22.

Input A

Input B

FIGURE 23–22   *Input waveforms for Example 23–14.*

Solution   The output waveform is shown in Figure 23–23.

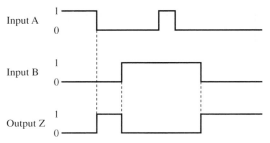

FIGURE 23–23 *Solution for Example 23–14.*

Section Review

1. A NOR gate is an OR gate with an inverter connected to the output (True/False).
2. When is the output of a NOR gate low?
3. When is the output of a NOR gate high?

— 23–13

EXCLUSIVE-OR GATES

The **exclusive-OR gate** shown in Figure 23–24(a) is a special type of OR gate that will produce an output only when one input is high. Unlike the OR gate, the output of an exclusive-OR will not be high when both inputs are high. The truth table of Figure 23–24(b) shows the input combinations and output signals for a two-input exclusive-OR gate. From the truth table, it is apparent that the output is 1 only when the inputs are unequal. The symbol $\oplus$ is used to represent the exclusive-OR function. The boolean equation for a two-input exclusive-OR gate is written as follows:

$$A \oplus B = Z \tag{23.7}$$

A	B	Z
0	0	0
0	1	1
1	0	1
1	1	0

FIGURE 23–24 *(a) Exclusive-OR gate. (b) Truth table.*

(a)

(b)

EXAMPLE 23–15

Input A

Input B

FIGURE 23–25 *Input waveforms for Example 23–15.*

Determine the output waveform for a two-input exclusive-OR gate with input waveforms shown in Figure 23–25.

Solution The output waveform is shown in Figure 23–26.

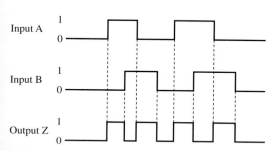

Input A

Input B

Output Z

FIGURE 23–26 *Solution for Example 23–15.*

A logic function similar to the exclusive-OR operation is the exclusive-NOR function. The output of an exclusive-NOR is simply the inverse of the exclusive-OR. That is, the output of the exclusive-NOR is low only when one input is high. The output is high whenever both inputs are high or low.

Section Review

1. When is the output of an exclusive-OR gate high?
2. If both inputs to an OR gate are high, what is the output?
3. The output of an exclusive-NOR is the opposite of an exclusive-OR (True/False).

23–14

LOGIC GATE COMBINATIONS

As logic circuits become more complex, the need to express these circuits in Boolean form becomes greater. A simple logic gate is quite straightforward in its operation. However, by grouping these gates into combinations, it becomes more difficult to determine which combinations of inputs will produce an output. Consider the circuit of Figure 23–27. Two OR gates and one AND gate form a logic gate combination that will cause output Z to become energized when a specific input combination is applied at A, B, C, and D.

The Boolean equation for the circuit of Figure 23–27 would be as follows:

$$(A + B) + (CD) = Z$$

The above Boolean equation tells us that if inputs A or B are high, the output will be high. Also, if inputs C and D are high, the output will be high. Output values for the truth table are derived from the Boolean equation, as shown in Figure 23–28.

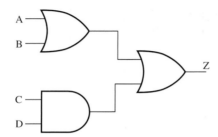

FIGURE 23–27 *Combination of logic gates.*

A	B	C	D	Z
0	0	0	0	0
0	0	0	1	0
0	0	1	0	0
0	0	1	1	1
0	1	0	0	1
0	1	0	1	1
0	1	1	0	1
0	1	1	1	1
1	0	0	0	1
1	0	0	1	1
1	0	1	0	1
1	0	1	1	1
1	1	0	0	1
1	1	0	1	1
1	1	1	0	1
1	1	1	1	1

FIGURE 23–28 *Truth table for logic circuit of Figure 23–27.*

___ **EXAMPLE 23–16** _____

For the circuit of Figure 23–29, determine the Boolean equation and truth table.

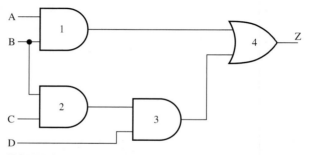

FIGURE 23–29 *Circuit for Example 23–16.*

Solution The output of gate 1 is AB, the output of gate 2 is BC, the output of gate 3 is BCD. Because gate 4 is the output of gates 1 and 3, the Boolean equation is

$$AB + BCD = Z$$

The truth table is shown in Figure 23–30.

A	B	C	D	Z
0	0	0	0	0
0	0	0	1	0
0	0	1	0	0
0	0	1	1	0
0	1	0	0	0
0	1	0	1	0
0	1	1	0	0
0	1	1	1	1
1	0	0	0	0
1	0	0	1	0
1	0	1	0	0
1	0	1	1	0
1	1	0	0	1
1	1	0	1	1
1	1	1	0	1
1	1	1	1	1

FIGURE 23–30 *Truth table for Example 23–16.*

_____ **EXAMPLE 23–17** _____

Solution The circuit of Figure 23–31 illustrates a logic gate combination that meets the requirements of the above Boolean equation. The inversion over B in the second portion of the equation is solved by connecting an inverter at the input of AND gate 2.

Draw the logic circuit for the following Boolean equation:

$$AB + \overline{B}C = Z$$

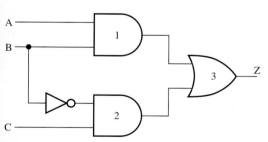

FIGURE 23–31 *Circuit for Example 23–17.*

EXAMPLE 23–18

For the circuit of Figure 23–32, determine the Boolean equation and truth table.

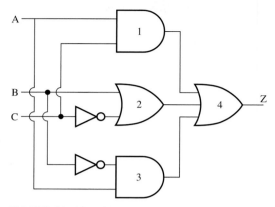

FIGURE 23–32 *Circuit for Example 23–18.*

Solution The output of gate 1 is AC, the output of gate 2 is $B + \overline{C}$, and the output of gate 3 is AB. Because the output of gate 4 depends on gates 1, 2, and 3, the Boolean equation is as follows:

$$AC + B + \overline{C} + A\overline{B} = Z$$

Figure 23–33 shows the truth table for the circuit of Figure 23–32.

A	B	C	Z
0	0	0	1
0	0	1	0
0	1	0	1
0	1	1	1
1	0	0	1
1	0	1	1
1	1	0	1
1	1	1	1

FIGURE 23–33 *Truth table for Example 23–18.*

Section Review

1. When logic gates are grouped together, it becomes less difficult to determine which combinations of inputs produce an output (True/False).
2. Why are logic gate combinations often expressed as Boolean equations?

— 23–15

BOOLEAN THEOREMS

Earlier in this chapter, Boolean equations for basic logic gates were derived. As Boolean equations increase in size and complexity, it becomes necessary to simplify these statements to their most basic form. Boolean theorems are used to reduce the size and com-

TABLE 23–4 *Single-variable theorems.*

Theorem	Equivalent Electric Circuit	Reduced Form	Boolean Form
1	A, A	A	$A \cdot A = A$
2	A, Open circuit	Open circuit	$A \cdot 0 = 0$
3	A, Short circuit	Short circuit	$A + 1 = 1$
4	A, $\bar{A}$	Short circuit	$A + \bar{A} = 1$
5	A, A	A	$A + A = A$
6	A, $\bar{A}$	Open circuit	$A \cdot \bar{A} = 0$
7	A, Short circuit	A	$A \cdot 1 = A$
8	A, Open circuit	A	$A + 0 = A$

plexity of Boolean equations. There are eight single-variable theorems and seven multi-variable theorems. The single-variable theorems with equivalent electric circuit components are shown in Table 23–4. Binary 0 is represented by an open circuit and binary 1 by a short circuit.

The seven multivariable theorems, given below, involve more than one variable.

Theorem 9 $x + y = y + x$
Theorem 10 $xy = yx$

Theorems 9 and 10 are referred to as the *commutative laws.* They indicate that the order in which variables are added or multiplied is unimportant.

Theorem 11 $x + (y + z) = (x + y) + z = x + y + z$
Theorem 12 $x(yz) = (xy)z = xyz$

Theorems 11 and 12 are the *associative laws,* which state that it is possible to group the terms of a sum or product in any order that is desired.

> Theorem 13 $x(y + z) = xy + xz$

Theorem 13 is the *distributive law,* which states that an expression can be expanded by multiplying term by term, as in ordinary algebra.

> Theorem 14 $x + xy = x$
> Theorem 15 $x + \bar{x}y = x + y$

EXAMPLE 23–19

Simplify the digital circuit shown in Figure 23–34.

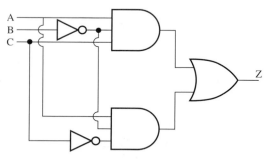

FIGURE 23–34   *Circuit for Example 23–19.*

Solution   First, write the Boolean equation for the circuit.

$$(A \cdot \bar{B} \cdot C) + A \cdot \bar{B} \cdot \bar{C}) = Z$$

Factor out the common variables $A\bar{B}$ using theorem 13.

$$A\bar{B}(C + \bar{C}) = Z$$

According to theorem 8, the term in parentheses is equal to 1. Therefore,

$$Z = A\bar{B} \cdot 1 = A\bar{B} \qquad \text{(according to theorem 2)}$$

The simplified digital circuit is shown in Figure 23–35.

FIGURE 23–35   *Simplified circuit for Example 23–19.*

EXAMPLE 23–20

Simplify $(\bar{A} + B)(A + \bar{B})$ $= Z$

Solution   Multiply out the following terms:

$$\bar{A}A + \bar{A}\bar{B} + BA + B\bar{B} = Z$$

$\bar{A}A$ and $B\bar{B}$ are equal to zero (theorem 4). Therefore,

$$Z = 0 + \bar{A}\bar{B} + BA + 0 = \bar{A}\bar{B} + BA$$

EXAMPLE 23–21

Solution Factor out common variables CD by

$$CD(A + \overline{A}B) = Z$$

Use theorem 15 to replace $A + \overline{A}B$ with $A + B$

$$Z = CD(A + B) = ACD + BCD$$

Simplify $ACD + \overline{A}BCD$
$= Z$

Section Review

1. What are Boolean theorems used for?
2. How many single-variable theorems are there?
3. There are eight multivariable theorems (True/False).
4. State the distributive law.

23–16

APPLYING DeMORGAN'S THEOREMS TO DIGITAL CIRCUITS

DeMorgan was a logician and mathematician who proposed two theorems that are an important part of Boolean algebra. **DeMorgan's theorems** are stated as follows:

> Theorem 1 The complement of a sum equals the product of the complements.
> Theorem 2 The complement of a product equals the sum of the complements.

In equation form,

$$\overline{A \cdot B} = \overline{A} + \overline{B} \qquad (23.8)$$
$$\overline{A + B} = \overline{A} \cdot \overline{B}$$

DeMorgan's theorems are used mainly for simplifying logic circuits. The following examples illustrate how DeMorgan's theorems can be used to reduce digital circuits.

EXAMPLE 23–22

Reduce the digital circuit in Figure 23–36 using DeMorgan's theorems.

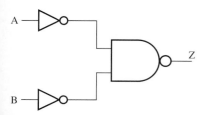

FIGURE 23–36 *Circuit for Example 23–22.*

Solution First, write the Boolean equation.

$$\overline{A} \cdot \overline{B} = \overline{Z}$$

Mathematically, whatever you do to one side of an equation, you can do to the other side. Therefore, if the entire output is inverted, the entire input could also be inverted.

$$\overline{\overline{A} \cdot \overline{B}} = Z$$

Apply DeMorgan's first theorem to the above equation.

$$\overline{\overline{A}} + \overline{\overline{B}} = Z$$

If an input is inverted twice, the double inversion cancels out.

$$A + B = Z$$

According to DeMorgan's first theorem, the logic circuit of Figure 23–36 reduces to the OR gate shown in Figure 23–37.

FIGURE 23–37
Simplified circuit for
Example 23–22.

EXAMPLE 23–23

Simplify the digital circuit shown in Figure 23–38 using DeMorgan's theorems.

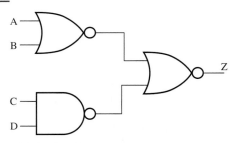

FIGURE 23–38 *Circuit for*
Example 23–23.

Solution First, write the Boolean equation for the circuit.

$$\overline{(A + B)} + \overline{(C \cdot D)} = \overline{Z}$$

$$\overline{\overline{(A + B)} + \overline{(C \cdot D)}} = Z$$

Apply DeMorgan's second theorem.

$$\overline{\overline{(A + B)}} \cdot \overline{\overline{(C \cdot D)}} = Z$$

Remember, the double inversions cancel out.

$$(A + B) \cdot (C \cdot D) = Z$$

The simplified logic circuit is shown in Figure 23–39.

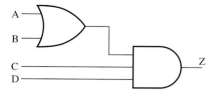

FIGURE 23–39 *Simplified circuit for Example 23–23.*

Section Review

1. State DeMorgan's two theorems.
2. DeMorgan's theorems are used mainly for simplifying logic circuits (True/False).

23–17

FLIP-FLOPS (FFs)

A **flip-flop (FF)** is a bistable multivibrator that has the ability to store information. The term *bistable* means that the device has two stable output states. Essentially, the flip-flop is a digital latch that will store a binary digit for a controlled length of time. Flip-flops are also known as *sequential circuits* and are special digital circuits in which the output depends on both the present inputs and previous outputs. Flip-flops have two outputs, Q and $\overline{Q}$, which are the inverse of each other. There are three basic types of flip-flops: set-reset (SR), D-type, and JK. The set-reset is the simplest type of FF. Figure 23–40(a) shows an **SR flip-flop** constructed with NAND gates. The two inputs are labeled *Set* and *Reset,* and the two outputs are Q and $\overline{Q}$. The schematic symbol for a NAND SR flip-flop is shown in Figure 23–40(b).

In Figure 23–40(a), if both inputs for the FF are zero, both outputs will be high. Because Q and $\overline{Q}$ are always the inverse of each other, the FF is said to be in an ambiguous state when set = 0 and reset = 0. The set and reset inputs are usually resting in the high state and one of the inputs will be pulsed low whenever the output is required to change. When both inputs to the NAND SR flip-flop are 1, the present output is determined by the previous output. Consider the circuit shown in Figure 23–41(a). In this circuit set = 0 and reset = 1. Because one input to NAND1 is low, the output will be high, and Q = 1. Because the Q output in Figure 23–41(a) is 1, both inputs to NAND2 will be high and the $\overline{Q}$ output will be 0. In Figure 23–41(b), the set input has been changed to 1. The Q output was previously 1, which remains fed back to the input of NAND2. The 1 at both inputs to NAND2 causes $\overline{Q}$ to remain in the low state. In other words, the Q output remains in its previous state when both inputs are high.

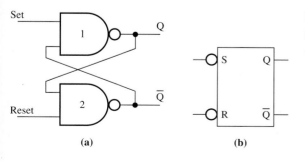

FIGURE 23–40 *(a) NAND SR flip-flop. (b) Schematic symbol.*

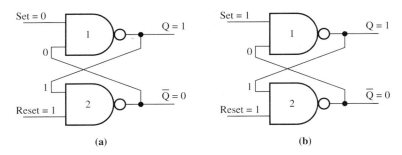

FIGURE 23-41   *(a) Q output is set. (b) Q output remains set.*

S	R	Q	$\overline{Q}$	State
0	0	1	1	Ambiguous
0	1	1	0	Set
1	0	0	1	Reset
1	1	No change		Previous

FIGURE 23-42   *Truth table for NAND SR flip-flop.*

Figure 23-42 shows a truth table for a NAND SR flip-flop. The operation of the NAND SR flip-flop can be summarized as follows:

1. Set = 1, Reset = 1. This condition is the normal resting state and it has no effect on the output state of the FF. The Q output will remain high or low, depending on the state it was in prior to this input condition occurring.
2. Set = 0, Reset = 1. This will always cause the Q output to go high. Whenever the Q output is high, the FF is said to be *set*. Once the FF has been set, the Q output will remain high until the reset input is changed to 0.
3. Set = 1, Reset = 0. This combination of inputs will cause the Q output to go low. When Q = 0, the FF is considered to be *reset*. When the FF is reset, the Q output will remain latched in the low state until the set input is 0.
4. Set = 0, Reset = 0. This condition tries to set and reset the FF simultaneously, producing an ambiguous result. Consequently, this combination cannot be used, which means the SR flip-flop has only three usable states.

_____ EXAMPLE 23-24 _____

For the NAND SR flip-flop waveforms shown in Figure 23-43, determine the Q output waveform. Assume Q = 0, initially.

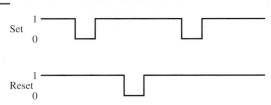

FIGURE 23-43   *Set and reset inputs for NAND SR flip-flop.*

Solution The Q output waveform is illustrated in Figure 23–44. The Q output is initially 0 and remains low until the set input goes low. When set = 0 and reset = 1, Q goes high. The Q output remains high until the reset is pulsed low. Q remains at 0 until the set is pulsed low.

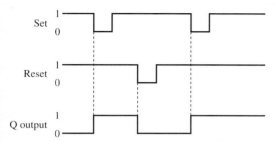

FIGURE 23–44   Solution for Example 23–24.

An SR flip-flop can also be constructed using NOR gates, as shown in Figure 23–45(a). The schematic symbol for a NOR SR flip-flop is shown in Figure 23–45(b). The NOR and NAND SR flip-flops both have only three usable states. The main difference between the NAND SR FF and the NOR SR FF is that the Q output is now taken from gate 2. The schematic symbols for the NAND and NOR SR are the same except that the NOR SR has no inverters shown at the inputs. In Figure 23–45(a), when set and reset are both high, the Q output and the $\overline{Q}$ output will both be 0 and the FF is said to be in an ambiguous state. When set = 1, and reset = 0, the Q output will be high and will remain high after the set input returns to 0. Because Q = 1 in this condition, the FF is set. When set = 0 and reset = 1, the output at Q will be 0 and will remain low after the reset input returns to 0. Because Q = 0, the FF is said to be reset. When set = 1 and reset = 1, the FF tries to set and reset itself at the same time. This causes an ambiguous result.

Figure 23–46 shows the truth table for a NOR SR flip-flop. The ambiguous state results when both inputs are high. The FF will remain latched when both inputs are 0.

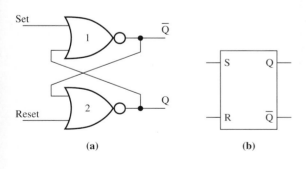

FIGURE 23–45   (a) NOR SR flip-flop. (b) Schematic symbol.

S	R	Q	$\overline{Q}$	State
0	0	No change		Previous
0	1	0	1	Reset
1	0	1	0	Set
1	1	0	0	Ambigious

FIGURE 23–46   Truth table for NOR SR flip-flop.

EXAMPLE 23-25

For the waveforms shown in Figure 23–47, assume Q = 0, initially, determine the Q output for a NOR SR flip-flop.

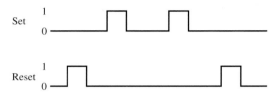

FIGURE 23–47 *Set and reset inputs for Example 23–25.*

Solution   The Q output waveform is illustrated in Figure 23–48. Q is initially 0, and remains low until the set input goes high. Q remains set until the reset input is pulsed high.

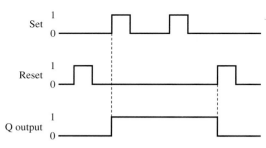

FIGURE 23–48 *Solution for Example 23–25.*

Section Review

1. A flip-flop is a monostable multivibrator (True/False).
2. Flip-flops are also known as sequential circuits (True/False).
3. What are the two outputs of a flip-flop labeled?

23-18

D-TYPE FLIP-FLOPS

The main disadvantage of the SR flip-flop is the fact that it has only three usable states. Figure 23–49(a) shows how the ambiguous state can be eliminated by adding two NAND gates and an inverter to the NAND SR flip-flop. This configuration is called a **D-type flip-flop.** The schematic symbol for a D-type FF is shown in Figure 23–49(b). The two inputs are labeled T and D. The T input represents a stable clock pulse. When the clock input is low, the D input will have no effect, because the 0 at the T input causes the outputs of both NAND1 and NAND2 to go high. When the clock input is high, the D input will produce a low at either set or reset, depending on whether the D input is high or low.

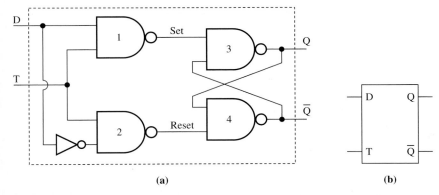

(a)   **(b)**

FIGURE 23–49 *(a) D-type flip-flop. (b) Schematic symbol.*

T	D	Q	$\overline{Q}$	State
0	0	No change		Previous
0	1	No change		Previous
1	0	0	1	Reset
1	1	1	0	Set

FIGURE 23–50 *Truth table for D-type flip-flop.*

Figure 23–50 shows the truth table for a D-type FF. The ambiguous state does not occur in this type of FF. Instead, there are two input conditions that cause the Q output to remain in its previous state. Whenever the clock input is low, the Q output will not change state. When T = 1, Q follows D. When T = 1 and D = 0, the FF is reset. The FF will be set when T = 1 and D = 1.

———— **EXAMPLE 23–26** ————

Determine the Q waveform for a D-type FF with T and D inputs as shown in Figure 23–51. Assume Q = 0, initially.

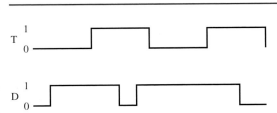

FIGURE 23–51 *D and T input waveforms for Example 23–26.*

Solution Figure 23–52 shows the Q output waveform. Q is initially 0 and goes high when D = 1 and T = 1. Q remains high until D returns to 0. When the D input goes high again, Q goes high. Q stays at 1 until D falls back to 0.

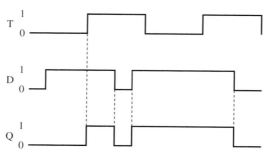

FIGURE 23–52 *Output waveform for Example 23–26.*

Section Review

1. What is the main disadvantage of the SR flip-flop?
2. What are the two inputs for a D-type flip-flop labeled?
3. How many usable states does a D-type flip-flop have?

___ 23–19 ___

JK FLIP-FLOPS

The **JK flip-flop** is the most common type of FF used in digital circuits. It has a clock input as well as J and K inputs. The J and K inputs are often referred to as control inputs because they determine what the FF does when the clock pulse is activated. There are two types of JK FF: positive-edge trigger and negative-edge trigger. The positive-edge trigger JK FF can only change state when the clock input goes from a 0 to a 1. Figure 23–53(a) shows an example of a positive- or leading-edge clock pulse. The arrow indicates the clock changing from a 0 to a 1. The clock pulse shown in Figure 23–53(b) is considered a negative- or trailing-edge clock pulse. The arrow indicates the clock changing from a 1 to a 0. Negative-edge trigger JK FFs will respond only to this input condition for the clock pulse.

Figure 23–54(a) illustrates how a positive-edge JK FF can be constructed using a NOR SR FF and two AND gates. The schematic symbol for a positive-edge JK FF is shown in Figure 23–54(b). The schematic symbol of Figure 23–54(c) represents a negative-edge trigger JK FF. The only difference between a positive-edge and a negative-edge trigger JK FF is where the clock pulse causes Q to change. Otherwise, the Q outputs are identical for any combination of J and K. Because the outputs are the same, the truth table for both types of JK FFs are also the same.

In Figure 23–54(a), when J and K are low, the outputs of both AND gates are zero, and the clock pulse will have no effect. When J is low and K is high, the output of AND gate 1 is low and the Q output is zero. When J is high and K is low, the Q output is initially zero, and the $\overline{Q}$ is 1. This places a 1 at the input of AND gate 1. When the clock pulse goes high, AND gate 1 is enabled and Q goes high. When the J and K are both high, the Q out-

(a)   (b)

FIGURE 23–53 *(a) Positive-edge clock pulse. (b) Negative-edge clock pulse.*

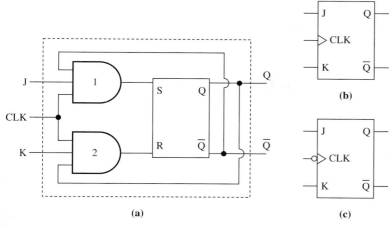

FIGURE 23–54   (a) Positive-edge trigger JK FF.
(b) Schematic symbol for positive-edge JK FF.
(c) Symbol for negative-edge trigger JK FF.

CLK	J	K	Q	$\overline{Q}$	State
X	0	0	No change		Previous
X	0	1	0	1	Reset
X	1	0	1	0	Set
X	1	1	Last state	Toggle	

FIGURE 23–55   Truth table for JK flip-flop.

put will go low when the clock input is high or high when the clock input is low. This is
known as the **toggle state.** Figure 23–55 shows the truth table for both types of JK FFs.
The clock transition is indicated by an X, which means either a negative- or positive-edge
trigger, depending on the type of FF.

EXAMPLE 23–27

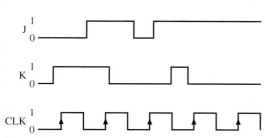

FIGURE 23–56   Input waveforms for
Example 23–27.

The waveforms shown in
Figure 23–56 are for a
positive-edge trigger JK
FF. Determine the Q out-
put, assuming Q is ini-
tially zero.

Solution   The output waveform is shown in Figure 23–57. The Q output can only
go high on the leading-edge clock pulse. For this particular combination of input
waveforms, the Q output will only change state when both J and K are high.

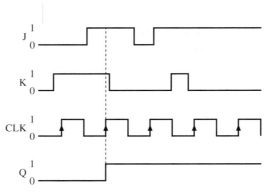

FIGURE 23–57   Solution for Example 23–27.

EXAMPLE 23–28

Determine the Q output for a negative-edge trigger JK FF with input waveforms shown in Figure 23–58. Assume Q = 0, initially.

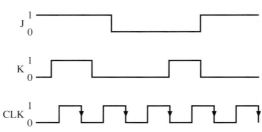

FIGURE 23–58   Input waveforms for Example 23–28.

Solution   Figure 23–59 shows the Q output waveform. Q is initially 0 and toggles to the high state when the first clock pulse is applied. Q remains high until the third clock pulse when J = 0 and K = 1. On the fourth clock pulse, Q returns high because J = 1 and K = 0 at this point.

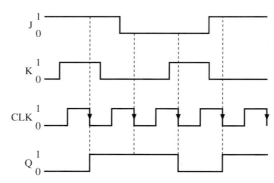

FIGURE 23–59   Solution for Example 23–28.

JK flip-flops are frequently combined in groups to store large amounts of data. In these types of combinational circuits, the Q outputs often are required to be initially all 0s or all 1s. To simultaneously set or reset all the Q outputs in a group of JK FFs, the FFs are provided with two additional inputs, as shown in Figure 23–60(a). The truth table for the

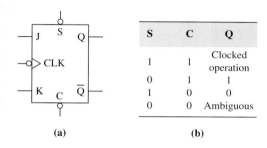

FIGURE 23–60 *(a) JK FF with set and clear inputs. (b) Truth table.*

S	C	Q
1	1	Clocked operation
0	1	1
1	0	0
0	0	Ambiguous

(a) (b)

set and clear inputs is shown in Figure 23–60(b). These two inputs are normally resting in the high state and are pulsed low when the Q output is required to change. When set = 1 and clear = 1, the FF operates normally and responds to different settings of the J, K, and clock inputs. When set = 0 and clear = 1, Q is immediately set to 1 regardless of the state of the J, K, and clock inputs. Set = 1 and clear = 0 will cause the Q output to be immediately reset to 0. If both the set and clear inputs were 0, the Q output would try to set and reset itself simultaneously, producing an ambiguous condition.

Section Review

1. What is the most common type of flip-flop used in digital circuits?
2. How many inputs does a JK flip-flop have?
3. What are the two types of JK flip-flops?
4. JK flip-flops have no ambiguous state (True/False).

23–20
SHIFT REGISTERS

A **shift register** is a device that is used for storing and transferring data, changing data from one format to another, sequencing control, and timing circuits. When the clock pulse shifts individual bits of data, the data are either shifted to the left or shifted to the right. Consequently, these registers are known as *shift-right* or *shift-left registers*. Figure 23–61 shows an example of a four-bit shift register constructed using four JK flip-flops. Because the JK FFs shown in Figure 23–61 are negative-edge triggered, when a trailing edge appears on the CLK line, it is applied to the CLK terminals of all four FFs simultaneously. At the instant the negative clock pulse is applied, all the FFs respond to the input levels at their previous J and K inputs. This type of clocked function is called *synchronous operation*. In Figure 23–61, the J and K inputs of one FF are the Q and $\overline{Q}$ outputs of the next FF. Each time a clock pulse is applied, the data is transferred, or shifted, one place to the right.

FIGURE 23–61 *Shift register using four JK FFs.*

Clock Pulses	Q_1	Q_2	Q_3	Q_4
0	0	0	0	0
1	1	0	0	0
2	1	1	0	0
3	1	1	1	0
4	1	1	1	1

FIGURE 23–62 *Four-bit shift register output table.*

Figure 23–62 shows the Q outputs produced by each successive clock pulse. If we assume that the clear line in Figure 23–61 has been pulsed, the Q outputs are initially 0 and the $\overline{Q}$ outputs are high. When the first negative clock pulse is applied, FF4 sees a 0 at the J input and a 1 at the K input. This combination causes Q_4 to go low. FF2 and FF3 will also produce a 0 at their respective Q outputs. FF1 will produce a 1 at its Q output after the first clock pulse because its J input is initially high and the K input is held low by an inverter. When the second negative clock pulse is applied, FF4 remains reset because $Q_3 = 0$ and $\overline{Q}_3 = 1$. FF3 remains reset because $Q_2 = 0$ and $\overline{Q}_2 = 1$. However, FF2 is now set because $Q_1 = 1$ and $\overline{Q}_1 = 0$. After the first clock pulse, FF1 remains set because its J and K inputs are fixed at 1 and 0, respectively. The data continues to be shifted through the third and fourth clock pulses. After four clock pulses, the shift register is full and will not change state when additional clock pulses are applied.

Section Review

1. What is a shift register?
2. Name the two basic types of shift registers.
3. Shift registers are synchronous circuits (True/False).

— 23–21

COUNTERS

Many situations arise in industry where there is a need for both simple and complex counting systems. A **counter** is a device that produces a numeric count each time an input clock pulse occurs. These clock pulses occur at known intervals, and, as a result, the counter can be used as an instrument for measuring time and frequency. There are two basic types of digital counters: *synchronous* and *asynchronous*. An asynchronous counter, or *ripple counter,* uses the output of one FF as the clock input to the next FF. As a result, the clock pulses pass through a ripple counter in a serial manner. The synchronous counter uses more complex circuitry to achieve basically the same result. In a synchronous circuit, the clock pulse is applied to the inputs of all the FFs at exactly the same instant, like a parallel circuit. The main advantage of synchronous counters is that they can operate at faster speeds.

Counting devices are classified as up-counters or down-counters. An up-counter will increase in numerical value each time a clock pulse is applied. These counters have a clear line that is used to reset the counter to its minimum value. Each successive clock pulse will cause the count stored by the device to increase by one. The maximum number stored by a digital counter is limited by the number of flip-flops used. The asynchronous up-counter shown in Figure 23–63 is capable of storing 2^4 possible combinations of binary data.

In the circuit of Figure 23–63, the Q output of one JK FF acts as the clock input for the next FF. This configuration is called asynchronous because all the FFs do not change

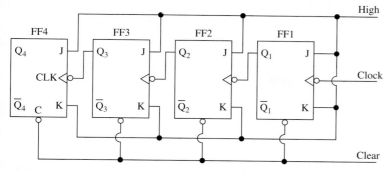

FIGURE 23–63 *Asynchronous up-counter.*

state in exact synchronism with the clock pulse. The clock pulse is applied only to the clock input of FF1. Because the J and K inputs for all the FFs are permanently high, the Q output of FF1 will toggle each time a negative clock pulse is applied. The Q output of FF1 acts as the clock input for FF2, so FF2 will toggle each time Q_1 goes from 1 to 0. Also, FF3 will toggle when Q_2 goes from 1 to 0, and FF4 will change state when Q_3 goes from a high to a low. Figure 23–64 shows the Q outputs for a four-bit up-counter. Q_1 represents the least significant digit in the counter and Q_4 represents the most significant digit. Because Q_1 will toggle on every clock pulse, it is shown to alternate between 0 and 1. Q_2 will change state on every second clock pulse, and Q_3 will change state on every fourth clock pulse. After 15 clock pulses have occurred, the counter's FFs are all set. When the sixteenth clock pulse occurs, FF1 changes from 1 to 0, causing FF2 to toggle low. As Q_2 changes from 1 to 0, this causes FF3 to change to 0, and FF4 is also reset.

Counters can be triggered by either a leading or trailing input pulse. A positive-edge triggered input causes the counter to increment or decrement when the input pulse goes from a low to a high state. In other words, every time a clock pulse occurs, the accumulated value of the counter changes. The circuit shown in Figure 23–63 uses negative-edge trig-

Clock Pulses	Q_4	Q_3	Q_2	Q_1
0	0	0	0	0
1	0	0	0	1
2	0	0	1	0
3	0	0	1	1
4	0	1	0	0
5	0	1	0	1
6	0	1	1	0
7	0	1	1	1
8	1	0	0	0
9	1	0	0	1
10	1	0	1	0
11	1	0	1	1
12	1	1	0	0
13	1	1	0	1
14	1	1	1	0
15	1	1	1	1
16	0	0	0	0

FIGURE 23–64 *Output states for four-bit up-counter.*

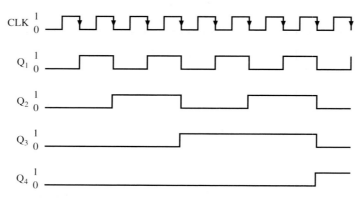

FIGURE 23–65 *Output waveforms for four-bit up-counter.*

gered JK FFs, and the counter would be classified as a negative-edge triggered device. The output waveforms for a four-bit asynchronous up-counter are shown in Figure 23–65.

Figure 23–66 shows an example of an asynchronous down-counter. This configuration is identical to the asynchronous up-counter, except the clock pulse is now supplied by the $\overline{Q}$ output instead of the Q output. The down-counter starts at its maximum count and decrements with each clock pulse. If we assume that all the Q outputs are initially 0, then all the $\overline{Q}$ outputs must be high. Because the FFs are negative-edge triggered in Figure 23–66, the first negative-going clock pulse will cause the Q output to change to 1. If Q_1 goes high, then $\overline{Q}_1$ will go low. FF2 sees the negative-going output of $\overline{Q}_1$ as a trigger pulse and, as a result, Q_2 goes high. FF3 toggles as a result of the $\overline{Q}_2$ output of FF2, which in turn causes Q_4 to go high.

Section Review

1. What is the basic operating principle of an asynchronous counter?
2. What are the two basic classifications of counters?
3. Clock pulses pass through a ripple counter in a parallel manner (True/False).

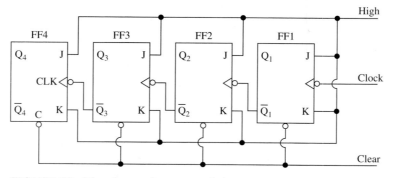

FIGURE 23–66 *Asynchronous down-counter.*

TROUBLESHOOTING DIGITAL CIRCUITS

Digital ICs generally consist of several logic gates contained in one package. Because these gates are isolated from each other, it is possible for one gate to malfunction without affecting the other gates in the IC. In other cases, the entire IC may not function. When troubleshooting a group of digital ICs, we are attempting to isolate the problem to a specific IC. The next step is to determine whether the problem is inside the IC or caused by something external, such as a faulty power supply. Other examples of external IC problems include cold solder joints and short circuits between IC pins. Solder bridges are often responsible for shorts between pins. Because the pins of an IC are very close together, a tiny amount of excess solder can bridge two pins together. If the problem is internal, and the IC is damaged, it is almost always replaced rather than repaired. IC chips are relatively inexpensive and impossible to repair in most cases.

Many of the troubleshooting techniques discussed previously in this book can be applied to digital circuits. The main difference between digital and analog testing is that we are now simply testing for ON or OFF conditions. Although general-purpose test instruments such as oscilloscopes, signal generators, and voltmeters can be used on digital equipment, special digital test equipment is available that greatly simplifies test procedures.

Figure 23–67(a) shows a *logic probe,* which is a test instrument designed to indicate the logic status at a single point in a digital circuit. A logic probe can be used to detect most problems associated with digital circuits. The logic probe indicates whether a logic level is high, low, or changing states. Some logic probes have three LEDs: one to indicate a high state, one for a low state, and one LED to indicate if the output is rapidly changing state. Other logic probes use a single LED to indicate various output conditions. When a single-lamp probe is touched to a high level or is open circuited, the LED is illuminated. When the probe is touched to a low level, the lamp goes out. In other words, the lamp flashes on or blinks off, depending on the polarity of the pulse. When the probe is connected to a pulse train, the lamp is partially illuminated. Although oscilloscopes are often used for pulse measurement, the logic probe is ideally suited for detecting pulses of short duration and low repetition rates. A **current tracer** is a device similar to a logic probe, except it is used to detect a changing current in a digital circuit. A typical current tracer is activated by the magnetic field originating from a current near its tip.

The **logic pulser** shown in Figure 23–67(b) is used to inject signals into a digital circuit. When testing digital devices such as logic gates and flip-flops, it is often necessary to trigger the device to determine if it is functioning properly or not. Logic pulsers are portable signal generators that are capable of supplying logic pulses of fixed duration to circuits under test. Logic pulsers generally are used in conjunction with logic probes to troubleshoot ICs. The pulser injects a signal and the probe monitors the output. Logic pulsers usually have a small push button that will supply a trigger pulse when pressed. Pulsers automatically determine the polarity of the pulse to be generated. When the push button is pressed, the pulser supplies a signal that is opposite in polarity to the signal presently at the input. For example, if the signal at the pin of an IC is low, the pulser will supply a high logic level when the button is pressed.

IC chips often contain a large number of pins that require simultaneous testing to obtain accurate results. Because the logic probe can only measure one output at a time, a **logic clip** is frequently employed in digital troubleshooting. The logic clip shown in Figure 23–68 clips onto standard dual-in line (DIP) ICs and instantly displays the logic states of

Lamp dim = open contact

Lamp on = HIGH

Lamp off = LOW

One flash = single pulse

Repetitive flashes = pulses

(a)

Pulser provides
repetitive pulses

Pulser

Current probe

Lamp on = current pulses

Lamp off = no current

(b)

FIGURE 23–67 *(a) Logic probe. (b) Logic pulser*

all the pins on the IC. Each of the clip's LEDs independently follow logic level fluctuations at the pin it is attached to. The logic clip is highly portable and much easier to use than an oscilloscope or voltmeter when testing for high and low states.

A **logic analyzer** is a more sophisticated version of the logic clip. Logic analyzers use CRT screens to display multiple waveforms simultaneously. A logic analyzer is basically a recording device that accepts multiple input channels of digital data. Figure 23–69 shows a typical logic analyzer. This type of device is capable of storing high-frequency digital signals and reproducing these signals on a CRT screen. The data in the analyzer's

FIGURE 23–68   *Logic clip.*

FIGURE 23–69   *Logic analyzer.*

memory can also be reproduced in binary form. This is quite useful when testing computer equipment. In computer systems, troubleshooting often involves checking the binary words that are passed along data buses and address buses. Logic analyzers are capable of providing continuous binary readouts of data being transmitted.

Section Review

1. What is a logic probe?
2. Logic pulsers are used to inject signals into a digital circuit (True/False).
3. What is a logic analyzer?

23–23

PRACTICAL APPLICATION SOLUTION

Your task described in the chapter opener was to design a digital control circuit that would activate an alarm when three combinations of pressure, level, and temperature occurred inside the mixing vat. The following steps illustrate the method of solution for this practical application of digital electronics.

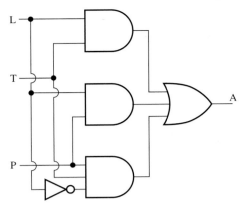

FIGURE 23-70 *Digital logic circuit.*

STEP 1   Derive a Boolean equation for the circuit. The three variables are labeled as follows: level = L, temperature = T, and pressure = P. The alarm is represented by A. The Boolean equation for the three combinations described in the chapter opener is as follows:

$$A = LT + LP + \overline{L}TP$$

STEP 2   Draw the digital circuit. Figure 23–70 shows a digital logic circuit for the Boolean equation derived in Step 1.

Summary

1. The binary number system is based on two digits: 1 and 0.
2. Logic gates are decision-making circuits that produce an output based on a certain combination of inputs.
3. Boolean theorems are used to reduce the size and complexity of Boolean equations.
4. A flip-flop is a bistable multivibrator that has the ability to store information.
5. The JK flip-flop is the most common type of FF used in digital circuits.
6. Shift registers are used for storing data, changing data from one format to another, sequencing control, and timing circuits.
7. A counter is a device that produces a numeric count each time an input clock pulse occurs.
8. Counters can be triggered by either a leading- or trailing-input pulse.
9. Logic probes are test instruments designed to indicate the logic status at a single point in a digital circuit.
10. A logic pulser is used to inject signals into a digital circuit.

Answers to Section Reviews

Section 23-1

1. True.
2. An illustration of the relationship between input and output logic levels of a gate circuit
3. A memory-storage device constructed of logic gates and capable of storing binary data

Section 23–2

1. The total number of individual symbols in the number system
2. False

Section 23–3

1. 1 and 0
2. A single binary digit
3. 8

Section 23–4

1. False
2. A device that converts binary numbers to decimal numbers
3. True

Section 23–5

1. 16
2. C
3. True

Section 23–6

1. False
2. 1000 0101
3. 39

Section 23–7

1. American Standard Code for Information Interchange
2. 7
3. True

Section 23–8

1. A dot ($\cdot$) or no symbol between Boolean operators
2. False
3. $N = 2^n$
4. 8

Section 23–9

1. False
2. +
3. High

Section 23–10

1. True
2. True

Section 23–11

1. True
2. Only when all inputs are high
3. True

Section 23–12

1. True
2. When any of the inputs are high
3. Only when all the inputs are low

Section 23–13

1. When only one input is high
2. Low
3. True

Section 23–14

1. False
2. To simplify the explanation of the circuit operation

Section 23–15

1. To reduce the size and complexity of Boolean equations
2. 8
3. False
4. An expression can be expanded by multiplying term by term, as in ordinary algebra

Section 23–16

1. The complement of a sum equals the product of the complements, and the complement of a product equals the sum of the complements.
2. True

Section 23–17

1. False
2. True
3. Q and $\overline{Q}$.

Section 23–18

1. Has only three usable states
2. D and T
3. 4

Section 23–19

1. JK flip-flops
2. 3
3. Positive-edge trigger and negative-edge trigger
4. True

Section 23–20

1. A device that is used for storing and transferring data
2. Shift-left and shift-right registers
3. True

Section 23–21

1. The output of one flip-flop acts as the clock input to the next flip-flop.
2. Up-counters and down-counters
3. False

Section 23–22

1. A test instrument designed to indicate the logic status at a single point in a digital circuit
2. True
3. A CRT screen used to display multiple waveforms simultaneously

Multiple Choice Questions

23–1 The radix, or base, of a number system represents the
(a) Largest digit   (b) Smallest digit
(c) Total number of individual symbols   (d) Binary equivalent of a decimal number

23–2 The binary number system is based on
(a) Ten digits   (b) Eight digits   (c) Two digits   (d) Sixteen digits

23–3 A single binary digit is often referred to as a
(a) Byte   (b) Bit   (c) Word   (d) Signal

23–4 The conversion of binary digits to decimal is called
(a) Decoding   (b) Encoding   (c) Cascading   (d) Inverting

23–5 The term *encoding* refers to
(a) Binary to decimal conversion
(b) Decimal to binary conversion
(c) Hexadecimal to decimal conversion
(d) Decimal to hexadecimal conversion

23–6 The hexadecimal number system uses
(a) Ten digits   (b) Twelve digits   (c) Eight digits   (d) Sixteen digits

23–7 The AND gate operates on the same principle as
(a) The series circuit   (b) The parallel circuit
(c) The series-parallel circuit   (d) None of the above

23–8 A parallel electrical circuit is equivalent to a(n):
(a) AND gate   (b) OR gate   (c) Inverter   (d) NAND gate

23–9 A digital circuit that produces an output signal that is the opposite of the input signal is called a(n)
(a) AND gate   (b) OR gate   (c) Inverter   (d) NAND gate

23–10 When the output of a gate is low only when all the inputs are high, it is called a(n)
(a) AND gate   (b) OR gate   (c) NOR gate   (d) NAND gate

23–11 Which mathematician proposed that the complement of a sum equals the product of the complements?
(a) George Boole   (b) Charles Shannon
(c) DeMorgan   (d) Georg Simon Ohm

23–12 A flip-flop is a(n)
(a) Astable multivibrator
(b) Circuit that can store decimal numbers
(c) Bistable multivibrator
(d) Decision-making circuit

23–13 The three most popular flip-flops are
(a) NAND, NOR, and JK   (b) NAND, NOR, and D-type
(c) NAND, SR, and JK   (d) SR, D-type, and JK

23–14 When the Q output of a flip-flop is high, the device is said to be
(a) Set   (b) Reset   (c) Cleared   (d) Ambiguous

23–15 The main disadvantage of the SR FF is the fact that it has
(a) Only two usable states   (b) No ambiguous state
(c) Only three usable states   (d) A toggle state

23–16 The JK FF is unique in that it has a(n)
(a) Ambiguous state **(b)** Previous state **(c)** Latch state **(d)** Toggle state

23–17 A counter with 9 JK FFs is capable of counting up to
(a) 128 **(b)** 256 **(c)** 512 **(d)** 1048

23–18 How many clock pulses are required to load an eight-bit word into a shift register?
(a) 2 **(b)** 4 **(c)** 8 **(d)** 16

23–19 A JK FF is reset; J = 1, K = 0. When a clock pulse occurs, the FF will
(a) Be set **(b)** Remain reset
(c) Produce an ambiguous state **(d)** Be reset

23–20 A JK FF has its J and K inputs = 0. When a clock pulse occurs, the state of the FF will be
(a) Reset **(b)** Set **(c)** Toggled **(d)** Unchanged

Practice Problems

23–21 Convert binary number 10101101 to decimal form.

23–22 Convert decimal number 37 to its equivalent binary form.

23–23 Change decimal number 479 to hexadecimal.

23–24 Convert decimal number 827 to BCD.

23–25 Change BCD number 1001100101 to decimal.

23–26 Determine the total possible combinations of input states for a three-input AND gate.

23–27 Figure 23–71 shows an example of an input waveform for an inverter. Determine the output waveform.

FIGURE 23–71

23–28 Determine the output waveform for a two-input NAND gate for the input waveforms shown in Figure 23–72.

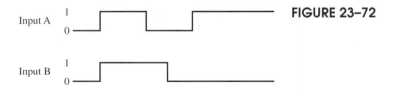

FIGURE 23–72

23–29 Derive the boolean equation for the logic gate combination shown in Figure 23–73.

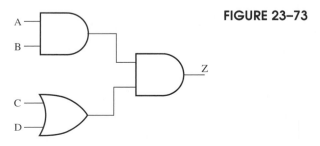

FIGURE 23–73

23–30 Determine the logic circuit for the following boolean equation:

$$(A + B) \cdot (C + \overline{D}) = Z$$

23–31 For the circuit of Figure 23–74, derive the boolean equation and truth table.

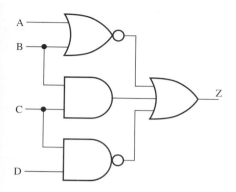

FIGURE 23–74

23–32 Simplify the following boolean equation:

$$A + \overline{A}B = Z$$

23–33 Reduce the digital circuit in Figure 23–75 using DeMorgan's theorems.

FIGURE 23–75

23–34 For the NAND SR FF waveform shown in Figure 23–76, determine the Q output. Assume Q = 0, initially.

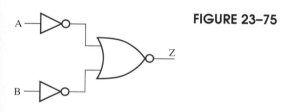

FIGURE 23–76

23–35 The waveforms shown in Figure 23–77 are for a positive-edge trigger JK FF. Determine the Q output, assuming Q is initially zero.

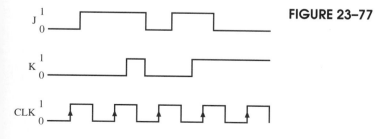

FIGURE 23–77

Essay Questions

23–36 What is a number system?

23–37 Define the term *radix*.

23–38 Explain the difference between decoding and encoding.

23–39 List the 16 digits used in the hexadecimal number system.

23–40 Describe the basic difference between binary coded decimal (BCD) and binary?

23–41 What is the ASCII code?

23–42 Why are truth tables used in digital circuits?

23–43 Name the five types of logic gates.

23–44 What is meant by a *logic level*?

23–45 Why use Boolean algebra in digital circuits?

23–46 What are the two types of Boolean theorems?

23–47 Name the three laws associated with Boolean theorems.

23–48 Explain the purpose of flip-flops in digital circuits.

23–49 List three types of flip-flops.

23–50 Describe the basic difference between a synchronous circuit and an asynchronous circuit.

23–51 Name four types of digital test instruments.

TABLE OF STANDARD RESISTOR VALUES

Resistance Tolerance ($\pm$%)

0.1% 0.25% 0.5%	1%	2% 5%	10%
10.0	10.0	10	10
10.1	—	—	—
10.2	10.2	—	—
10.4	—	—	—
10.5	10.5	—	—
10.6	—	—	—
10.7	10.7	—	—
10.9	—	—	—
11.0	11.0	11	—
11.1	—	—	—
11.3	11.3	—	—
11.4	—	—	—
11.5	11.5	—	—
11.7	—	—	—
11.8	11.8	12	12
12.0	—	—	—
12.1	12.1	—	—
12.3	—	—	—
12.4	12.4	—	—
12.6	—	—	—
12.7	12.7	—	—
12.9	—	—	—
13.0	13.0	13	—
13.2	—	—	—
13.3	13.3	—	—
13.5	—	—	—
13.7	13.7	—	—
13.8	—	—	—
14.0	14.0	—	—
14.2	—	—	—
14.3	14.3	—	—
14.5	—	—	—

0.1% 0.25% 0.5%	1%	2% 5%	10%
14.7	14.7	—	—
14.9	—	—	—
15.0	15.0	15	15
15.2	—	—	—
15.4	15.4	—	—
15.6	—	—	—
15.8	15.8	—	—
16.0	—	16	—
16.2	16.2	—	—
16.4	—	—	—
16.5	16.5	—	—
16.7	—	—	—
16.9	16.9	—	—
17.2	—	—	—
17.4	17.4	—	—
17.6	—	—	—
17.8	17.8	18	18
18.0	—	—	—
18.2	18.2	—	—
18.4	—	—	—
18.7	18.7	—	—
18.9	—	—	—
19.1	19.1	—	—
19.3	—	—	—
19.6	19.6	—	—
19.8	—	—	—
20.0	20.0	20	—
20.3	—	—	—
20.5	20.5	—	—
20.8	—	—	—
21.0	21.0	—	—
21.3	—	—	—

0.1% 0.25% 0.5%	1%	2% 5%	10%
21.5	21.5	—	—
21.8	—	—	—
22.1	22.1	22	22
22.3	—	—	—
22.6	22.6	—	—
22.9	—	—	—
23.2	23.2	—	—
23.4	—	—	—
23.7	23.7	—	—
24.0	—	24	—
24.3	24.3	—	—
24.6	—	—	—
24.9	24.9	—	—
25.2	—	—	—
25.5	25.5	—	—
25.8	—	—	—
26.1	26.1	—	—
26.4	—	—	—
26.7	26.7	27	27
27.1	—	—	—
27.4	27.4	—	—
27.7	—	—	—
28.0	28.0	—	—
28.4	—	—	—
28.7	28.7	—	—
29.1	—	—	—
29.4	29.4	—	—
29.8	—	—	—
30.1	30.1	30	—
30.5	—	—	—
30.9	30.9	—	—
31.2	—	—	—

0.1% 0.25% 0.5%	1%	2% 5%	10%
31.6	31.6	—	—
32.0	—	—	—
32.4	32.4	—	—
32.8	—	—	—
33.2	33.2	33	33
33.6	—	—	—
34.0	34.0	—	—
34.4	—	—	—
34.8	34.8	—	—
35.2	—	—	—
35.7	35.7	36	—
36.1	—	—	—
36.5	36.5	—	—
37.0	—	—	—
37.4	37.4	—	—
37.9	—	—	—
38.3	38.3	—	—
38.8	—	—	—
39.2	39.2	39	39
39.7	—	—	—
40.2	40.2	—	—
40.7	—	—	—
41.2	41.2	—	—
41.7	—	—	—
42.2	42.2	—	—
42.7	—	—	—
43.2	43.2	43	—
43.7	—	—	—
44.2	44.2	—	—
44.8	—	—	—
45.3	45.3	—	—
45.9	—	—	—

0.1% 0.25% 0.5%	1%	2% 5%	10%
46.4	46.4	—	—
47.0	—	—	—
47.5	47.5	47	47
48.1	—	—	—
48.7	48.7	—	—
49.3	—	—	—
49.9	49.9	—	—
50.5	—	—	—
51.1	51.1	51	—
51.7	—	—	—
52.3	52.3	—	—
53.0	—	—	—
53.6	53.6	—	—
54.2	—	—	—
54.9	54.9	—	—
56.2	—	—	—
56.6	56.6	56	56
56.9	—	—	—
57.6	57.6	—	—
58.3	—	—	—
59.0	59.0	—	—
59.7	—	—	—
60.4	60.4	—	—
61.2	—	—	—
61.9	61.9	62	—
62.6	—	—	—
63.4	63.4	—	—
64.2	—	—	—
64.9	64.9	—	—
65.7	—	—	—
66.5	66.5	—	—
67.3	—	—	—

0.1% 0.25% 0.5%	1%	2% 5%	10%
68.1	68.1	68	68
69.0	—	—	—
69.8	69.8	—	—
70.6	—	—	—
71.5	71.5	—	—
72.3	—	—	—
73.2	73.2	—	—
74.1	—	—	—
75.0	75.0	75	—
75.9	—	—	—
76.8	76.8	—	—
77.7	—	—	—
78.7	78.7	—	—
79.6	—	—	—
80.6	80.6	—	—
81.6	—	—	—
82.5	82.5	82	82
83.5	—	—	—
84.5	84.5	—	—
85.6	—	—	—
86.6	86.6	—	—
87.6	—	—	—
88.7	88.7	—	—
89.8	—	—	—
90.9	90.9	91	—
92.0	—	—	—
93.1	93.1	—	—
94.2	—	—	—
95.3	95.3	—	—
96.5	—	—	—
97.6	97.6	—	—
98.8	—	—	—

Note: These values are generally available in multiples of 0.1, 1, 10, 100, 1 k, and 1 M.

MAGNETIC PARAMETER CONVERSIONS		
SI (MKS)	CGS	English
Φ webers (Wb)	maxwells	lines
1 Wb	$= 10^8$ maxwells	$= 10^8$ lines
B Wb/m^2	gauss (maxwells/cm^2)	lines/in.^{2}
1 Wb/m^2	$= 10^4$ gauss	$= 6.452 \times 10^4$ lines/in^2
A 1 m^2	$= 10^4$ cm^2	$= 1550$ in.^{2}
μ_o $4\pi \times 10^{-7}$ Wb/Am	$= 1$ gauss/oersted	$= 3.20$ lines/Am
$\mathcal{F}$ NI(At)	$= 0.4\pi NI$ (gilberts)	NI (At)
1 At	$= 1.257$ gilberts	1 gilbert $= 0.7958$ At
H NI/l (at/m)	$0.4\pi NI/l$ (oersteds)	NI/l (At/in.)
1 At/m	$= 1.26 \times 10^{-2}$ oersted	$= 2.54 \times 10^{-2}$ At/in.
H$_g$ 7.97×10^5 B$_g$ (At/m)	B$_g$ (oersteds)	0.313B$_g$ (At/in.)

APPENDIX C

THE GREEK ALPHABET AND COMMON MEANINGS

Name	Capital	Lower case	Commonly used to designate
Alpha	A	α	Angles, area, coefficients
Beta	B	β	Angles, flux density, coefficients
Gamma	Γ	γ	Conductivity, specific gravity
Delta	Δ	δ	Variation, density
Epsilon	E	ϵ	Base of natural logarithms
Zeta	Z	ζ	Impedance, coefficients, coordinates
Eta	H	η	Hysteresis, coefficients, efficiency
Theta	Θ	θ	Phase angle, temperature
Iota	I	ι	
Kappa	K	κ	Dielectric constant, susceptibility
Lambda	Λ	λ	Wavelength
Mu	M	μ	Micro, amplification factor, permeability reluctivity
Nu	N	ν	
Xi	Ξ	ξ	
Omicron	O	o	
Pi	Π	π	Ratio of circumference to diameter = 3.1416
Rho	P	ρ	Resistivity
Sigma	Σ	σ	Sign of summation
Tau	T	τ	Time constant, time phase displacement
Upsilon	Υ	ν	
Phi	Φ	ϕ	Magnetic flux, angles
Chi	X	χ	
Psi	Ψ	ψ	Dielectric flux, phase difference
Omega	Ω	ω	Capital, ohms; lowercase, angular velocity

DEVICE DATA SHEETS

- Silicon rectifier
- Zener diode
- LED lamp
- Single digit, seven segment LED
- Amplifier transistor
- General purpose transistor
- General purpose JFET
- Unijunction transistor
- Triode thyristor
- Operational amplifier
- Timer
- Quad 2-input AND gate
- Hex inverting gate
- Quad 2-input NAND gate
- Quad 2-input NOR gate

MOTOROLA SEMICONDUCTORS

PO BOX 20912 • PHOENIX, ARIZONA 85036

Designers Data Sheet

LEAD MOUNTED
SILICON RECTIFIERS

50-1000 VOLTS
DIFFUSED JUNCTION

"SURMETIC" RECTIFIERS

. . . subminiature size, axial lead mounted rectifiers for general-purpose low-power applications.

Designers Data for "Worst Case" Conditions

The Designers Data Sheets permit the design of most circuits entirely from the information presented. Limit curves – representing boundaries on device characteristics – are given to facilitate "worst case" design.

*MAXIMUM RATINGS

Rating	Symbol	1N4001	1N4002	1N4003	1N4004	1N4005	1N4006	1N4007	Unit
Peak Repetitive Reverse Voltage Working Peak Reverse Voltage DC Blocking Voltage	V_{RRM} V_{RWM} V_R	50	100	200	400	600	800	1000	Volts
Non-Repetitive Peak Reverse Voltage (halfwave, single phase, 60 Hz)	V_{RSM}	60	120	240	480	720	1000	1200	Volts
RMS Reverse Voltage	$V_{R(RMS)}$	35	70	140	280	420	560	700	Volts
Average Rectified Forward Current (single phase, resistive load, 60 Hz, see Figure 8, T_A = 75°C)	I_O				1.0				Amp
Non-Repetitive Peak Surge Current (surge applied at rated load conditions, see Figure 2)	I_{FSM}				30 (for 1 cycle)				Amp
Operating and Storage Junction Temperature Range	T_J, T_{stg}				-65 to +175				°C

*ELECTRICAL CHARACTERISTICS

Characteristic and Conditions	Symbol	Typ	Max	Unit
Maximum Instantaneous Forward Voltage Drop (I_F = 1.0 Amp, T_J = 25°C) Figure 1	v_F	0.93	1.1	Volts
Maximum Full-Cycle Average Forward Voltage Drop (I_O = 1.0 Amp, T_L = 75°C, 1 inch leads)	$V_{F(AV)}$	–	0.8	Volts
Maximum Reverse Current (rated dc voltage) T_J = 25°C T_J = 100°C	I_R	0.05 1.0	10 50	μA
Maximum Full-Cycle Average Reverse Current (I_O = 1.0 Amp, T_L = 75°C, 1 inch leads	$I_{R(AV)}$	–	30	μA

* Indicates JEDEC Registered Data.

MECHANICAL CHARACTERISTICS

CASE: Transfer Molded Plastic
MAXIMUM LEAD TEMPERATURE FOR SOLDERING PURPOSES: 350°C, 3/8" from case for 10 seconds at 5 lbs. tension
FINISH: All external surfaces are corrosion-resistant, leads are readily solderable
POLARITY: Cathode indicated by color band
WEIGHT: 0.40 Grams (approximately)

DIM	MILLIMETERS		INCHES	
	MIN	MAX	MIN	MAX
A	5.97	6.60	0.235	0.260
B	2.79	3.05	0.110	0.120
D	0.76	0.86	0.030	0.034
K	27.94	–	1.100	–

CASE 59-04
Does Not Conform to DO-41 Outline.

MOTOROLA

Designers Data Sheet

500 MILLIWATT HERMETICALLY SEALED GLASS SILICON ZENER DIODES

- Complete Voltage Range – 2.4 to 110 Volts**
- DO-35 Package – Smaller than Conventional DO-7 Package
- Double Slug Type Construction
- Metallurgically Bonded Construction
- Nitride Passivated Die

Designer's Data for "Worst Case" Conditions

The Designer's Data sheets permit the design of most circuits entirely from the information presented. Limit curves – representing boundaries on device characteristics – are given to facilitate "worst case" design.

*MAXIMUM RATINGS

Rating	Symbol	Value	Unit
DC Power Dissipation @ $T_L \leq 75^oC$	P_D		
Lead Length = 3/8"		500	mW
Derate above $T_L = 75^oC$		4.0	mW/^oC
Operating and Storage Junction Temperature Range	T_J, T_{stg}	–65 to +200	^oC

*Indicates JEDEC Registered Data

**See 1N5273 thru 1N5281 for devices > 110 volts.

MECHANICAL CHARACTERISTICS

CASE: Double slug type, hermetically sealed glass

MAXIMUM LEAD TEMPERATURE FOR SOLDERING PURPOSES: 230^oC, 1/16" from case for 10 seconds

FINISH: All external surfaces are corrosion resistant with readily solderable leads

POLARITY: Cathode indicated by color band. When operated in zener mode, cathode will be positive with respect to anode

MOUNTING POSITION: Any

GLASS ZENER DIODES

500 MILLIWATTS
2.4-110 VOLTS

NOTES:
1. PACKAGE CONTOUR OPTIONAL WITHIN A AND B. HEAT SLUGS, IF ANY, SHALL BE INCLUDED WITHIN THIS CYLINDER, BUT NOT SUBJECT TO THE MINIMUM LIMIT OF B.
2. LEAD DIAMETER NOT CONTROLLED IN ZONE F TO ALLOW FOR FLASH, LEAD FINISH BUILDUP AND MINOR IRREGULARITIES OTHER THAN HEAT SLUGS.
3. POLARITY DENOTED BY CATHODE BAND.
4. DIMENSIONING AND TOLERANCING PER ANSI Y14.5, 1973.

DIM	MILLIMETERS		INCHES	
	MIN	MAX	MIN	MAX
A	3.05	5.08	0.120	0.200
B	1.52	2.29	0.060	0.090
D	0.46	0.56	0.018	0.022
F	–	1.27	–	0.050
K	25.40	38.10	1.000	1.500

All JEDEC dimensions and notes apply.

CASE 299-02
DO-204AH
(DO-35)

STEADY STATE POWER DERATING

ELECTRICAL CHARACTERISTICS

(T_A = 25°C unless otherwise noted. Based on dc measurements at thermal equilibrium; lead length = 3/8''; thermal resistance of heat sink = 30°C/W) V_F = 1.1 max @ I_F = 200 mA for all types.

JEDEC Type No (Note 1)	Nominal Zener Voltage V_Z @ I_{ZT} Volts (Note 2)	Test Current I_{ZT} mA	Max Zener Impedance A and B Suffix only		Max Reverse Leakage Current				Max Zener Voltage Temperature Coeff. (A and B Suffix only) θ_{VZ} (%/°C) (Note 3)
					A and B Suffix only			Non-Suffix	
			Z_{ZT} @ I_{ZT} Ohms	Z_{ZK} @ I_{ZK} = 0.25 mA Ohms	I_R µA	@ V_R Volts		I_R @ V_R Used for Suffix A µA	
						A	B		
1N5221	2.4	20	30	1200	100	0.95	1.0	200	−0.085
1N5222	2.5	20	30	1250	100	0.95	1.0	200	−0.085
1N5223	2.7	20	30	1300	75	0.95	1.0	150	−0.080
1N5224	2.8	20	30	1400	75	0.95	1.0	150	−0.080
1N5225	3.0	20	29	1600	50	0.95	1.0	100	−0.075
1N5226	3.3	20	28	1600	25	0.95	1.0	100	−0.070
1N5227	3.6	20	24	1700	15	0.95	1.0	100	−0.065
1N5228	3.9	20	23	1900	10	0.95	1.0	75	−0.060
1N5229	4.3	20	22	2000	5.0	0.95	1.0	50	·0.055
1N5230	4.7	20	19	1900	5.0	1.9	2.0	50	±0.030
1N5231	5.1	20	17	1600	5.0	1.9	2.0	50	·0.030
1N5232	5.6	20	11	1600	5.0	2.9	3.0	50	+0.038
1N5233	6.0	20	7.0	1600	5.0	3.3	3.5	50	+0.038
1N5234	6.2	20	7.0	1000	5.0	3.8	4.0	50	+0.045
1N5235	6.8	20	5.0	750	3.0	4.8	5.0	30	+0.050
1N5236	7.5	20	6.0	500	3.0	5.7	6.0	30	+0.058
1N5237	8.2	20	8.0	500	3.0	6.2	6.5	30	+0.062
1N5238	8.7	20	8.0	600	3.0	6.2	6.5	30	+0.065
1N5239	9.1	20	10	600	3.0	6.7	7.0	30	+0.068
1N5240	10	20	17	600	3.0	7.6	8.0	30	+0.075
1N5241	11	20	22	600	2.0	8.0	8.4	30	+0.076
1N5242	12	20	30	600	1.0	8.7	9.1	10	+0.077
1N5243	13	9.5	13	600	0.5	9.4	9.9	10	+0.079
1N5244	14	9.0	15	600	0.1	9.5	10	10	+0.082
1N5245	15	8.5	16	600	0.1	10.5	11	10	+0.082
1N5246	16	7.8	17	600	0.1	11.4	12	10	+0.083
1N5247	17	7.4	19	600	0.1	12.4	13	10	+0.084
1N5248	18	7.0	21	600	0.1	13.3	14	10	+0.085
1N5249	19	6.6	23	600	0.1	13.3	14	10	+0.086
1N5250	20	6.2	25	600	0.1	14.3	15	10	+0.086
1N5251	22	5.6	29	600	0.1	16.2	17	10	+0.087
1N5252	24	5.2	33	600	0.1	17.1	18	10	+0.088
1N5253	25	5.0	35	600	0.1	18.1	19	10	+0.089
1N5254	27	4.6	41	600	0.1	20	21	10	+0.090
1N5255	28	4.5	44	600	0.1	20	21	10	+0.091
1N5256	30	4.2	49	600	0.1	22	23	10	+0.091
1N5257	33	3.8	58	700	0.1	24	25	10	+0.092
1N5258	36	3.4	70	700	0.1	26	27	10	+0.093
1N5259	39	3.2	80	800	0.1	29	30	10	+0.094
1N5260	43	3.0	93	900	0.1	31	33	10	+0.095
1N5261	47	2.7	105	1000	0.1	34	36	10	+0.095
1N5262	51	2.5	125	1100	0.1	37	39	10	+0.096
1N5263	56	2.2	150	1300	0.1	41	43	10	+0.096
1N5264	60	2.1	170	1400	0.1	44	46	10	+0.097
1N5265	62	2.0	185	1400	0.1	45	47	10	+0.097
1N5266	68	1.8	230	1600	0.1	49	52	10	+0.097
1N5267	75	1.7	270	1700	0.1	53	56	10	+0.098
1N5268	82	1.5	330	2000	0.1	59	62	10	+0.098
1N5269	87	1.4	370	2200	0.1	65	68	10	+0.099
1N5270	91	1.4	400	2300	0.1	66	69	10	+0.099
1N5271	100	1.3	500	2600	0.1	72	76	10	+0.110
1N5272	110	1.1	750	3000	0.1	80	84	10	+0.110

NOTE 1. Tolerance — The JEDEC type numbers shown indicate a tolerance of ±10% with guaranteed limits on only V_Z, I_R and V_F as shown in the electrical characteristics table. Units with guaranteed limits on all six parameters are indicated by suffix "A" for ±10% tolerance and suffix "B" for ±5.0% units.

†For more information on special selections contact your nearest Motorola representative.

NOTE 2. Special Selections† Available Include:

1. Nominal zener voltages between those shown.

2. Two or more units for series connection with specified tolerance on total voltage. Series matched sets make zener voltages in excess of 200 volts possible as well as providing lower temperature coefficients, lower dynamic impedance and greater power handling ability.

3. Nominal voltages at non-standard test currents.

 MOTOROLA *Semiconductor Products Inc.*

FLV110·FLV140·FLV150·FLV160

RED GaAsP LED LAMPS

OPTOELECTRONICS PRODUCT GROUP

GENERAL DESCRIPTION — The FLV110, FLV140, FLV150 and FLV160 are red light emitting diodes encapsulated in diffused plastic. These LED devices provide an intense large area light source with wide angle viewing. Visual light emission is in the 600 nm to 700 nm range.

- SOLID STATE THUS NO REPLACEMENT REQUIRED
- NO SOCKET REQUIRED
- HIGH ON/OFF CONTRAST
- FLEXIBLE PINS ON FLV110, FLV140 AND FLV150
    - FOR GOOD HEAT SINKING
    - FOR RIGHT ANGLE BENDING
    - FITS STANDARD SOCKETS AND DRILLED HOLES
- HEAVY COPPER LEADS ON FLV160
    - FOR WIRE WRAPPING
    - FOR RIGID STANDOFF FROM PC BOARD
- SINGLE MOLDED BODY ELIMINATES THERMAL CYCLING PROBLEMS
- HIGH TEMPERATURE EPOXY ENCAPUSLATION WITHSTANDS SEVERE ENVIRONMENTAL TEMPERATURES
- LOW POWER CONSUMPTION MEANS IC COMPATIBILITY

PIN CONNECTIONS

ANODE   CATHODE

FLV110
FLV140
FLV150

ANODE   CATHODE

FLV160

ABSOLUTE MAXIMUM RATINGS

Maximum Temperature and Humidity

Storage Temperature	$-55°C$ to $+100°C$
Operating Temperature	$-55°C$ to $+100°C$
Pin Temperature (Soldering, 5 s)	260°C
Relative Humidity at 85°C	85%

Maximum Power Dissipation

Total Dissipation at $T_A = 25°C$	120 mW
Derate Linearly from 25°C	1.6 mW/°C

Maximum Voltage and Currents

V_R	Reverse Voltage	3.0 V
I_F	Forward dc Current	50 mA
i_f	Peak Forward Current (1.0 μs pulse width)	1.0 A

ELECTRICAL AND RADIANT CHARACTERISTICS: $T_A = 25°C$

SYMBOL	CHARACTERISTIC	MIN	TYP	MAX	UNITS	TEST CONDITIONS
V_F	Forward Voltage		1.7	2.0	V	$I_F = 20$ mA
BV_R	Reverse Breakdown Voltage	3.0	8.0		V	$I_R = 10\,\mu$A
I_O	Axial Luminous Intensity	0.8	2.0		mcd	$I_F = 20$ mA
$\theta_{1/2}$	Angle of Half Intensity		± 35		degrees	$I_F = 20$ mA
λ_{pk}	Peak Wavelength		665		nm	$I_F = 20$ mA

TYPICAL ELECTRICAL CHARACTERISTIC CURVES

EMISSION SPECTRUM

FORWARD CURRENT VERSUS FORWARD VOLTAGE

INTENSITY VERSUS FORWARD CURRENT

INTENSITY VERSUS VIEWING ANGLE

INTENSITY VERSUS TEMPERATURE

PEAK WAVELENGTH VERSUS TEMPERATURE

FND350 · FND360 · FND357 · FND367
RED GaAsP 0.362 — INCH SINGLE DIGIT
NUMERIC LED DISPLAY
OPTOELECTRONICS PRODUCT GROUP

GENERAL DESCRIPTION — The FND350, FND360, FND357 and FND367 are red GaAsP single digit 7-segment displays with a 3/8 inch character height. FND357 and FND367 are common cathode configuration. These displays are designed for applications in which the viewer is within fifteen feet of the display. Each digit has a brightness code (05, 06, 07 . . .) for brightness consistency within a display. The FND360 and FND367 are suitable for applications in high-ambient light.

- **COMPACT — 10 DIGITS IN 3" PANEL WIDTH**
- **LOW CURRENT REQUIREMENTS 2-20 mA/SEGMENT**
- **LOW VOLTAGE 1.7 V**
- **INTENSITY CODE MARKING FOR UNIFORM DISPLAYS**
- **RIGHT-HAND DECIMAL**
- **MAXIMIZED CONTRAST RATIO WITH INTEGRAL LENS CAP**
- **FND360, FND367 SUITABLE FOR USE IN HIGH AMBIENT LIGHT**

ABSOLUTE MAXIMUM RATINGS

Maximum Temperature and Humidity

Junction Temperature	-25°C to +85°C
Storage Temperature	-25°C to +85°C
Pin Temperature (Soldering, 5 s)	260°C
Relative Humidity at 65°C	98%

Maximum Voltage and Currents

V_R	Reverse Voltage	3.0 V
$I_{F(Avg)}$	Average Forward Current/Segment or Decimal Point	25 mA
	Derate from 25°C Ambient Temperature	0.3 mA/°C
I_{pk}	Peak Current/Segment or Decimal Point (100 μs pulse) 1000 pps, $T_A = 25$°C	200 mA

PIN CONNECTIONS (FRONT VIEW)

PIN	FND357/367	FND350/360
1	Common Cath.	Common Anode
2	Segment F	Segment F
3	Segment G	Segment G
4	Segment E	Segment E
5	Segment D	Segment D
6	Common Cath.	Common Anode
7	Decimal Point	Decimal Point
8	Segment C	Segment C
9	Segment B	Segment B
10	Segment A	Segment A

ELECTRICAL CHARACTERISTICS: $T_A = 25$°C

SYMBOL	CHARACTERISTICS	MIN	TYP	MAX	UNITS	TEST CONDITIONS
V_F	Forward Voltage		1.7	2.0	V	$I_F = 20$ mA
BV_R	Reverse Breakdown Voltage	3.0	12		V	$I_R = 1.0$ mA
I_0	Axial Luminous Intensity					
	Each Segment (Note 1)					
	FND350, 357	240	450		μcd	$I_F = 20$ mA
	FND360, 367	590	900			
ΔI_0	Intensity Matching, Segment to Segment (Note 3)		±33		%	$I_F = 20$ mA
	Intensity Matching Within One Intensity Class		±20		%	$I_F = 20$ mA, all segments at once
L_0	Average Segment Luminance (Note 2)					
	FND350, 357		26		ftL	$I_F = 20$ mA
	FND360, 367		52			
$\theta_{1/2}$	Viewing Angle to Half Intensity		±27		degrees	
λ_{pk}	Peak Wavelength		665		nm	$I_F = 20$ mA

NOTES:
1. Typical operating current under single digit (dc) drive is approximately 2 to 20 mA average per segment for most ambient light conditions. The 125 mA specification is a representative peak current.
2. Measured on mechanical axis of package. See Average Luminous Intensity curve for other forward currents.
3. Segment-to-segment from average segment intensity.

TYPICAL ELECTRICAL CHARACTERISTIC CURVES

**FORWARD CURRENT VERSUS
FORWARD VOLTAGE**

**MAXIMUM AVERAGE CURRENT
RATING VERSUS
AMBIENT TEMPERATURE**

**ANGULAR DISTRIBUTION
OF LUMINOUS INTENSITY**

**AVERAGE LUMINOUS INTENSITY
VERSUS AVERAGE
FORWARD CURRENT**

**RELATIVE LUMINOUS EFFICIENCY
(mcd PER mA) VERSUS PEAK
CURRENT PER SEGMENT**

**RELATIVE LUMINOUS INTENSITY
VERSUS JUNCTION
TEMPERATURE**

EMISSION SPECTRUM

MAXIMUM RATINGS

Rating	Symbol	Value	Unit
Collector-Emitter Voltage	V_{CEO}	40	Vdc
Collector-Base Voltage	V_{CBO}	4.0	Vdc
Collector Current — Continuous	I_C	100	mAdc
Total Device Dissipation @ $T_A = 25°C$ Derate above 25°C	P_D	625 5.0	mW mW/°C
Total Device Dissipation @ $T_C = 25°C$ Derate above 25°C	P_D	1.5 12	Watt mW/°C
Operating and Storage Junction Temperature Range	T_J, T_{stg}	− 55 to + 150	°C

THERMAL CHARACTERISTICS

Characteristic	Symbol	Max	Unit
Thermal Resistance, Junction to Ambient	$R_{\theta JA}$(1)	200	°C/W
Thermal Resistance, Junction to Case	$R_{\theta JC}$	83.3	°C/W

MPSA20

CASE 29-04, STYLE 1
TO-92 (TO-226AA)

3 Collector
2 Base
1 Emitter

AMPLIFIER TRANSISTOR

NPN SILICON

Refer to MPS3904 for graphs.

ELECTRICAL CHARACTERISTICS (T_A = 25°C unless otherwise noted.)

Characteristic	Symbol	Min	Max	Unit
OFF CHARACTERISTICS				
Collector-Emitter Breakdown Voltage(2) (I_C = 1.0 mAdc, I_B = 0)	$V_{(BR)CEO}$	40	—	Vdc
Emitter-Base Breakdown Voltage (I_E = 100 μAdc, I_C = 0)	$V_{(BR)EBO}$	4.0	—	Vdc
Collector Cutoff Current (V_{CB} = 30 Vdc, I_E = 0)	I_{CBO}	—	100	nAdc
ON CHARACTERISTICS				
DC Current Gain(2) (I_C = 5.0 mAdc, V_{CE} = 10 Vdc)	h_{FE}	40	400	—
Collector-Emitter Saturation Voltage (I_C = 10 mAdc, I_B = 1.0 mAdc)	$V_{CE(sat)}$	—	0.25	Vdc
SMALL-SIGNAL CHARACTERISTICS				
Current-Gain — Bandwidth Product(2) (I_C = 5.0 mAdc, V_{CE} = 10 Vdc, f = 100 MHz	f_T	125	—	MHz
Output Capacitance (V_{CB} = 10 Vdc, I_E = 0, f = 1.0 MHz)	C_{obo}	—	4.0	pF

(1) $R_{\theta JA}$ is measured with the device soldered into a typical printed circuit board.
(2) Pulse Test: Pulse Width ≤ 300 μs, Duty Cycle ≤ 2.0%.

MAXIMUM RATINGS

Rating	Symbol	Value	Unit
Collector-Emitter Voltage	V_{CEO}	−40	Vdc
Collector-Base Voltage	V_{CBO}	−40	Vdc
Emitter-Base Voltage	V_{EBO}	−5.0	Vdc
Collector Current — Continuous	I_C	−200	mAdc
Total Device Dissipation @ T_A = 25°C Derate above 25°C	P_D	625 5.0	mW mW/°C
Total Device Dissipation @ T_C = 60°C	P_D	450	mW
Total Device Dissipation @ T_C = 25°C Derate above 25°C	P_D	1.5 12	Watts mW/°C
Operating and Storage Junction Temperature Range	T_J, T_{stg}	−55 to +150	°C

THERMAL CHARACTERISTICS

Characteristic	Symbol	Max	Unit
Thermal Resistance, Junction to Ambient	$R_{\theta JA}$	200	°C/W
Thermal Resistance, Junction to Case	$R_{\theta JC}$	83.3	°C/W

MPS8093

CASE 29-04, STYLE 1
TO-92 (TO-226AA)

3 Collector

2 Base

1 Emitter

1
2
3

GENERAL PURPOSE TRANSISTOR

PNP SILICON

Refer to 2N4402 for graphs.

ELECTRICAL CHARACTERISTICS (T_A = 25°C unless otherwise noted.)

Characteristic	Symbol	Min	Max	Unit
OFF CHARACTERISTICS				
Collector-Emitter Breakdown Voltage (I_C = −10 mAdc)	$V_{(BR)CEO}$	−40	—	Vdc
Collector-Base Breakdown Voltage (I_C = −100 µAdc)	$V_{(BR)CBO}$	−40	—	Vdc
Emitter-Base Breakdown Voltage (I_E = −100 µAdc)	$V_{(BR)EBO}$	−5.0	—	Vdc
Collector Cutoff Current (V_{CB} = −20 V)	I_{CBO}	—	−100	nAdc
Emitter Cutoff Current (V_{EB} = −3.0 V)	I_{EBO}	—	−100	nAdc
ON CHARACTERISTICS				
DC Current Gain (I_C = −50 mAdc, V_{CE} = −2.0 Vdc)	h_{FE}	100	300	—
Collector-Emitter Saturation Voltage (I_C = −50 mAdc, I_B = −5.0 mAdc)	$V_{CE(sat)}$	—	−0.25	Vdc
Base-Emitter On Voltage (I_C = −50 mAdc, V_{CE} = −2.0 V)	$V_{BE(on)}$	−0.6	−1.0	Vdc

2N5457
thru
2N5459★

CASE 29-04, STYLE 5
TO-92 (TO-226AA)

JFETs
GENERAL PURPOSE

N-CHANNEL — DEPLETION

★These are Motorola
designated preferred devices.

MAXIMUM RATINGS

Rating	Symbol	Value	Unit
Drain-Source Voltage	V_{DS}	25	Vdc
Drain-Gate Voltage	V_{DG}	25	Vdc
Reverse Gate-Source Voltage	V_{GSR}	− 25	Vdc
Gate Current	I_G	10	mAdc
Total Device Dissipation @ T_A = 25°C Derate above 25°C	P_D	310 2.82	mW mW/°C
Junction Temperature Range	T_J	125	°C
Storage Channel Temperature Range	T_{stg}	− 65 to + 150	°C

ELECTRICAL CHARACTERISTICS (T_A = 25°C unless otherwise noted.)

Characteristic		Symbol	Min	Typ	Max	Unit
OFF CHARACTERISTICS						
Gate-Source Breakdown Voltage (I_G = − 10 μAdc, V_{DS} = 0)		$V_{(BR)GSS}$	− 25	—	—	Vdc
Gate Reverse Current (V_{GS} = − 15 Vdc, V_{DS} = 0) (V_{GS} = − 15 Vdc, V_{DS} = 0, T_A = 100°C)		I_{GSS}	— —	— —	− 1.0 − 200	nAdc
Gate Source Cutoff Voltage (V_{DS} = 15 Vdc, I_D = 10 nAdc)	2N5457 2N5458 2N5459	$V_{GS(off)}$	− 0.5 − 1.0 − 2.0	— — —	− 6.0 − 7.0 − 8.0	Vdc
Gate Source Voltage (V_{DS} = 15 Vdc, I_D = 100 μAdc) (V_{DS} = 15 Vdc, I_D = 200 μAdc) (V_{DS} = 15 Vdc, I_D = 400 μAdc)	2N5457 2N5458 2N5459	V_{GS}	— — —	− 2.5 − 3.5 − 4.5	— — —	Vdc
ON CHARACTERISTICS						
Zero-Gate-Voltage Drain Current* (V_{DS} = 15 Vdc, V_{GS} = 0)	2N5457 2N5458 2N5459	I_{DSS}	1.0 2.0 4.0	3.0 6.0 9.0	5.0 9.0 16	mAdc
SMALL-SIGNAL CHARACTERISTICS						
Forward Transfer Admittance Common Source* (V_{DS} = 15 Vdc, V_{GS} = 0, f = 1.0 kHz)	2N5457 2N5458 2N5459	$|Y_{fs}|$	1000 1500 2000	— — —	5000 5500 6000	μmhos
Output Admittance Common Source* (V_{DS} = 15 Vdc, V_{GS} = 0, f = 1.0 kHz)		$|Y_{os}|$	—	10	50	μmhos
Input Capacitance (V_{DS} = 15 Vdc, V_{GS} = 0, f = 1.0 MHz)		C_{iss}	—	4.5	7.0	pF
Reverse Transfer Capacitance (V_{DS} = 15 Vdc, V_{GS} = 0, f = 1.0 MHz)		C_{rss}	—	1.5	3.0	pF

*Pulse Test: Pulse Width ≤ 630 ms; Duty Cycle ≤ 10%.

FIGURE 1 — NOISE FIGURE versus FREQUENCY

FIGURE 2 — NOISE FIGURE versus SOURCE RESISTANCE

FIGURE 3 — TYPICAL DRAIN CHARACTERISTICS
$V_{GS(off)} \cong -1.2$ VOLTS

FIGURE 4 — COMMON SOURCE TRANSFER CHARACTERISTICS
$V_{GS(off)} \cong -1.2$ VOLTS

MOTOROLA

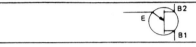

SILICON UNIJUNCTION TRANSISTORS

. . .designed for pulse and timing circuits, sensing circuits, and thyristor trigger circuits. These devices feature:

● Low Peak Point Current — 1.0 μA Typical
● Low Emitter Reverse Current — 5.0 nA Typical
● Passivated Surface for Reliability and Uniformity
● One-Piece Injection-Molded Unibloc‡ Plastic Package for Economy and Reliability
● High η for greater bandwith.

**PN UNIJUNCTION
TRANSISTORS**

MAXIMUM RATINGS ($T_A = 25^{\circ}$ unless otherwise noted)

Rating	Symbol	Value	Unit
RMS Power Dissipation*	P_D*	300	mW
RMS Emitter Current	I_e	50	mA
Peak-Pulse Emitter Current**	I_e**	1.5	Amp
Emitter Reverse Voltage	V_{B2E}	30	Volts
Interbase Voltage†	V_{B2B1}†	35	Volts
Operating Junction Temperature Range	T_J	-55 to +125	$^{\circ}$C
Storage Temperature Range	T_{stg}	-55 to +150	$^{\circ}$C

*Derate 3.0 mW/$^{\circ}$C increase in ambient temperature.
**Duty cycle ≤ 1%, PRR = 10 PPS (see Figure 5).
†Based upon power dissipation at $T_A = 25^{\circ}$C.

**FIGURE 1 – UNIJUNCTION
TRANSISTOR SYMBOL
AND NOMENCLATURE**

**FIGURE 2 – STATIC EMITTER
CHARACTERISTICS CURVES**

STYLE 9
PIN 1 BASE 1
 2 EMITTER
 3 BASE 2

DIM	MILLIMETERS		INCHES	
	MIN	MAX	MIN	MAX
A	4.32	5.33	0.170	0.210
B	4.44	5.21	0.175	0.205
C	3.18	4.19	0.125	0.165
D	0.41	0.56	0.016	0.022
F	0.41	0.48	0.016	0.019
G	1.14	1.40	0.045	0.055
H	–	2.54	–	0.100
J	2.41	2.67	0.095	0.105
K	12.70	–	0.500	–
L	6.35	–	0.250	–
N	2.03	2.92	0.080	0.115
P	2.92	–	0.115	–
R	3.43	–	0.135	–
S	0.36	0.41	0.014	0.016

All JEDEC dimensions and notes apply.

CASE 29-02

2N4870, 2N4871

ELECTRICAL CHARACTERISTICS (T$_A$ = 25°C unless otherwise noted)

Characteristic		Fig. No.	Symbol	Min	Typ	Max	Unit
Intrinsic Standoff Ratio*		4, 7	η*				—
(V$_{B2B1}$ = 10 V) 2N4870				0.56	—	0.75	
2N4871				0.70	—	0.85	
Interbase Resistance		10,11	R$_{BB}$				k ohms
(V$_{B2B1}$ = 3.0 V, I$_E$ = 0)				4.0	6.0	9.1	
Interbase Resistance Temperature Coefficient		11	αR$_{BB}$				%/°C
(V$_{B2B1}$ = 3.0 V, I$_E$ = 0, T$_A$ = –65 to +125°C)				0.10	—	0.90	
Emitter Saturation Voltage**			V$_{EB1(sat)}$**				Volts
(V$_{B2B1}$ = 10 V, I$_E$ = 50 mA)				—	2.5	—	
Modulated Interbase Current			I$_{B2(mod)}$				mA
(V$_{B2B1}$ = 10 V, I$_E$ = 50 mA)				—	15	—	
Emitter Reverse Current		6	I$_{EB2O}$				μA
(V$_{B2E}$ = 30 V, I$_{B1}$ = 0)				—	0.005	1.0	
Peak-Point Emitter Current		8, 9	I$_P$				μA
(V$_{B2B1}$ = 25 V)				—	1.0	5.0	
Valley-Point Current**		12, 13	I$_V$**				mA
(V$_{B2B1}$ = 20 V, R$_{B2}$ = 100 ohms) 2N4870				2.0	5.0	—	
2N4871				4.0	7.0	—	
Base-One Peak Pulse Voltage 2N4870		3, 16	V$_{OB1}$	3.0	6.0	—	Volts
2N4871				5.0	8.0	—	

* η, Intrinsic standoff ratio, is defined in terms of the peak-point voltage, Vp, by means of the equation: $V_p = \eta\, V_{B2B1} + V_F$, where V$_F$ is about 0.49 volt at 25°C @ I$_F$ = 10 μA and decreases with temperature at about 2.5 mV/°C. The test circuit is shown in Figure 4. Components R$_1$, C$_1$, and the UJT form a relaxation oscillator; the remaining circuitry serves as a peak-voltage detector. The forward drop of Diode D$_1$ compensates for V$_R$. To use, the "cal" button is pushed, and R$_3$ is adjusted to make the current meter, M$_1$, read full scale. When the "cal" button is released, the value of η is read directly from the meter, if full scale on the meter reads 1.0.

** Use pulse techniques: PW ≈ 300 μs, duty cycle ≤ 2.0% to avoid internal heating, which may result in erroneous readings.

FIGURE 3 – V$_{OB1}$ TEST CIRCUIT

FIGURE 4 – η TEST CIRCUIT

FIGURE 5 – PRR TEST CIRCUIT AND WAVEFORM

2N4441 thru 2N4444

A O—|◁—O K   G

REVERSE BLOCKING TRIODE THYRISTORS

. . . designed for high-volume consumer phase-control applications such as motor speed, temperature, and light controls and for switching applications in ignition and starting systems, voltage regulators, vending machines, and lamp drivers requiring:

- Small, Rugged, Thermopad Construction — for Low Thermal Resistance, High Heat Dissipation, and Durability.

- Practical Level Triggering and Holding Characteristics @ 25°C
  I_{GT} = 7.0 mA (Typ)
  I_H = 6.0 mA (Typ)

- Low "On" Voltage — V_{TM} = 1.0 Volt (Typ) @ 5.0 Amp @ 25°C

- High Surge Current Rating — I_{TSM} = 80 Amp

SILICON CONTROLLED RECTIFIERS

8.0 AMPERES RMS
50 thru 600 VOLTS

MAXIMUM RATING (T_J = 100°C unless otherwise noted.)

Rating	Symbol	Value	Unit
Peak Repetitive Forward and Reverse Blocking Voltage (Note1) 2N4441	V_{DRM} V_{RRM}	50	Volts
2N4442		200	
2N4443		400	
2N4444		600	
*Non-Repetitive Peak Reverse Blocking Voltage (t = 5.0 ms (max) duration)	V_{RSM}		Volts
2N4441		75	
2N4442		300	
2N4443		500	
2N4444		700	
*RMS On-State Current (All Conduction Angles)	$I_{T(RMS)}$	8.0	Amp
Average On-State Current, T_C = 73°C	$I_{T(AV)}$	5.1	Amp
*Peak Non-Repetitive Surge Current (1/2 cycle, 60 Hz preceded and followed by rated current and voltage)	I_{TSM}	80	Amp
Circuit Fusing (T_J = −40 to +100°C; t = 1.0 to 8.3 ms)	I^2t	25	A^2s
*Peak Gate Power	P_{GM}	5.0	Watts
*Average Gate Power	$P_{G(AV)}$	0.5	Watt
*Peak Forward Gate Current	I_{GM}	2.0	Amp
*Peak Reverse Gate Voltage	V_{RGM}	10	Volts
*Operating Junction Temperature Range	T_J	−40 to +100	°C
*Storage Temperature Range	T_{stg}	−40 to +150	°C
Mounting Torque (6-32 screw) (Note 2)	—	8.0	in. lb.

THERMAL CHARACTERISTICS

Characteristic	Symbol	Typ	Max	Unit
*Thermal Resistance, Junction to Case	$R_{\theta JC}$	—	2.5	°C/W
Thermal Resistance, Junction to Ambient	$R_{\theta JA}$	40	—	°C/W

*Indicates JEDEC Registered Data.

Notes 1, 2, See page 2

STYLE 1:
PIN 1. CATHODE
2. ANODE
3. GATE

NOTES:
1. DIM "D" UNCONTROLLED IN ZONE "H"
2. DIM "F" DIA THRU
3. HEAT SINK CONTACT AREA (BOTTOM)
4. LEADS WITHIN 0.005" RAD OF TRUE POSITION (TP) AT MAXIMUM MATERIAL CONDITION.

DIM	MILLIMETERS		INCHES	
	MIN	MAX	MIN	MAX
A	16.13	16.38	0.635	0.645
B	12.57	12.83	0.495	0.505
C	3.18	3.43	0.125	0.135
D	1.09	1.24	0.043	0.049
F	3.51	3.76	0.138	0.148
G	4.22 BSC		0.166 BSC	
H	2.67	2.92	0.105	0.115
J	0.813	0.864	0.032	0.034
K	15.11	16.38	0.595	0.645
M	9° TYP		9° TYP	
Q	4.70	4.95	0.185	0.195
R	1.91	2.16	0.075	0.085
U	6.22	6.48	0.245	0.255
V	2.03	—	0.080	—

CASE 90-06

2N4441 thru 2N4444

ELECTRICAL CHARACTERISTICS ($T_C = 25°C$ unless otherwise noted)

Characteristic	Symbol	Min	Typ	Max	Unit
Peak Forward Blocking Current (V_D = Rated V_{DRM}, $T_J = 100°C$, gate open)	I_{DRM}	–	–	2.0	mA
Peak Reverse Blocking Current (V_D = Rated V_{RRM}, $T_J = 100°C$, gate open)	I_{RRM}	–	–	2.0	mA
Gate Trigger Current (Continuous dc) (V_D = 7.0 Vdc, R_L = 100 Ohms) $\quad\quad T_C = 25°C$ $^*\quad T_C = -40°C$	I_{GT}	– 	7.0 	30 60	mA
Gate Trigger Voltage (Continuous dc) (V_D = 7.0 Vdc, R_L = 100 Ohms) $\quad\quad T_C = 25°C$ (V_D = 7.0 Vdc, R_L = 100 Ohms) $\quad\quad T_C = -40°C$ (V_D = Rated V_{DRM}, R_L = 100 Ohms) $\quad\quad T_J = 100°C$	V_{GT}	– – 0.2	0.75 – –	1.5 2.5 –	Volts
Peak On-State Voltage (Pulse Width = 1.0 to 2.0 ms, Duty Cycle ≤ 2.0%) $\quad\quad (I_{TM}$ = 5.0 A peak) $^*\quad (I_{TM}$ = 15.7 A peak)	V_{TM}	– –	1.0 –	1.5 2.0	Volts
Holding Current (V_D = 7.0 Vdc, gate open) $\quad\quad T_C = 25°C$ $^*\quad T_C = -40°C$	I_H	– –	6.0 –	40 70	mA
Gate Controlled Turn-On Time (I_{TM} = 5.0 A, I_{GT} = 20 mA, V_D = Rated V_{DRM})	t_{gt}	–	1.0	–	μs
Circuit Commutated Turn-Off Time (I_{TM} = 5.0 A, I_R = 5.0 A) (I_{TM} = 5.0 A, I_R = 5.0 A, $T_J = 100°C$)	t_q	– –	15 20	– –	μs
Critical Rate of Rise of Off-State Voltage (V_D = Rated V_{DRM}, Exponential Waveform, $T_J = 100°C$, Gate Open)	dv/dt	–	50	–	V/μs

*Indicates JEDEC Registered Data

Note 1.   Ratings apply for zero or negative gate voltage but positive gate voltage shall not be applied concurrently with a negative potential on the anode. When checking forward or reverse blocking capability, thyristor devices should not be tested with a constant current source in a manner that the voltage applied exceeds the rated blocking voltage.

Note 2.   Torque rating applies with use of torque washer (Shakeproof WD19522 #6 or equivalent). Mounting torque in excess of 8 in. lbs. does not appreciably lower case-to-sink thermal resistance. Anode lead and heatsink contact pad are common.

For soldering purposes (either terminal connection or device mounting), soldering temperatures shall not exceed +225°C.

MC1741, MC1741C
MC1741N, MC1741NC

OPERATIONAL AMPLIFIER
SILICON MONOLITHIC
INTEGRATED CIRCUIT

INTERNALLY COMPENSATED, HIGH PERFORMANCE OPERATIONAL AMPLIFIERS

. . . designed for use as a summing amplifier, integrator, or amplifier with operating characteristics as a function of the external feedback components.

- No Frequency Compensation Required
- Short-Circuit Protection
- Offset Voltage Null Capability
- Wide Common-Mode and Differential Voltage Ranges
- Low-Power Consumption
- No Latch Up
- Low Noise Selections Offered — N Suffix

G SUFFIX
METAL PACKAGE
CASE 601

(Top View)

P1 SUFFIX
PLASTIC PACKAGE
CASE 626
(MC1741C,MC1741NC)

U SUFFIX
CERAMIC PACKAGE
CASE 693

(Top View)

MAXIMUM RATINGS (T_A = +25°C unless otherwise noted)

Rating	Symbol	MC1741C	MC1741	Unit
Power Supply Voltage	V_{CC}	+18	+22	Vdc
	V_{EE}	-18	-22	Vdc
Input Differential Voltage	V_{ID}	±30		Volts
Input Common Mode Voltage (Note 1)	V_{ICM}	+15		Volts
Output Short Circuit Duration (Note 2)	t_S	Continuous		
Operating Ambient Temperature Range	T_A	0 to +70	-55 to +125	°C
Storage Temperature Range	T_{stg}			°C
Metal, Flat and Ceramic Packages		-65 to +150		
Plastic Packages		-55 to +125		

Note 1   For supply voltages less than + 15 V, the absolute maximum input voltage is equal to the supply voltage.
Note 2   Supply voltage equal to or less than 15 V.

L SUFFIX
CERAMIC PACKAGE
CASE 632
TO 116

P2 SUFFIX
PLASTIC PACKAGE
CASE 646
(MC1741C,MC1741NC)

NC		NC
NC		NC
Offset Null		NC
Inputs		V_{CC}
		Output
V_{EE}		Offset Null
NC		NC

(Top View)

EQUIVALENT CIRCUIT SCHEMATIC

F SUFFIX
CERAMIC PACKAGE
CASE 606-04
TO-91

NC
Offset Null
Inputs
V_{EE}

NC
NC
V_{CC}
Output
Offset Null

(Top View)

ELECTRICAL CHARACTERISTICS (V_{CC} = 15 V, V_{EE} = 15 V, T_A = 25°C unless otherwise noted).

Characteristic	Symbol	MC1741 Min	MC1741 Typ	MC1741 Max	MC1741C Min	MC1741C Typ	MC1741C Max	Unit
Input Offset Voltage ($R_S \leq$ 10 k)	V_{IO}	–	1.0	5.0	–	2.0	6.0	mV
Input Offset Current	I_{IO}	–	20	200	–	20	200	nA
Input Bias Current	I_{IB}	–	80	500	–	80	500	nA
Input Resistance	r_i	0.3	2.0	–	0.3	2.0	–	MΩ
Input Capacitance	C_i	–	1.4	–	–	1.4	–	pF
Offset Voltage Adjustment Range	V_{IOR}	–	±15	–	–	±15	–	mV
Common Mode Input Voltage Range	V_{ICR}	±12	±13	–	±12	±13	–	V
Large Signal Voltage Gain (V_O = ±10 V, $R_L \geq$ 2.0 k)	A_v	50	200	–	20	200	–	V/mV
Output Resistance	r_o	–	75	–	–	75		Ω
Common Mode Rejection Ratio ($R_S \leq$ 10 k)	CMRR	70	90	–	70	90		dB
Supply Voltage Rejection Ratio ($R_S \leq$ 10 k)	PSRR		30	150	–	30	150	µV/V
Output Voltage Swing	V_O							V
($R_L \geq$ 10 k)		±12	±14		±12	±14		
($R_L \geq$ 2 k)		±10	±13	–	±10	±13	–	
Output Short-Circuit Current	I_{os}	–	20	–	–	20	–	mA
Supply Current	I_D	–	1.7	2.8	–	1.7	2.8	mA
Power Consumption	P_C	–	50	85	–	50	85	mW
Transient Response (Unity Gain — Non-Inverting)								
(V_I = 20 mV, $R_L \geq$ 2 k, $C_L \leq$ 100 pF) Rise Time	t_{TLH}	–	0.3	–	–	0.3		µs
(V_I = 20 mV, $R_L \geq$ 2 k, $C_L \leq$ 100 pF) Overshoot	os	–	15	–		15	–	%
(V_I = 10 V, $R_L \geq$ 2 k, $C_L \leq$ 100 pF) Slew Rate	SR	–	0.5	–		0.5	–	V/µs

ELECTRICAL CHARACTERISTICS (V_{CC} = 15 V, V_{EE} = 15 V, T_A = *T_{high} to T_{low} unless otherwise noted.)

Characteristic	Symbol	MC1741 Min	MC1741 Typ	MC1741 Max	MC1741C Min	MC1741C Typ	MC1741C Max	Unit
Input Offset Voltage ($R_S \leq$ 10 kΩ)	V_{IO}		1.0	6.0	–	–	7.5	mV
Input Offset Current	I_{IO}							nA
(T_A = 125°C)		–	7.0	200	–			
(T_A = -55°C)			85	500	–	–	–	
(T_A = 0°C to +70°C)		–	–	–	–		300	
Input Bias Current	I_{IB}							nA
(T_A = 125°C)		–	30	500	–	–	–	
(T_A = -55°C)		–	300	1500	–	–	–	
(T_A = 0°C to +70°C)		–	–	–	–		800	
Common Mode Input Voltage Range	V_{ICR}	±12	±13	–	–	–	–	V
Common Mode Rejection Ratio ($R_S \leq$ 10 k)	CMRR	70	90	–	–		–	dB
Supply Voltage Rejection Ratio ($R_S \leq$ 10 k)	PSRR		30	150	–	–	–	µV/V
Output Voltage Swing	V_O							V
($R_L \geq$ 10 k)		±12	±14	–	–	–	–	
($R_L \geq$ 2 k)		±10	±13	–	±10	±13	–	
Large Signal Voltage Gain ($R_L \geq$ 2 k, V_{out} = ±10 V)	A_v	25	–	–	15	–		V/mV
Supply Currents	I_D							mA
(T_A = 125°C)		–	1.5	2.5				
(T_A = -55°C)		–	2.0	3.3				
Power Consumption (T_A = +125°C)	P_C	–	45	75				mW
(T_A = -55°C)			60	100				

*T_{high} = 125°C for MC1741 and 70°C for MC1741C
 T_{low} = -55°C for MC1741 and 0°C for MC1741C

MOTOROLA *Semiconductor Products Inc.*

TYPICAL CHARACTERISTICS

($V_{CC} = +15$ Vdc, $V_{EE} = -15$ Vdc, $T_A = +25°C$ unless otherwise noted)

FIGURE 6 – POWER BANDWIDTH
(LARGE SIGNAL SWING versus FREQUENCY)

FIGURE 7 – OPEN LOOP FREQUENCY RESPONSE

FIGURE 8 – POSITIVE OUTPUT VOLTAGE SWING
versus LOAD RESISTANCE

FIGURE 9 – NEGATIVE OUTPUT VOLTAGE SWING
versus LOAD RESISTANCE

FIGURE 10 – OUTPUT VOLTAGE SWING versus
LOAD RESISTANCE (Single Supply Operation)

FIGURE 11 – SINGLE SUPPLY INVERTING AMPLIFIER

MOTOROLA *Semiconductor Products Inc.*

NOISE CHARACTERISTICS (Applies for MC1741N and MC1741NC only, V_{CC} = 15 V, V_{EE} = –15 V, T_A = +25°C)

Characteristic	Symbol	MC1741N			MC1741NC			Unit
		Min	Typ	Max	Min	Typ	Max	
Burst Noise (Popcorn Noise) (BW = 1.0 Hz to 1.0 kHz, t = 10 s, R_S = 100 k) (Input Referenced)	E_n	–	–	20	–	–	20	μV/peak

FIGURE 1 — BURST NOISE versus SOURCE RESISTANCE

FIGURE 2 — RMS NOISE versus SOURCE RESISTANCE

FIGURE 3 — OUTPUT NOISE versus SOURCE RESISTANCE

FIGURE 4 — SPECTRAL NOISE DENSITY

FIGURE 5 — BURST NOISE TEST CIRCUIT (N Suffixed Devices Only)

For applications where low noise performance is essential, selected devices denoted by an N suffix are offered. These units have been 100% tested for burst noise pulses on a special noise test system. Unlike conventional peak reading or RMS meters, this system was especially designed to provide the quick response time essential to burst (popcorn) noise testing.

The test time employed is 10 seconds and the 20 μV peak limit refers to the operational amplifier input thus eliminating errors in the closed-loop gain factor of the operational amplifier under test

 MOTOROLA *Semiconductor Products Inc.*

SE555F,H,N,N-14 • SE555C,F,H,N,N-14 • NE555F,H,N,N-14

FEATURES

- Turn off time less than 2μs
- Maximum operating frequency greater than 500kHz
- Timing from microseconds to hours
- Operates in both astable and monostable modes
- High output current
- Adjustable duty cycle
- TTL compatible
- Temperature stability of 0.005% per °C
- SE555 Mil std 883A,B,C available M38510 (JAN) approved, M38510 processing available.

PIN CONFIGURATIONS

N PACKAGE

GROUND	1	8	V_CC
TRIGGER	2	7	DISCHARGE
OUTPUT	3	6	THRESHOLD
RESET	4	5	CONTROL VOLTAGE

APPLICATIONS

- Precision timing
- Pulse generation
- Sequential timing
- Time delay generation
- Pulse width modulation
- Pulse position modulation
- Missing pulse detector

ABSOLUTE MAXIMUM RATINGS

PARAMETER	RATING	UNIT
Supply voltage		
SE555	+18	V
NE555, SE555C,	+16	V
Power dissipation	600	mW
Operating temperature range		
NE555	0 to +70	°C
SE555, SE555C	–55 to +125	°C
Storage temperature range	–65 to +150	°C
Load temperature (soldering, 60sec)	300	°C

EQUIVALENT SCHEMATIC

F,N-14 PACKAGE

GND	1	14	V_CC
NC	2	13	NC
TRIGGER	3	12	DISCHARGE
OUTPUT	4	11	NC
NC	5	10	THRESHOLD
RESET	6	9	NC
NC	7	8	CONTROL VOLTAGE

H PACKAGE

BLOCK DIAGRAM

SE555F,H,N,N-14 • SE555C,F,H,N,N-14 • NE555F,H,N,N-14

DC ELECTRICAL CHARACTERISTICS $T_A = 25°C$, V_{CC} = +5V to +15 unless otherwise specified.

PARAMETER	TEST CONDITIONS	SE555			NE555/SE555C			UNIT
		Min	Typ	Max	Min	Typ	Max	
Supply voltage		4.5		18	4.5		16	V
Supply current (low state)[1]	V_{CC} = 5V R_L = ∞		3	5		3	6	mA
	V_{CC} = 15V R_L = ∞		10	12		10	15	mA
Timing error (monostable)	R_A = 2KΩ to 100KΩ							
Initial accuracy[2]	C = 0.1μF		0.5	2.0		1.0	3.0	%
Drift with temperature			30	100		50		ppm/°C
Drift with supply voltage			0.05	0.2		0.1	0.5	%/V
Timing error (astable)	R_A, R_B = 1kΩ to 100kΩ							
Initial accuracy[2]	C = 0.1μF		1.5			2.25		%
Drift with temperature	V_{CC} = 15V		90			150		ppm/°C
Drift with supply voltage			0.15			0.3		%/V
Control voltage level	V_{CC} = 15V	9.6	10.0	10.4	9.0	10.0	11 0	V
	V_{CC} = 5V	2.9	3.33	3.8	2.6	3.33	4.0	V
Threshold voltage	V_{CC} = 15V	9.4	10.0	10.6	8.8	10.0	11.2	V
	V_{CC} = 5V	2.7	3.33	4.0	2.4	3.33	4.2	V
Threshold current[3]			0.1	0.25		0.1	0.25	μA
Trigger voltage	V_{CC} = 15V	4.8	5.0	5.2	4.5	5.0	5.6	V
	V_{CC} = 5V	1.45	1.67	1.9	1.1	1.67	2.2	V
Trigger current	V_{TRIG} = 0V		0.5	0.9		0.5	2.0	μA
Reset voltage[4]		0.4	0.7	1.0	0.4	0.7	1.0	V
Reset current			0.1	0.4		0.1	0.4	mA
Reset current	V_{RESET} = 0V		0.4	1.0		0.4	1.5	mA
Output voltage (low)	V_{CC} = 15V							
	I_{SINK} = 10mA		0.1	0.15		0.1	0.25	V
	I_{SINK} = 50mA		0.4	0.5		0.4	0.75	V
	I_{SINK} = 100mA		2.0	2.2		2.0	2.5	V
	I_{SINK} = 200mA		2.5			2.5		V
	V_{CC} = 5V							
	I_{SINK} = 8mA		0.1	0.25		0.3	0.4	V
	I_{SINK} = 5mA		0.05	0.2		0.25	0.35	V
Output voltage (high)	V_{CC} = 15V							
	I_{SOURCE} = 200mA		12.5			12.5		V
	I_{SOURCE} = 100mA	13.0	13.3		12.75	13.3		V
	V_{CC} = 5V							
	I_{SOURCE} = 100mA	3.0	3.3		2.75	3.3		V
Turn off time[5]	V_{RESET} = V_{CC}		0.5	2.0		0.5		μs
Rise time of output			100	200		100	300	ns
Fall time of output			100	200		100	300	ns
Discharge leakage current			20	100		20	100	na

NOTES

1. Supply current when output high typically 1mA less.
2. Tested at V_{CC} = 5V and V_{CC} = 15V.
3. This will determine the maximum value of $R_A + R_B$, for 15V operation, the max total
   R = 10 megohm, and for 5V operation, the max total R = 3.4 megohm.
4. Specified with trigger input high.
5. Time measured from a positive going input pulse from 0 to 0 8xV_{CC} into the threshold
   to the drop from high to low of the output. Trigger is tied to threshold.

DM54S08/DM74S08 Quad 2-Input AND Gates

General Description
This device contains four independent gates each of which performs the logic AND function.

Connection Diagram

Dual-In-Line Package

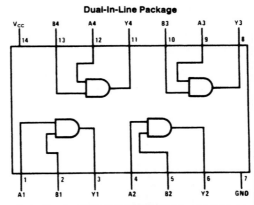

TL/F/6444-1

Order Number DM54S08J, DM54S08W or DM74S08N
See NS Package Number J14A, N14A or W14B

Function Table

$Y = AB$

Inputs		Output
A	B	Y
L	L	L
L	H	L
H	L	L
H	H	H

H = High Logic Level
L = Low Logic Level

Absolute Maximum Ratings (Note)

If Military/Aerospace specified devices are required, please contact the National Semiconductor Sales Office/Distributors for availability and specifications.

Supply Voltage	7V
Input Voltage	5.5V
Operating Free Air Temperature Range	
DM54S	−55°C to +125°C
DM74S	0°C to +70°C
Storage Temperature Range	−65°C to +150°C

Note: *The "Absolute Maximum Ratings" are those values beyond which the safety of the device cannot be guaranteed. The device should not be operated at these limits. The parametric values defined in the "Electrical Characteristics" table are not guaranteed at the absolute maximum ratings. The "Recommended Operating Conditions" table will define the conditions for actual device operation.*

Recommended Operating Conditions

Symbol	Parameter	DM54S08			DM74S08			Units
		Min	Nom	Max	Min	Nom	Max	
V_{CC}	Supply Voltage	4.5	5	5.5	4.75	5	5.25	V
V_{IH}	High Level Input Voltage	2			2			V
V_{IL}	Low Level Input Voltage			0.8			0.8	V
I_{OH}	High Level Output Current			−1			−1	mA
I_{OL}	Low Level Output Current			20			20	mA
T_A	Free Air Operating Temperature	−55		125	0		70	°C

Electrical Characteristics over recommended operating free air temperature (unless otherwise noted)

Symbol	Parameter	Conditions		Min	Typ (Note 1)	Max	Units
V_I	Input Clamp Voltage	V_{CC} = Min, I_I = −18 mA				−1.2	V
V_{OH}	High Level Output Voltage	V_{CC} = Min, I_{OH} = Max V_{IL} = Max	DM54	2.5	3.4		V
			DM74	2.7	3.4		
V_{OL}	Low Level Output Voltage	V_{CC} = Min, I_{OL} = Max V_{IH} = Min				0.5	V
I_I	Input Current @ Max Input Voltage	V_{CC} = Max, V_I = 5.5V				1	mA
I_{IH}	High Level Input Current	V_{CC} = Max, V_I = 2.7V				50	µA
I_{IL}	Low Level Input Current	V_{CC} = Max, V_I = 0.5V				−2	mA
I_{OS}	Short Circuit Output Current	V_{CC} = Max (Note 2)	DM54	−40		−100	mA
			DM74	−40		−100	
I_{CCH}	Supply Current with Outputs High	V_{CC} = Max			18	32	mA
I_{CCL}	Supply Current with Outputs Low	V_{CC} = Max			32	57	mA

Switching Characteristics at V_{CC} = 5V and T_A = 25°C (See Section 1 for Test Waveforms and Output Load)

Symbol	Parameter	R_L = 280Ω				Units
		C_L = 15 pF		C_L = 50 pF		
		Min	Max	Min	Max	
t_{PLH}	Propagation Delay Time Low to High Level Output	2.5	7	3	9	ns
t_{PHL}	Propagation Delay Time High to Low Level Output	2.5	7.5	3	11	ns

Note 1: All typicals are at V_{CC} = 5V, T_A = 25°C.
Note 2: Not more than one output should be shorted at a time, and the duration should not exceed one second.

National Semiconductor

DM54S04/DM74S04 Hex Inverting Gates

General Description

This device contains six independent gates each of which performs the logic INVERT function.

Connection Diagram

Dual-In-Line Package

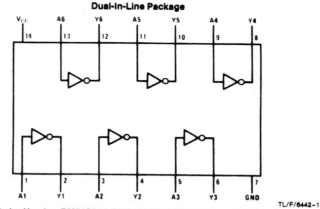

TL/F/6442–1

Order Number DM54S04J, DM54S04W, DM74S04M or DM74S04N
See NS Package Number J14A, M14A, N14A or W14B

Function Table

$Y = \overline{A}$

Input	Output
A	Y
L	H
H	L

H = High Logic Level
L = Low Logic Level

Absolute Maximum Ratings (Note)

If Military/Aerospace specified devices are required, please contact the National Semiconductor Sales Office/Distributors for availability and specifications.

Supply Voltage	7V
Input Voltage	5.5V
Operating Free Air Temperature Range	
DM54S	$-55°C$ to $+125°C$
DM74S	$0°C$ to $+70°C$
Storage Temperature Range	$-65°C$ to $+150°C$

Note: *The "Absolute Maximum Ratings" are those values beyond which the safety of the device cannot be guaranteed. The device should not be operated at these limits. The parametric values defined in the "Electrical Characteristics" table are not guaranteed at the absolute maximum ratings. The "Recommended Operating Conditions" table will define the conditions for actual device operation.*

Recommended Operating Conditions

Symbol	Parameter	DM54S04			DM74S04			Units
		Min	Nom	Max	Min	Nom	Max	
V_{CC}	Supply Voltage	4.5	5	5.5	4.75	5	5.25	V
V_{IH}	High Level Input Voltage	2			2			V
V_{IL}	Low Level Input Voltage			0.8			0.8	V
I_{OH}	High Level Output Current			-1			-1	mA
I_{OL}	Low Level Output Current			20			20	mA
T_A	Free Air Operating Temperature	-55		125	0		70	°C

Electrical Characteristics over recommended operating free air temperature range (unless otherwise noted)

Symbol	Parameter	Conditions		Min	Typ. (Note 1)	Max	Units
V_I	Input Clamp Voltage	V_{CC} = Min, I_I = -18 mA				-1.2	V
V_{OH}	High Level Output Voltage	V_{CC} = Min, I_{OH} = Max V_{IL} = Max	DM54	2.5	3.4		V
			DM74	2.7	3.4		
V_{OL}	Low Level Output Voltage	V_{CC} = Min, I_{OL} = Max V_{IH} = Min				0.5	V
I_I	Input Current @ Max Input Voltage	V_{CC} = Max, V_I = 5.5V				1	mA
I_{IH}	High Level Input Current	V_{CC} = Max, V_I = 2.7V				50	μA
I_{IL}	Low Level Input Current	V_{CC} = Max, V_I = 0.5V				-2	mA
I_{OS}	Short Circuit Output Current	V_{CC} = Max (Note 2)	DM54	-40		-100	mA
			DM74	-40		-100	
I_{CCH}	Supply Current with Outputs High	V_{CC} = Max			15	24	mA
I_{CCL}	Supply Current with Outputs Low	V_{CC} = Max			30	54	mA

Switching Characteristics at V_{CC} = 5V and T_A = 25°C (See Section 1 for Test Waveforms and Output Load)

Symbol	Parameter	R_L = 280Ω				Units
		C_L = 15 pF		C_L = 50 pF		
		Min	Max	Min	Max	
t_{PLH}	Propagation Delay Time Low to High Level Output	2	4.5	2	7	ns
t_{PHL}	Propagation Delay Time High to Low Level Output	2	5	2	8	ns

Note 1: All typicals are at V_{CC} = 5V, T_A = 25°C.
Note 2: Not more than one output should be shorted at a time, and the duration should not exceed one second.

DM54S00/DM74S00 Quad 2-Input NAND Gates

General Description

This device contains four independent gates each of which performs the logic NAND function.

Connection Diagram

Dual-In-Line Package

TL/F/6489-1

Order Number DM54S00J, DM54S00W, DM74S00M or DM74S00N
See NS Package Number J14A, M14A, N14A or W14B

Function Table

$$Y = \overline{AB}$$

Inputs		Output
A	**B**	**Y**
L	L	H
L	H	H
H	L	H
H	H	L

H = High Logic Level
L = Low Logic Level

Absolute Maximum Ratings (Note)

If Military/Aerospace specified devices are required, please contact the National Semiconductor Sales Office/Distributors for availability and specifications.

Supply Voltage	7V
Input Voltage	5.5V
Operating Free Air Temperature Range	
DM54S	−55°C to +125°C
DM74S	0°C to +70°C
Storage Temperature Range	−65°C to +150°C

Note: *The "Absolute Maximum Ratings" are those values beyond which the safety of the device cannot be guaranteed. The device should not be operated at these limits. The parametric values defined in the "Electrical Characteristics" table are not guaranteed at the absolute maximum ratings. The "Recommended Operating Conditions" table will define the conditions for actual device operation.*

Recommended Operating Conditions

Symbol	Parameter	DM54S00			DM74S00			Units
		Min	Nom	Max	Min	Nom	Max	
V_{CC}	Supply Voltage	4.5	5	5.5	4.75	5	5.25	V
V_{IH}	High Level Input Voltage	2			2			V
V_{IL}	Low Level Input Voltage			0.8			0.8	V
I_{OH}	High Level Output Current			−1			−1	mA
I_{OL}	Low Level Output Current			20			20	mA
T_A	Free Air Operating Temperature	−55		125	0		70	°C

Electrical Characteristics over recommended operating free air temperature range (unless otherwise noted)

Symbol	Parameter	Conditions		Min	Typ (Note 1)	Max	Units
V_I	Input Clamp Voltage	V_{CC} = Min, I_I = −18 mA				−1.2	V
V_{OH}	High Level Output Voltage	V_{CC} = Min, I_{OH} = Max V_{IL} = Max	DM54	2.5	3.4		V
			DM74	2.7	3.4		
V_{OL}	Low Level Output Voltage	V_{CC} = Min, I_{OL} = Max V_{IH} = Min				0.5	V
I_I	Input Current @ Max Input Voltage	V_{CC} = Max, V_I = 5.5V				1	mA
I_{IH}	High Level Input Current	V_{CC} = Max, V_I = 2.7V				50	μA
I_{IL}	Low Level Input Current	V_{CC} = Max, V_I = 0.5V				−2	mA
I_{OS}	Short Circuit Output Current	V_{CC} = Max (Note 2)	DM54	−40		−100	mA
			DM74	−40		−100	
I_{CCH}	Supply Current with Outputs High	V_{CC} = Max			10	16	mA
I_{CCL}	Supply Current with Outputs Low	V_{CC} = Max			20	36	mA

Switching Characteristics at V_{CC} = 5V and T_A = 25°C (See Section 1 for Test Waveforms and Output Load)

Symbol	Parameter	R_L = 280Ω				Units
		C_L = 15 pF		C_L = 50 pF		
		Min	Max	Min	Max	
t_{PLH}	Propagation Delay Time Low to High Level Output	2	4.5	2	7	ns
t_{PHL}	Propagation Delay Time High to Low Level Output	2	5	2	8	ns

Note 1: All typicals are at V_{CC} = 5V, T_A = 25°C.

Note 2: Not more than one output should be shorted at a time and the duration should not exceed one second.

DM54S02/DM74S02 Quad 2-Input NOR Gates

General Description

This device contains four independent gates each of which performs the logic NOR function.

Connection Diagram

Dual-In-Line Package

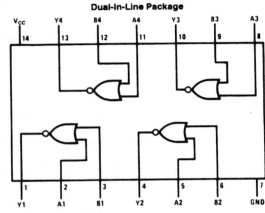

TL/F/6490-1

Order Number DM54S02J, DM54S02W or DM74S02N
See NS Package Number J14A, N14A or W14B

Function Table

$$Y = \overline{A + B}$$

Inputs		Output
A	**B**	**Y**
L	L	H
L	H	L
H	L	L
H	H	L

H = High Logic Level

L = Low Logic Level

Absolute Maximum Ratings (Note)

If Military/Aerospace specified devices are required, please contact the National Semiconductor Sales Office/Distributors for availability and specifications.

Supply Voltage	7V
Input Voltage	5.5V
Operating Free Air Temperature Range	
DM54S	$-55°C$ to $+125°C$
DM74S	$0°C$ to $+70°C$
Storage Temperature Range	$-65°C$ to $+150°C$

Note: *The "Absolute Maximum Ratings" are those values beyond which the safety of the device cannot be guaranteed. The device should not be operated at these limits. The parametric values defined in the "Electrical Characteristics" table are not guaranteed at the absolute maximum ratings. The "Recommended Operating Conditions" table will define the conditions for actual device operation.*

Recommended Operating Conditions

Symbol	Parameter	DM54S02			DM74S02			Units
		Min	Nom	Max	Min	Nom	Max	
V_{CC}	Supply Voltage	4.5	5	5.5	4.75	5	5.25	V
V_{IH}	High Level Input Voltage	2			2			V
V_{IL}	Low Level Input Voltage			0.8			0.8	V
I_{OH}	High Level Output Current			-1			-1	mA
I_{OL}	Low Level Output Current			20			20	mA
T_A	Free Air Operating Temperature	-55		125	0		70	°C

Electrical Characteristics over recommended operating free air temperature (unless otherwise noted)

Symbol	Parameter	Conditions		Min	Typ (Note 1)	Max	Units
V_I	Input Clamp Voltage	$V_{CC} = $ Min, $I_I = -18$ mA				-1.2	V
V_{OH}	High Level Output Voltage	$V_{CC} = $ Min, $I_{OH} = $ Max $V_{IL} = $ Max	DM54	2.5	3.4		V
			DM74	2.7	3.4		
V_{OL}	Low Level Output Voltage	$V_{CC} = $ Min, $I_{OL} = $ Max $V_{IH} = $ Min				0.5	V
I_I	Input Current @ Max Input Voltage	$V_{CC} = $ Max, $V_I = 5.5$V				1	mA
I_{IH}	High Level Input Current	$V_{CC} = $ Max, $V_I = 2.7$V				50	μA
I_{IL}	Low Level Input Current	$V_{CC} = $ Max, $V_I = 0.5$V				-2	mA
I_{OS}	Short Circuit Output Current	$V_{CC} = $ Max (Note 2)	DM54	-40		-100	mA
			DM74	-40		-100	
I_{CCH}	Supply Current with Outputs High	$V_{CC} = $ Max			17	29	mA
I_{CCL}	Supply Current with Outputs Low	$V_{CC} = $ Max			26	45	mA

Switching Characteristics at $V_{CC} = 5$V and $T_A = 25°C$ (See Section 1 for Test Waveforms and Output Load)

Symbol	Parameter	$R_L = 280\Omega$				Units
		$C_L = 15$ pF		$C_L = 50$ pF		
		Min	Max	Min	Max	
t_{PLH}	Propagation Delay Time Low to High Level Output	1.5	5.5	2	7.5	ns
t_{PHL}	Propagation Delay Time High to Low Level Output	1.5	5.5	2	7.5	ns

Note 1: All typicals are at $V_{CC} = 5$V, $T_A = 25°C$.

Note 2: Not more than one output should be shorted at a time, and the duration should not exceed one second.

ANSWERS FOR ODD-NUMBERED QUESTIONS AND PROBLEMS

Chapter 1	**1–1**	**(b)** 600 B.C.
	1–3	**(c)** Charge
	1–5	**(c)** Electromagnetism
	1–7	**(b)** Michael Faraday
	1–9	**(a)** Walter Schottky
	1–11	**(c)** Computer
	1–13	**(d)** 52.4×10^3
	1–15	**(b)** Inertia
	1–17	**(d)** Matter
	1–19	**(c)** J.J. Thomson
	1–21	**(a)** Bound electrons
	1–23	**(d)** Permissible energy levels
	1–25	**(b)** A positive charge and a negative charge brought close together
	1–27	**(a)** 10^{-4} **(b)** 10^5 **(c)** 10^{-8} **(d)** 10^8
	1–29	3.64×10^{-4} m
	1–31	**(a)** 6,770,000 **(b)** 0.00325 **(c)** 846,000,000 **(d)** 0.000173
	1–33	**(a)** 10^3 **(b)** 10^5 **(c)** 10^{-9} **(d)** 10^{12} **(e)** 10^3
	1–35	**(a)** 3.40×10^{10} **(b)** 1.08×10^{-2} **(c)** 2.17×10^{-8} **(d)** 2.76×10^{16}
	1–37	**(a)** $1.248 \times 10^{20}/6.24 \times 10^{18} = 20$ C **(b)** $1.56 \times 10^{18}/6.24 \times 10^{18} = 0.25$ C
		(c) $1.872 \times 10^{19}/6.24 \times 10^{18} = 3$ C **(d)** $3.12 \times 10^{19}/6.24 \times 10^{18} = 5$ C
Chapter 2	**2–1**	**(b)** Electrons move in a random pattern
	2–3	**(c)** 41 mA
	2–5	**(d)** From the negative terminal of a power supply to the positive
	2–7	**(c)** Joule
	2–9	**(d)** Voltage rise
	2–11	**(d)** Good conductor
	2–13	**(c)** The reciprocal of resistance
	2–15	**(c)** Used as jumpers on printed circuit boards
	2–17	**(a)** Taper
	2–19	**(a)** Between 5% and 10%
	2–21	$t = 0.375$ s
	2–23	$R = 3.39 \ \Omega$
	2–25	$\rho = 0.16 \ \Omega\cdot$m
	2–27	**(a)** 0.0164 in. **(b)** 0.0742 in. **(c)** 0.0346 in. **(d)** 0.0286 in.
	2–29	$R_{60°} = 5.15 \ \Omega$
	2–31	$R_{20°} = 2.08 \ \Omega$

2–33	$R = 1.61\ \Omega$	
2–35	$20{,}000 \pm 5\% = 19{,}000 - 21{,}000\ \Omega$	
2–37	$6.4\ \Omega \pm 20\%$	
2–39	Blue, gray, green, gold	

3–1	**(c)** Voltage and current	**Chapter 3**
3–3	**(a)** Voltage, current, and resistance	
3–5	**(c)** Ability to do work	
3–7	**(b)** 100%	
3–9	$55.8\ \Omega$	
3–11	6 V	
3–13	9 V	
3–15	The current also doubles	
3–17	1440 W	
3–19	17,600 W or 17.6 kW	
3–21	$I = 0.83$ A, $R = 144\ \Omega$	
3–23	0.5 A	
3–25	0.83 hp	
3–27	3,600,000 J	
3–29	72,000 kW·h	
3–31	**(a)** $2.88 **(b)** $3.12 **(c)** $8.064 **(d)** $0.90	

4–1	**(d)** Only one current path	**Chapter 4**
4–3	**(b)** Decreases	
4–5	**(b)** With high internal resistance	
4–7	**(d)** All of the above	
4–9	**(d)** Series aiding	
4–11	**(c)** Allows the load to operate at a lower value of voltage than the source	
4–13	**(a)** Internal resistance	
4–15	**(b)** Current increases	
4–17	$110\ \Omega$	
4–19	18.46 μA	
4–21	26.13 V	
4–23	42.15 V	
4–25	$P_1 = 117.76$ W, $P_2 = 80.64$ W, $P_3 = 57.6$ W, $P_T = 256$ W	
4–27	$V_{R1} = 41.32$ V, $V_{R2} = 19.38$ V	
4–29	$R_s = 3\ k\Omega$, $P_s = 21.36$ mV	
4–31	$1.25\ \Omega$	

5–1	**(c)** More than one current path	**Chapter 5**
5–3	**(b)** Doubles	
5–5	**(b)** Two resistors in parallel	
5–7	**(c)** Is 10 times greater than the other resistor	
5–9	**(c)** A nodal equation	
5–11	**(a)** Total current increases	
5–13	**(a)** The greatest amount of heat	
5–15	$156.7\ \Omega$	
5–17	$10\ k\Omega$	
5–19	0.62 A	
5–21	$I_1 = 13.33$ A, $I_2 = 6.67$ A	
5–23	$781.25\ \Omega$	
5–25	$27.7\ \Omega$	

5–27	$P_1 = 28.8$ W, $P_2 = 72$ W	
5–29	$P_1 = 12$ W, $P_2 = 4.8$ W, $P_T = 16.8$ W	
5–31	**(a)** $I_1 = 0.21$ mA, $I_2 = 0.5$ A, $I_3 = 0.833$ A	**(b)** 64.17 Ω **(c)** 225 W
5–33	**(a)** $P_1 = 25$ mW, $P_2 = 11.36$ mW	**(b)** 36.36 mW

Chapter 6

6–1	**(c)** Combinations of series and parallel resistors
6–3	**(b)** Two series networks connected in parallel
6–5	**(d)** All of the above
6–7	**(a)** 107 mA
6–9	**(b)** 4.5 V
6–11	**(c)** Bleeder current
6–13	**(a)** Positive and negative voltages
6–15	**(b)** Current
6–17	**(b)** Bleeder resistors
6–19	1920 Ω
6–21	30 mA
6–23	4 V
6–25	121.4 mW
6–27	$R_3 = 50$ Ω, $R_1 = 4.55$ Ω, $R_2 = 8.33$ Ω
6–29	40.82 kΩ
6–31	765 mA
6–33	$V_{R1} = 1.12$ V, $V_{R2} = 8.88$ V

Chapter 7

7–1	**(b)** Two magnetic fields
7–3	**(a)** Shunt
7–5	**(c)** Low-voltage, low-resistance
7–7	**(b)** 0 A
7–9	**(d)** Ω/V
7–11	**(b)** Decreases
7–13	**(a)** Center
7–15	**(d)** Parallax
7–17	**(c)** 3 1/2 digit display
7–19	184.8 mA
7–21	$R_1 = 3.95$ kΩ, $R_2 = 19.95$ kΩ, $R_3 = 39.95$ kΩ
7–23	85 kΩ/V

Chapter 8

8–1	**(b)** Current or voltage
8–3	**(c)** Resistors
8–5	**(c)** Add the voltage in the KVL equation
8–7	**(a)** 1 less
8–9	**(c)** Resistance
8–11	**(d)** All of the above
8–13	**(c)** Sum
8–15	**(a)** Voltage sources are shorted and current sources are open circuited
8–17	**(d)** Low efficiency
8–19	$I_1 = 17.14$ mA, $I_2 = 7.5$ mA
8–21	2.29 A
8–23	53.33 Ω
8–25	4.01 V
8–27	115.2 W
8–29	708.93 μA
8–31	25.67 mW

8–33 $V_{R1} = 2.86$ V, $V_{R2} = 22.86$ V, $V_{R3} = 17.14$ V
8–35 33.1 W
8–37 $I_1 = 16$ mA, $I_2 = 6.42$ mA, $I_3 = 10.04$ mA, $I_4 = 16.46$ mA
8–39 $I_1 = 4.08$ A, $I_2 = 2.14$ A, $I_3 = 2.77$ A, $I_4 = 1.94$ A, $I_5 = 0.63$ A
8–41 1.39 A
8–43 1.22 mA
8–45 $E_{TH} = 6$ V, $R_{TH} = 3$ Ω
8–47 (a) $I_L = 0.41$ mA, $V_{RL} = 19.27$ V   (b) $I_L = 0.33$ mA, $V_{RL} = 22.44$ V
8–49 7.72 mA
8–51 1.01 A
8–53 (a) $R_L = 2.8$ Ω   (b) 12.86 W
8–55 (a) 0.8 Ω   (b) 4 Ω   (c) 0.8 Ω   (d) 83.33%

9–1 (b) Iron   **Chapter 9**
9–3 (c) Tiny magnets
9–5 (b) Crystals
9–7 (a) Magnetic flux
9–9 (b) High permeability
9–11 (d) Is a permanent magnet
9–13 (a) Oersted
9–15 (b) Permanent magnet
9–17 0.03 T
9–19 1.74×10^{-9} Wb
9–21 2.75 T

10–1 (b) Weber   **Chapter 10**
10–3 (c) Reluctance
10–5 (a) Permeance
10–7 (a) Magnetizing force increases
10–9 (b) Increase reluctance
10–11 (b) Sum of individual reluctances
10–13 (a) Increase reluctance
10–15 (c) Fringing decreases
10–17 186.57 At/Wb
10–19 38.82×10^4 At/Wb
10–21 10.53×10^{-3} T
10–23 3.1 T
10–25 (a) 1704.55 At/m   (b) 0.47 T   (c) 29.37×10^{-5} Wb

11–1 (d) Can be easily raised or lowered in value   **Chapter 11**
11–3 (d) Generators and oscillators
11–5 (a) Peak value
11–7 (a) Radians per second
11–9 (c) Radio waves
11–11 (a) Time period
11–13 (c) Amplitude
11–15 (d) rms
11–17 (a) Pulse train
11–19 75.40 rad
11–21 99.7°
11–23 30.53 μs
11–25 250 m

	11–27	2.61 A
	11–29	8.455 V
	11–31	(a) 2.12 A (b) 1.5 A
	11–33	45 μs
	11–35	77.8%

Chapter 12

12–1	(c) Signal-measuring and signal-generating instruments
12–3	(d) The peak voltage
12–5	(c) Both AC and DC voltages
12–7	(d) Graticule
12–9	(b) Provide a stable waveform display
12–11	(c) Oscillator
12–13	(d) Audio and radio
12–15	(a) 30 V (b) 15 V (c) 10.6 V

12–17 (a) $E_{p\text{-}p} = (10 \text{ V/division})(4/1) = 40 \text{ V}$ (b) $E_p = 1/2E_{p\text{-}p} = 0.5(40 \text{ V}) = 20 \text{ V}$
(c) $E_{rms} = 0.707E_p = 0.707(20 \text{ V}) = 14.14 \text{ V}$

12–19 (a) +20 V, −30 V   (b) $T = 0.025 \text{ s}, f = 40 \text{ Hz}$

Chapter 13

13–1	(a) Voltage
13–3	(d) Charge
13–5	(b) Doubles
13–7	(c) Dielectric constant
13–9	(c) Transient state
13–11	(c) Coupling
13–13	(b) Dielectric leakage
13–15	816 μC
13–17	16.25 V
13–19	4.68×10^{-9} C
13–21	0.35×10^{-12} F = 0.35 nF
13–23	2000 μF
13–25	1.215×10^{-12} F = 1.215 pF
13–27	0.55 mm
13–29	680,000 pF = 0.68 μF ± 10%, 100 V
13–31	22 μs
13–33	220 μF
13–35	0.89 s
13–37	30 μF
13–39	1.65 s
13–41	8.87 ms
13–43	0.24 s
13–45	(a) 1.852 μF (b) $V_1 = 55.56$ V, $V_2 = V_3 = 4.44$ V

Chapter 14

14–1	(b) Michael Faraday
14–3	(a) Faraday's law
14–5	(b) Low self-inductance
14–7	(d) Henry
14–9	(c) Coefficient of coupling
14–11	(b) Flux
14–13	355 μH
14–15	3.5 H
14–17	310.24 mH

14–19	307.2 H
14–21	8.33 H
14–23	**(a)** 0.75 H **(b)** 0.125
14–25	299.72 mH
14–27	33.96 mA
14–29	16.24 V
14–31	2.3 J

15–1	**(c)** The same frequency
15–3	**(b)** Core type and shell type
15–5	**(c)** The high-voltage winding is the primary
15–7	**(c)** Directly proportional to
15–9	**(a)** Load resistance equals the internal resistance of the source
15–11	**(d)** Eddy currents
15–13	**(c)** Variacs
15–15	**(a)** 10% and 90% of the pulse amplitude
15–17	373.85 V
15–19	7.51 mWb
15–21	3.2 kΩ
15–23	2.83
15–25	87.5%

Chapter 15

16–1	**(d)** Impedance
16–3	**(a)** Inductive reactance
16–5	**(b)** Capacitive reactance
16–7	**(d)** Ohms
16–9	**(a)** Resistance
16–11	**(d)** Real power
16–13	**(d)** Apparent power
16–15	**(b)** Directly proportional to frequency
16–17	58.76 V
16–19	8.89 H
16–21	$15 \angle{-45°}$ A
16–23	173.06 V
16–25	**(a)** 205.88 Ω **(b)** 0.59 A **(c)** 0.42 A **(d)** 1.01 A
16–27	1.16 μF
16–29	**(a)** 442.1 Ω **(b)** 54.29 mA **(c)** $V_{C1} = 14.4$ V, $V_{C2} = 9.6$ V
16–31	$400 \angle{35°}$ Ω
16–33	**(a)** $2234.5 \angle{57.5°}$ Ω **(b)** $V_R = 111.76$ V, $V_L = 175.43$ V
16–35	2.2 μF
16–37	$2235 \angle{10.2°}$ Ω
16–39	$R_s = 8.3$ Ω, $X_s = 5.54$ Ω
16–41	120.64 VAR

Chapter 16

17–1	**(c)** Said to be damped
17–3	**(b)** Low impedance, high current
17–5	**(c)** Selectivity
17–7	**(a)** Resistance
17–9	**(d)** Flywheel effect
17–11	**(b)** Steep response curve
17–13	**(d)** Level of sound and reference level

Chapter 17

17–15	(a) 5.07 µF	(b) 10 A
17–17	63.67 kHz	
17–19	(a) 50.33 kHz	(b) 3.16	(c) 15.93 kHz
17–21	37 dB	
17–23	20 W	

Chapter 18

18–1	(b) Removes unwanted signals
18–3	(b) Inefficient power transfer
18–5	(a) Insertion loss
18–7	(b) 50% of the input power
18–9	(a) At its minimum
18–11	(c) A doubling of frequency
18–13	(a) Inductors
18–15	(a) Tenfold change in frequency
18–17	1496 Hz
18–19	Assume $R = 1$ kΩ, $L = 19.89$ mH
18–21	$f_r = 1220.7$ Hz, $BW = 127.3$ Hz
18–23	5.03 kHz

Chapter 19

19–1	(c) A crystal lattice
19–3	(a) Zero
19–5	(d) Extrinsic conductivity
19–7	(a) Majority carriers
19–9	(b) Barrier potential
19–11	(a) Junction resistance will vary inversely as the current through the PN junction varies
19–13	(d) Anode and cathode
19–15	(d) Acts like an open switch
19–17	(d) Alternating current to direct current for 360° per cycle
19–19	(b) Double the input frequency
19–21	(a) Equal the peak voltage
19–23	(d) Variable-voltage capacitors
19–25	9.3 mA
19–27	6.53 V
19–29	(a) 6.36 V	(b) 12.73 V
19–31	6.75 V
19–33	14.1%
19–35	20%

Chapter 20

20–1	(b) Vacuum tubes
20–3	(a) Emitter
20–5	(d) $I_E = I_C + I_B$
20–7	(c) Alpha
20–9	(b) Cutoff
20–11	(d) Drain, source, gate
20–13	(c) N-channel and P-channel
20–15	(d) Transconductance
20–17	(d) SCRs and triacs
20–19	(c) Triacs allow current flow regardless of the biasing voltage and polarity of gate current
20–21	(d) Used primarily in switching circuits

20–23	(a) 200	(b) 0.995		
20–25	(a) 2.19 V	(b) 1.49 V	(c) 1.49 mA	(d) 5.08 V	(d) 3.59 V
20–27	85.8			
20–29	118.5 Hz			

21–1 (b) Stage
21–3 (a) AC input signals and DC biasing voltages
21–5 (c) Distorted
21–7 (a) Emitter current and the output is the collector
21–9 (c) Very high
21–11 (c) Common-collector amplifier
21–13 (d) Buffers
21–15 (c) Common-collector
21–17 (a) Product of the individual gains
21–19 (c) Crossover distortion
21–21 77.6
21–23 30.92 kΩ
21–25 9.3 mV
21–27 10,201
21–29 2.65 V

22–1 (c) High-gain, direct-coupled multistage amplifier
22–3 (b) Differential amplifier
22–5 (a) Open loop
22–7 (c) Common-emitter amplifier
22–9 (c) Unity gain, high input impedance, and low output impedance
22–11 (a) Comparator
22–13 (b) Single-pole filter
22–15 (a) $18\ V_{rms}$ (b) $9.6\ V_{DC}$
22–17 +9.39 V
22–19 4.82 kHz
22–21 1053.8 Hz

23–1 (c) Total number of individual symbols
23–3 (b) Bit
23–5 (b) Decimal to binary conversion
23–7 (a) Series circuit
23–9 (c) Inverter
23–11 (c) DeMorgan
23–13 (d) SR, D-type, and JK
23–15 (c) Only three usable states
23–17 (c) 512
23–19 (a) Be set
23–21 173
23–23 1DF
23–25 265
23–29 $(\overline{AB}) \cdot (C + D) = Z$
23–31 $(A + B) + (BC) + (\overline{CD}) = Z$

A	B	C	D	Z
0	0	0	0	1
0	0	0	1	1
0	0	1	0	1
0	0	1	1	1
0	1	0	0	1
0	1	0	1	1
0	1	1	0	1
0	1	1	1	1
1	0	0	0	1
1	0	0	1	1
1	0	1	0	1
1	0	1	1	0
1	1	0	0	1
1	1	0	1	1
1	1	1	0	1
1	1	1	1	1

23–33 $A \cdot B = Z$

GLOSSARY

Absolute permittivity (ϵ) The flux produced with a vacuum as dielectric; also known as absolute capacity.

AC resistance The effective resistance of a conductor; includes factors such as skin effect and radiation loss.

Active device A device, such as a transistor, capable of controlling voltage or current.

Active filter A filter network that uses an active device to obtain the desired filtering effect.

Additive polarity A winding connection that causes the direction of counter emf in both windings to be the same.

Admittance (Y) The ease at which an AC current flows in a circuit; the reciprocal of impedance, measured in siemens (S).

Admittance triangle A right-angle triangle that relates conductance, susceptance, and admittance.

Air-core inductor An inductor that contains no magnetic iron and generally is wound on a tubular insulating material.

Air gap A part of the magnetic circuit of electromagnets, and used to increase reluctance.

Alternating current (AC) A current in which the magnitude and direction varies with time.

Alternating emf A voltage in which the magnitude and direction varies with time.

American Wire Gauge (AWG) A standard for manufacturing and numbering wires.

Ammeter A measuring instrument used to indicate electrical current in amperes.

Ampere (A) The SI base unit of electric current; the rate of electric charge flow when 1 C of charge passes a given point in 1 s.

Ampere's circuit law A law that states the algebraic sum of the rises and drops of the mmf around a closed loop of a magnetic circuit is equal to zero.

Amplifier A device capable of increasing the voltage, current, or power of an electrical signal.

Amplitude The maximum positive or negative value of an alternating current, voltage, or power; also known as peak value.

Analog meter A moving-coil measuring instrument.

AND gate A logic gate that only will produce an output voltage when all inputs receive a signal; equivalent to an electrical series circuit.

Angular velocity The rate of change of a quantity, such as voltage, in an AC circuit.

Anode The positive element of an electrical device or battery.

Apparent power The product of the total rms voltage and current in a circuit, expressed in volt-amperes (VA).

Arcing A phenomenon caused by interrupting current, such as opening a switch, that produces a very high induced voltage because of the rapidly collapsing field.

Argument The angular displacement between two quantities such as between current and voltage.

Atom The smallest particle of an element that can exist alone or in combination.

Attenuation A reduction in the magnitude of a quantity, such as voltage or current.

Autotransformer A transformer having one winding that is common to both primary and secondary sides.

Avalanche breakdown A condition produced when the reverse-bias voltage across a PN junction is increased beyond the tolerances of the semiconductor material.

Average power The average of instantaneous power over a complete cycle, expressed in watts (W).

Average value The mean value over an integral number of repetitions. When calculating average values of sine waves, only one-half of the cycle is used.

Band A specific range of frequencies.

Bandpass The width of a band of frequencies.

Band-pass filter A circuit used to allow a specific range of frequencies to pass and to attenuate all other frequencies.

Band-stop filter A circuit designed to attenuate a specific range of frequencies; also known as band-reject filter and notch filter.

Bandwidth *(BW)* The total number of cycles below and above resonant frequency for which the current is equal to or greater than 70.7% of its resonant value.

Barrier potential The difference of potential, or charge, at the PN junction.

Base *See* Radix.

Battery A series connection of voltaic primary cells or secondary cells.

Bel The unit of measure for the logarithm of the ratio of two voltages, currents, or power levels.

B–H curve A curve that illustrates the various stages of magnetization of ferrous material; also that illustrates the demagnetization of material.

Bias voltage Any voltage that is connected across a PN junction.

Binary coded decimal (BCD) A modification of the decimal number system where the decimal digits are independently coded as four-bit binary numbers.

Binary number system A number system based on two digits: 1 and 0.

Bipolar junction transistor (BJT) A three-terminal device whose emitter-to-collector conduction is controlled by its base terminal current.

Bit A single binary digit.

Bleeder resistor A resistor that divides the voltage into different values from a single source.

Bode plot A straight-line approximation of the magnitude and phase angle response to frequency.

Bound electrons The electrons very close to the nucleus and tightly held in their orbit.

Breakdown The point at which current begins to flow in an insulator.

Breakdown voltage The amount of reverse-biased voltage required to produce a sharp increase in reverse current flow.

Breakover voltage The amount of voltage required to switch a device into a conducting state.

Breakpoints The points at which the gain drops to 3 dB below the mid-frequency gain.

Bridge rectifier An electronic circuit that converts alternating current to direct current.

Buffer An electronic circuit that performs impedance matching or isolation functions.

Butterworth filter A two-pole filter that produces a roll-off rate of −40 dB/decade.

Bypass capacitor A capacitor connected in an electronic circuit to block DC signals and pass AC signals.

Byte A group of eight binary digits.

Capacitance *(C)* The property of an electric circuit to oppose any change in voltage across the circuit; measured in farads (F).

Capacitive reactance (X_C) The opposition of capacitance to alternating current.

Capacitor *(C)* A device capable of storing electrical energy, constructed of two conductor materials separated by an insulator.

Cascading An amplifier configuration where the output of one amplifier is connected to the input of another amplifier.

Cathode (1) The negative terminal of an electrical device or battery. (2) The part of an electronic device from which electrons are emitted.

Cathode ray tube (CRT) An electron-beam tube in which the beam, varied in position and intensity to produce a visible pattern, is focused to a small cross section on a luminescent screen.

Cell A chemical source of electrical energy consisting of two electrodes in an electrolyte.

Center-tap rectifier A rectifier circuit that uses two diodes and a center-tap transformer to produce full-wave rectification.

Center-tap transformer A transformer with two separate secondary voltages that are 180° out of phase with each other.

Channel The path connecting the source and drain in a FET.

Charge An accumulation or deficiency of electrons.

Charge curve An exponential curve showing the increase in charge across an inductor or capacitor from zero to maximum.

Circuit breaker A device designed to physically interrupt the flow of current in a circuit.

Circular mil (CM) The area of a circular cross section having a diameter of 1 mil.

Class A amplifier An amplifier that is capable of responding to positive and negative signals without becoming saturated or cutoff.

Class AB amplifier An amplifier that is biased partially on to reduce crossover distortion.

Class B amplifier An amplifier with its Q point biased at cutoff.

Class C amplifier An amplifier that produces the highest efficiency and highest distortion of the three classes of operation.

Clipping Amplitude distortion that is primarily caused by improper biasing and excessive input signals.

Coefficient of coupling *(k)* The degree of closeness with which the primary and secondary windings are coupled.

Coercive force The demagnetizing force necessary to remove the residual flux from the magnetic material.

Color code band A method of determining resistance, capacitance, and tolerance of electrical and electronic components such as resistors and capacitors.

Common-base amplifier A BJT amplifier with a low voltage gain and high current gain.

Common-collector amplifier A BJT amplifier in which the signal is introduced into the base-collector portion of the transistor and extracted at the emitter; also known as an emitter follower amplifier.

Common-drain amplifier A FET amplifier with similar characteristics to a common-collector amplifier; also known as a source follower amplifier.

Common-emitter amplifier A BJT amplifier capable of providing high values of both voltage and current gain.

Common-gate amplifier A FET amplifier used primarily to transform a low input impedance to a high output impedance.

Common mode rejection ratio (CMRR) A measure of a differential amplifier's ability to eliminate, or reject, common signals such as unwanted noise or hum.

Common-source amplifier A FET amplifier with high input impedance, medium output impedance, and a voltage gain greater than 1.

Common winding The winding of an autotransformer that has a physical connection between the primary and secondary.

Comparator A circuit that is used to detect whether the voltage at one input is higher or lower than the voltage at another input.

Complex circuit A circuit that contains two or more sources in different branches; or a circuit that combines series and parallel elements in various interconnections.

Conductance *(G)* The ability of resistance to pass alternating current; the reciprocal of resistance, measured in siemens (S).

Conduction band A band that contains electrons with an energy level so high that the electrons are not attached or bound to any atom.

Conductivity The reciprocal of resistivity; the conductance per unit length and cross section of a material.

Conductor A material that offers a low resistance to the passage of electric current.

Conventional current A current flow, adopted by the IEEE, in which current is defined as the direction in which positive charge carriers flow through a circuit.

Copper loss A term loosely applied to the I^2R loss that occurs in a conductor because of the resistance of the conductor.

Coulomb *(C)* The SI unit of charge: the quantity of electric charge possessed by 6.24×10^{18} electrons.

Coulomb constant *(k)* A proportionality constant whose value depends on what material fills the volume of space in which the bodies are located.

Coulomb's law A law that states the force of attraction or repulsion between two charged bodies is directly proportional to the square of the distance between them.

Counter A device that produces a numeric count each time an input clock pulse occurs.

Counter emf The emf that is generated in an electric circuit by a change of current within the circuit itself.

Coupling A method of connecting two circuits so that they have a common impedance that allows the transfer of energy from one circuit to another.

Coupling capacitor A capacitor that will pass AC signals and block DC signals.

Critical frequency *See* Cutoff frequency.

Crossover distortion Distortion caused by the breakover voltage of a PN junction.

CRT *See* Cathode ray tube.

Crystal An assembly of molecules having a definite internal structure and the external form of a solid enclosed by a number of symmetrical plane faces.

Current A movement of electric charge, the direction of which is taken arbitrarily as the movement of positive charges and opposite to the movement of negative charges.

Current divider rule A rule that states the amount of current in one of two parallel resistances is calculated by multiplying their total current by the other resistance and dividing by their sum.

Current limiting resistor A resistor used specifically to limit the flow of current in a circuit.

Current magnification factor *See* Q factor.

Current source A device capable of supplying current.

Current tracer A test instrument used to detect a changing current in a digital circuit.

Cutoff frequency A specific frequency where the signal is attenuated.

Cycle One complete set of positive and negative values of alternating current or voltage; the smallest nonrepetitive portion of a periodic wave.

Damping (1) Any means employed to keep moving parts from oscillating. (2) The reduction in Q factor by reducing the resistance of the parallel-resonant circuit.

Damping resistor A variable resistor used in parallel-resonant circuits to reduce the Q of the circuit.

Darlington amplifier A high-gain, two-stage BJT amplifier.

D'Arsonval galvanometer A very sensitive electrical instrument used in detecting extremely small currents.

D'Arsonval principle A principle that states when current is fed through a moving coil, the resulting magnetic field reacts with the magnetic field of a permanent magnet and causes the coil to rotate.

Decade A tenfold change in frequency.

Decade box A laboratory device with a set of dials that can be rotated to provide a desired value of resistance.

Decibel (dB) A logarithmic unit used to express gain or loss in signal level; equal to ten times the common logarithm of a power ratio.

Decoding The conversion of binary digits to decimal.

DeMorgan's theorems Theorems that state the complement of a sum equals the product of the complements, and the complement of a product equals the sum of the complements.

Depletion mode A biasing condition for a FET in which the device is designed to operate at drain current levels below the drain current at the pinchoff point.

Depletion region The interchange of holes and electrons that cancels out the current carriers in the vicinity of the PN junction.

Derived units The units that are not selected as fundamental but are derived from the fundamental.

Diac A two-terminal bidirectional thyristor that is capable of conducting current in both directions.

Diamagnetic material A material that exhibits a very slight opposition to magnetic lines of force.

Dielectric Insulating material used to store electrical charges.

Dielectric constant (ϵ_r) The property that determines the electrostatic energy stored per unit volume for unit potential gradient.

Dielectric field *See* Electrostatic field.

Dielectric flux (ψ) The total strength of an electric field.

Dielectric hysteresis An effect in a dielectric material caused by changes in orientation of electron orbits in the dielectric.

Dielectric strength The voltage per unit thickness at which breakdown occurs.

Differential amplifier An amplifier consisting of two direct-coupled transistors designed to amplify the difference between the two input terminals.

Differentiator An electronic circuit that produces an output waveform whose value is equal to the rate of change of the input at any instant in time.

Digital counter A sophisticated frequency meter that is capable of measuring frequencies over an extremely wide range.

Digital multimeter (DMM) A solid state electronic instrument that measures electrical quantities and displays the measured value in decimal numeric form.

Diode An electronic device that permits current flow in one direction only.

Diode ratings The limiting values assigned by a manufacturer to various parameters of a diode.

Direct current (DC) An unchanging, unidirectional current.

Discharge curve An exponential curve showing the decrease in charge across an inductor or capacitor from maximum to zero.

Dissipation factor The ratio of energy dissipated to the energy stored in an element for one cycle.

DMM *See* Digital multimeter.

Domain A microscopic needle-shaped crystal that contains a large number of spinning electrons.

Doping The process of adding impurities to semiconductors.

Dot convention (1) A notation used to represent the direction of flux produced by a coil. (2) A method of identifying the terminals of transformer windings where terminals with the same instantaneous polarity are identified as dots and the other terminals are left blank.

Dot matrix display A visual display device designed to provide alphanumeric readouts.

Double-subscript notation A notation in which the first subscript designates the point at which a value is measured with respect to a reference point, which is designated by a second subscript.

D-type flip-flop A flip-flop with four usable states, two of which are the same.

Dual A parallelism between the equations and theorems of electric circuits.

Duty cycle A characteristic of a pulse waveform that is a ratio of the conducting time to the period of one cycle.

Dynamic resistance The internal emitter resistance of a BJT, which is not a fixed value but is dependent on the current flow in the circuit.

Eddy current A circulating current caused by induced emf within a magnetic material.

Effective resistance The total of all resistive effects for an AC circuit; obtained by dividing the total losses caused by current (i.e., hysteresis loss) by the square of the rms value of current.

Effective value *See* rms value.

Efficiency A measure of how completely the power put into a device is used as output.

Elastance The opposition to the setting up of electric lines of force in an electric insulator or dielectric.

Electric current *See* Current.

Electric flux density (*D*) *See* Flux density.

Electric potential The amount of work required to move a unit charge from one point to another.

Electrodynamometer A meter that is capable of measuring both AC and DC voltages and currents.

Electromagnet A magnet excited by a current in a coil surrounding a ferromagnetic core.

Electromagnetic induction The process by which an electromotive force is induced, or generated, in an electric circuit when there is a change in the magnetic flux linking the circuit.

Electromotive force (emf) The force that tends to cause a movement of charge carriers in a conductor; commonly known as voltage.

Electron An elementary particle containing the smallest negative electric charge.

Electron flow The movement of electrons past a given point from the negative terminal of the power source, through the load, to the positive terminal of the source.

Electron shell A group of several closely spaced permissible energy levels that may be occupied by orbiting electrons.

Electrostatic field The field surrounding a charged body.

Electrostatic force The force that exists between charged bodies.

Element A substance that cannot be chemically decomposed and contains atoms of one kind only.

Emitter follower *See* Common-collector amplifier.

Encoding The conversion of decimal digits to binary.

Energy The ability to do work.

Engineering notation A form of scientific notation in which a number is shown with a power of 10 that is divisible by 3 so that the number can be directly related to an SI prefix.

Enhancement mode A biasing condition for a FET in which the device is designed to operate at drain current levels above the drain current at the pinchoff point.

Equivalent circuit A circuit that, under certain conditions of use, may replace another circuit without substantial effect on electrical performance.

Exchange interaction An electrostatic force that maintains a magnetic alignment up to a certain temperature.

Excluse-OR gate A logic gate that will produce an output only when one input is high.

Extrinsic conductivity The conductivity of a material containing impurities.

Farad (F) The unit of capacitance.

Faraday's law A law that states the generation of voltage increases with additional flux or a faster rate of change of the flux field.

Feedback Defined as taking a portion of the output signal and returning, or feeding it back, to the input.

Ferromagnetic materials Materials possessing pronounced magnetic properties.

Ferromagnetism Phenomenon exhibited by materials having a permeability that is considerably greater than unity and that varies with flux density.

Field effect transistor (FET) A three-terminal, voltage-controlled semiconductor device.

Field intensity A measurement of the magnetomotive force needed to establish a certain flux density in a unit length of the magnetic circuit.

Filter A circuit designed to pass certain frequencies and block other frequencies.

Flip-flop (FF) A bistable multivibrator that is capable of storing binary data.

Flux density *(D)* The charge-inducing capability of an electric field; the number of lines of force per unit area.

Flux linkage The product of the number of lines of magnetic flux and the number of turns of a coil through which they pass. The unit of linkage is one unit of magnetic flux passing through one turn of a coil.

Flywheel effect The oscillations of charge and energy that occur at the natural *LC* resonant frequency.

Focus control An oscilloscope adjustment that allows the focus of a waveform to be modified.

Form factor A ratio between the average and effective values of a signal.

Forward bias A condition where the internal resistance of a diode is very low.

Fourier series A method of mathematical analysis of a recurring nonsinusoidal waveform.

Free electron The valence electrons of a material that are free to move among the positive material ions forming a crystal lattice structure.

Frequency The number of cycles per second.

Frequency counter An instrument designed to measure frequency.

Fringing The spreading of magnetic lines of force as they cross an air gap in a ferromagnetic material.

Full-wave rectifier A device that converts alternating current to direct current for 360° per cycle.

Function generator A signal generator used to generate a wide variety of waveshapes.

Fundamental frequency The lowest frequency component in the harmonic or Fourier expansion of a nonsinusoidal quantity.

Fundamental units The units selected to serve as a basis from which other units of the system may be derived.

Fuse A series-connected protection device designed to create an open circuit in the event of excess current flow in the circuit.

Fusible resistor A resistor designed to burn open when the power rating of the resistor is exceeded.

Gain A ratio of output to input; also called amplification.

Galvanometer A very sensitive electrical instrument used in detecting and measuring extremely small currents.

Gate trigger current The amount of gate current required to switch on an SCR.

Gauss's law A law that states that the net number of electric lines of force crossing any closed surface in an outward direction is numerically equal to the net total charge within that surface.

General transformer equation The equation describing the voltage induced into a transformer winding in terms of the peak mutual flux, frequency, and number of turns.

Generated voltage An induced voltage that is developed by mechanical motion between a conductor and a magnetic field.

Germanium A tetravalent semiconductor element having low conductivity at room temperature and increasing conductivity with rising temperature.

Grain-oriented steel Sheets of steel that are cut so that the magnetic flux flows in the direction of the structural grain of the material.

Graticule A transparent ruled screen mounted in front of the fluorescent screen in an oscilloscope.

Ground An electrical connection between a circuit and the earth, or between a circuit and a metal object that takes the place of the earth.

Half-power frequency *See* Half-power point.

Half-power point The point(s) on a response curve that represents 50% of the input power; also known as the 3 dB point.

Half-wave rectifier A device that allows current to flow in one direction during the positive half-cycle of input voltage and blocks reverse flow during the negative half-cycle.

Hall-effect sensor A device that produces an output voltage in the presence of a magnetic field.

Harmonic An integral multiple of the repetition (or fundamental) frequency of the waveform.

Harmonic distortion A change in waveform because of the inclusion of additional frequency components.

Heat losses Losses that occur in electronic circuits primarily because of eddy currents and hysteresis losses.

Henry (H) The SI unit of inductance.

Hertz (Hz) The SI unit of frequency.

Hexadecimal number system A number system with 16 digits: 0, 1, 2, 3, 4, 5, 6, 7, 8, 9, A, B, C, D, E, F.

High-pass filter A circuit that is designed to pass high frequencies and to attenuate low frequencies.

Holding current The minimum value of current required to maintain conduction.

Hole Positive electrical charge existing in a semiconductor material.

Hole flow The current consisting of electrons jumping from one position to another.

Horsepower (hp) A practical unit of power, equal to 746 watts.

Hypotenuse The longest side of a right-angle triangle.

Hysteresis The lagging of the magnetization of a ferromagnetic material behind the magnetomotive force that produces it.

Hysteresis loss A heat loss that occurs in any magnetic material in which the magnetic field continually reverses direction.

IEEE The Institute of Electrical and Electronic Engineers.

Impedance *(Z)* The total opposition to the flow of alternating current.

Impedance matching The process of matching the output impedance of one circuit to the input impedance of another circuit to achieve maximum power transfer. *See also* Matched circuit.

Impedance triangle A vector triangle formed by the resistance, reactance, and impedance of a circuit.

Inductance *(L)* The property of a circuit that opposes any change in the amount of current in that circuit.

Induced emf The emf associated with a changing magnetic field.

Inductive reactance (X_L) The opposition of an inductor, such as a coil, to the flow of alternating current.

Inductive time constant *See* Time constant.

Inductor A device that introduces inductance into an electric circuit.

Inferred zero-resistance temperature The temperature at which the extrapolation of a resistance versus temperature curve intersects the temperature axis.

In phase A condition where there is no angular displacement between quantities; e.g., between voltage and current.

Input offset voltage A compensating voltage applied to an op-amp to reference the output voltage to zero.

Insertion loss The difference between power received at the load before a device is inserted and the power after insertion.

Instantaneous value The magnitude of a voltage, current, or power at an exact instant in time.

Insulator A material that is a poor conductor and has electrons that are tightly bound to individual atoms.

Integrator An op-amp circuit that uses an *RC* network to provide a linearly increasing output voltage that is inversely proportional to the *RC* time constant.

Intensity control An oscilloscope adjustment that varies the brightness of the displayed waveform.

Internal resistance The resistance to the flow of an electric current within a source.

Internal sync An oscilloscope function that applies the output of the vertical amplifier to the sweep generator.

Intrinsic semiconductor A semiconductor in its purest form.

Intrinsic standoff ratio The internal resistance ratio of a UJT, represented by the Greek letter η.

Inverter A circuit that produces an output signal that is opposite to the input signal.

Inverting amplifier An op-amp configuration that performs the same basic function as a common-emitter BJT amplifier.

Ion An atom having an electrical charge.

Iron-core inductor An inductor with a core made of various alloys of iron and other materials to give the required characteristics for a particular application.

Isolation transformer A transformer in which there is no physical connection between the secondary and ground.

JFET Junction field effect transistor.

JK flip-flop A flip-flop with four usable states; each one being unique to the other.

Joule (J) The SI unit of work or energy; the work done by a force of 1 N acting through the distance of 1 m; or the work done by the transfer of an electric charge of 1 C through a potential difference of 1 V.

Kilogram (kg) The SI base unit of mass.

Kilowatt-hour (kW · h) The amount of work done in a specific period of time.

Kinetic energy The energy that a mechanical system possesses by virtue of its motion.

Kirchhoff's current law (KCL) A law that states the algebraic sum of the currents entering and leaving a node is zero.

Kirchhoff's voltage law (KVL) A law that states, in any closed loop, the algebraic sum of the voltage drops and rises equals zero.

Lagging A term used to describe a waveform that is delayed in reference to another waveform.

Law of conservation of electric charge A law that states the algebraic sum of all electric charges in any isolated system is a constant.

Laws of magnetic attraction and repulsion Like magnetic poles repel each other; unlike magnetic poles attract each other.

LCD *See* Liquid crystal display.

Leading A term used to describe a waveform that is advanced in reference to another waveform.

Leakage current (1) The DC current that flows through a capacitor because of imperfections in the dielectric or that flows to surface paths from one plate to the other. (2) The small amount of reverse-biased current that flows through a PN junction, caused by minority carriers.

Leakage resistance The relatively high resistance representing the dielectric in a capacitor.

Leak-off The process by which a capacitor discharges through its leakage resistance.

Least significant bit (LSB) The digit of a binary number that has the lowest position weight.

LED *See* Light-emitting diode.

Left-hand rule A rule in which the left hand determines the direction of the magnetic field surrounding a conductor and the current flowing through the conductor.

Length (l) The SI unit of length, in meters (m).

Lenz's law A law that states the current that flows as a result of an induced emf is in such a direction that the magnetic field established by the current reacts to stop the motion that generates the emf.

Light-emitting diode (LED) A diode that radiates energy in the form of light when conducting current.

Linear circuit A circuit made up of resistors and driven by sources of constant voltage and current.

Line of force The path along which an electric charge moves in an electric field.

Liquid crystal display (LCD) A low-power indicating device used to produce alphanumeric displays.

Load The name given to any device connected across an energy source.

Loading effect (1) A condition producing an inaccurate reading by an ammeter caused by the meter's resistance being too high. (2) A condition producing an inaccurate voltmeter reading because of the voltmeter acting as a shunt resistor.

Load line A graphical representation of the operating range of a transistor between saturation and cutoff.

Load resistance The amount of opposition to current flow offered by a load.

Load resistor A resistor used to represent the ohmic value of a load.

Logarithmic scale A graph scale in which the linear displacement along the axis is proportional to the logarithm of the numbers represented.

Logic analyzer A recording device that displays multiple waveforms simultaneously.

Logic clip A troubleshooting instrument used to instantly display the logic states of all the pins on an IC.

Logic pulser A troubleshooting device used to inject signals into a digital circuit.

Loop Any closed path in an electric circuit.

Loop analysis A circuit analysis method in which Kirchhoff's voltage law equations are written to solve for loop currents in a circuit.

Loop current The current flowing in a closed path.

Loose coupling The coefficient of coupling between two magnetic circuits that produces a small value of coupling.

Low-pass filter A circuit designed to pass low frequencies and attenuate high frequencies.

Magnetic circuit The closed path around which magnetic flux passes.

Magnetic field The space near a magnet in which magnetic forces are present.

Magnetic field intensity *See* Field intensity.

Magnetic flux The total number of lines of force in a given region.

Magnetic pole The regions where a magnet's strength is concentrated.

Magnetic saturation A condition where a magnetic material is saturated to the point that to produce an increase in the flux density, an extremely large increase in the magnetizing force is necessary.

Magnetism A property associated with materials that attract iron and iron alloys.

Magnetization curve *See* B–H curve.

Magnetomotive force (mmf) The force that causes a magnetic flux to exist in any magnetic circuit.

Majority carrier The current carrier that is present in the largest quantity.

Mass The property that determines the acceleration the body will have when acted upon by a given force.

Matched circuit A circuit where the source and load impedances are equal.

Matter A substance of which all physical objects are composed.

Maximum efficiency A condition that occurs in transformers when the copper losses are equal to the core losses.

Maximum power transfer theorem A theorem that states maximum power is drawn from a source when the load resistance equals the internal resistance of the source.

Maximum value The greatest value a signal may obtain; also known as peak value.

Mean An average value obtained by adding successive, equally spaced values and then dividing the sum by the number of values.

Meter (m) The SI base unit of length or distance.

Millman's theorem A procedure that combines the source transformation theorem with both the Thévenin and Norton theorems.

Minority carrier The current carrier that is present in the smallest quantity.

mmf *See* Magnetomotive force.

Molecule The smallest particle of a substance that has all the chemical and physical properties of the substance.

MOSFET A semiconductor device in which the gate is electrically insulated from the source-drain channel.

Most significant bit (MSB) The digit of a binary number that has the highest position weight.

Multimeter *See* VOM.

Multiplier The resistance connected in series with the moving coil of a voltmeter to extend the basic range of measurement.

Multistage amplifier An electronic circuit that combines more than one amplifier.

Multivibrator An oscillator that produces a nonsinusoidal waveform.

Mutual inductance The inductance between two separate coils where the change of current in one coil will induce a voltage in the other coil.

NAND gate A logic gate that produces a low output only when all inputs are high.

Negative ground A power supply connection where the negative terminal is connected to ground.

Neutral The common return conductor in a single phase of a polyphase system.

Neutron An elementary particle with no electrical charge.

Nodal analysis A circuit analysis method in which Kirchhoff's current law equations are written to solve for unknown voltages in a circuit.

Node Any point in a circuit where two or more circuit paths intersect.

Node voltage The voltage at a node with respect to a common reference point.

Noninverting amplifier An op-amp configuration that has no phase displacement between the input and output signals.

NOR gate A logic gate that produces a low output when any input is high.

North pole The region on a magnet from which magnetic lines of force exit.

Norton equivalent circuit A circuit determined by applying Norton's theorem.

Norton's theorem A procedure for reducing a complex electrical circuit to one having a single current source and a parallel resistance.

Nucleus The core of an atom composed mainly of protons and neutrons.

Octave A doubling of frequency.

Ohm (Ω) The unit of measurement applied to resistance and impedance; a conducting path that has a resistance of 1 Ω when the passage of a current of 1 A requires a potential difference across the path of 1 V.

Ohmmeter A measuring instrument used to indicate resistance in ohms.

Ohm's law A law that states the current produced in a given conductor is directly proportional to the difference of potential between its end points.

One-to-one transformer A transformer that receives energy at one voltage and delivers it at the same voltage.

Open circuit An open loop with infinite resistance.

Operational amplifier (op-amp) An integrated circuit amplifier.

Optoelectronic device A photoelectric device that converts light to electrical energy or vice versa.

OR gate A logic gate that will produce an output voltage when a voltage is applied at any input; equivalent to a electrical parallel circuit.

Oscillator A circuit that continuously generates a periodic time-varying wave that can be either sinusoidal or nonsinusoidal in shape.

Oscilloscope A measuring instrument consisting of a cathode ray tube (CRT) and various associated electronic circuitry; commonly used to automatically plot a particular voltage variation versus time.

Out of phase A condition where there is an angular displacement between quantities; e.g., between voltage and current.

Parallax error An error in reading measuring instruments caused by looking at a meter from an angle that will cause the pointer to appear left or right of the true position.

Parallel circuit A circuit with two or more common points so that the same voltage is across all elements.

Parallel resonant circuit A circuit in which the inductance and capacitance have equal values of reactance.

Paramagnetic material A material that becomes only slightly magnetized when under the influence of a strong magnetic field.

Passband The band of frequencies that are passed through a circuit such as a filter.

Passive device A device, such as a resistor, whose value does not change as a result of the application of voltage or current.

Passive filter A filter network consisting of passive devices.

Peak-to-peak value The amplitude of a waveform from its positive peak to its negative peak value.

Peak value The maximum instantaneous positive or negative value of a voltage, current, or power during a cycle.

Pentavalent impurity An impurity with five valence electrons.

Percent ripple The ratio between the rms of the AC component and the average DC output voltage.

Percent voltage regulation A ratio, expressed in percentage form, between the no-load and the full-load voltage of a device.

Period The duration of one cycle of a sustained oscillation or alternation.

Periodic waveform A wave that repeats itself after given time intervals.

Permeability Permeance per unit length and cross-sectional area of magnetic materials.

Permeance The property of a magnetic circuit that permits the passage of magnetic flux.

Permittivity The ratio, in an insulating material, of the electric flux density to the electric field intensity, compared with the same ratio for free space.

Phase angle The angle by which a sinusoidally varying quantity is displaced in time from another quantity of the same frequency.

Phasor The representation of a steady-state, sine-varying quantity as a complex number.

Phasor diagram A two-dimensional drawing that shows the magnitude and phase relationships of two or more sinusoidally varying quantities.

Photodiode A diode in which conductivity is controlled by light intensity.

Photoresistor A resistor made of semiconductor material that changes resistivity as the level of light around the semiconductor changes.

Pinchoff point The point where an FET becomes saturated; represents the point where the rate of current change approaches zero.

Pinchoff voltage The amount of gate voltage in a FET circuit that is required to produce pinchoff.

PN junction A single piece of semiconductor material formed with both P-type and N-type extrinsic materials.

Pole (1) The boundary where magnetic lines of force enter or leave a ferromagnetic material. (2) A term used to describe how an electrical contact arrangement operates. (3) A filter network that produces a −20 dB/decade rate of attenuation; *see also* Single-pole filter.

Positive ground A power supply connection where the positive terminal is connected to ground.

Positive ion An atom that has lost one or more valence electrons and has a net positive charge.

Potential energy Energy possessed by a system by virtue of position or condition.

Potentiometer A variable resistor.

Power The rate at which work is done or energy is converted from one form to another.

Power factor (PF) The ratio of the active power to the apparent power in an electrical system.

Power factor correction The addition of capacitors or inductors to a circuit in order to reduce the total current drawn from an AC source by reducing the power factor angle of the circuit.

Power supply A device that converts one type of electric potential or current to another.

Power triangle A right-angle triangle relating apparent power, real power, and reactive power.

Precision resistor A resistor with a tolerance of under 1%.

Primary cell An electrolytic cell in which two electrodes of different conducting materials associated with an electrolyte generate an emf.

Primary winding The transformer winding that receives energy from the supply circuit.

Product-over-sum rule A rule that states, when only two resistances are connected in parallel, the resistance of the parallel combination of two resistances is equal to the product of the individual resistances divided by their sum.

Proton A component of the nucleus of an atom with a positive charge equal in magnitude to that of an electron.

Pulsating current A unidirectional current whose magnitude varies with time.

Pulse A brief but sharp rise or fall of electrical current or voltage.

Pulse amplitude The maximum signal produced by a pulse waveform.

Pulse repetition rate (PRR) The frequency at which pulses occur.

Pulse train A group of consecutive pulses.

Pulse width The time between the start of the rise and the start of the fall of the input pulse.

Push button A momentary contact device.

Push-pull amplifier A class B amplifier circuit that uses a PNP and an NPN transistor to drive a load.

Pythagorean theorem A method of solving for an unknown in a right-angle triangle when two quantities are known.

Q factor A figure of quality or merit of a resonant circuit.

Q point The quiesescent point on the load line that indicates the steady-state condition of the V_{CE} of a BJT when no input signal is applied to the circuit.

Radian (rad) The SI unit of angular measure, approximately equal to 57.3°.

Radiation loss A power loss associated with high-frequency AC circuits such as antennas.

Radio wave An electromagnetic wave of radio frequency.

Radix The total number of individual symbols in a number system.

Ramp voltage A linearly increasing voltage.

Reactive power The rate at which energy is stored and alternately returned to the source by a reactive component such as an inductor or capacitor.

Reactive volt-ampere (VAR) The unit of reactive power.

Real power *See* Average power.

Rectifier An electronic circuit that converts AC into pulsating DC.

Rectifier filter A device used to reduce the AC component of a pulsating DC signal.

Reference node Common point for all sources in a network.

Reference phasor A phasor lying on the reference axis.

Reference voltage A voltage used as a standard of reference.

Reflected load The value of a load resistance reflected from the secondary into the primary of a transformer that is equal to the load resistance multiplied by the square of the transformation ratio.

Refresh A method of maintaining a charge across a capacitor.

Relative permeability The ratio of the flux in a material to the flux that would exist if the material were replaced with air.

Relative permittivity The property that determines the electrostatic energy stored per unit volume for unit potential gradient; dimensionless because it is a ratio.

Relaxation oscillator An electronic circuit used to create a repetitive signal voltage.

Relay An electromagnetic device that will open or close one or more sets of electrical contacts.

Reliability factor An indication of the failure rate of devices such as resistors.

Reluctance The opposition that a magnetic path offers to magnetic flux when a magneto-motive force is applied.

Residual flux density *See* Residual magnetism.

Residual magnetism The amount of flux density remaining in the material after the magnetizing force has been removed.

Resistance (R) The property of a device or a circuit that opposes the movement of current through it; measured in ohms (Ω).

Resistivity (specific resistance) (ρ) The resistance of a conductor having unit length and unit cross-sectional area.

Resistor A device used to insert electrical resistance into a circuit.

Resonance A condition in an AC circuit where the inductive reactance equals the capacitive reactance.

Resonant frequency The frequency at which the inductive reactance is numerically equal to the capacitive reactance of an AC circuit.

Response curve A plotted curve indicating the reaction of a circuit to a given input.

Reverse bias A condition where the internal resistance of a diode is very high.

Rheostat A device to control circuit current by varying the amount of resistance in the resistance element.

rms value The value of a sine wave that indicates its heating effect; represents 0.707 of the peak value.

Root-mean-square *See* rms value.

Sawtooth waveform A waveform in which the magnitude increases uniformly with time for a period and then falls rapidly to zero.

Schmitt trigger An electronic switching device.

Scientific notation A format of representing very large or very small numerical values using powers of 10.

SCR *See* Silicon-controlled device.

Second (s) The SI base unit of time.

Secondary cell A cell that is designed to be charged after discharge.

Secondary winding The winding of a transformer that receives energy by induction from the primary.

Second harmonic A harmonic with a frequency that is double that of the fundamental frequency.

Selectivity A measure of the bandwidth of a resonant circuit in terms of the Q factor and resonant frequency.

Self-inductance The ability of an inductor to induce a voltage into itself.

Semiconductor A material having electrical properties intermediate between those of good electrical conductors and insulators.

Semilogarithmic scale A graph scale in which linear graduations are indicated on the vertical axis and logarithmic graduations are indicated on the horizontal axis.

Series aiding A connection of two or more voltage sources where the flow of electrons inside the sources flows from the positive terminal of one source to the negative terminal of another source.

Series circuit A circuit in which there is only one current path and all components are connected end to end along this path.

Series-dropping resistor A resistor used to reduce the supply voltage to a value suitable for a load.

Series opposing A connection of two or more voltage sources where the sources "subtract" from each other.

Series-parallel circuit A circuit containing both series- and parallel-connected devices.

Seven-segment display An indicating device made up of segments that can be controlled to produce alphanumeric displays.

Shift register A device that is used for storing and transferring data, changing data from one format to another, sequencing control, and timing circuits.

Short circuit A closed-loop with zero-ohms resistance.

Shunt resistor A resistor connected in parallel with another branch of a circuit; used in measuring instruments to provide a path around meter movement.

SI Abbreviation for International System of Units (Systéme Internationale d'Unités).

Siemens (S) The SI unit of conductance; the reciprocal of the ohm (Ω).

Signal generator A test instrument designed to provide alternating voltage at a certain frequency and amplitude.

Silicon A tetravalent semiconductor element used in the manufacture of semiconductor devices.

Silicon-controlled rectifier (SCR) A three-terminal PNPN device that acts as a gated diode; that is, the gate terminal switches the device on, allowing current to flow from anode to cathode.

Sine wave An electrical waveform created when magnitude varies in proportion to the sine function of the angles of rotation.

Single-ended input amplifier A differential amplifier connected so that only one input is utilized.

Single-ended output amplifier A differential amplifier connected so that only one output is utilized.

Single-pole filter A filter with one RC network that produces a -20 dB/decade rate of attenuation; also known as a first-order filter.

Slew rate (SR) An indication of how fast the output voltage of an op-amp can change; the base unit is volts/second.

Solenoid An electromagnet constructed by winding a coil of wire around a cylindrical form.

Sound pressure level (SPL) The relationship between the level of sound and the reference level.

Source conversion theorem A procedure for replacing a constant-current source in a circuit diagram with an equivalent constant-voltage source, and vice versa.

Source-follower *See* Common-drain amplifier.

South pole The region on a magnet that magnetic lines of force enter.

Space permeability The permeability of air or vacuum.

Square wave The waveform of alternating current or voltage that is approximately square or rectangular.

SR flip-flop A flip-flop made of NAND gates or NOR gates with only three usable states.

Stage A single amplifier circuit.

Standoff voltage The amount of emitter voltage required to switch on a UJT.

Static charge A charge that is transferred by causing friction between materials.

Static electricity The movement of static charges between materials.

Steady state A condition where the signal is constant and does not change with time.

Step-down transformer A transformer that receives energy at one voltage and delivers it at a lower voltage.

Step-up transformer A transformer that receives energy at one voltage and delivers it at a higher voltage.

Stopband The band of frequencies that are attenuated.

Storage element A device that receives electrical energy, stores it in the form of a magnetic field, and returns the energy to the circuit at a later point in time.

Stray capacitance Capacitance arising from proximity of component parts, wire, and ground.

Subtractive polarity A winding connection that causes the direction of counter emf in each winding to be opposite.

Summing amplifier An op-amp circuit that is used to combine several input signals together while maintaining a common ground.

Superposition theorem A procedure that is used to analyze networks containing multiple power sources by isolating sources, solving for individual currents, and adding the results.

Susceptance (*B*) The ability of a reactive component, such as a coil or capacitor, to permit current flow; the reciprocal of reactance, measured in siemens (S).

Sweep generator Used in oscilloscopes to develop a voltage at the horizontal deflection plate that increases linearly with time.

Switch A device that mechanically interrupts the flow of current in a circuit by opening a set of contacts.

Tank circuit *See* Parallel resonant circuit.

Taper The relationship between angle and resistance in a variable resistor.

Temperature coefficient of resistance (α) The change in resistance per ohm for each degree of change in temperature from the reference temperature of 20° C.

Ten-to-one rule A rule that states, when two resistors are connected in parallel, if one resistor is ten or more times greater than the other resistor, then the greater value resistor may be ignored.

Tesla (T) The SI unit of flux density.

Thermistor A resistor whose resistance varies with temperature.

Thévenin equivalent circuit The reduction of a circuit to its Thévenin values.

Thévenin's theorem A procedure for reducing a complex electrical circuit to one having a single voltage source and a series resistance.

Threshold voltage The amount of gate voltage required to produce a channel in a FET.

Throw The total number of individual circuits that each pole of a switch is capable of controlling.

Thyristor A semiconductor device that exhibits the regenerative switching characteristic of a four-layer or PNPN arrangement.

Tight coupling When the coefficient of coupling between two magnetic circuits is 1, or unity.

Time *(t)* The SI unit of time, in seconds (s).

Time constant The time required for the voltage across a capacitor or inductor to rise to 63.2% of maximum or to fall to 36.8% of its maximum value.

Time period The amount of time required for one cycle to change.

Toggle state The opposite of the previous state.

Tolerance Allowable deviation from the marked value of a component.

Toroid A coil wound on a circular core.

Transconductance The relative ability of a gate to control the drain current of a FET.

Transformation ratio *See* Turns ratio.

Transformer A device that transfers energy from one circuit to another by electromagnetic induction.

Transformer polarity The relative direction of the induced voltages in the primary and secondary windings with respect to the winding terminals.

Transient The part of the change in a variable that disappears when going from one steady-state condition to another; also known as transient state.

Transistor A three-terminal semiconductor device used for amplification or switching.

Triac A gated diac capable of conducting current in both directions when the gate terminal is triggered.

Triangular wave A periodic ramp waveform where the voltage level changes from one level to another at a constant rate producing a symmetrical alternating triangle.

Trigger control An oscilloscope adjustment to provide a stable waveform display.

Trimmer resistor A variable resistor used where small and infrequent adjustments of a resistance are necessary to maximize circuit performance.

Trivalent impurities Impurities that have only three electrons.

True power The power developed by a purely resistive component.

Truth table A graphical illustration of the relationship between input and output logic levels of a gate circuit.

Tuned circuit A resonant circuit.

Tuning The adjustment of the frequency of a circuit or of a system to secure optimum performance.

Turns ratio The ratio of the number of secondary turns to the number of primary turns in a transformer winding.

Two-pole filter A filter using two *RC* networks to produce a roll-off rate of −40 dB/decade.

UJT relaxation oscillator *See* Relaxation oscillator.

Unijunction transistor (UJT) A three-terminal device that acts as a diode with its own internal voltage divider biasing circuit.

Unit hypotenuse A right-angle triangle used for deriving sine values.

Valence band A band of electron energies that contains all the energy levels available to the valence electrons.

Valence orbit The outermost shell occupied by orbiting electrons.

VAR The unit of reactive power.

Varactor diode A PN junction semiconductor diode that is designed for low losses at high frequencies.

Varistor A voltage-dependent, metal-oxide material that has the property of rapidly decreasing resistance with increasing voltage.

Vector A quantity that has both magnitude and direction.

Vector diagrams A diagram that displays quantities that consist of magnitude and direction.

Vertical controls The selector switches on an oscilloscope that allow for an appropriate volts/division range to be selected; also move the waveform up or down by using the position knob.

Virtual ground A node that is at ground potential but is not physically connected to the system or chassis ground.

Volt (V) The SI unit of voltage or potential difference.

Voltage divider A device to permit a fixed or variable fraction of a given supply voltage to be obtained.

Voltage divider rule A rule that states the ratio between any two voltage drops in a series circuit is the same as the ratio of the two resistances across which these voltage drops occur.

Voltage drop *(V)* (1) The difference of voltages at the two terminals of a passive element. (2) The energy used by free electrons when engaged in current flow.

Voltage follower A circuit in which the output voltage tracks, or follows, the input voltage.

Voltage regulation The voltage variation of a source as it changes from no load to full load as a ratio of full-load voltage. *See* Percent voltage regulation.

Voltage regulator A device that, supplied a given input voltage, will provide an adjustable output voltage.

Voltage rise The energy imparted to the free electrons for the development and maintenance of electric current.

Volt-ampere (VA) The product of the current and voltage; used to determine power in a DC circuit.

Volt-ampere-reactive *See* VAR.

Voltmeter A measuring instrument used to indicate the magnitude of voltage in a circuit.

Voltmeter loading *See* Loading effect.

Voltmeter sensitivity An indication of the shunting action of a voltmeter, measured in ohms/volt.

VOM An instrument used to measure voltage, resistance, and current.

Watt (W) The power expended when 1 A of direct current flows through a resistance of 1 Ω; unit of active or average power.

Wattmeter An instrument used for measuring power in watts.

Waveform The pattern of variations of a current or voltage.

Wavelength The distance that a radio wave travels in the time of one cycle; its symbol is lambda (λ).

Weber (Wb) The SI unit of magnetic flux.

Weber's theory A theory that states the molecules of a magnetic material are tiny magnets, each with a north and south pole and a surrounding magnetic field.

Wheatstone bridge An electronic instrument that measures resistance with a high degree of accuracy.

Wien-bridge oscillator A circuit that uses both positive and negative feedback to produce a sinusoidal waveform.

Wiper The movable contact in a potentiometer.

Wire table A table of sizes and properties of round copper wire, manufactured to AWG standards.

Work The amount of energy converted from one form to another as a result of motion or conversion of energy.

Zener diode A diode that is designed to conduct in the reverse direction when its value of breakdown voltage is reached.

Zero-crossing detector An electronic device used to detect the instant a signal crosses the zero axis.

Zero-ohm resistor A resistor with an ohmic value of 0 Ω.

INDEX